Smart Sensors, Measurement and Instrumentation

Volume 50

Series Editor

Subhas Chandra Mukhopadhyay, School of Engineering, Macquarie University, Sydney, NSW, Australia

The Smart Sensors, Measurement and Instrumentation series (SSMI) publishes new developments and advancements in the fields of Sensors, Instrumentation and Measurement technologies. The series focuses on all aspects of design, development, implementation, operation and applications of intelligent and smart sensors, sensor network, instrumentation and measurement methodologies. The intent is to cover all the technical contents, applications, and multidisciplinary aspects of the field, embedded in the areas of Electrical and Electronic Engineering, Robotics, Control, Mechatronics, Mechanical Engineering, Computer Science, and Life Sciences, as well as the methodologies behind them. Within the scope of the series are monographs, lecture notes, selected contributions from specialized conferences and workshops, special contribution from international experts, as well as selected PhD theses.

Indexed by SCOPUS and Google Scholar.

Biswajeet Pradhan · Subhas Mukhopadhyay
Editors

IoT Sensors, ML, AI and XAI: Empowering A Smarter World

Springer

Editors
Biswajeet Pradhan
Centre for Advanced Modelling
and Geospatial Information Systems
(CAMGIS)
University of Technology Sydney
Ultimo, NSW, Australia

Subhas Mukhopadhyay
School of Engineering
Macquarie University
North Ryde, NSW, Australia

ISSN 2194-8402 ISSN 2194-8410 (electronic)
Smart Sensors, Measurement and Instrumentation
ISBN 978-3-031-68601-6 ISBN 978-3-031-68602-3 (eBook)
https://doi.org/10.1007/978-3-031-68602-3

This Springer imprint is published by the registered company Springer Nature Switzerland AG
The registered company address is: Gewerbestrasse 11, 6330 Cham, Switzerland

Preface

As editors of this volume, it is our pleasure to present *IoT Sensors, ML, AI and XAI: Empowering A Smarter World*. In this book, we have curated a collection of chapters that explore the intersection of IoT sensors, machine learning (ML), artificial intelligence (AI), and explainable AI (XAI), showcasing how these technologies are shaping a smarter, more interconnected world.

The rapid advancement of technology has enabled the development of smart sensors capable of capturing real-time environmental data with unprecedented accuracy and granularity. Coupled with the power of AI and ML algorithms, these sensors have the potential to revolutionize various domains, from agriculture and healthcare to transportation and environmental monitoring.

The chapters in this volume delve into diverse applications of IoT sensors, ML, and AI, ranging from autonomous quadcopters and smart irrigation systems to medical diagnostics and environmental monitoring. Each chapter presents innovative research and practical implementations that highlight the transformative impact of these technologies on our daily lives.

Furthermore, this volume emphasizes the importance of explainable AI (XAI) in enhancing the interpretability and trustworthiness of AI systems. By demystifying the decision-making process of AI algorithms, XAI enables stakeholders to better understand and validate the outcomes, fostering greater acceptance and adoption of AI-powered solutions.

Chapter Summaries:

Chapter "Sensors, ML and AI for Real World Applications" emphasizes the integration of smart sensors, AI, and ML for accurate real-time environmental monitoring. By combining sensor data with AI algorithms, the chapter demonstrates how to bridge gaps in data reliability and derive valuable insights for various applications.

Chapter "Flying IoT: Sensor Fusion Performance Analysis for UAV Applications in Indoor Spaces" focuses on the fusion of unmanned aerial vehicles (UAVs) with IoT platforms, this chapter explores the development of autonomous quadcopters for indoor sensing networks. By leveraging vision-based and laser sensors, the chapter highlights the potential of UAVs in 3D mapping and localization within IoT frameworks.

Chapter "Machine Learning Empowered IoT Devices, Analysis of Indoor and Outdoor Temperature and Health Risks" investigates the application of machine learning in analyzing temperature data from IoT devices to assess health risks associated with indoor and outdoor temperature variations. It discusses the preprocessing of data and the classification of temperature datasets using machine learning models.

Focusing on network security, Chapter "Identification of IoT Devices Through Machine Learning and Hardware Fingerprints Based on Clock-Skew" proposes a novel solution for identifying devices on local networks using machine learning algorithms. It discusses the design of the solution architecture and the development of machine learning models for automated device identification based on clock-skew analysis.

Addressing agricultural challenges, Chapter "Development of IoT Soil Sensor Node: Installation and Modeling" presents the design and development of an IoT soil sensor node for real-time field data communication. It discusses calibration, testing, and field deployment of the sensor node, highlighting its potential applications in weather forecasting and irrigation scheduling.

Chapter "A Novel Determination of Chemical Oxygen Demand (COD) Using Thermal Sensor: A Key Initiative in Wastewater Treatment" introduces a novel method for portable chemical oxygen demand (COD) detection using thermal sensors. It discusses the principle of COD detection, experimental setup, and comparison of machine learning algorithms for COD prediction, emphasizing environmental monitoring applications.

Focusing on public safety, Chapter "Developing a Cloud-Based Weapon Detection System Using Transfer Learning and Generative Adversarial Networks" proposes a cloud-based object detection system for baggage security using deep learning techniques. It discusses the integration of transfer learning and generative adversarial networks (GANs) for threat detection during X-ray scanning, emphasizing real-time surveillance capabilities.

Chapter "Implementation of IoT and Machine Learning Techniques in Smart Irrigation Systems" addresses water management challenges in agriculture through an IoT-based smart irrigation system. It discusses the implementation of wireless sensor networks and machine learning algorithms for optimizing irrigation designs based on crop types and weather conditions.

Exploring the architecture of autonomous vehicles, Chapter "Autonomous Driving from an Architectural Perspective" reviews technical aspects of self-driving cars, including sensor technology, obstacle detection techniques, and collaborative safety frameworks. It highlights challenges and future research directions in perception, planning, and control of autonomous vehicles.

Focusing on road safety, Chapter "A Prototype Design for Effective Traffic Navigation" summarizes pothole detection and navigation techniques using various sensors and machine learning algorithms. It discusses real-time detection methods and navigation strategies to mitigate road hazards and improve vehicle safety.

Chapter "Integrating Topology and Geospatial Knowledge for Mapping Road Network Layers from High-Resolution Remote Sensing Images" presents a deep learning-based pipeline for road extraction and vectorization from high-resolution remote sensing images. It discusses the development of convolutional neural network (CNN) models for road feature extraction, emphasizing the significance of shadow removal for accurate road mapping.

Addressing medical diagnostics, Chapter "Unlocking the Potential: Machine Learning and Deep Learning in Leukemia Diagnosis with Explainable AI" explores AI-driven approaches for leukemia diagnosis using microscopic blood cell imaging. It discusses segmentation and classification techniques for disease detection and highlights the importance of explainable AI in enhancing trust and interpretability of diagnostic models.

Focusing on healthcare, Chapter "Non-invasive Stress Recognition Framework Using Consumer Internet of Things in Smart Healthcare Applications" introduces a wearables-based stress recognition framework for smart healthcare applications. It discusses the integration of machine learning models at the edge layer and cloud layer for real-time stress prediction and analysis, emphasizing personalized healthcare solutions.

Chapter "Explainable Multi-task Learning Approach for Skin Lesion Classification" presents a multi-task learning approach for skin lesion classification using deep learning techniques. It discusses the integration of classification and segmentation tasks to improve model accuracy and interpretability, highlighting the potential for enhanced diagnostic performance in dermatology.

Addressing operational efficiency, Chapter "IoT Operational Intelligence" explores the concept of operational intelligence in IoT systems. It discusses sensor clusters, machine learning, and inference models for optimizing data processing and improving outcomes in real-time domains such as health and emergencies.

Chapter "Techniques to Implement Blockchain in Internet of Medical Things (IoMT)" explores methods for integrating blockchain with IoMT to enhance security and privacy. Four major techniques are classified and compared based on their contributions to addressing data security, privacy, and interoperability issues. The chapter provides a comparative analysis of these methods, offering insights into their implementations and contributions to enhancing security in IoMT.

Chapter "Offloading Strategies and Computing Paradigms in IoT: A Survey" explores various offloading mechanisms and computing paradigms for optimizing data processing in IoT. It discusses standard data offloading methods like cellular offloading and computing paradigms such as cloudlet computing, fog computing, and serverless computing. Offers insights into their strengths and limitations and concludes by summarizing key findings and emphasizing the significance of offloading strategies in IoT.

Chapter "Predicting Forest Canopy Height Using GEDI LiDAR Based Machine Learning Technique Over Similipal Biosphere, India" utilizes machine learning models with diverse data sources like Sentinel and GEDI LiDAR to estimate forest canopy height. Results indicate enhanced prediction accuracy, aiding in sustainable forest resource management and carbon storage tracking.

Chapter "AI-Driven Environmental Monitoring Using Google Earth Engine" explores AI's role in automating environmental data collection and analysis, highlighting applications like air quality assessment and deforestation detection.

Offers insights into AI's opportunities and challenges, emphasizing transparency and stakeholder participation.

Chapter "A Fusion Approach for Kannada Speech Recognition Using Audio and Visual Cue" introduces a novel fusion approach for Kannada speech recognition, leveraging both audio and visual cues. Communication, a fundamental aspect of human interaction, is facilitated through the proposed audiovisual speech recognition system, particularly beneficial for the deaf and hearing-impaired population. By integrating audio and visual models, the system achieves higher accuracy rates compared to individual models.

Chapter "Analysis of Machine Learning Approach for Spamming Electronic Mail Detection" delves into the analysis of machine learning techniques for spam email detection. With the exponential growth of electronic mail, the proliferation of spam emails poses significant challenges. Through the application of various machine learning algorithms, including logistic regression, decision trees, random forest, SVM, naive Bayes, and BERT, the study achieves impressive accuracy rates in identifying spam emails, thus mitigating potential risks associated with communication overload and malware exposure.

Chapter "An Enhanced Caries Detection and Prediction Using DentSU_Net" presents an enhanced caries detection and prediction system using DentSU_Net. Segmenting dental images is vital for accurate clinical diagnosis, and the proposed DentSU_Net addresses limitations in existing segmentation techniques. By incorporating Swin_Unet and Squeeze and Excitation (SE) attention layers, DentSU_Net achieves superior performance in accurately identifying caries regions, providing dentists with an efficient tool for clinical analysis.

In Chapter "Skin Lesions Classification of Dermoscopy Images Using Deep Learning Technique", the authors delve into the pressing issue of skin cancer, particularly melanoma, which affects a significant number of individuals worldwide. They explore the challenges in accurately identifying and classifying different types of skin lesions, highlighting the importance of early detection for improved prognosis. Through innovative deep learning-based methods, the authors propose a solution aimed at enhancing the accuracy and efficiency of skin lesion analysis, offering promising insights into the field of medical image analysis.

We hope that this book serves as a valuable resource for researchers, practitioners, and policymakers alike, inspiring further innovation and collaboration in harnessing the potential of IoT, ML, AI, and XAI for the betterment of society.

We would like to express our sincere gratitude to all the contributors for their valuable insights and contributions to this volume. Their dedication and expertise have enriched the discourse on IoT sensors, ML, AI, and XAI, paving the way for a smarter and more sustainable future.

North Ryde, Australia — Prof. Subhas Mukhopadhyay
Ultimo, Australia — Prof. Biswajeet Pradhan

Contents

Sensors, ML and AI for Real World Applications

Subhas C. Mukhopadhyay and Krishanthi P. Jayasundera

Abstract Accurate measurement of real time environmental parameters with smart sensors along with application of artificial intelligence (AI) and machine learning (ML) will open incredible opportunities in many fields. Application of AI and ML on data that is not genuine will provide very limited useful information. The current gaps are surrounding the application of ML and AI on data in which the sources are not properly known or verifiable in many cases. The proposed combo-sensor will bridge the gap by exploring the real strength of AI and ML while it is applied in combination with sensors which are fabricated for real time measurement of health, the environment and many other parameters of interest. A few typical sensors based applications are considered in which the combo-sensor has derived important parameters within the working environments and provide very useful information.

1 Introduction

Sensors is an essential member of a family of sensors network or internet of things. Sensors measure parameters of interest and provide useful information to the system. The measured data by the sensors is analysed to extract necessary information for post-operation which may be as simple as just to display the value of the measured data or as complex as taking some action to correct certain events. The analysis of measured data from sensors form a different branch of science and engineering and known as Data Science. Different types of algorithms are used and are developed to analyse data which are commonly known as Machine Learnings (ML) and introduction of extra knowledge to extract more useful information to the system for some useful action is known as Artificial Intelligence (AI). Both ML and AL are

S. C. Mukhopadhyay (✉)
School of Engineering, Macquarie University, Sydney, NSW 2109, Australia
e-mail: Subhas.Mukhopadhyay@mq.edu.au

K. P. Jayasundera
University Technology Sydney, Ultimo, NSW 2200, Australia

B. Pradhan and S. Mukhopadhyay (eds.), *IoT Sensors, ML, AI and XAI: Empowering A Smarter World*, Smart Sensors, Measurement and Instrumentation 50,
https://doi.org/10.1007/978-3-031-68602-3_1

an emerging field of current research and significantly new developments are taking place in almost every fields of science and engineering.

The design, development, fabrication and applications of interdigital sensors have found numerous applications in the last few decades [1]. The configuration of electrodes for the interdigital sensors plays a significant role as described in [2] as well spatial resolution needs to be properly considered for its performance [3]. Being a capacitive sensor, the excitation and measurement for an interdigital sensor need some design consideration while making it an embedded sensor [4]. Planar electromagnetic sensors, a combination of interdigital and meander type magnetic sensor has been used to determine nitrate and other contaminants in water [5, 6]. Polymer coated Interdigital sensor has been used to determine nitrate contamination is water [7]. Interdigital sensor is affected due to change of temperature and humidity. If the sensor is used in water dipping environment, humidity will not influence the performance but change of operating temperature will affect the output signals. Introducing intelligent algorithms to compensate the effect of temperature will allow the sensor to determine correct value of nitrate concentration at outside uncontrolled environments [8]. Recent developments in the last decade has shown more and more sensing system with an ability to upload the measured data into cloud allowing the data to be available through internet. The extension of smart nitrate sensor to an internet of things (IoT) enabled system has been achieved using LoRa based communication system [9]. Interdigital based sensing system to detect leaching of phthalate from plastic bottle in water and beverages has been developed and reported in [10, 11].

The measured data from sensors are to be analysed to extract the correct information and it can be very critical if the outcome/s is/are used to take an important decision. Delays in taking the decision and accuracy in correctness can be fatal in many situations. Combo-sensor which is combination of smart sensors, intelligent decision-making system with machine learning and artificial intelligence is very helpful for this type of situation.

This chapter has been divided into a few sections. The section II describes fabrication of sensors and the section III presents interfacing of sensors to embedded processor. The next section will discuss processing of data and wireless communication to gateway. The following section will present a few case studies before the chapter is concluded.

2 Design and Fabrication of Sensor

The design and implementation of sensor which are fabrication in-house are of interdigital type. Figure 1 shows a few typical interdigital sensors fabricated in-house, (a) FR4 as the base substrate, (b) Glass substrate and (c) Micro-Electro Mechanical System (MEMS) based sensor [12, 13]. The electrode configuration, pitch resolution and overall size will determine sensitivity of the sensor [14–16].

In recent times, interest on flexible sensors has been in upward swing as a very fast rate due to many disadvantages of rigid sensors such as (i) difficulty in fabrication

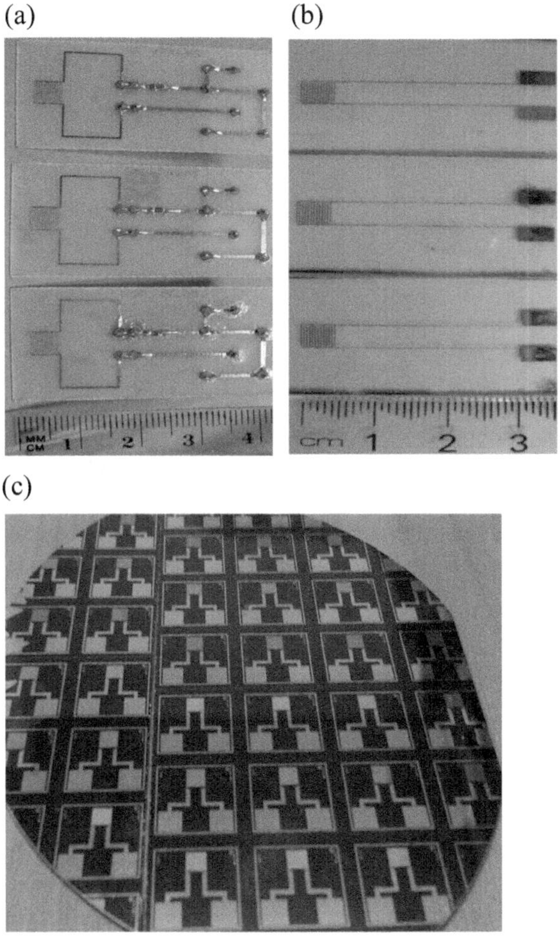

Fig. 1 Fabrication of interdigital sensor; **a** FR4 based **b** Glass based and **c** MEMS based

process, (ii) the cost of fabrication of sensors with rigid substrates is high, (iii) difficult to attach sensors with rigid substrates, (iv) Electrode material comes off on attachment of the chips, and (v) difficult to work with liquids due to hydrophilicity. The flexible sensors are fabricated with different substrate materials such as PDMS (Polydimethylsiloxane), PET (Polyethylene terephthalate) or Polyimide film [17]. The electrodes are fabricated from aluminium, silver, carbon nanotubes, graphene and mixture with PDMS with other materials. The selection of substrate as well as electrode materials depends on typical application as there are constraints on cost, fabrication complexity, life time, compliant with environment and disposal. A typical flexible sensor is shown in Fig. 2 [18, 19].

Different types of sensors have been fabricated over the years, sensors for detection of LPG gases are reported in [20, 21], a novel planar coil-based inductive-capacitive water level sensor has been described in [22], Eddy-Current-Based Angle Sensor

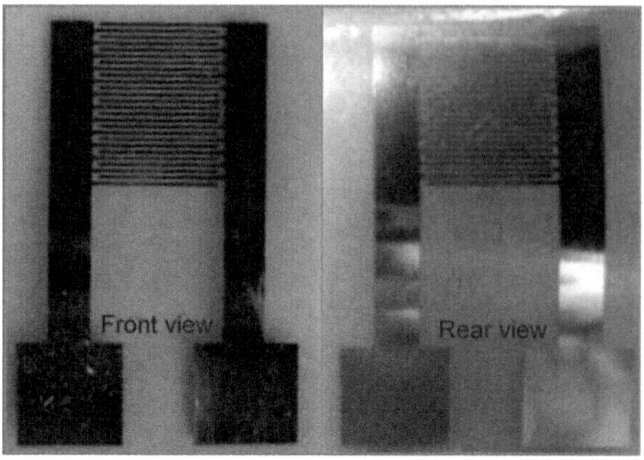

Fig. 2 Fabricated flexible sensor [18]

has been developed in [23], Capacitive Sensor for Combined Angular and Linear Displacement Sensing has been designed and developed in [24] and electrochemical method based detection of calcium as well as magnesium in water bodies has been reported in [25] respectively.

3 Interfacing Electronic Circuits

Interfacing electronics circuits play important roles on the design and development of IoT enabled smart sensing system. The raw sensor needs to be connected with proper signal conditioning circuits so that the output signals of the sensor can be interfaced with the embedded processor for optimum utilization. The interfacing electronic circuit can be as simple like as shown in Fig. 3, an inverting amplifier [26] or can be complex as shown in Fig. 4 for measurement of angle [23, 27]. The interfacing of voltage and current signal for monitoring electrical appliance for smart home application is shown in [28]. In many applications, the output of the sensor may not have linear relationship with the measured variable so different types of linearization circuits which can also be part of the interfacing electronics are needed [29]. Linearization are also implemented in the software.

Interfacing electronics is very helpful to utilise the maximum capacity of the system. Sometimes, the electronics circuits may be designed carefully so that there is no need of any special circuits to interface to embedded controller as reported in [30–32]. The system will allow interfacing of resistive, inductive as well as capacitive type of sensors. Towards the development of low-cost interfacing electronics system, the complexity of the working principle of the sensors need to be properly understood. Inductive sensors in many applications are used and the response is usually non-linear in nature. To obtain output directly from the sensors without using any kind of embedded processors, innovative approach needs to be adopted as reported in

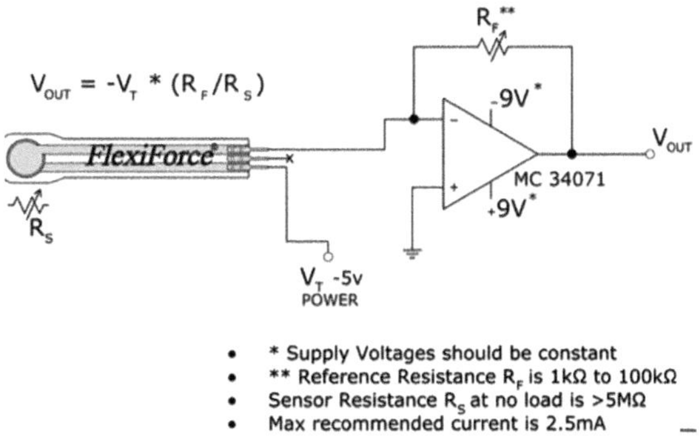

Fig. 3 A simple inverting amplifier based interfacing electronic circuit [26]

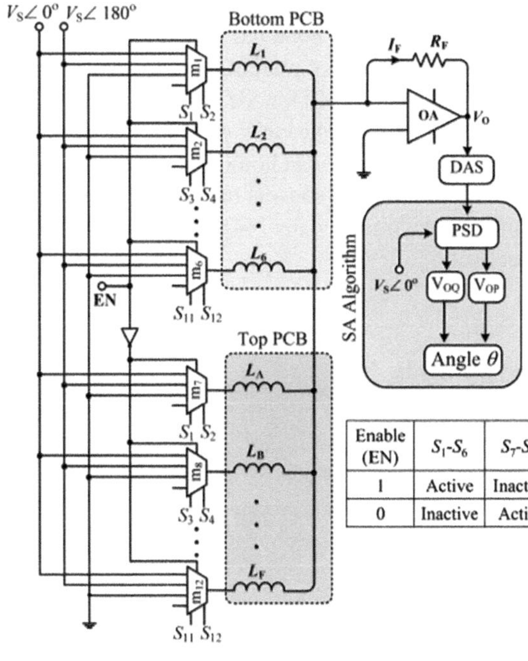

Enable (EN)	S_1-S_6	S_7-S_{12}
1	Active	Inactive
0	Inactive	Active

Fig. 4 A interfacing electronic circuit for angle measurement [23]

[33–36] for the inductive types of sensors. Similar approach can be extended for the capacitive type of sensors and useful literature are available in [37–39].

4 IoT Enabled Sensing System

The measured signals by sensors are conditioned by interfacing electronics and connected to embedded processor at the input pins. The microcontroller then processed the data and transmitted to gate-way using a suitable wireless communication system. The gate-way are connected to cloud and the data are finally saved in suitable servers. The data can be accessed from the server for processing, visualization and used for any corrective actions. A complete architecture for the IoT enabled sensing system is shown in Fig. 5. The selection of gate-way as well as other components depends on many factors, availability of components and cost can be most important consideration.

In recent times, design, development and applications of IoT enabled sensing systems are being reported at a very fast rate. A step by step design method to develop an IoT sensor node has been presented in [40], the paper is very useful for early career researchers working in the area. A similar method has been adopted to develop a portable IoT node for measurement of sulphur content in water has been reported in [41]. An early work on IoT enabled sensing system for monitoring environmental condition at home is available in [42]. IoT based phosphate detection using Graphite/PDMS sensors has been presented in [43], MWCNTs/PDMS sensor for nitrate detection in water [44] and Point-of-Care Monitoring of Bone Loss in [45] respectively.

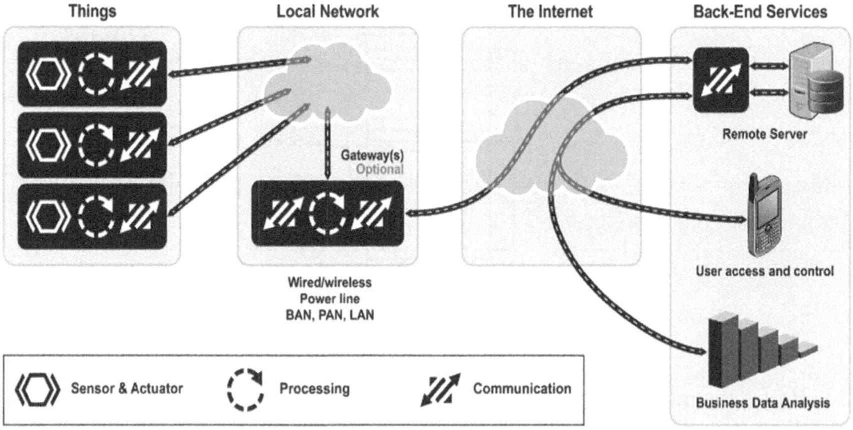

Fig. 5 An architecture of an IoT enabled sensing system

5 Off-Line Data Analytics

Data measured by sensors are analysed and are extremely useful to determine the cause and effect relationship in an environment. In normal sensing situation, the collection of data and analysis are done as a sequential process. The results are used to take some corrective action immediately, if the sensor/s is/are used in a control system [46]. In many other situations, the measured data are used for visualization and used for different other applications after post-processing [47]. Off line data analytics are applied on the data which are already available and are not connected with the source of data in real time [48]. Though off-line data and their analytics are very useful in many applications but the focus of this book chapter is not on off-line data rather sensing system where the actions on data are take place in real time.

6 Combo-Sensors

The theme of the combo-sensor is slightly different with respect to conventional sensors. In this concept, the data captured by the sensor is analysed immediately and based on the outcomes, the system is instructed to take corrective action. The analytical technique includes some kind of machine learning (ML) algorithm. In [49], a novel drone delivery system for multi-story apartment buildings with balconies has been proposed. The developed drone has a frame size of 295 mm and is equipped with a stereo camera and a ranging sensor. The captured images (video streams) by the camera are analysed on the embedded processor carried by the drone. It employs two methods for Vertical Grid Screening (VGS), i.e., Grid Screening (GS) and Square Screening (SS), to detect unique markers to identify the precise balcony that needs to receive the product. Based on the analysis, it explores the scanning and trajectory methods required for autonomous flight to accurately approach the marker location. The proposed machine learning system is trained on a YOLOv5 model for image recognition of the marker, and four different models and batch sizes are compared. Based on the correct identification of the target destination, the item is delivered. The ML algorithm plays an important role on the success of the delivery of the items by the drones.

Wireless-sensor-network-based home monitoring system has been implemented as described in [50]. The purpose of the system is to monitor daily activities of an elderly living inside the smart home. The behaviour involves functional assessment of daily activities. The research has reported a mechanism for estimation of elderly well-being condition based on usage of house-hold appliances connected through various sensing units. Two wellness functions to determine the status of the elderly on performing essential daily activities have been defined. The wellness functions are determined based on the collected real time data from the sensors. The developed intelligent algorithms are able to take decision to generate warning text message to caregiver when the wellness value drops below a certain threshold value. The warning message is extremely important to save the life of the monitored elderly person and off-line data analysis will not be of much use in this type of applications. There are many incidents when an elderly has fallen down at home which lead to death of the person. To save the life of the person it is extremely important that the data is analysed immediately and the decision to provide help is instructed without any time delay. The delay is associated with the real time data analysis consisting of data collection and decision making. This type of system is a perfect example of combination of sensor, data analysis involving machine learning and artificial intelligence to save the life of human beings under monitoring.

The paper in [51] proposes applications of deep learning in detecting cancerous cells inside patients via laparoscopic camera on da Vinci Xi surgical robots. Currently, the surgeons rely on their experience to identify the tumors. Once the location of the tumor has been approximated, da Vinci Xi provides intra operative ultrasound and Indocyanine Green (ICG) with Fluorescence Imaging to further assist the surgeon. Intra operative ultrasound shows the depth of the tumor and makes a 3D reconstruction of the organ on a tablet computer beside the surgeon's console. Injecting ICG and turning on fluorescent light makes the kidney green and the tumor grey. If the location of the tumor cannot be identified then intra operative ultrasound will not work. Incorrect doses of ICG will either make the whole field of view green or will not change color. ICG also comes with side effects which makes it necessary to keep the injection of ICG minimum. This creates a huge problem to the surgeon even though they are highly experienced. Under this situation, deep learning technique based on Convolution Neural Network (CNN) helps to identify the cancerous cells while running the surgery. This will provide a second opinion besides the surgeon's experience in identifying tumors during surgery. Detection of tumor is done using object detection and classification/localizing using Gradient-weighted Class Activation Mapping (Grad-CAM) technique. Localization means heat map is drawn on the image highlighting the classified class. The following steps are followed in the research.

Step#1: Deep learning on a live surgical video to show the locations of the tumors on a global range.

Step#2: Classifying among cancerous tissue, non-cancerous tissue, fatty tissue and localizing the identified class in close range.

Step#3: If it is preferred to have two class classification, the third step is for classifying between cancerous and non-cancerous tissue with localization with Gradient-based Class Activation mapping (GRADCAM) in close range.

The idea is that once the more aggressive tumor had been identified, the close range classification will be used to identify if there are any more tumors left inside the patient before closing the wounds. The system generates a result with a possibility of cancer, fatty-tissue and non-cancerous option. The medical surgeon will use his own expertise to take a final call. The combined system is very useful for successful operation on human subject.

7 Conclusions

The chapter has presented usefulness of real time application of data collected from sensor, different analytical techniques including machine learning (MLs), intelligence including Artificial Intelligence (AI) and finally decision making for the success of a sensing system. Real time data analysis and use of results from it can be extremely useful in many applications involving fast decision making, quick recovery and finally life savings.

References

1. S.C. Mukhopadhyay, C.P. Gooneratne, J.K. Roy, Progress of interdigital sensors over last two decades: a researcher's perspective, in *Smart Sensors, Measurement and Instrumentation*, vol. 36, ed. by S.C. Mukhopadhyay, B. George, J.K. Roy, T. Islam (Interdigital Sensors Springer-Verlag, 2021), pp. 1–22. ISBN 978-3-030-6683-9
2. A. Ranjan, B. George, S.C. Mukhopadhyay, Interdigital proximity sensor: electrode configuration, interfacing and an application, in *Smart Sensors, Measurement and Instrumentation*, vol. 36, ed. by S.C. Mukhopadhyay, B. George, J.K. Roy, T. Islam (Interdigital Sensors Springer-Verlag, 2021), pp. 23–34. ISBN 978-3-030-6683-9
3. A. Nag, S.C. Mukhopadhyay, C.P. Gooneratne, fabrication of interdigitated sensors: issues and resolution, in *Smart Sensors, Measurement and Instrumentation*, vol. 36, ed. by S.C. Mukhopadhyay, B. George, J.K. Roy, T. Islam (Interdigital Sensors Springer-Verlag, 2021), pp. 35–69. ISBN 978-3-030-6683-9
4. M.E.E. Alahi, A. Nag, N. Afsarimanesh, S.C. Mukhopadhyay, J.K. Roy, A simple embedded sensor: excitation and interfacing, in *Smart Sensors, Measurement and Instrumentation. Advanced Interfacing Techniques for Sensors*, vol. 25, ed. by B. George, J.K. Roy, V. Jagadeesh Kumar, S.C. Mukhopadhyay (Springer, 2017), pp. 111–138. ISBN 978-3-319-55368-9
5. M.A. Yunus, S.C. Mukhopadhyay, S. Ibrahim, Planar electromagnetic sensors based estimation of nitrates contamination in water sources using independent component analysis. IEEE Sens. J. **12**(6), 2024–2033 (2012)
6. M.A. Yunus, S.C. Mukhopadhyay, Novel planar electromagnetic sensors for detection of nitrates and contamination in natural water sources. IEEE Sens. J. **11**(6), 1440–1447 (2011)
7. M.E.E. Alahi, S.C. Mukhopadhyay, L. Burkitt, Imprinted polymer coated impedimetric nitrate sensor for real-time water quality monitoring. Sens. Actuators B **259**, 753–761 (2018)
8. M.E.E. Alahi, L. Xie, S.C. Mukhopadhyay, L. Burkitt, A temperature compensated smart nitrate-sensor for agricultural industry. IEEE Trans. Ind. Electron. **64**(9), 7333–7341 (2017)
9. M.E.E. Alahi, N. Pereira-Ishak, S.C. Mukhopadhyay, L. Burkitt, An internet-of-things enabled smart sensing system for nitrate monitoring. IEEE Internet Things J. **5**(6), 4409–4417 (2018)
10. A.I. Zia, M.S.A. Rahman, S.C. Mukhopadhyay, P.-L. Yu, I.H. Al-Bahadly, C.P. Gooneratne, J. Kosel, T.-S. Liao, Technique for rapid detection of phthalates in water and beverages. J. Food Eng. **116**, 515–523 (2013)
11. A.I. Zia, S.C. Mukhopadhyay, P.-L. Yu, I.H. Al-Bahadly, C.P. Gooneratne, J. Kosel, Rapid and Molecular selective electrochemical sensing of phthalates in aqueous solution. Biosens. Bioelectron. **67**, 342–349 (2015). https://doi.org/10.1016/j.bios.2014.08.050
12. A. Nag, M. Alahi, E. Eshrat, S.C. Mukhopadhyay, Z. Liu, Multi-walled carbon nanotubes-based sensors for strain sensing applications. MDPI Sens. **21**(4), 1261 (2021)
13. N. Anindya, S.C. Mukhopadhyay, J. Kosel, Flexible carbon nanotube nanocomposite sensor for multiple physiological parameter monitoring. Sens. Actuators A **251**, 148–155 (2016)

14. A. Nag, S. Feng, S.C. Mukhopadhyay, J. Kosel, D. Inglis, 3-D printed mould-based Graphite/PDMS sensor for low force applications. Sens. Actuators A: Phys. **280**, 525–534 (2018)
15. N. Afsarimanesh, S.C. Mukhopadhyay, M. Kruger, Performance assessment of interdigital sensor for varied coating thicknesses to detect CTX-I. IEEE Sens. J. **18**(10), 3524–3531 (2018)
16. X. Wang, Y. Wang, H. Leung, S.C. Mukhopadhyay, M. Tian, J. Zhou, Mechanism and experiment of planar electrode sensors in water pollutant measurement. IEEE Trans. Instrum. Meas. **64**(2), 516–523 (2015)
17. A. Nag, B.V.B.R. Simorangkir, E. Valentin, T. BJorninen, L. Ukkonen, M.R. Hashmi, S.C. Mukhopadhyay, A transparent strain sensor based on PDMS-embedded conductive fabric for wearable sensing applications. IEEE Access **6**, 71020–71027 (2018)
18. A. Nag, B. Menzies, S.C. Mukhopadhyay, Performance analysis of flexible printed sensors for robotic arm applications. Sens. Actuators A: Phys. **276**, 226–236 (2018)
19. X. Wang, Y. Wang, H. Leung, S.C. Mukhopadhyay, S. Chen, Y. Cui, A self-adaptive and wide-range conductivity measurement method based on planar interdigital electrode array. IEEE Access, 173157–173165 (2019)
20. A. Nag, A.I. Zia, X. Li, S.C. Mukhopadhyay, J. Kosel, Novel sensing approach for LPG leakage detection: Part I: operating mechanism and preliminary results. IEEE Sens. J. **16**(2), 996–1003 (2016)
21. A. Nag, A.I. Zia, X. Li, S.C. Mukhopadhyay, J. Kosel, Novel sensing approach for LPG leakage detection: Part II: effects of particle size, composition and coating layer thickness. IEEE Sens. J. **16**(2), 1088–1094 (2016)
22. L. Areekath, G. Lodha, S.K. Sahana, B. George, L. Philip, S.C. Mukhopadhyay, Feasibility of a planar coil-based inductive-capacitive water level sensor with a quality-detection feature: an experimental study. Sensors, MDPI **22**(15), 5508 (2022)
23. A.S. Anil Kumar, B. George, S.C. Mukhopadhyay, An eddy-current-based angle sensor with a minimally modified shaft as a sensing element. IEEE Open J. Instrum. Meas. **1**, 1–9 (2022)
24. A.S. Anil Kumar, N. Anandan, B. George, S.C. Mukhopadhyay, Improved capacitive sensor for combined angular and linear displacement sensing. IEEE Sens. J. **19**(22), 10253–10261 (2019)
25. F. Akhter, A. Nag, M.E.E. Alahi, H. Liu, S.C. Mukhopadhyay, Electrochemical detection of calcium and magnesium in water bodies. Sens. Actuators A: Phys. **305**, 111949 (2020)
26. K. Kaur, S.C. Mukhopadhyay, J. Schnepper, M. Haefke, H. Ewald, A zigbee based wearable physiological parameters monitoring system. IEEE Sens. J. **12**(3), 423–430 (2012)
27. A.S. Anil Kumar, B. George, S.C. Mukhopadhyay, Design and development of a variable reluctance-based thin planar angle sensor. IEEE Trans. Ind. Electron. **70**(9), 9653–9662 (2022)
28. N.K. Suryadevara, S.C. Mukhopadhyay, S.D.T. Kelly, S.P.S. Gill, WSN-based smart sensors and actuator for power management in intelligent buildings. IEEE Trans. Mechatron. **20**(2), 564–571 (2015). https://doi.org/10.1109/TMECH.2014.2301716
29. T. Islam, S.C. Mukhopadhyay, Linearization of the sensors characteristics: a review. Int. J. Smart Sens. Intell. Syst. **12**(1) (2019). https://doi.org/10.21307/ijssis-2019-007
30. R.N. Ponnalagu, B. George, V.J. Kumar, An improved direct digital converter for bridge connected resistive sensors. IEEE Sens. J. **16**(10), 3679–3688 (2016)
31. L. Areekath, B. George, F. Reverter, Analysis of a direct microcontroller interface for capacitively-coupled resistive sensors. IEEE Trans. Instrum. Meas. **70**, 1–10 (2021). Art. no. 1501010. https://doi.org/10.1109/TIM.2020.3034969
32. R. Nandagopal, B. George, A simple microcontroller based digitizer for differential inductive sensors, in *Proceedings of IEEE International Instrumentation and Measurement Technology Conference, May 11–14, 2015, Pisa, Italy* (2015), pp 148–153
33. C.S. Anoop, B. George, Electronic scheme for computing inverse-cosine and its application to a GMR based angle sensor. IEEE Trans. Instrum. Meas. **61**(7), 1991–1999 (2012)
34. N. Philip, B. George, Design and analysis of a dual-slope inductance-to-digital converter for differential reluctance sensors. IEEE Trans. Instrum. Meas. **63**(5), 1364–1371 (2014)
35. A. Babu, B. George, An efficient readout scheme for simultaneous measurement from multiple wireless passive LC sensors. IEEE Trans. Instrum. Meas. **67**(5), 1161–1168 (2018)

36. G. Harikumar, B. George, A. Sankaran, S. Haneefa, A closed-loop signal conditioning scheme for core-less planar LVDT. IEEE Trans. Instrum. Meas. **60** (2023). Art. no. 9505010
37. B. George, V.J. Kumar, Switched-capacitor triple slope capacitance to digital converter. IEE Circuits Devices Syst. **153**(2), 148–152 (2006)
38. V. Sreenath, B. George, An improved closed-loop switched capacitor capacitance to frequency converter and its evaluation. IEEE Trans. Instrum. Meas. **67**(5), 1028–1035 (2018)
39. V. Prashanth, B. George, An improved capacitance-to-digital converter for leaky capacitive sensors. IEEE Sens. J. **15**(11), 6238–6247 (2015)
40. A. James, A. Seth, S.C. Mukhopadhyay, IoT enabled sensor node: a tutorial paper. Int. J. Smart Sens. Intell. Syst. **13**(1), 1–18 (2020)
41. B. Shearan, F. Akhter, S.C. Mukhopadhyay, Development of an IoT-enabled portable sulphur sensor: a tutorial paper. IEEE Sens. J. **22**(11), 10075–10088 (2022)
42. S.D.T. Kelly, N.K. Suryadevara, S.C. Mukhopadhyay, Towards the implementation of IoT for environmental condition monitoring in homes. IEEE Sens. J. **13**(10), 3846–3853 (2013)
43. A. Nag, M.E.E. Alahi, S. Feng, S.C. Mukhopadhyay, IoT-based sensing system for phosphate detection using Graphite/PDMS sensors. Sens. Actuators A: Phys. **286**, 43–50 (2019)
44. F. Akhter, H.R. Siddiquei, M.E.E. Alahi, S.C. Mukhopadhyay, *An IoT-Enabled Portable Sensing System with MWCNTs/PDMS Sensor for Nitrate Detection in Water, Measurement*, vol. 178 (Elsevier, 2021), p. 109424
45. N. Afsarimanesh, M.E.E. Alahi, S.C. Mukhopadhyay, M. Kruger, Development of IoT-based impedometric biosensor for point-of-care monitoring of bone loss. IEEE J. Emerg. Sel. Top. Circuits Syst. **8**(2), 211–220 (2018)
46. A. Seth, A. James, S.C. Mukhopadhyay, 1/10th scale autonomous vehicle based on convolutional neural network. Int. J. Smart Sens. Intell. Syst. **13**(1), 1–17 (2020)
47. F. Akhter, S. Khadivizand, H.R. Siddiquei, M.E.E. Alahi, S.C. Mukhopadhyay, IoT enabled intelligent sensor node for smart city: pedestrian counting and ambient monitoring. Sensors **19**, 3374(19 pages) (2019). https://doi.org/10.3390/s19153374
48. W. Zhang, Y. Cao, X. Hu, J. Mi, P. Zhang, S.C. Mukhopadhyay, Y. Li, Z. Liu, IAT: a full-scale IVUS analysis toolbox based on deep learning algorithms for clinical diagnosis. IEEE Trans. Instrum. Meas. **72**, 1–15 (2023)
49. A. Seth, A. James, E. Kuantama, S.C. Mukhopadhyay, R. Han, Drone high-rise aerial delivery with vertical grid screening. Drones, MDPI **7**(5), 300 (2023)
50. N.K. Suryadevara, S.C. Mukhopadhyay, Wireless sensor network based home monitoring system for wellness determination of elderly. IEEE Sens. J. **12**(6), 1965–1972 (2012)
51. R.I. Azad, S.C. Mukhopadhyay, M. Asadnia, Using explainable deep learning in da Vinci Xi robot for tumor detection. Int. J. Smart Sens. Intell. Syst. **14**, 16 (2021), https://doi.org/10.21307/ijssis-2021-017

Flying IoT: Sensor Fusion Performance Analysis for UAV Applications in Indoor Spaces

Alice James, Avishkar Seth, Natalia Chatrath, Endrowednes Kuantama, Richard Han, and Subhas Mukhopadhyay

Abstract In recent times, the integration of Unmanned Aerial Vehicles (UAVs) with Internet of Things (IoT) platforms has brought a new dimension of mobility, significantly enhancing the capabilities of complex sensor networks. This chapter focuses on the development of an autonomous quadcopter, tailored as a core component of an AI-driven sensing network in indoor settings. We explore the integration of various vision-based and laser sensors with an onboard computer, utilizing the Robot Operating System (ROS) and WiFi technology for real-time wireless data collection and communication. Our system achieves significant milestones in 3D mapping and localization, demonstrating the tracking camera T265 sensor's superiority in localization and computational efficiency over traditional SLAM algorithms. Through rigorous testing and calibration, including initial flight trials, we highlight the vast potential of these autonomous systems within an IoT framework. Moreover, we delve into the edge computing capabilities of the onboard Jetson Nano processor, optimizing its performance for enhanced functionality. These optimizations led to full utilization of the processor's capabilities, reducing the mapping time of a 3 m $\times$ 3 m area from 4 to 5 min to under 3 min, with improved accuracy and reduced feature drift. Utilizing advanced AI and computer vision techniques, we test and compare SLAM algorithms, further cementing the role of UAVs in revolutionizing IoT applications.

Keywords Robotics · Sensor applications · UAV · SLAM · Computer vision

---

A. James (✉) · A. Seth (✉) · N. Chatrath · S. Mukhopadhyay
School of Engineering, Macquarie University, Macquarie Park, Sydney, NSW 2109, Australia
e-mail: alice.james@mq.edu.au

A. Seth
e-mail: avishkar.seth@mq.edu.au

E. Kuantama · R. Han
School of Computing, Macquarie University, Macquarie Park, Sydney, NSW 2109, Australia

B. Pradhan and S. Mukhopadhyay (eds.), *IoT Sensors, ML, AI and XAI: Empowering A Smarter World*, Smart Sensors, Measurement and Instrumentation 50,
https://doi.org/10.1007/978-3-031-68602-3_2

1 Introduction

The recent advancements and research in IoT systems have witnessed significant growth, expanding their presence not only in the consumer market but also in critical domains such as military applications [1], logistics [2], fire detection [3], and more [4, 5]. Internet of Flying Things (IoFT) or Flying IoT is a relatively new research domain exploring the integration of drones with IoT expanding applications in various domains due to their mobility, size and capability to access hard-to-reach areas [6]. This convergence of technologies holds the promise of revolutionizing data acquisition, enabling real-time insights, and enhancing decision-making processes across a wide spectrum of industries and sectors.

The fusion of aerial robotics, offering mobile sensing, actuation, and communication capabilities, ushers in a new era for IoT. We focus on micro aerial robots, particularly Vertical Take Off and Landing (VTOL) quadcopters used in complex indoor environments presenting new challenges in sensing and navigation. These aerial sensing machines have shown versatile applications in emergency services [7, 8], warehouse operations [9], and indoor monitoring [10]. Autonomous navigation and sensing in GPS-denied scenarios is challenging [11] due to various reasons such as limited environmental awareness and lack of precise location data.

Communication methods most commonly used in IoT devices are either through WiFi [12, 13], Bluetooth [14], Low-Power Wide-Area Network (LPWAN) [15, 16], cellular, Zigbee, Radio-Frequency Identification (RFID), or Near Field Communication (NFC) [6]. However, due to the space and energy constraints inherent to micro UAVs, the available communication methods are often constrained, with a preference for low-power and short-range options such as Zigbee and Bluetooth. These technologies enable efficient data exchange within the confined and resource-limited environments of micro UAVs. Also, usually a ground control station (GCS) is involved which helps using WiFi and radio connections to communicate with the drone. In our case, we utilise a WiFi connection to the GCS for real time connection and transfer of data.

In this study, we have constructed a prototype UAV and conducted rigorous testing of various sensors and control algorithms to validate its autonomous capabilities. While commercial off-the-shelf UAVs serve as suitable entry points for beginners, they often present limitations, primarily due to their closed-system nature, which restricts the flexibility for additional enhancements and in-depth testing of both software and hardware components [17].

The remainder of the paper is organized as follows. Section II presents literature reviewed in the similar domain and section III describes the proposed system overview. Methods of research are presented in section IV and experimental results are discussed in section V. Finally, section VI presents our conclusions.

2 Literature Review

A combination of sensors is typically employed to enable effective performance and communication in UAVs operation. There are different types of flight controller variations with sensors that are specifically designed to meet the requirements of autonomous UAVs [18, 19]. Among the variations used in UAVs, the most commonly used are inertial measurement units which consist of a 9DoF accelerometer, gyroscope, and magnetometer (compass) with a barometer for altitude estimation [20]. Out of the multiple open-source flight controller options such as Pixhawk series, Parallax Elev-8 [21], Hackflight [22], and Paparazzi [23] are most commonly used for research purposes. We use the Pixhawk 2.4.8 flight controller with Ardupilot autopilot firmware due to its versatile software and hardware functionality. The sensor readings obtained from the flight controller are used in real-time to update the actuation system. Integrating these sensors and the sophisticated algorithms running on the flight controller enables the drones to work together seamlessly, perform precise grasping maneuvers, ensure payload stability, and maintain coordinated flight throughout the operation. A combination of visual and ranging sensors are most commonly used for autonomous navigation of drones, particularly in GPS-denied environments.

Key considerations in sensor choice encompass cost-effectiveness, energy consumption implications on system efficiency, adaptability to environmental conditions, and the constraints imposed by size and weight. Additionally, the update frequency and latency of the sensors are scrutinized in relation to the specific demands of the operational environment and application scenarios. We categorically examine these selection parameters across three sensor types: vision-based, localization-based, and point cloud sensors.

Vision-based sensors like depth cameras, which utilize Time-of-Flight (ToF) or structured infrared light, provide depth information essential for 3D mapping and spatial orientation but can be affected by light conditions and environmental factors [24]. Stereo cameras use binocular disparity to compute depth, offering good perception for mapping despite potential issues with occlusions [10]. Monocular cameras, costeffective for capturing 2D imagery, lack direct depth information [25, 26]. RGBD cameras combine color with depth data, enhancing robustness but with limitations in range and depth accuracy compared to some depth cameras [27].

Location-based sensors provide positioning and orientation data for systems. Inertial Measurement Units (IMUs) offer motion and orientation data by measuring linear acceleration and angular velocity but can drift over time. Visual inertial tracking cameras combine visual data with IMU feedback for enhanced pose estimation, offering high accuracy especially where GPS is unreliable, yet they need frequent calibration [10]. GPS provides broad positioning by triangulating satellite data, effective in open outdoor environments but less accurate indoors or where signals are obstructed [28]. Furthermore, point-cloud based ranging lidar sensors are useful in detection and avoidance close objects in many environments [29].

Sensor fusion methods are categorized into particle filter maps, graph-based maps, and 3D maps for applications in Simultaneous Localization and Mapping (SLAM). Particle filter based algorithms can be advantageous in several scenarios where dealing with uncertainty and noise is crucial [30]. However, graph-based techniques such as Real-Time Appearance-Based Mapping (RTABMap) can be better in scalability, accuracy and superior memory management algorithms [31].

Our research involves developing an indoor aerial robot prototype using open-source software and hardware. We explore the challenges and methodologies of its development. The study also examines the performance of few popular sensors such as RPLidar A1, Intel Realsense D435i, and T265 in varied environmental conditions to assess their suitability for drones. Results from preliminary flight tests are analyzed, and we demonstrate optimization techniques to enhance the prototype's functionality.

3 System Overview

In this study, the aim was to design an autonomous indoor UAV using open source software, enabling the development of custom software and hardware programs. To achieve this, an open source flight controller and a companion computer option were selected as the primary components. The system configuration used in this project is partially adapted from [32]. Figure 1 presents a block diagram divided into two main sections: Hardware and Software, detailing the components and connections within an Unmanned Aerial Vehicle (UAV) system. Figure 2 shows the prototype UAV developed to run the experiments in this project.

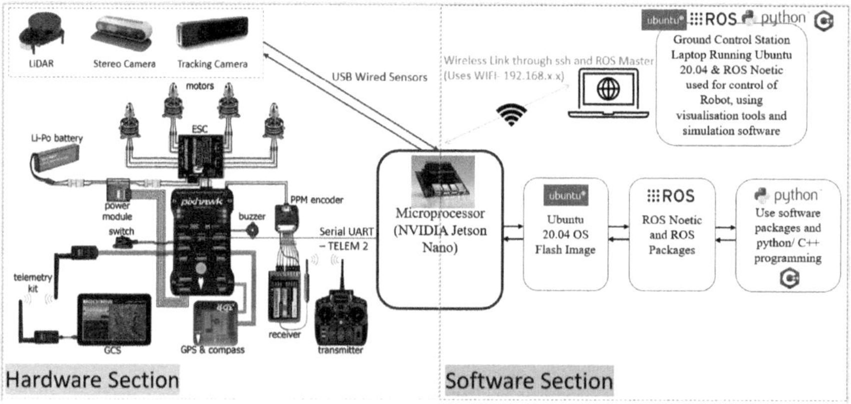

Fig. 1 The block diagram of the flying IoT system showing Pixhawk, NVIDIA Jetson Nano, and other electronics and sensors

Fig. 2 The development of flying IoT UAV prototype **a** T265 + D435i **b** RPLidar + D435i

3.1 Quadcopter Avionics (VTOL)

Referring to Figs. 1 and 2, the quadcopter avionics are built for VTOL capabilities to maneuver indoor spaces. The main element is the flight controller (Pixhawk 2.4.8) running Ardupilot firmware used to connect the main electronic components. The motors are controlled by 30 amp Electronic Speed Controllers (ESC), which regulate the speed of the 1750 kV motors. A 3S 3200 mAH Li-Po (Lithium Polymer) battery provides power to the system, along with a power module with a switch and a buzzer for alerts. The UAV includes a GPS and compass module for positioning and navigation outdoors. Radio telemetry is enabled outdoors mainly through a kit connected to a Ground Control Station (GCS), allowing for remote monitoring and control. Additionally, a PPM encoder (Pulse Position Modulation) is present, for transforming the receiver signals into a format usable by the Pixhawk, and Serial UART (Universal Asynchronous Receiver-Transmitter) communication is indicated, facilitating serial communication to the companion computer. A 295 mm carbon fiber quadrotor frame and 3 bladed 5 inch propellers are used. With a maximum flight duration of eight minutes and a total payload of roughly 1.5 kg, this provides the UAV with a balance of stability and control.

3.2 Computing Platform

Central to the hardware is a microprocessor, the GPU enabled NVIDIA Jetson Nano, which serves as the brain of the UAV. The computer enables complex flight behaviour to a drone. The software is built around an Ubuntu 20.04 Operating System, which is flashed onto the Jetson Nano. The Robot Operating System (ROS) Noetic is used, which is a middleware suite for writing robot software. Python and C++ are the programming languages utilized, indicating the use of software packages and custom programming for UAV operations. The major packages used are OpenCV for computer vision and mavros for communication messages to the flight controller via UART serial mavlink protocol.

3.3 Sensors

The UAV is equipped with various sensors including LiDAR (Light Detection and Ranging) for measuring distances using laser light, a stereo camera for depth perception, and a tracking camera for motion tracking and navigation. We use the 360° RPLiDAR, Intel D435i camera, and Intel T265 tracking camera to perform various autonomous operations. These sensors are connected though a USB 3.0 cable to the Jetson Nano. Additionally, the Pixhawk flight controller also has inbuilt IMU and Barometer sensors enabling it for positioning and altitude estimations.

3.4 IoT Network

The Jetson nano has a wireless link is established using Wi-Fi within the 192.168.x.x IP range, enabling SSH (Secure Shell) and ROS Master connections to the onboard computer. This setup connects the UAV to a laptop, which acts as the Ground Control Station. The laptop runs Ubuntu 20.04 and ROS Noetic, and is utilized for controlling the robot, using visualization tools such as RQT and RVIZ, and running simulation software. The ROS network enables distributed computing allowing it to add multiple machines on the same network and communicate with each node via messages. Usually the Jetson Nano acts as the master and GCS connects to this node via "ROS MASTER URI".

3.5 Simultaneous Localization and Mapping (SLAM)

In indoor settings, GPS is often ineffective, leading UAVs to depend on Simultaneous Localization and Mapping (SLAM) techniques for navigation. SLAM allows a UAV to simultaneously map an unfamiliar environment and track its own position within that map in real-time. This technology is essential for accurate and reliable navigation in indoor spaces, making it a valuable tool for a wide range of applications, including inspection, surveillance, and search and rescue operations. When used in conjunction with ROS, the RTAB-Map (Real-Time Appearance-Based Mapping) makes loop closure detection and map production easier [31]. To do this, it subscribes to different ROS topics that offer sensor and odometry data. RTAB-Map then publishes the generated 2D or 3D maps as well as localisation data throughout the ROS network. The flight controller receives these commands for attitude and position control once this data has been analysed and converted.

4 Methodology

This section delves into the key sensor technologies involved in the sensor fusion process for SLAM. We showcase the initial setup and experimental techniques used to study the UAVs behaviours with the described prototype and environment.

4.1 *Camera Calibration and Depth Estimation*

The RealSense SDK, essential for applications using Intel's depth cameras, interfaces with sensors to collect data, while the wrapper, a higher-level abstraction, facilitates data format conversion and integration with existing frameworks, ensuring crossplatform compatibility. The Intel RealSense T265 Tracking Camera, equipped with fisheye cameras, an IMU module, and a Movidius VPU, enables precise 6DoF tracking through SLAM. It along with the D435i depth camera valued in drone systems for its precise, low-latency localization capabilities, efficiently interfacing with the ROS ecosystem via the librealsense library.

Before calibrating the D435i and D455 depth cameras, issues like noisy images and false depth readings were noted, with the D435i showing more image gaps and distortion due to its narrower field of view. Calibration included tuning both cameras and their integrated IMU, which consists of a Magnetometer, Gyroscope, and Accelerometer, using a Madgwick filter for precise orientation. Calibration of the IMU required data collection from six distinct positions using a Librealsense SDK Python script, considering orientation, temperature, humidity, and magnetic fields. Parameters for the accelerometer involved scale factors (Sx, Sy, Sz), bias (a bias), and off-axis terms (Cxy, Cyx, Czx, Cxz, Czy, Cyz) to ensure accuracy and orthogonality of the axes. The calibration process for the IMU sensor corrected its raw gravitational constant reading from 9.633 to 9.805, but tests showed a real-sense viewer value of 9.32, indicating over a 5% error at position 5, likely due to slight angular misalignment. However, apart from position 5, the calibrated gravitational constant displayed in the viewer was within 1% of the actual 9.81 value, meeting system requirements for IMU accuracy and alignment, with deviations attributed to sensor noise. The calibration of Intel.

RealSense D435i and D455 depth cameras was performed using a checkerboard with a ROS package, considering both intrinsic and extrinsic camera parameters as seen in Fig. 3. Adequate lighting and stable camera mounting minimized external calibration influences, while the ROS topics "/camera/camera info" and "/camera/image raw" verified image capture. Calibration involved a rectification matrix to correct depth maps and a projection matrix to convert 3D to 2D images, with tangential distortion being the most significant error observed, likely due to manufacturing tolerances.

The D435i camera was evaluated for its depth point cloud performance in complex indoor settings, considering distances, surface types, angles, and lighting. Utilizing

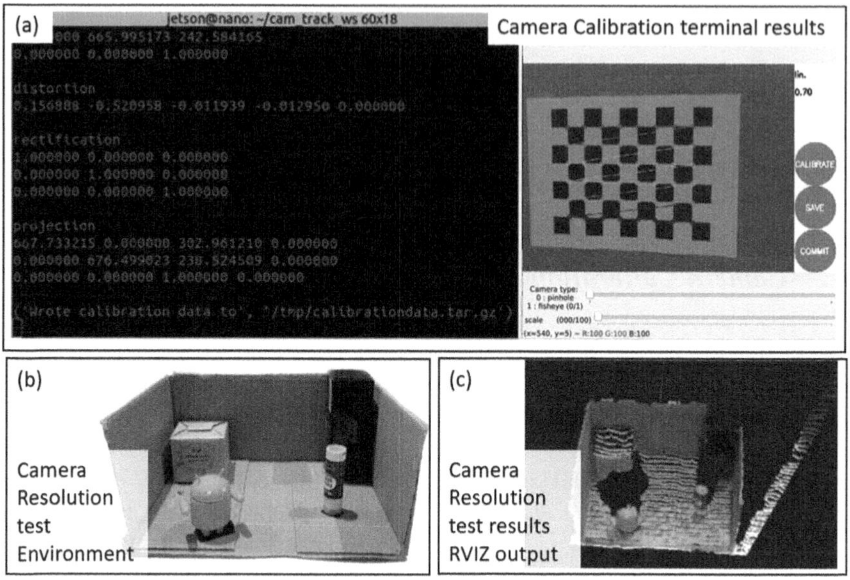

Fig. 3 **a** Standard stereo camera calibration procedure for ROS **b** Test environment and **c** Camera resolution test result of simulated environment

the librealsense SDK or the ROS wrapper, the camera's RGBD data is transformed into useful point clouds and disparity images. Tests on various materials such as Fig. 3 showed the D435i's minimum detection distance was approximately 35 cm, with noticeable declines in object detection clarity, especially with glass, and underperformed compared to another model in low light and resolution. This analysis informs efficient flight path planning and mapping accuracy for 3D spatial reconstructions [33]. The fisheye stereo calibration approach, discussed in method [33] and implemented in OpenCV, incorporates a mechanism designed for fish-eye lens cameras. Initially, images are taken of a predetermined plane with known target positions, such as a checkerboard pattern demonstrated previously in Fig. 3. It is critical to use raw images from the T265, as the standard images are precalibrated. These images should cover the complete field of vision from various distances. This step is the foundation for fisheye calibration, which is mathematically expanded in the next sections.

The basic equation for transforming a point p in the world coordinate frame to a point $p(c)$ in a camera reference frame using rotation R and translation T is given by:

$$p(c) = RP + T \tag{1}$$

with coordinate representations,

$$p = Pc(1), q = Pc(2), r = Pc(3). \tag{2}$$

the coordinates of the pinhole projection of the point p that is constrained within the pinhole model is expressed by, (3)

$$\begin{aligned} g &= \frac{p}{r} \\ h &= \frac{q}{r} \\ Rad_c^2 &= g^2 + h^2 \\ \theta &= \mathrm{g}\tan(Rad_c) \end{aligned} \tag{3}$$

The radial distortion of the fisheye lens is represented by,

$$Rad_c(\theta) = k_1\theta + k_2\theta^3 + k_3\theta^5 + k_3\theta^7 + \cdots$$

further, the points $[p^{'}; q^{'}]$ are calculated by,

$$p' = (\theta(d)/Rad_c)g \tag{4}$$

$$p' = (\theta(d)/Rad_c)h. \tag{5}$$

Lastly, the coordinates are converted into the appropriate pixel coordinates. The description is $[s;t]$ with,

$$\begin{aligned} s &= f(x)(p' + \alpha q') + c(p) \\ s &= f(x)q' + c(q). \end{aligned} \tag{6}$$

It is now possible to calibrate the camera using the estimated points from the bundle adjustment; the following parameters may be determined at the same time: Camera intrinsic parameters, Extrinsic parameters, Stereo parameters (relative), and Estimated point coordinates. Before calibration, the T265 visual inertial camera exhibited significant noise and error accumulation, affecting distance accuracy, relocalisation, and pose tracking. Issues stemmed from IMU sensor biases, scale factor instabilities, and misalignment, particularly integration and drift errors from the gyroscope. Integration errors arose from discretized rotational readings leading to cumulative inaccuracies, despite accelerometers and magnetometers aiding in reducing drift. Initial observations showed the z-axis most prone to drift and overestimation, the x-axis exhibited irregular fluctuations, and all axes faced measurable deviations from the truth as seen in Fig. 4.

Calibration aimed to rectify these errors using either on-chip calibration via librealsense SDK or externally with the Kalibr toolbox, this study chose the latter. Calibration, conducted under controlled conditions (no abrupt movements, adequate lighting, optimal temperature, and increased framerate to reduce motion blur), honed in on intrinsic camera values and IMU noise. Post-calibration results indicated a more consistent drift, unlike the prior erratic behavior, and improved closed-loop accuracy

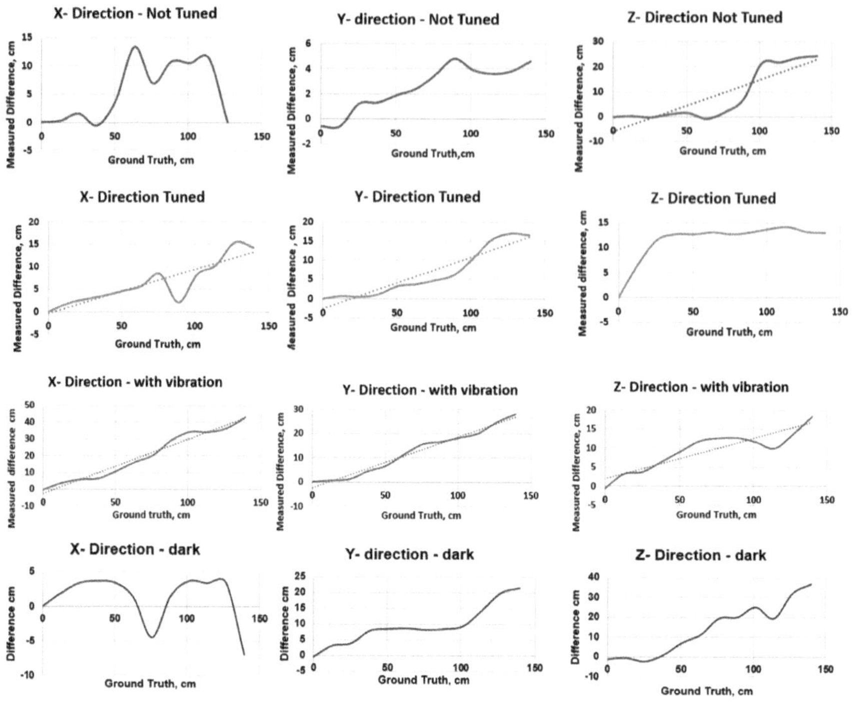

Fig. 4 T265 tracking camera untuned and tuned axis readings and vibration and dark environment testing distance readings

enabling the camera to relocalize to the origin. The z-axis drift became more linear, though the x-axis remained sensitive, exhibiting spikes in measured value, suggesting a refinement of sensor offsets and reduced warping.

The graphs in Fig. 4 depict the measured difference from ground truth for the T265 camera across the X, Y, and Z axes, both before and after tuning. Pre-tuning, the X-axis shows significant fluctuations, the Y-axis demonstrates minor variations, and the Z-axis reveals a consistent overestimation. Post-tuning, while the X and Z axes exhibit a more stable and linear drift, the Y-axis shows an increased drift, indicating a marked improvement in overall measurement accuracy but with a noted increase in Y-axis deviation from the untuned state.

The T265 camera's performance under dark lighting conditions and vibrations was tested to assess its tracking accuracy, which is crucial for indoor use and during the typical vibrations experienced in drone flight. The dark environment tests revealed significant z-axis drift and variable x-axis measurements, while vibration tests showed a linear error rate with up to 45 cm drift across all axes, and a notable tracking halt when the camera faced downwards.

These errors during vibration tests are likely due to forces impacting the IMU's accelerometer, resulting in inaccurate linear acceleration measurements. The T265's

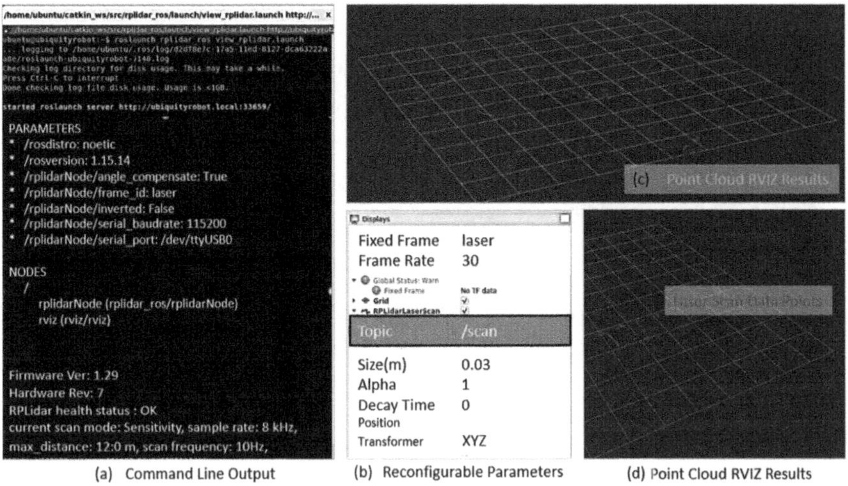

Fig. 5 RPLiDAR RVIZ result point cloud data **a** Command line output **b** Screeshot of reconfigurable parameters **c** Point cloud RVIZ results **d** Zoomed In PCL results

update rate and adaptability to such conditions were also evaluated, emphasizing the need for robust pose data handling under random and abrupt drone movements.

4.2 *LiDAR Point Cloud Data*

The RPlidar A1's spin rate can be modulated via PWM to alter point cloud data frequency, assessed through ROS' rplidar ros package. The lidar's output, covering range, angle, and intensity, is broadcast across three topics and visualized in Rviz after proper leveling shown in Fig. 5. Initial tests showed inconsistencies at maximum frequency, with distant detections flickering, possibly due to reflective interference. Detected range diminishes with distance, and the angular resolution is less suited for detailed indoor mapping necessary for the project.

4.3 *Sensor Fusion*

The proposed algorithm synthesizes camera, IMU, and LiDAR inputs for precise realtime mapping and localization, employing RTAB-Map and SLAM. Sensor fusion melds camera visuals with LiDAR depth, enhancing detection and mapping. RTAB-Map, an open-source SLAM framework, relies on diverse sensors for 3D mapping and localization, optimizing via loop closures. SLAM, central to navigation, balances mapping with self-localization, though costs vary with sophistication. The process

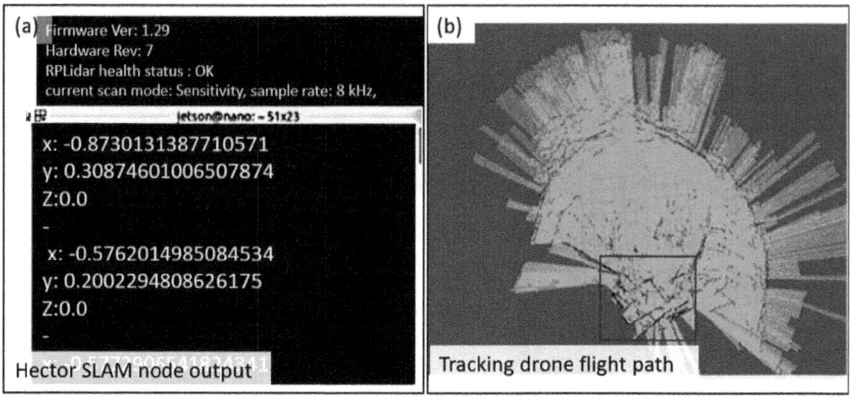

Fig. 6 RVIZ result of HECTOR SLAM LiDAR point cloud data **a** terminal output **b** tracking path output

involves mapping the environment, aligning data with a model, and estimating position to reconcile new and known data, success hinging on the tools and algorithms used.

The code discussed pertains to the extraction of LiDAR point cloud data for mapping, employing ROS functionalities to interpret and visualize spatial data from the surroundings. The 'rostopic map' command processes this data to form an environmental map, while 'rostopic echo' displays the data on a given topic for user inspection. Additionally, the 'hector slam' package's 'hector mapping' algorithm is highlighted for its role in real-time SLAM, creating a 2D occupancy grid map from fused sensor data, which is instrumental for autonomous navigation and accurate self-localization in robotics (see Fig. 6).

The article "ROS: SLAM with RTAB-Map" [34] discusses how the RTAB-Map algorithm within the ROS ecosystem achieves simultaneous mapping and localization by combining data from various sensors like RGB-D cameras and laser range finders. This fusion results in a detailed 3D point cloud and a graph-based map, offering a comprehensive representation of the environment and robust localization by comparing real-time sensor data with the map.

RTAB-Map's sensor fusion system overview is shown in Fig. 7 enhances navigation in diverse conditions, with RGB-D cameras providing depth for dynamic scenes and laser range finders ensuring precision over large areas. The algorithm's ability to fuse sensor information leads to improved mapping and localization, making it versatile for various robotic applications and capable of handling environmental challenges and sensor discrepancies. The camera and Lidar send the messages through ROS topics to the OBC. Using this information, the computer creates the RTABMap and trajectory which is sent back as position and altitude commands to the flight controller. This is done though mavros and mavlink communication through the computer.

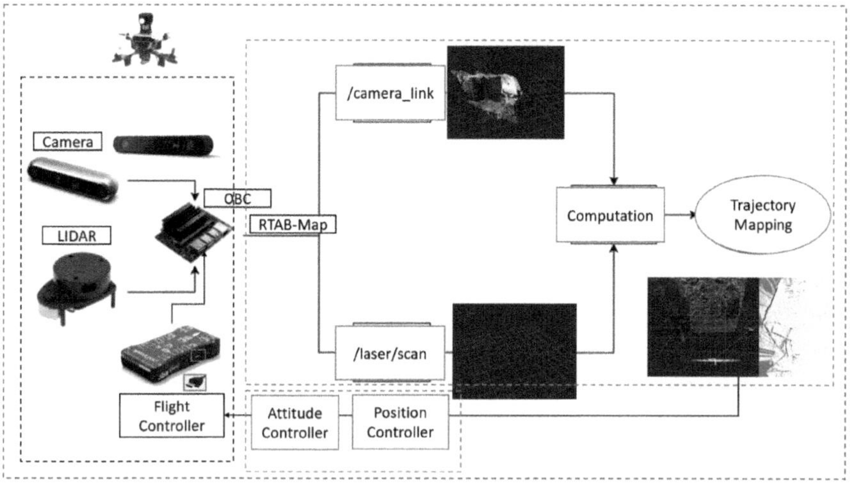

Fig. 7 Overview of RTAB mapping using depth camera and LiDAR

5 Experimental Results

In our experimental arrangement, the Jetson Nano computer is used on board the UAV to perform all computations. Prior to installing propellers, we confirm and test the computational needs and functioning. We also correct any calibration and setup stages that are required for the connection between various sensors, controllers, and actuators on the UAV. A real-world experimental setting is developed to evaluate the performance of the model and algorithm.

5.1 Preliminary Flight Tests

The drones developed in Fig. 2 is used for the tests performed in this section. Conducting flight tests was pivotal in amalgamating data from isolated component evaluations into a single, unified analysis, offering a chance to synergize different system elements previously tested in isolation. These collective trials yielded insightful revelations, offering a complete perspective on the system's operational behavior and efficacy in an authentic setting. Such integrated testing was crucial for confirming the system's functional integrity, pinpointing possible deficiencies, and honing the design for enhanced performance.

Figures 8 and 9 display, on one side, the UAV in its operational habitat, illustrating the practical application during the mapping exercise. On the other, it portrays the LiDAR-generated mapping visual in RVIZ, stemming from the drone's LiDAR readings and manifesting as a detailed point cloud delineating the physical features of the environment. Additionally, there is an image showing the camera's mapping output, synthesizing visual data into a coherent depiction of the surveyed area, thereby facilitating an enriched visual comprehension. Collectively, these images offer a multi-faceted view of the drone's navigational and mapping capabilities, crucial for tasks across surveying, mapping, and related fields.

The two configurations used are the D435i + T265 and the D435i + RPLidar for testing purposes. A comparative analysis of the T265, and Rplidar A1 sensors

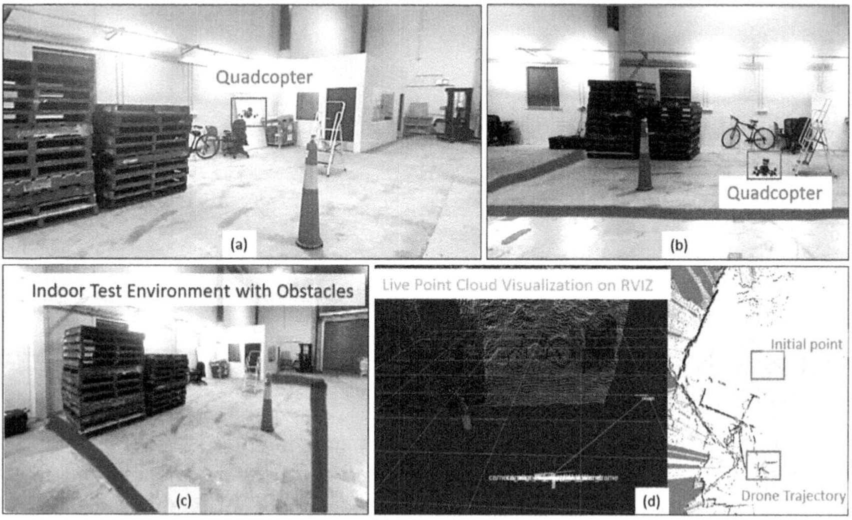

Fig. 8 Indoor flight tests showing the testing area and ROS visualizations for the D435i and RPLidar test **a** and **b** Flying quadcopter **c** Test environment **d** RVIZ live point cloud visualization results showing the trajectory of the drone

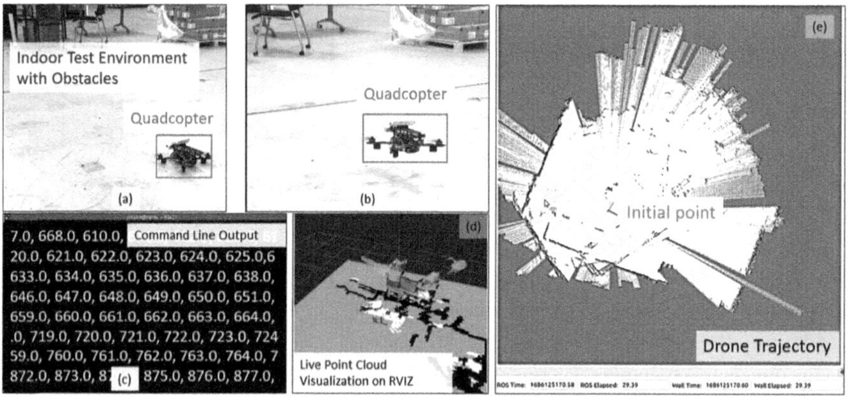

Fig. 9 Indoor flight tests showing the testing area and ROS visualizations for the D435i and T265 test **a** and **b** Flying quadcopter **c** Terminal output **d** and **e** RVIZ live point cloud visualization results showing the trajectory of the drone

Table 1 RPLidar A1 versus T265 performance result comparison

Quality	T265	RPLidar A1
Accuracy	Affected by light and fast motion	Affected by fast motion, low scan
Range	<2 m before drift	<2 m point cloud
Field of view	360° horizontal 180° vertical (cant face down)	360° horizontal 0–10° vertical
Size	108 mm × 25 mm × 13 mm	102 mm × 72 mm × 84 mm
Data Output	2D point cloud	6DoF
Weight	55 g	115 g

(with Rplidar A1), respectively, on these metrics to determine their suitability for the system

shown in Table 1 is conducted to select the most suitable options for a sensor fusion algorithm, based on their performance in various calibration, testing scenarios, and flight tests. The selection criteria hinge on the system's demands for 3D mapping, localization, onboard computing, which include spatial resolution, coverage area, depth accuracy, lighting dependency, power requirements, localization robustness, and operational range. The D435i and T265 are assessed against the D435i and Hector Slam.

The T265 outperforms Hector Slam in terms of localization robustness and computational efficiency, with superior tracking capabilities and built-in processing. Given these evaluations, the D435i and T265 are chosen for the sensor fusion algorithm, aligning with the system's requirement for high-quality data acquisition and efficient processing.

5.2 *Edge Computing Optimisation*

The initial phase of 3D mapping involved handheld navigation through indoor spaces, gauging environmental adaptability. Linked to a Jetson Nano were several components: a D455, T265, Ethernet cable, keyboard, mouse, and monitor. All sensors operated together within the librealsense SDK framework, with data processing and subsampling delivered to the ROS via the realsense2 camera node. The T265 supplies camera streams and inertial data, while the D435i offers depth and color data, along with point clouds. For the 3D map creation, RTabmap and associated tools (GTSAM, g2o, libpointmatcher, SURF/SIFT) were employed. RTABMAP's launch files were modified for dual-camera data interpretation.

The Rtabmap node processes camera data and conducts feature-based tracking, with initial mapping results, taking about 4–5 min to map a 3 m x 3 m area as seen in Fig. 10. Observations revealed issues like minimal texture despite correct colors, latency evident in camera frames, time-consuming structural replication, object and feature drift creating ghost points, double-imaging, and slow odometry recovery. Understanding how the Jetson Nano's hardware impacts algorithm performance is key. Rtabmap's main tasks—feature extraction, matching, bundle adjustment, graph optimization—all demand substantial CPU, GPU, memory, and network capabilities. Optimization depends on the GPU, while memory handles large data sets vital for processing. During initial mapping shown in Fig. 11, only 20% of the Jetson's capabilities were used. Enhancements suggested include utilizing multiple CPU cores, enlarging USB buffer size, and increasing RAM. Adjustments to the Jetson Nano's configuration aimed to optimize processing power and data transfer rates. They were changed to utilize all 4 cores of the Jetson nano, the USB buffer size was increased to 1000 to allow for quick data rate transfer and the RAM was increased using SWAP file in the Ubuntu workspace as seen in Fig. 12. In a warehouse setting, the system setup mirrored previous arrangements but yielded an enhanced map from RTabMap using the D435i and T265 cameras within a 3–4 min span. This map showed notable improvements: dynamic objects were dynamically removed from the map, the system could recalibrate if it went off track, and there were no redundant object edges. The texture and color representation on objects were distinct and precise, with more accurate object placement. However, the surroundings appeared somewhat blurry, although the latency was minimal. The point cloud exhibited clear detection in new areas, with low noise and well-distributed points. Table 2 in the chapter illustrates that these advancements and the efficient use of Jetson Nano's resources contributed to a significantly optimized 3D mapping process.

Table 2 in the chapter illustrates that these advancements and the efficient use of Jetson Nano's resources contributed to a significantly optimized 3D mapping process.

6 Conclusion

In our study, the UAV prototype developed, showed advancements in 3D mapping and localization. Initial tests with the D435i and T265 sensors yielded promising spatial resolution and depth accuracy, crucial for detailed environmental mapping. In comparative analyses, the T265 demonstrated superior localization robustness and computational efficiency over Hector Slam, making it the preferred choice for sensor fusion. The edge computing optimizations on the Jetson Nano were transformative. Initially, the system utilized only 20% of the NVIDIA Jetson's capabilities. Postoptimization, we achieved full utilization of its four CPU cores, expanded the USB buffer size to 1000, and increased RAM through a SWAP file. These enhancements significantly reduced the time to map a 3 m $\times$ 3 m area from 4–5 min to under 3 min, with improved structural replication and reduced feature drift.

The proposed system has several advantages over existing commercial UAVs. First, it is open-source, which allows for customization and modification to meet specific needs. Second, it is cost-effective, as it uses off-the-shelf components. Third, it is lightweight and portable, making it suitable for indoor operations. These advancements have potential implications for various sectors, including surveillance, inspection, and emergency response, where precise and efficient indoor mapping is crucial.

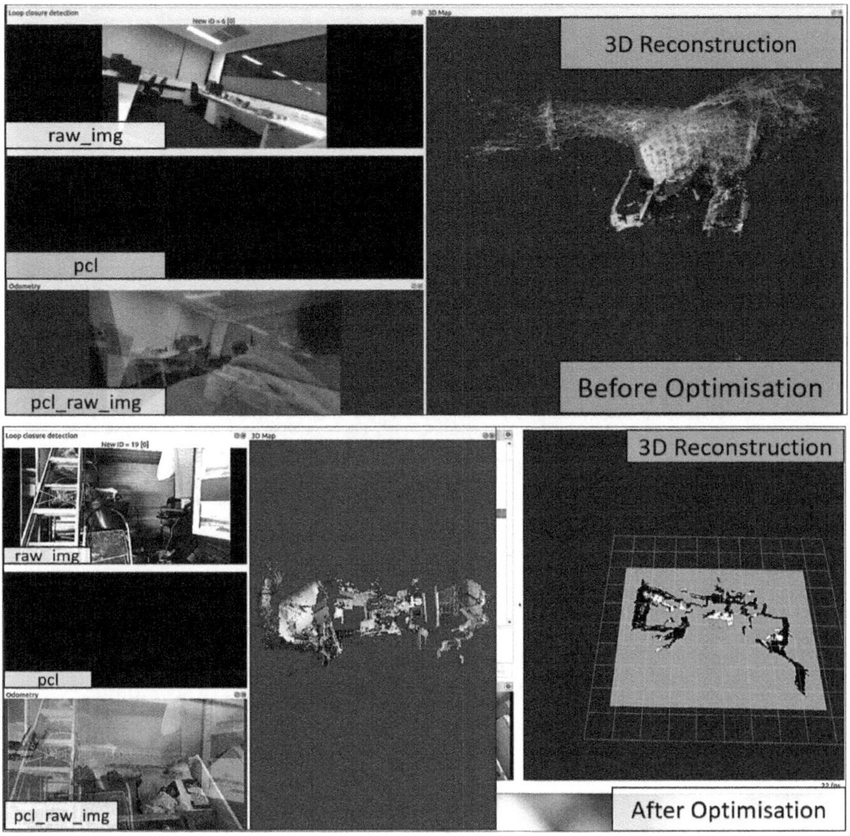

Fig. 10 3D mapping results **a** before and **b** after optimizations performed on the Jetson Nano

Future work could focus on improving the system's autonomy by developing algorithms for obstacle avoidance and path planning. Furthermore, the system offers scalability potential, allowing integration into a network of multiple flying IoT devices. Each UAV in this network would be interconnected through a ROS network, facilitating seamless communication between ground stations and individual UAVs. This architecture enables each UAV to operate as an independent ROS node, ensuring robust and coordinated communication across the network.

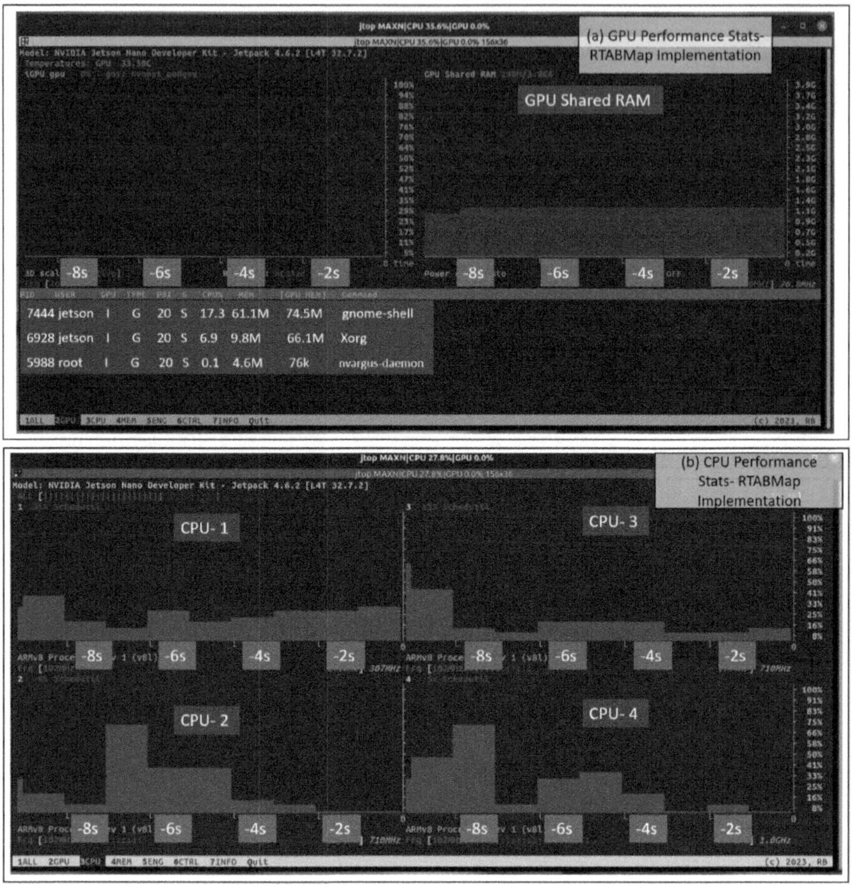

Fig. 11 Initial Jetson Nano Statistics—RTAB implementation **a** GPU performance **b** CPU performance

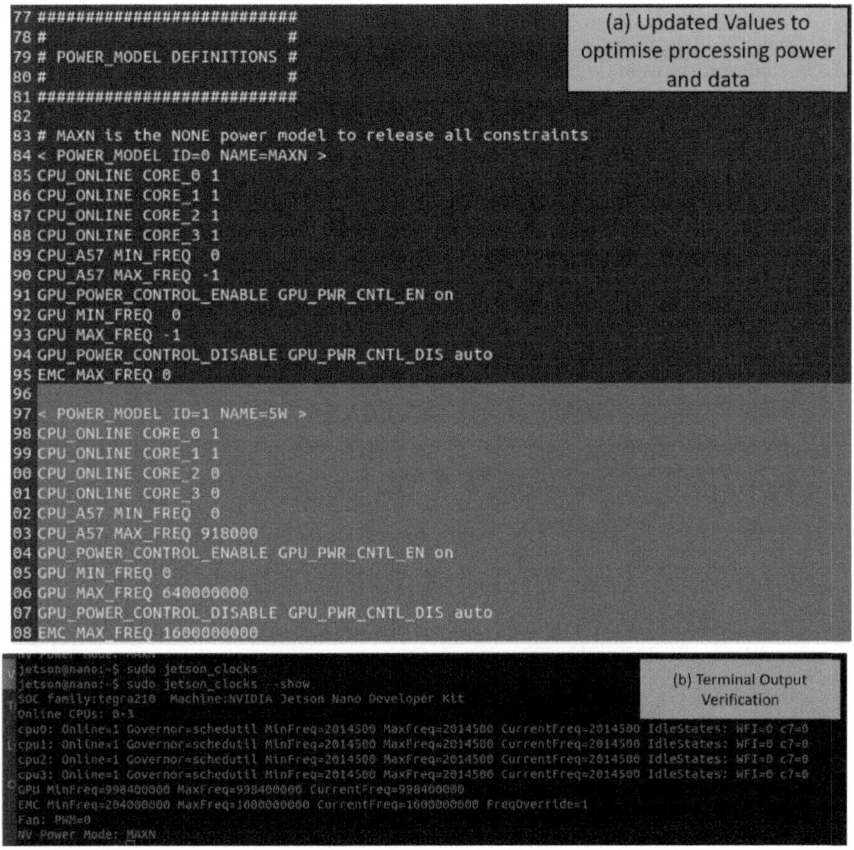

Fig. 12 Jetson nano optimization **a** modified code structure **b** terminal output

Table 2 3D map reconstruction performance comparison

Quality	Not optimised	Optimised
Processing time	~4 min for 3 m	<3 min for 4 m
Map quality	A lot of shifts reducing accuracy	Objects correctly placed—improved accuracy and dynamic object handling
Loop closure	Slow rate of loop detection	Higher and quicker loop detection

References

1. F.T. Johnsen, Z. Zieliński, K. Wrona, N. Suri, C. Fuchs, M. Pradhan, J. Furtak, B. Vasilache, V. Pellegrini, M. Dyk, M. Marks, M. Krzysztoń, Application of iot in military operations in a smart city, in *2018 International Conference on Military Communications and Information Systems (ICMCIS)* (2018), pp. 1–8. https://doi.org/10.1109/ICMCIS.2018.8398690

2. Y. Song, F.R. Yu, L. Zhou, X. Yang, Z. He, Applications of the internet of things (IoT) in smart logistics: a comprehensive survey. IEEE Internet Things J. **8**(6), 4250–4274 (2020)
3. M. Anbarasan, B. Muthu, C.B. Sivaparthipan, R. Sundarasekar, S. Kadry, S. Krishnamoorthy, R. Samuel, A.A. Dasel, Detection of flood disaster system based on iot, big data and convolutional deep neural network. Comput. Commun. **150**, 150–157 (2020) https://doi.org/10.1016/j.comcom.2019.11.022
4. A. James, A. Seth, S.C. Mukhopadhyay, *IoT System Design—A Project Based Approach* (Springer, Cham, 2022), pp. 9–33. https://doi.org/10.1007/978-3-030-85863-6-2
5. W. Shafik, S.M. Matinkhah, F. Shokoor, Cybersecurity in unmanned aerial vehicles: a review. Int. J. Smart Sens. Intell. Syst. **16** (2023)
6. S. Zaidi, M. Atiquzzaman, C.T. Calafate, Internet of flying things (IoFT): a survey. Comput. Commun. **165**, 53–74 (2021). https://doi.org/10.1016/j.comcom.2020.10.023
7. Y. Bi, M. Lan, J. Li, S. Lai, B.M. Chen, A lightweight autonomous MAV for indoor search and rescue. Asian J. Control **21**(4), 1732–1744 (2019). https://doi.org/10.1002/asjc.2162
8. E.M. Moreira, J.C.D. Oliveira, P.F.F. Rosa, A flying iot network to help in disaster recovery, in *2019 Latin American Robotics Symposium (LARS), 2019 Brazilian Symposium on Robotics (SBR) and 2019 Workshop on Robotics in Education (WRE)* (2019), pp. 352–357. https://doi.org/10.1109/LARS-SBR-WRE48964.2019.00068
9. L. Wawrla, O. Maghazei, T. Netland, Applications of drones in warehouse operations. Whitepaper. ETH Zurich, D-MTEC (2019), p. 212
10. Y. Tao, E. Iceland, B. Li, E. Zwecher, U. Heinemann, A. Cohen, A. Avni, O. Gal, A. Barel, V. Kumar, *Learning to Explore Indoor Environments using Autonomous Micro Aerial Vehicles* (2023)
11. T. Elmokadem, A.V. Savkin, Towards fully autonomous UAVs: A survey. Sensors **21**(18) (2021) https://doi.org/10.3390/s21186223
12. A. James, A. Seth, S.C. Mukhopadhyay, *WiFi Based IoT System* (Springer, Cham, 2022), pp. 105–136. https://doi.org/10.1007/978-3-030-85863-6
13. N. Ayadi, N. Derbel, N. Morette, C. Novales, G. Poisson, Simulation and experimental evaluation of the ekf simultaneous localization and mapping algorithm on the wifibot mobile robot. J. Artif. Intell. Soft Comput. Res. **8**, 101–191 (2018)
14. A. James, A. Seth, S.C. Mukhopadhyay, *Bluetooth Based IoT System* (Springer, Cham, 2022), pp. 137–166. https://doi.org/10.1007/978-3-030-85863-6
15. A. James, A. Seth, S.C. Mukhopadhyay, *LoRa Communication Based IoT System* (Springer, Cham, 2022), pp. 167–191. https://doi.org/10.1007/978-3-030-85863-6
16. A. James, A. Seth, S.C. Mukhopadhyay, Iot enabled sensor node: a tutorial paper. Int. J. Smart Sens. Intell. Syst. **13**, 1–18 (2020)
17. H. Lim, J. Park, D. Lee, H.J. Kim, Build your own quadrotor: open-source projects on unmanned aerial vehicles. IEEE Rob. Autom. Mag. **19**(3), 33–45 (2012) https://doi.org/10.1109/MRA.2012.2205629
18. E. Ebeid, M. Skriver, K.H. Terkildsen, K. Jensen, U.P. Schultz, A survey of open-source uav flight controllers and flight simulators. Microprocess. Microsyst. **61**, 11–20 (2018)
19. A. Eltayeb, M.F. Rahmat, M.A.M. Basri, Sliding mode control design for the attitude and altitude of the quadrotor uav. Int. J. Smart Sens. Intell. Syst. **13**, 1–13 (2020)
20. L. Meier, P. Tanskanen, F. Fraundorfer, M. Pollefeys, Pixhawk: a system for autonomous flight using onboard computer vision, in *2011 Ieee International Conference on Robotics and Automation* (IEEE, 2011), pp. 2992–2997
21. D. Norris, *Build Your Own Quadcopter: Power Up Your Designs with the Parallax Elev-8* (McGraw-Hill Education, New York, NY, USA, 2014)
22. S.D. Levy, Robustness through simplicity: a minimalist gateway to neurorobotic flight. Front. Neurorobot. **14**, 16 (2020)
23. G. Hattenberger, M. Bronz, M. Gorraz, Using the paparazzi uav system for scientific research, in *IMAV 2014, International Micro Air Vehicle Conference and Competition 2014* (2014), p. 247

24. J.A. Paredes, F.J. Alvarez, T. Aguilera, F.J. Aranda, Precise drone location´ and tracking by adaptive matched filtering from a top-view ToF camera. Expert Syst. Appl. **141**, 112989 (2020). https://doi.org/10.1016/j.eswa.2019.112989
25. S. Krul, C. Pantos, M. Frangulea, J. Valente, Visual SLAM for indoor livestock and farming using a small drone with a monocular camera: a feasibility study. Drones **5**(2), 41 (2021). https://doi.org/10.3390/drones5020041
26. A. Seth, A. James, E. Kuantama, S. Mukhopadhyay, R. Han, Vertical trajectory analysis using QR code detection for drone delivery application, in *Sensing Technology* (Springer, Cham, Switzerland, 2023), pp. 476–483. https://doi.org/10.1007/978-3-031-29871-4
27. Z. Xu, X. Zhan, B. Chen, Y. Xiu, C. Yang, K. Shimada, A real-time dynamic obstacle tracking and mapping system for uav navigation and collision avoidance with an rgb-d camera, in *2023 IEEE International Conference on Robotics and Automation (ICRA)* (IEEE, 2023), pp. 10645–10651
28. N. Gyagenda, J.V. Hatilima, H. Roth, V. Zhmud, A review of GNSSindependent UAV navigation techniques. Rob. Auton. Syst. **152**, 104069 (2022). https://doi.org/10.1016/j.robot.2022.104069
29. E. Aldao, L.M. González-de Santos, H. González-Jorge, LiDAR based detect and avoid system for UAV navigation in UAM corridors. Drones **6**(8), 185 (2022). https://doi.org/10.3390/drones6080185
30. G. Raja, S. Suresh, S. Anbalagan, A. Ganapathisubramaniyan, N. Kumar, PFIN: an efficient particle filter-based indoor navigation framework for UAVs. IEEE Trans. Veh. Technol. **70**(5), 4984–4992 (2021). https://doi.org/10.1109/TVT.2021.3072727
31. M. Labbé, F. Michaud, Rtab-map as an open-source lidar and visual simultaneous localization and mapping library for large-scale and long-term online operation. J. Field Robot. **36**(2), 416–446 (2019). https://doi.org/10.1002/rob.21831
32. A. Seth, A. James, E. Kuantama, S. Mukhopadhyay, R. Han, Drone highrise aerial delivery with vertical grid screening. Drones **7**(5), 300 (2023). https://doi.org/10.3390/drones7050300
33. T. Appelt, J. Lucht, M. Bleier, A. Nüchter, Calibration and validation of the intel t265 for visual localisation and tracking underwater, in *The International Archives of the Photogrammetry, Remote Sensing and Spatial Information Sciences* ***XLIII-B2-2021*** (2021), pp. 635–641. https://doi.org/10.5194/isprs-archives-XLIII-B2-2021-635-2021
34. Sebastian: ROS: Simultaneous Mapping and Localization with RTABmap. Medium (2022). Accessed 09 Feb 2024

Machine Learning Empowered IoT Devices, Analysis of Indoor and Outdoor Temperature and Health Risks

Md. Aamir Sohail and Naushad Ahmad

Abstract The Internet of Things (IoT) is a widely used technology due to its high applicability. Since it is capable of generating data at high velocity, this data can also be beneficial in machine learning. This paper pre-processed data about temperature generated through IoT devices, followed by analysis, consequences on health, and ML applications. In order to establish a maximum indoor and outdoor temperature threshold for health, this study aims to identify and evaluate the evidence regarding the direct and indirect health effects of Indoor and Outdoor temperatures, as well as the temperature thresholds at which those effects are noticed. Additionally, the data generated by IoT devices are fed into machine learning models for classification as indoor or outdoor based on the given temperature and comparative analysis of predictability for the given dataset. The analysis of data generated by the IOT device is done using available tools in Python. Some valuable insights about data related to temperature and its associated parameters are extracted, and their effect is estimated. Different machine learning models have been applied to the given dataset and measured using available metrics. Overall, this paper contains the analytical part of the data to measure the health risks based on the indoor and outdoor temperature and their classification using the best supporting Machine Learning models.

Keywords Indoor temperature · Machine learning · Internet of Things (IoT) · SVM

1 Introduction

To-days and coming era we are much interested to work in a free environment i.e., without any other human interaction we instruct the device to perform some specific tasks. The IoT provides an omnipresence connection between smart devices with sensors and communication capabilities that are doing your work in cyberspace

Md. A. Sohail · N. Ahmad (✉)
Department of CS and IT, Mahatma Gandhi Central University, Motihari 845401, India
e-mail: naushad13bhu@gmail.com

B. Pradhan and S. Mukhopadhyay (eds.), *IoT Sensors, ML, AI and XAI: Empowering A Smarter World*, Smart Sensors, Measurement and Instrumentation 50,
https://doi.org/10.1007/978-3-031-68602-3_3

i.e., the transformation of data from smart devices to other smart devices over a network. Day by day information and communication technology take new place in an environment. When the Internet of Things appears then information and communication technology have found a special boost to cover all the fields like Agriculture Industries, Automobile Industries, Medical Industries etc. The field of agricultural activities has found the most boost to effective utilization of hard-to-reach resources. These kinds of technologies can give maximum advantage to remotely monitor the preventive measures to detect crop threats and damages, crop growth, soil moisture, soil fertility, and soil erosion [19]. The Internet of Things or sometimes called the Internet of All is a specific paradigm that combines some technologies like as Radio Frequency Identification (RFID), wireless communication technology, sensors network, internet, and embedded systems to shape a system that links the real worlds with cyber worlds [26]. The internet is a collection and combination of more than one network to transfer information/data to each other. In the 1845s, the Russian army started to use Machine to Machine Short (M2M) for data transmission on that time only cables/wired data transfer over the network uses the non-IP protocol. M2M system is hardware-based and used for vertical applications deployed in a closed system. The main aims of M2M communication are automatically triggered actions performed by sequences of events. The next wave of the Internet is the Internet of Things in short (IoT). Kevin Ashton changed the "Machine-to-Machine" communication name to "Internet of Things" in the year 1999. At that time Internet was became to Internet of Things (IoT) [14]. M2M Communication was changed into IoT with reduces the effort of human interaction. IoT always plays a very important role in effective operations for data management [15]. IoT connects devices or sensors that observe and record the physical activity of an environment [23].

To measure the numerous interior environmental parameters, loggers were installed in the various rooms and connected to temperature and relative humidity sensors. A logger roamed the area, taking measurements every hour. The reading material for this study was gathered in both the kitchen and the living room [24]. This article offers a critical assessment of the different investigations into the development of monitoring tools and the various applications of a data logger. This document also offers further details on how to build a simple data logger to monitor indoor and outdoor temperatures [7]. There is proof that the environment affects how common some viral respiratory infections are. Since most people spend their time indoors, markers of the interior environment may be more accurate predictors of the occurrence and spread of disease than those of the external climate. Thus, understanding the interactions between indoor and outdoor surroundings might provide insight into how they affect the seasonality of illnesses caused by respiratory viruses [21]. This work proposes an indoor/outdoor categorization system that determines whether the sensor device is indoors or outdoors using light measurements and machine learning methods. An Arduino-based measuring device is used to take measurements of UV light, color temperature, brightness, and the red, green, blue, and clear components of light every minute [27].

The daily mean inside temperatures are typically 10 °C lower or 10 °C higher than the outside temperatures. Further evidence that the external temperature is not a trustworthy predictor of the inside temperature for the analyzed dwellings is provided by the fact that there is little connection between interior and exterior temperatures in all residences [36]. The mean radiant temperature plays a significant role in our understanding of the radiant heat exchange between the human body and the environment. The evolution of the idea will be covered in this article, along with its qualitative characterization, quantitative evaluation methods, and related challenges [8].

1.1 Research Issues and Motivation

Data Acquisition and Integration: One of the key challenges in machine learning-empowered IoT devices for analysing indoor and outdoor temperature and health risks is the acquisition and integration of diverse data sources. These devices need to collect temperature data from various sensors and sources, both indoors and outdoors, and integrate it with other relevant health-related data such as air quality, humidity, and personal health records [16]. Data Quality and Reliability: Ensuring the quality and reliability of the acquired data is crucial for accurate analysis. IoT devices may encounter issues like sensor inaccuracies, data transmission errors, and missing data, which can affect the reliability of the temperature and health data. Developing techniques to address data quality issues and ensuring data integrity is essential for reliable analysis and risk assessment [2]. Data Privacy and Security: IoT devices collect sensitive personal information, including health-related data. Protecting the privacy and security of this data is a significant concern [29]. There is a need for robust security measures to prevent unauthorized access, data breaches, and misuse of personal health information [39]. Real-Time Analysis and Decision-Making: Analysing temperature data and health risks in real-time is essential for timely intervention and decision-making. Developing efficient algorithms and models that can process and analyse large volumes of data in real time is a challenge. It requires balancing accuracy, computational efficiency, and the ability to provide actionable insights promptly [4]. Interpretability and Explain ability: Machine-learning models used for analysing temperature and health risks in IoT devices may often be complex, making it difficult to interpret and explain the results. Ensuring the interpretability and explain ability of the models is crucial for gaining trust and acceptance from end-users and stakeholders [1].

Health Risk Assessment: Analysing indoor and outdoor temperature and health risks using machine learning-empowered IoT devices can provide valuable insights into the potential health risks associated with temperature variations. By identifying patterns and correlations between temperature data and health outcomes, it becomes possible to assess the risks posed by extreme temperatures, heat waves, or cold spells on vulnerable populations [9]. Early Warning Systems: Real-time analysis of temperature data combined with health-related information can enable the development of early warning systems for health risks [28]. By continuously monitoring temperature levels and analysing health indicators, it becomes possible to detect and predict potential health emergencies, such as heatstroke, hypothermia, or respiratory illnesses, allowing for timely interventions and mitigations [6].

Personalized Health Monitoring: Machine learning-empowered IoT devices can enable personalized health monitoring by integrating temperature data with individual health profiles. This integration can help in assessing the impact of temperature on individuals with specific health conditions, identifying triggers for health exacerbations, and providing personalized recommendations for managing temperature-related health risks [38]. Urban Planning and Infrastructure Design: analysing indoor and outdoor temperature data can contribute to urban planning and infrastructure design. Understanding temperature variations in different locations and environments can help optimize building designs, HVAC systems, and urban green spaces. It can also aid in the development of climate-responsive cities and energy-efficient systems [25]. Public Health Policies: The insights gained from analysing temperature and health risks can inform the development of public health policies and interventions. By understanding the relationship between temperature, health outcomes, and vulnerable populations, policymakers can implement targeted interventions to mitigate risks, improve public health, and reduce healthcare costs [17]. Overall, the combination of machine learning, IoT devices [20], and temperature data analysis holds significant potential for enhancing health monitoring, risk assessment, and decision-making, ultimately leading to improved public health and well-being.

Organization of the paper Sect. 2 describe the literature review with Tables 1, 2 and 3. The Sect. 3 Internet of Things (IoT) and Machine learning analysis. Section 4 Indoor and Outdoor temperature analysis, Sect. 5 materials methods and methodology with Tables 4 and 5 and Figs. 1, 2, 3, 4 and 5. The Sect. 6 results and subsection contains the analysis of the results. The Sects. 7 and 8 contains the future perspective and advancement, and conclusion.

2 Literature Review

In Kumar et al. [12] this paper author develops mobile Internet of Things (IoT) system that enabled monitoring of quality in nature by analyzing Carbon Monoxide, Particulate Matter (PM), and Smoke to measure real-time surrounding data. This system can measure the local area air quality and generate alerts through a buzzer device that was fabricated into the system. The proposed system uses MQ2, MQ9 gas sensor, and PMS3003G3 particle sensor. The author wrote that some major advantages of the proposed system are cost-efficient, small, and portable. The system uses "ThingSpeak", ThingSpeak is a data collection method in the cloud with analysis of live data using MATLAB. The implementation of the paper is a low-cost Internet of Things and low power consumption-based air pollution monitoring system. It can install ubiquity, it can smoothly provide operational efficiency anywhere and everywhere than conventional wired models. In the future scope, the author wrote we can make that system more usable in the other sectors and more user friendly.

Table 1 Related studies and progress

References	YOP	Research objective	Component
Picallo-Perez et al. [22]	2021	Ventilation of buildings with heat recovery systems	Temperature, CO_2
Tham et al. [32]	2020	Indoor temperature and health	Temperature
Dzien et al. [5]	2020	Measurements of the forehead temperature recorded in the frigid environment	Temperature
Wang and Norbäck [35]	2022	Adult perceptions of the visible and measured interior environment's thermal comfort and interior air quality	Relative Humidity (RH), Temp
Ullah et al. [34]	2021	Internet of Things-based intelligent greenhouse climate control that consumes less energy	Temp, RH, CO_2, SR, WS
Tashakkori et al. [31]	2021	An IoT-based system for monitoring beehives	Heat, humidity, and weight
Mohamed et al. [18]	2021	Air quality indoors and overheating	Temp, CO_2, PM2.5
Xie et al. [37]	2020	A pig barn's interior environment is modeled using mechanical ventilation	H2S, CO_2, Temp, NH_3

Table 2 Common daily activity classification

Different types of outdoor events	Different types of indoor events
Adventure/mountaineering	Dinner
Sports/javelin throw	Entertainment/film
Environmental protection	Fitness/gym
Bicycle/walking	Fashion
Swimming	Banquet
Corporate meeting	Amusement
Sky diving	Lecture

Table 3 Pollutant gases and sensors with pollutant causes and disease

Sensors	Pollutant	Diseases	Causes
MQ131	Ozone gas	Asthma	Smog
MQ2	Methane	Arrhythmia	Mobile combustion
MQ135	NO_2	Cough and wheezing	Cars, trucks, and buses
MQ7	CO	Chest pain	Incomplete combustion
DHT11	Temp	Heatstroke	Cutting down forests

Table 4 Mean temperature and yearly mean temperature difference

S. no.	IN/OUT	Mean temperature (T_m)	Yearly mean temperature difference (Y_t)
1	IN	30.6999	30.2679
2	OUT	39.7658	38.0553

Table 5 Month wise mean indoor-outdoor temperature

S. no.	Month	IN temperature	OUT temperature
1	January	30.5060	40.2723
2	February	31.1428	40.0873
3	March	31.1359	36.9694
4	April	29.8434	38.5498
5	May	29.3707	39.5369
6	June	27.2981	33.9758
7	July	29.0086	35.1966
8	August	29.5671	33.9621
9	September	31.2535	**42.4040**
10	October	31.3138	41.0003
11	November	30.8531	38.5675
12	December	**31.9220**	36.1413

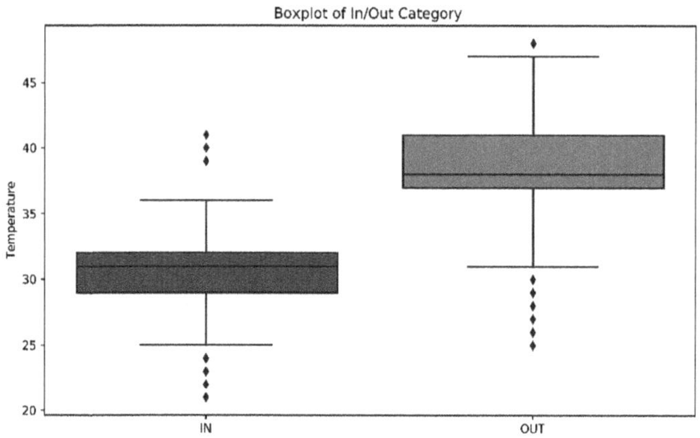

Fig. 1 Box plot of mean indoor-outdoor temperature

In Idrees and Zheng [10] this paper the author gives a very comprehensive review but in short of the (APMS) Air Pollution Monitoring Systems. The author categorizes the recent research work into two different parts first one is static and the second one is the mobile air monitoring system. The monitoring system on the protocols and emerging technology like as Internet of Things devoted system, portable monitoring

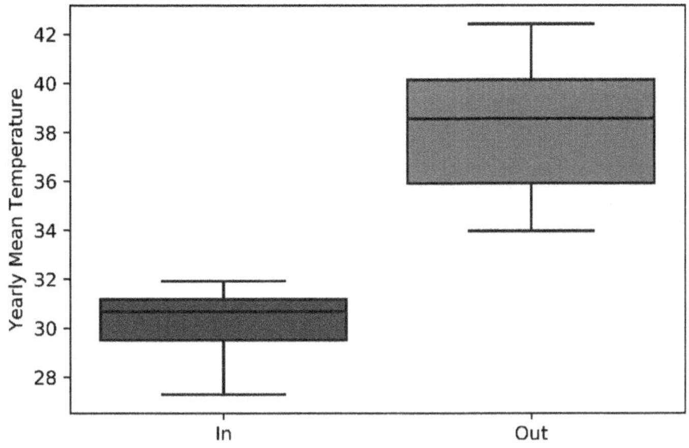

Fig. 2 Box plot of yearly mean indoor-outdoor temperature

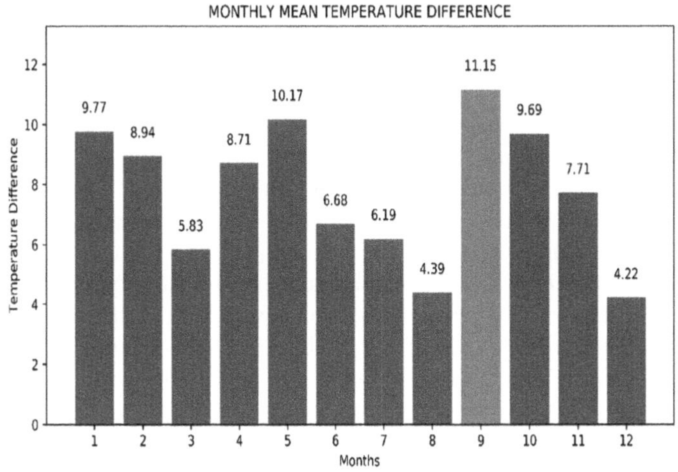

Fig. 3 Monthly mean temperature difference of Indoor and Outdoor temperature

devices, Wireless Sensor Network based systems, and other more. The main objective of this research review to need to enlist an air pollution monitoring approach to the formation of more realistic, reliable, and consistent in the future. In Kamgueu et al. [11] this paper's contribution of the author divided the whole article into two parts. The first part contains the selection of the most suitable for connecting the internet with WSN in such a way that low-cost method and efficient, to develop a mathematical model using integer linear programming. The second fold contains the integration of WSN to the internet with the network topology where a place is built. The proposed method has been conducted to indicate the successful production of desired results of real deployments of systems and research experiments. In Srivastava

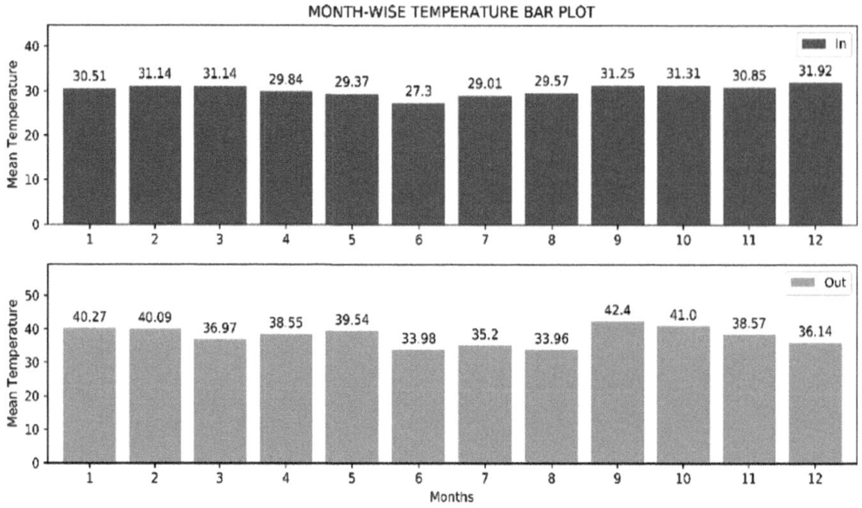

Fig. 4 Month wise Bar plot of indoor-outdoor temperature

et al. [30] this article researchers used the Internet of Things to develop an air quality monitoring system using the Arduino system. Researchers collect the data from different monitoring stations. This paper uses Big Data concepts because of the large value of data collected through different monitoring stations and how data affects air pollution quality. These concepts demonstrate data analytics with an in-depth study of air quality monitoring systems. The authors [33], the variances in the interior temperature and humidity had a significant influence on the subjects' physiological and subjective responses, which laid the groundwork for the disparity in cognitive performance. These results showed that by lowering indoor humidity at very high temperatures, impaired cognitive performance might be improved.

In Table 1 authors give reference of recent year of publication, research objective of the same literature with base component temperature are listed.

In Table 2 researchers classify the outdoor and indoor events in vary systemetic manners. Outdoor activity like sports, walking, etc. and indoor activity like dinner, lecture etc.

The Table 3 author listed base Internet of Things sensors like MQ131, MQ135 and DHT11 use for temperature sensing. Further listed diseases and causes due to the same sensors and pollutant concentrations. Therefor diseases heatstroke to begin with consequences of high temperature because of cutting down the forests.

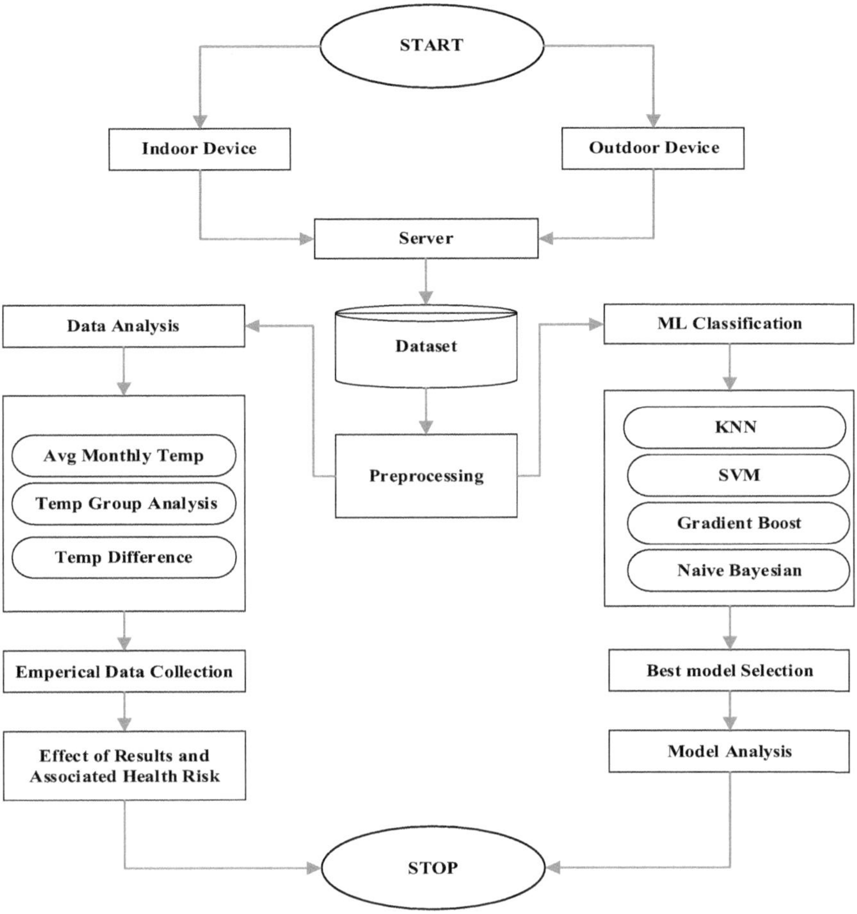

Fig. 5 Architecure of the indoor-outdoor temperature operation

3 IoT Devices and Machine Learning Integration

The integration of IoT devices and machine learning is a rapidly evolving field, and there are numerous research papers and articles that discuss its various aspects. Atakan et al. highlights the combination of IoT and machine learning for data analysis and decision-making in their paper "Machine Learning for IoT Analytics: A Survey". They provide a comprehensive overview of how machine-learning techniques can be applied to IoT data to extract meaningful insights [16]. Bonomi et al. introduce the concept of fog computing and its role in the Internet of Things in their paper "Fog Computing and Its Role in the Internet of Things". They discuss how fog computing, which extends cloud computing capabilities to the edge of the network, can enable efficient data processing and machine learning inference on IoT devices

[3]. The paper "Edge Intelligence: Paving the Last Mile of Artificial Intelligence with Edge Computing" by Duan et al. explores the concept of edge intelligence and its significance in IoT and machine learning integration. They emphasize how edge computing can facilitate real-time analytics and decision-making at the network edge [40].

4 Indoor and Outdoor Temperature Analysis

4.1 Data Collection and Sensor Integration

Collecting indoor temperature data typically involves the use of sensors or data loggers placed strategically within the indoor environment.

4.2 Energy Consumption Optimization

Energy consumption optimization involves identifying and implementing strategies to reduce energy usage while maintaining desired levels of comfort, productivity, or operational requirements.

4.3 Thermal Comfort Enhancement

Enhancing thermal comfort involves creating indoor conditions that promote the satisfaction and well-being of occupants in terms of thermal conditions.

4.4 Mitigation of Heat Stress and Hypothermia Risks

Mitigating heat stress and hypothermia risks involves implementing measures to prevent or minimize the adverse effects of extreme heat and cold temperatures on individuals.

5 Materials and Methodology

Internet-of-Things data was collected from the Kaggle (https://www.kaggle.com/datasets/atulanandjha/temperature-readings-iot-devices) that have 5 columns id, room_id, noted_data, temp, out/in. Temperature readings in and out of a room in an office building (admin). This was timed at random and was recorded. The pace of recording was one second. IoT devices were installed outside and inside a room at an organization for a pilot testing project to measure the heat index. Because the device was still in the alpha testing stage, the data has a random quality in the middle and outliers, which makes the analysis more interesting. Data have been taken from Kaggle. Temperature reading of an enterprise building room, both inside and outside. This was recorded at random intervals. The recording speed was per second. Data is also pre-processed and redundancy was removed making data looks more concise [13].

N_{in} = Number of samples in indoor category

N_{out} = Number of samples in outdoor category

N_m = Number of months samples for the given year.

$$T_m = \sum_{i=1}^{i=n} \frac{t_i}{N_{in/out}} \tag{1}$$

$$Y_t = \sum_{j=1}^{j=12} \frac{T_{mj}}{N_m} \tag{2}$$

Indoor Category, $T_m > Y_t$ Outdoor Category, $T_m > Y_t$.

In Fig. 1 the box plot of Indoor and outdoor temperature is eventually very clearly.

In the same manner Fig. 2 the box plot of yearly mean Indoor and outdoor temperature is eventually very clearly.

In this table authors listed month with monthly mean indoor and outdoor temperature.

As we can see in Fig. 5, mean temperature difference i.e., T_{md} is maximum during the month of September, depicted as "9". We can see that the mean temperature difference between indoor and outdoor is maximum during the month of September (9).

6 Results

The use of machine learning-enabled IoT sensors for the analysis of indoor and outdoor temperature data and health hazards have produced extremely positive results. We have made important progress in comprehending and mitigating temperature-related health hazards.

Assuming the indoor temperatures to be constant, since the mean monthly differences are not so much. Outdoor temperature during September month is relatively high resulting in high temperature difference. In that scenario, there is a chance of more health risks.

In terms of classification of temperature, Support Vector Machine (SVM) is able to classify the temperature from the given data, with more accuracy, precision and recall value. Other models like Gradient Boost Classifier, Ridge Classifier and AdaBoost Classifier are not comparatively performing well (see Table 6).

6.1 *Analysis of the Results*

In Fig. 6, the Ridge Classifier has the lowest accuracy and a downward trending line, while the SVM has the highest accuracy of the other five models at 0.99988. In Fig. 7, the Ridge Classifier has the lowest accuracy and a falling trending line, while the SVM has the highest precision (0.99988), outperforming the other five models. Figure 8 shows that the SVM has the highest recall (0.99988), surpassing the other five models, while the Ridge Classifier has the lowest accuracy and a decreasing trending line. In Fig. 9, the SVM has the greatest f1-score of 0.99988 among the other five models, while the Ridge Classifier has the lowest accuracy and a downward sloping line.

Table 6 Models classification results with accuracy, precision, recall and f1-score

Performance evaluations	SVM	KNN	Naïve Bayes	Ridge classifier	Gradient boosting classifier	AdaBoost classifier
Accuracy	**0.99988**	0.91117	0.95974	0.83008	0.87521	0.87521
Precision	**0.99988**	0.91117	0.95974	0.83008	0.87521	0.87521
Recall	**0.99976**	0.91117	0.95974	0.83008	0.87521	0.87521
F1-score	**0.99984**	0.94214	0.97287	0.83008	0.87521	0.87521

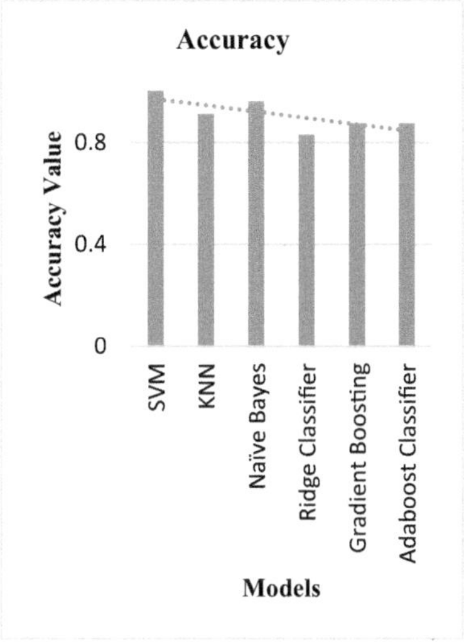

Fig. 6 Accuracy of all models

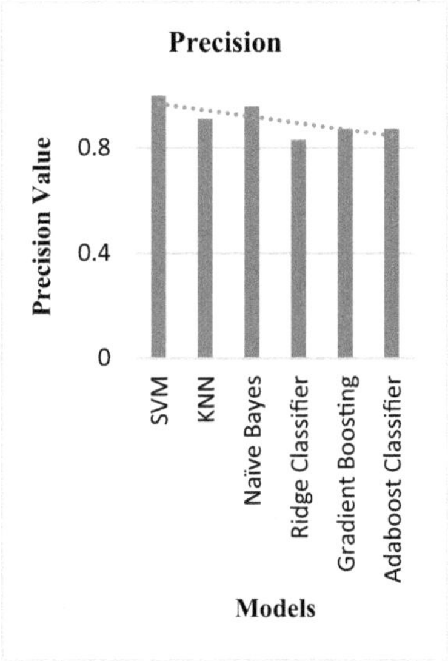

Fig. 7 Precision of all models

7 Future Perspectives and Advancement

Future perspectives and advancements in the field of thermal comfort and temperature regulation include the integration of smart building technologies for real-time monitoring and control, personalized thermal comfort solutions using wearable technologies and personalized sensing, predictive analytics and machine learning for accurate forecasting and optimization, advancements in insulation materials and passive

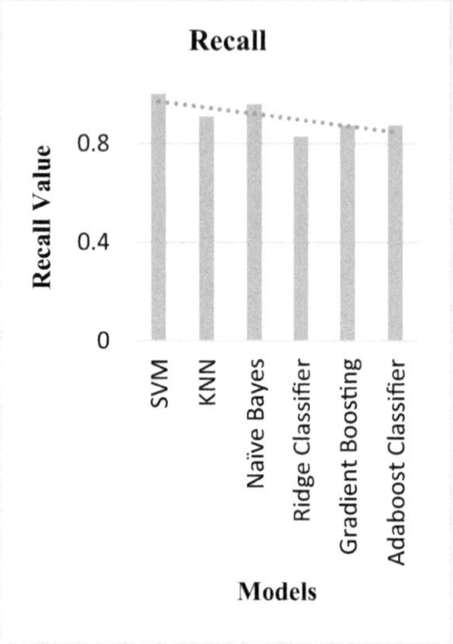

Fig. 8 Recall of all models

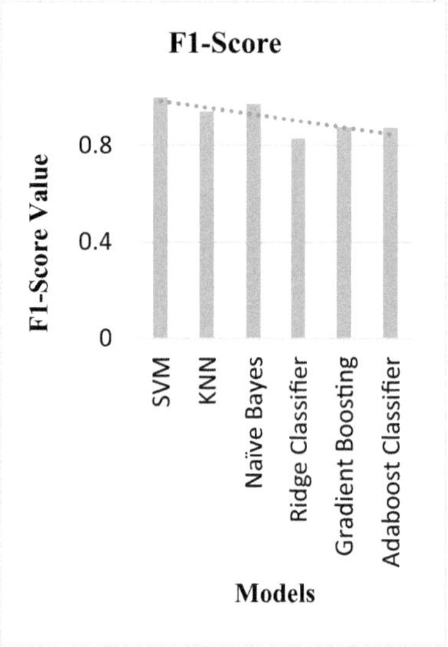

Fig. 9 F1-Score of all models

design strategies for improved energy efficiency, renewable energy integration for sustainable heating and cooling, human-centric design approaches prioritizing occupant well-being, and global sustainability initiatives driving green building standards and certifications. Collaboration among researchers, engineers, architects, policymakers, and industry professionals will be crucial in translating these advancements into practical applications that enhance comfort, reduce energy consumption, and promote sustainability.

8 Conclusion

As a result, machine learning-enabled IoT sensors have revolutionized the analysis of indoor and outdoor temperature data and its consequences for human health. These developments give us important information on how to manage temperature-related health hazards, optimize energy use, improve thermal comfort, and forecast extreme weather occurrences. Machine learning and IoT integration have significant promise for developing smarter, more resilient settings that prioritize human well-being in the face of changing climatic circumstances as technology continues to advance.

In terms of outdoor temperature, September month has the highest average temperature which may give an insight about the geographical location from where data has been taken and further inference can be drawn followed by an in depth location centric study and analysis can be done.

References

1. I. Ahmad, K. Chang, Mission critical user priority-based random access scheme for collision resolution in coexisting PS-LTE and LTE-M networks. IEEE Access **7**, 115505–115517 (2019). https://doi.org/10.1109/ACCESS.2019.2934778
2. N. Ahmed, D. De, I. Hussain, Internet of Things (IoT) for smart precision agriculture and farming in rural areas. IEEE Internet Things J. **5**(6), 4890–4899 (2018). https://doi.org/10.1109/JIOT.2018.2879579
3. F. Bonomi, et al., Fog computing and its role in the internet of things, in *Proceedings of the First Edition of the MCC Workshop on Mobile Cloud Computing*. (ACM, New York, NY, USA, 2012), pp. 13–16. https://doi.org/10.1145/2342509.2342513
4. M. Domb, Smart home systems based on Internet of Things, in *IoT and Smart Home Automation [Working Title]* (IntechOpen, 2019), pp. 2073–2075. https://doi.org/10.5772/intechopen.84894
5. C. Dzien et al., Covid-19 screening: are forehead temperature measurements during cold outdoor temperatures really helpful? Wien. Klin. Wochenschr. **133**(7–8), 331–335 (2021). https://doi.org/10.1007/s00508-020-01754-2
6. K.L. Ebi et al., Heat watch/warning systems save lives: estimated costs and benefits for Philadelphia 1995–98. Bull. Am. Meteor. Soc. **85**(8), 1067–1074 (2004). https://doi.org/10.1175/BAMS-85-8-1067
7. V. Essien, et al., Application of data logger for monitoring indoor and outdoor temperature of buildings: a review, in *IOP Conference Series: Materials Science and Engineering*, vol. 1107(1) (2021), p. 012181. https://doi.org/10.1088/1757-899X/1107/1/012181
8. H. Guo et al., On the understanding of the mean radiant temperature within both the indoor and outdoor environment, a critical review. Renew. Sustain. Energy Rev. **117**(June), 109207 (2020). https://doi.org/10.1016/j.rser.2019.06.014
9. P.D. Hunter et al., The effect of risk perception on public preferences and willingness to pay for reductions in the health risks posed by toxic cyanobacterial blooms. Sci. Total. Environ. **426**, 32–44 (2012). https://doi.org/10.1016/j.scitotenv.2012.02.017
10. Z. Idrees, L. Zheng, Low cost air pollution monitoring systems: a review of protocols and enabling technologies. J. Ind. Inf. Integr. **17**, 100123 (2020). https://doi.org/10.1016/j.jii.2019.100123
11. P.O. Kamgueu, E. Nataf, T. Djotio, Architecture for an efficient integration of wireless sensor networks to the Internet through Internet of Things gateways. Int. J. Distrib. Sens. Netw. **13**(11), 155014771774473 (2017). https://doi.org/10.1177/1550147717744735

12. A. Kumar, M. Kumari, H. Gupta, Design and Analysis of IoT based Air Quality Monitoring System', in *2020 International Conference on Power Electronics & IoT Applications in Renewable Energy and its Control (PARC)*. (IEEE, 2020), pp. 242–245. https://doi.org/10.1109/PARC49193.2020.236600
13. V. Kumar, N. Ahmad, Deep learning for air quality prediction after COVID-19 pandemic based on pollutant and meteorological data. SSRN Electron. J. (2022). https://doi.org/10.2139/ssrn.4292346
14. L. Tan, N. Wang, Future internet: the Internet of Things, in *2010 3rd International Conference on Advanced Computer Theory and Engineering(ICACTE)* (IEEE, 2010), pp. V5-376–V5-380. https://doi.org/10.1109/ICACTE.2010.5579543
15. Ma M, Wang P (2013) Data management for Internet of Things: challenges, approaches and opportunitieshttps://doi.org/10.1109/GreenCom-iThings-CPSCom.2013.199
16. M.S. Mahdavinejad et al., Machine learning for internet of things data analysis: a survey. Digit. Commun. Netw. **4**(3), 161–175 (2018). https://doi.org/10.1016/j.dcan.2017.10.002
17. A.J. McMichael et al., International study of temperature, heat and urban mortality: the "ISOTHURM" project. Int. J. Epidemiol. **37**(5), 1121–1131 (2008). https://doi.org/10.1093/ije/dyn086
18. S. Mohamed et al., Overheating and indoor air quality in primary schools in the UK. Energy Build. **250**, 111291 (2021). https://doi.org/10.1016/j.enbuild.2021.111291
19. S. Navulur, S.C.S. Sastry, M.N. Giri Prasad, Agricultural management through wireless sensors and Internet of Things. Int. J. Electr. Comput. Eng. (IJECE)**7**(6), 3492 (2017)https://doi.org/10.11591/ijece.v7i6.pp3492-3499
20. A. Ojha, S.P. Sahu, D.K. Dewangan, *VDNet: Vehicle Detection Network Using Computer Vision and Deep Learning Mechanism for Intelligent Vehicle System* (2022), pp. 101–113. https://doi.org/10.1007/978-981-16-3097-2_9
21. J. Pan et al., Correlating indoor and outdoor temperature and humidity in a sample of buildings in tropical climates. Indoor Air **31**(6), 2281–2295 (2021). https://doi.org/10.1111/ina.12876
22. A. Picallo-Perez et al., Ventilation of buildings with heat recovery systems: thorough energy and exergy analysis for indoor thermal wellness. J. Build. Eng. **39**, 102255 (2020). https://doi.org/10.1016/j.jobe.2021.102255
23. E.S. Pramukantoro, H. Anwari, An event-based middleware for syntactical interoperability in Internet of Things. Int. J. Electr. Comput. Eng. (IJECE) **8**(5), 3784 (2018). https://doi.org/10.11591/ijece.v8i5.pp3784-3792
24. H. Rau, et al., Changing energy cultures? Household energy use before and after a building energy efficiency retrofit. Sustain. Cities Soc.**54**(November 2019), 101983 (2020). https://doi.org/10.1016/j.scs.2019.101983
25. I. Raya-Moreno et al., Comparing current chemical methods to assess biochar organic carbon in a Mediterranean agricultural soil amended with two different biochars. Sci. Total. Environ. **598**, 604–618 (2017). https://doi.org/10.1016/j.scitotenv.2017.03.168
26. A. Rghioui, A. Oumnad, Internet of Things: surveys for measuring human activities from everywhere. Int. J. Electr. Comput. Eng. (IJECE) **7**(5), 2474 (2017). https://doi.org/10.11591/ijece.v7i5.pp2474-2482
27. M.B. Rhudy, et al., Indoor and outdoor classification using light measurements and machine learning. Appl. Artif. Intell. **36**(1) (2022). https://doi.org/10.1080/08839514.2021.2012001
28. S. Sahu, S.P. Sahu, D.K. Dewangan, *Pedestrian Detection Using MobileNetV2 Based Mask R-CNN* (2023a), pp. 299–318. https://doi.org/10.1007/978-981-19-5845-8_22
29. S. Sahu, S.P. Sahu, D.K. Dewangan, *Pedestrian Detection Using ResNet-101 based Mask R-CNN* (2023b), p. 020008. https://doi.org/10.1063/5.0134276
30. C. Srivastava, S. Singh, A.P. Singh, *Intelligent Systems, Technologies and Applications*, ed by S.M. Thampi, et al. (Singapore, Springer Singapore (Advances in Intelligent Systems and Computing, 2020). https://doi.org/10.1007/978-981-13-6095-4
31. R. Tashakkori, A.S. Hamza, M.B. Crawford, Beemon: an IoT-based beehive monitoring system. Comput. Electron. Agric. **190**(August), 106427 (2021). https://doi.org/10.1016/j.compag.2021.106427

32. S. Tham et al., Indoor temperature and health: a global systematic review. Public Health **179**, 9–17 (2020). https://doi.org/10.1016/j.puhe.2019.09.005
33. X. Tian, Z. Fang, W. Liu, Decreased humidity improves cognitive performance at extreme high indoor temperature. Indoor Air **31**(3), 608–627 (2021). https://doi.org/10.1111/ina.12755
34. I. Ullah et al., An optimization scheme for IoT based smart greenhouse climate control with efficient energy consumption. Computing **104**(2), 433–457 (2022). https://doi.org/10.1007/s00607-021-00963-5
35. J. Wang, D. Norbäck, Subjective indoor air quality and thermal comfort among adults in relation to inspected and measured indoor environment factors in single-family houses in Sweden-the BETSI study. Sci. Total. Environ. **802**, 149804 (2022). https://doi.org/10.1016/j.scitotenv.2021.149804
36. D.W. Waugh et al., Indoor heat exposure in Baltimore: does outdoor temperature matter? Int. J. Biometeorol. **65**(4), 479–488 (2021). https://doi.org/10.1007/s00484-020-02036-2
37. Q. Xie, et al., Correlations, variations, and modelling of indoor environment in a mechanically-ventilated pig building. J. Clean. Prod. **282**(xxxx), 124441 (2021). https://doi.org/10.1016/j.jclepro.2020.124441
38. X. Ye et al., Ambient temperature and morbidity: a review of epidemiological evidence. Environ. Health Perspect. **120**(1), 19–28 (2012). https://doi.org/10.1289/ehp.1003198
39. A. Zanella et al., Internet of Things for smart cities. IEEE Internet Things J. **1**(1), 22–32 (2014). https://doi.org/10.1109/JIOT.2014.2306328
40. Z. Zhou et al., Edge intelligence: paving the last mile of artificial intelligence with edge computing. Proc. IEEE **107**(8), 1738–1762 (2019). https://doi.org/10.1109/JPROC.2019.2918951

Identification of IoT Devices Through Machine Learning and Hardware Fingerprints Based on Clock-Skew

Pedro Miguel Sánchez Sánchez, Alberto Huertas Celdrán, Sergio Marín Sánchez, Kallol Krishna Karmakar, Gérôme Bovet, and Gregorio Martínez Pérez

Abstract This chapter proposes a novel solution for identifying devices on a local network by analyzing differences in their internal clocks relative to a reference clock. The study consists of four stages: designing the solution architecture, obtaining data from five identical Raspberry Pi 4 Model B devices, conducting statistical analysis, and developing a machine learning model to automate device identification based on previously obtained statistical data. A client–server architecture was employed to control packet sending rates, and data captures were performed sequentially and in parallel. Deviations between timestamps and the reference device were analyzed, and only parallel data samples were used for further analysis due to their expected behavior. These samples were processed using a sliding window of 1 min, generating various statistical variables for machine learning. Various supervised classification and unsupervised anomaly detection algorithms were evaluated. Anomaly detection algorithms were trained individually for each device, and their performance was assessed separately. Tree-based models, specifically Decision Trees (95.92% Accuracy) and Random Forest (99.22% Accuracy), exhibited the best performance, while anomaly detection models showed lower results. Consequently, the Random Forest algorithm was selected and trained using all available data and previously utilized hyperparameters. The final model achieved 99.38% accuracy, 99.39% recall, and 99.38% f-score, suggesting the potential for automating the identification of identical devices remotely.

P. M. Sánchez Sánchez · S. M. Sánchez · G. M. Pérez
University of Murcia, Murcia, Spain

A. H. Celdrán
University of Zurich, Zürich, Switzerland

K. K. Karmakar (✉)
University of Newcastle, Newcastle, Australia
e-mail: kallolkrishna.karmakar@newcastle.edu.au

G. Bovet
armasuisse S+T, Bern, Switzerland

B. Pradhan and S. Mukhopadhyay (eds.), *IoT Sensors, ML, AI and XAI: Empowering A Smarter World*, Smart Sensors, Measurement and Instrumentation 50,
https://doi.org/10.1007/978-3-031-68602-3_4

1 Introduction

In recent years the number of devices connected to the Internet has increased greatly [1]. This is due to the use of smartphones, tablets and other devices that require internet connection to perform most (or all) of the tasks for which they have been designed. In particular, the number of Internet of Things (IoT) devices using new technologies such as 5G [2] has developed to a large extent.

Each device connected to the Internet has several identifiers associated with it, such as IP or MAC address. These identifiers should serve to uniquely identify a device, but in practice this is not the case. IP addresses can change automatically due to dynamic IP addressing (via DHCP servers), but they can also be modified by individuals themselves.

These modifications can be purely privacy or performance related, but are often related to cybercrime. Criminals may try to spoof their identifiers so that people, seeking to connect to a legitimate service, end up connecting to their malicious equipment.

These types of devices, despite being widely used, lack strong security measures and for this reason there have been several cyberattacks in this sector [3, 4]. One of the main problems they present is the intrusion into the IoT device network of a malicious device with the same software and hardware as legitimate devices [5]. In addition to gaining access to such a network, it may be that the device attempts to impersonate one of the legitimate devices, thus perpetrating a spoofing attack [6, 7]. It can also encrypt the data on a computer (ransomware attack), or introduce viruses on many computers in order to carry out Distributed Denial of Service (DDoS) attacks using thousands of infected computers (distributed denial of service attacks) [8], among others.

Due to this threat, it is necessary to know whether the equipment with which you want to establish is the right one or not. At this point it is necessary to talk about the identification of devices remotely [9]. There are several alternatives for this process. On the one hand, you can identify the class of the device, i.e., whether it is a computer, a printer, an Internet of Things (IoT) device, etc. On the other hand, the model of the device can be identified, for example, distinguishing a device of brand A and model B from one that is brand X and model Y. Finally, and the most difficult part, you can distinguish devices on an individual level, even if they have the same hardware.

For all the reasons mentioned above, the question arises as to how it is possible to know with which equipment a communication is being established, or what differentiates one device from another on the Internet if they both have the same identifiers [10]. Currently there are ways to ensure the identity of a user or device, such as digital certificates issued by certification authorities, in Spain, the Fábrica Nacional de Moneda y Timbre (FNMT). However, these options can be cloned along with the rest of a device's software [11].

One answer to these questions is that even if the devices present the same hardware, have the same hardware identifiers (such as IP or MAC address) or software (such as digital certificates or keys), and run the same software, they will never be the same. This is because in the manufacturing process of the devices there will always be differences (however small) that will make the devices distinguishable from each other, for example, a device executes a function in 1.2 ns and another in 1.4 ns. The differences are minimal, but they do exist [12].

In contrast, several of the solutions found and discussed in the State of the Art section (Chap. 2) work differently. Some of them are based on identifying a device model, but not an individual device.

In the current landscape of Big Data and Machine Learning, we can exploit the differences discussed above in such a way that fingerprints of each device are generated and thus know if we are really connecting to the right device or not.

This is the framework in which this work is focused. The aim is to create a system that, starting from a common synchronization point, compares the deviations of the clocks of the different devices and thus creates a statistical fingerprint of the behavior of each one. Subsequently, the process of analyzing these statistical values will be automated by means of a Machine Learning model. In order to achieve the main objective of the work, the automatic identification of identical devices, several intermediate goals have been established:

- To present the IoT architecture, as well as its various applications.
- To present different solutions within the Machine Learning field that can be applied to our problem.
- To design our own working architecture, which will be followed throughout the development of the project.
- To analyze the different ways of obtaining a time stamp, with sufficient accuracy, from a device.
- To generate a dataset with the different clock deviations of the devices under analysis.
- To statistically analyze the differences between the different clocks of the devices, in order to see if they are statistically distinguishable.
- To generate a new dataset with different statistical variables of the previous deviations. Where correlated variables will be eliminated.
- Evaluate different Machine Learning algorithms for the task of distinguishing between devices.
- To compare the results obtained with existing solutions and to describe future avenues in this line of research. describe future avenues for this line of research.

The rest of this paper is structured as follows. In the State of the Art section (Sect. 2) a presentation of the IoT architecture and its applications is made, as well as, a presentation of Machine Learning and some of its algorithms. In the Analysis of objectives and methodology section (Sect. 3), the proposed objectives are analyzed and a working methodology is established. In the Design and resolution section (Sect. 4) we will discuss our proposal to address this problem. Several datasets will be obtained, and several Machine Learning models will be trained with them. In the Results section (Sect. 5) the results obtained will be analyzed, in particular, the different classifiers used will be evaluated. Finally, the Conclusions and future directions section presents the final conclusions of the work and discusses possible future avenues for this line of research.

2 State of the Art

This section studies the state of the art regarding the application of different techniques that aim to identify a device through the network independently of an identifier that can be identified. Machine Learning algorithms are widely used for this purpose.

To begin the literature review, the work developed by Pascal Oser et al. [13] will be analyzed. In this work, a system is developed to identify devices among a total of 562 at the *Conseil Européen pour la Recherche Nucléaire* (CERN) center. For this purpose, periodic samples are taken from the timestamps contained in TCP packets sent to and received from these devices. From these timestamps we obtain the variables that will be used to train the different machine learning models. The variables obtained are the increment between two consecutive timestamps, the timestamp counter overflow, the median of all timestamps, etc. An interesting variable is the counter overflow. RFC 1323 [14] defines that the timestamp counter (TSval field) has a length of 4 bytes (32 bits), therefore, the largest value it may be assigned to 232, since it is an unsigned integer. It could be thought that all devices overflow at the same time, but each manufacturer sets the value at which the counter overflows. Once he has all the variables registered for all the devices, he compares several machine learning models to classify each of the devices. The results obtained are an accuracy of 99.22% with MLP, 99.14% with SVM and finally 99.67% with Random Forest. With this data, he decides to keep the Random Forest model, which he trains with all the training data, finally obtaining a precision of 97.03% and an accuracy of 99.76%.

Another related work has been developed by Salma Abdalla Hamad et al. [15]. In this work a system is developed that identifies a device that wants to connect to the network and decides whether to allow it to perform this action or deny it. In addition, this system will periodically check the devices that are already on the network for malicious behavior. For devices that are already connected to the network, a random sample is taken that will check if they have been classified correctly. In case it has been misclassified but is still on the whitelist, its authorization matrix will be updated. On the other hand, if it is not on the whitelist, it will be quarantined for further analysis. Device fingerprints are obtained from the payload of messages sent by that device. From them, 67 characteristics such as Ethernet packet size, header size, destination IP, Time To Live (TTL), etc. are obtained. With these traces it trains different models such as AdaBoost, Decision Trees, Linear Discriminant Analysis (LDA), KNN, Naive Bayes, SVM, Random Forest and XGBoost. It obtained a final accuracy of 89%.

The next paper to be analyzed is by Ahmet Aksoy et al. [16]. This paper designs a system (called SysID) that is able to identify devices through network traffic. This system only needs a network packet to identify the device. Given a packet in the trace, n headers are selected from the packet using a genetic algorithm, which implements a fitness function that tries to reduce this number of headers. The traces that have been used to classify these devices come from a database of another paper [17], from which 20 measurements of 23 IoT devices are obtained. For the classification, different algorithms from the Waikato Environment for Knowledge Analysis (WEKA) tool [18] have been used, such as decision tables, J48 decision trees, OneR and PART. It is decided to use a two-level classification, they are first classified between their vendor or the device itself, depending on whether there are more than one same device from the same vendor. Then the devices from the same vendor are sorted against each other. This scheme helps to have better results as each vendor can be better identified by a different algorithm. The results obtained are 82% average accuracy among all classifiers of each vendor. The individual values are between 42.2% and 100%.

The next work to be seen is that done by Hossein Jafari et al. [19]. This paper does not use packets from the network or transport layer, but from the physical layer. To perform this work the footprint of each device will be obtained using radio frequencies, for this reason, this study will only focus on wireless connections, in this case 6 ZigBee devices [20]. For each of the devices a 5-min capture is performed. This capture will be of the SNR at 5 different levels, resulting in 10 GB per device and SNR level, which in total would be 300 GB of data. It then proceeds to train three Deep Learning algorithms, DNN, CNN and LSTM (the three models belong to the TensorFlow library programmed for Python [21]), with which it obtains results of 96.3%, 94.7% and 76% respectively.

Fabian Lanze et al. [22] have developed a work in which they create a device identification system based on clock skew. Their goal is to generate fingerprints for each access point, so that if a device can know if that access point is legitimate, or if it can be a potential attacker. The first thing they do is to define the skew of a clock as the first derivative of the offset of that clock. Practically they will use the offset stored by the NTP service daemon to approximate the skew of the device connected to the access point. It will use wireless connection and collect up to 6000 packets per device in each sample (10 min). If a sample does not reach 500 packets, due to connection problems or any other reason, that sample is deleted. With this process, 388 access points are measured. Finally, he tries to generate a fingerprint for each access point statistically, based mainly on the clock skew distribution. As a result, he finds that his method is not sufficiently accurate and that many devices have a similar fingerprint, so he concludes that it is not possible to identify them in this way alone.

In the work developed by Yair Meidan et al. [23], a system is developed that is able to identify IoT devices through their network traffic. The tests are performed with traces over wireless connection and with IoT devices as well as PCs, smartphones, etc. From these traces they focus specifically on TCP sessions, where each is represented as a set of features of each packet. First, they run a binary classifier with which they obtain the probability that a device is IoT, using only information from one TCP session of each device. In this section they obtain a 100% accuracy rate. Finally, they train different classification models, such as Gradient Boost Machine (GBM), Random Forest, XGBoost, only on the devices previously labeled as IoT, finally obtaining a result of 99.281% accuracy in the identification of the IoT device.

As a last article, the proposal made by Loh Chin Choong Desmond et al. [24] will be analyzed. In this work they have built a device identification system based on probe request messages from different devices. For this purpose, they will distinguish different combinations of the tuple: "Machine", "Wireless Network Interface Card (NIC) Driver", "Operating System" as distinct devices. Next, they check that in the broadcast time of these messages they see several behaviors repeating. Then, they observe that the latency between bursts of these messages also exhibits these distinctive behaviors. It is with these latencies that they will train a model that is able to distinguish the devices. The model they choose is based on unsupervised learning, Maximum Variance Clustering [25]. This model is like the k-means algorithm, the distinction is that k-means needs to know the number of k clusters beforehand, while Maximum Variance Clustering needs to know the maximum variance within a cluster σmax. With this algorithm they obtain accuracy results between 70 and 80%.

From this literature review, it has been concluded that although there are several solutions for device identification, only one of them uses hardware mechanisms, the proposal by Hossein Jafari et al. [19]. The rest of the proposals focus on software aspects, in particular on various features of TCP forwarding. Another of the shortcomings found was the desired level of identification; several proposals do not seek individual identification of the devices, but only classify them by manufacturer, model and other characteristics that do not identify a device univocally. Due to these reasons, the solution proposed in this work proposes a lower-level identification method that can be used to increase the security of various environments that must remain secure against possible attacks (Table 1).

3 Objectives and Methodology

This section will discuss the objectives of the work and analyze the decisions made in the process of identifying devices.

The main objective of this work is to identify identical devices. The proposed solution is able to identify the patterns related to the deviation of their internal clocks and use them to recognize each device by means of a Machine Learning model. To achieve this goal, several intermediate goals have been established:

- To use what has been seen in the state of the art, to know and select the challenges of this project.
- To obtain a record of the internal clocks of each of the devices that will participate in the analysis. For what we will define a client–server structure that will collect a record of the data.
- Generate a statistical footprint of these records by using a sliding window that allows grouping the data and extracting common statistics. Eliminating correlated variables.
- Analyze different proposals of Machine Learning algorithms that are able to take as input the statistical fingerprints and, with them, identify each device.

Table 1 State-of-the-art comparison

Work	Identification	Data source	Approach	Supervision	Algorithms	Results
[13]	Device type	TCP header	Machine learning	Supervised	MLP, SVM, Random Forest	99.76% accuracy, 97.03% precision
[15]	Individual	Payload features	Machine learning	Supervised	AdaBoost, k-NN, Decision Tree, SVM, Random Forest, XGBoost	89% accuracy
[16]	Device type and model	TCP header	Machine learning	Supervised	J48 Decision Tree, OneR, PART	42.2–100% accuracy, 82% average
[19]	Individual	Radio frequency	Machine learning	Supervised	DNN, CNN, LSTM	96.3% accuracy DNN, 94.7% accuracy CNN, 76% accuracy LSTM
[22]	Device model	NTP registers	Statistical analysis	–	–	Method not able to uniquely identify devices
[23]	Device type and model	TCP sessions	Machine learning	Supervised	GBM, Random Forest and XGBoost	99.281% accuracy
[24]	Individual	*Probe request* messages	Machine learning	Unsupervised	Maximum Variance Clustering	70–80% accuracy
This work	Individual	CPU cyles	Machine learning	Supervised and Unsupervised	Random Forest, XGBoost, Decision Tree, Isolation Forest, LOF, OC-SVM	99.38% accuracy, 99.39% recall, and 99.38% f1-score

In order to achieve the proposed objectives, the following work methodology is presented:

1. Bibliographic review. Review of the works related to the subject of this work. Specifically, we seek to understand how they obtain their data and their method of classifying devices.
2. Design of the work architecture. A generic procedure to be followed is designed. Based on this procedure, the rest of the work is carried out.
3. Data collection. Generation of a dataset containing the timestamps of both the reference device and the device under analysis. This process is performed for each device to be analyzed.
4. Generation of statistical fingerprint. For each device a sliding window is used to group the data in groups of n samples. From each group a set of statistical variables are obtained and labeled with the device from which they have been obtained.
5. Search for correlated variables. A correlation test is performed on the data in order to reduce the dimensionality of the data. The correlated variables do not provide information and therefore have to be eliminated.
6. Partitioning of the data set. In order to train Machine Learning algorithms, the data set is divided into two smaller ones (training and test).
7. Hyperparameter tuning. To decide which of the algorithms is better to classify the data of this work, different Machine Learning algorithms are compared. These algorithms must be adjusted correctly to obtain the best results. To perform this adjustment, a partition similar to the one performed with the set of all the data is performed, but with a reduced set so that the training takes less time.
8. Evaluation of the different models. Once all the algorithms to be compared have been trained, their results are compared by identifying devices to choose the best candidate.
9. Training of the final model. Once the algorithm to be used and its hyperparameters have been decided, a final training is performed with the complete set of training data.
10. Evaluation of the final model. Analysis of the results obtained.

4 Design and Implementation

This section describes the experiments performed, their analysis and the search for a Machine Learning model, as well as the results of the work.

In this work we will use a network topology as illustrated in Fig. 1. In this topology we have an observer (RockPro64 model) and 5 devices to be analyzed (Raspberry Pi 4 Model B).

The idea is to send messages from the observer to each of the devices and have each of the devices respond to those messages, thus obtaining a timestamp of that device.

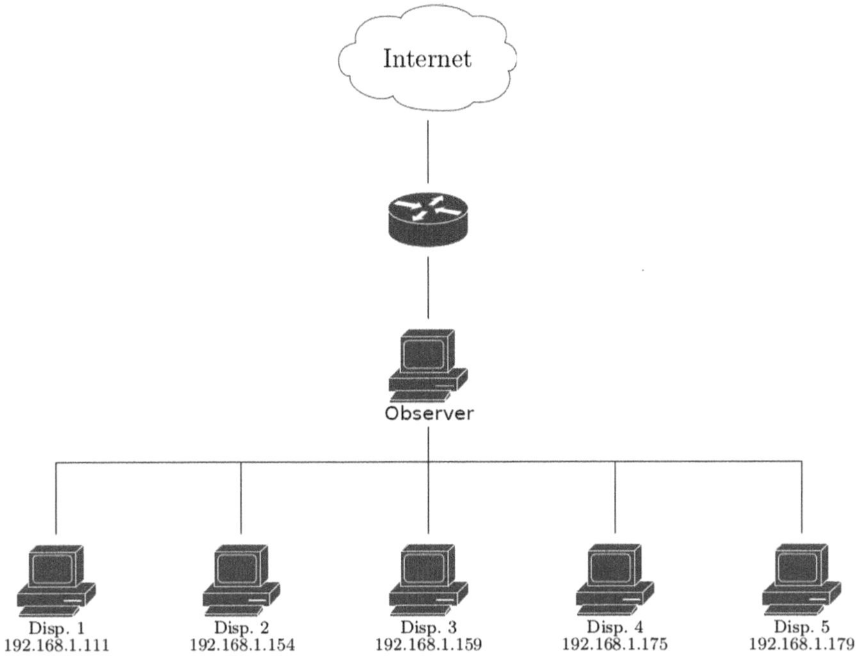

Fig. 1 Network topology

a. **Proposed architecture**

In this work, an architecture has been proposed on which the entire development of this work will be based. This architecture can be seen in Fig. 2.

This architecture is composed of two parts. On the one hand, the devices that generate and share the data, and on the other hand, the external device that oversees both analyzing and processing the data, as well as adjusting the Machine Learning algorithms used. This device is also in charge of evaluating the results obtained by the trained models and comparing them with each other. Each of these parts will be explained below.

- (A) Establishment of the stability conditions. In this phase, the devices are prepared for sample collection. Fixed CPU frequencies are set, to avoid affecting processor cycles. The NTP service that modifies the internal clock of the device is also disabled.
- (B) Data collection. In this phase data is obtained from the different devices to be analyzed. Every second these devices obtain the value of their internal clock and send it to the client. This process is repeated during the collection of the entire sample, which should be a sufficiently long period of time.
- (C) Data analysis and processing. In this phase the client obtains each of the timestamps from the devices. With these values he calculates the increment between each pair of marks of a device and with this he builds a data set. On this data set,

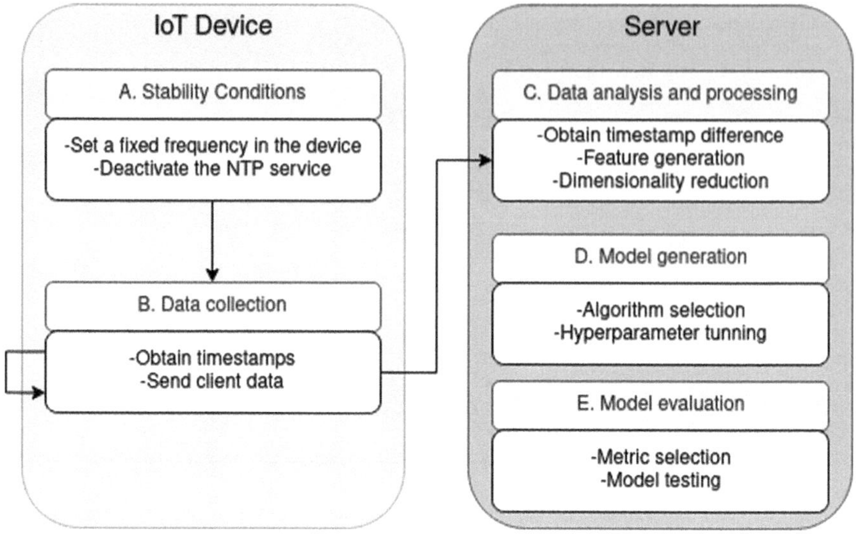

Fig. 2 Proposed architecture

and by using a sliding window, various statistical values are obtained that will be representative of the window in question. Some of these values may be correlated with each other, therefore, all of them will be ignored except one, which will serve as a representative.

- (D) Model generation. In this phase the different Machine Learning algorithms to be used in this work are selected. In this project we have chosen to use supervised learning algorithms oriented to classification, and unsupervised learning algorithms oriented to anomaly detection. Once the algorithms to be used are known, we proceed to adjust their hyperparameters to make the resulting models fit the data of this work in the best possible way.
- (E) Model evaluation. In this phase, the metrics to be used to compare the different models with each other are chosen. Different metrics are selected to compare the classification algorithms and the anomaly detection algorithms. The results of the desired metrics are obtained for each of the trained models, and once all of them are in possession, the models are compared with each other.

b. **Protocol selection**

The options considered were Transmission Control Protocol (TCP) and Internet Control Message Protocol (ICMP). Initially the idea was to use the timestamps contained in the headers of these protocols, but this idea was discarded since they only allowed the times to be obtained in milliseconds. The objective was to obtain the times in nanoseconds, so the data was sent in the body of the packet, which does not have such a short space limitation. To keep the connection persistent throughout the sample, the TCP protocol was chosen, since once the connection is established, it will be maintained until the end.

c. **Data gathering**

A client–server structure will be used to obtain the data. The devices to be analyzed will be the servers and the observer will be the client. With this idea in mind, a server program is created that listens to any IP address on a given port and responds with a time stamp in nanoseconds. This program will run on each of the devices under analysis. When capturing the data, it was necessary to establish the internal clock on which the data will be taken, since there are several data will be taken, as there are different clocks within a device.

- system_clock which contains the current time [26]. This clock is modifiable by services that keep the system time up to date, such as the Network Time Protocol (NTP) service.
- steady_clock which contains the number of CPU cycles since device power-up [27].

For the purposes of this project the steady_clock will be used.

Once the devices are all listening, the observer will run a client program that will be in charge of sending the messages to the corresponding server. This program saves a timestamp at the beginning of the execution t_start, which will be our reference point. Then it will send *n* messages equispaced in 1 second intervals. At each execution of the loop we will get a timestamp at the observer ti, and a timestamp from the device t′i. This provides several pieces of data:

- The timestamp relative to each message from the start, ti-tstart. Theoretically should give exact values, since a message is sent every second, but there is a certain delay. some delay.
- The absolute timestamp of the observer ti.
- The absolute timestamp of the device t′i.
- The deviation of the device clock from that of the observer, ti-t′i.

An example of how this data is stored can be seen in Table 2.

This process will be performed for each device in both a sequential sample and a parallel sample. In the sequential sample, timestamps will be received from one device for 2 hours. Once this time has elapsed, timestamps will be received from the next device, and this structure is repeated for each device, obtaining a final data sample of about 10 hours. In the parallel sample, on the other hand, timestamps from all devices will be received simultaneously, resulting in an overall capture of 12 hours.

Table 2 Example of the collected data

Time	TSrock	TSr	Offset	Device
292	119,238,112,796,030	104,592,709,716,803	−14,645,403,079,227	192.168.1.111
1,001,191,222	119,239,113,986,960	104,593,710,167,425	−14,645,403,819,535	192.168.1.111
2,001,485,862	119,240,114,281,600	104,594,710,453,699	−14,645,403,827,901	192.168.1.111
…	…	…	…	…

d. **Data analysis**

In this section, we will study the incremental clock drift in each of the samples, as well as the cumulative drift at each point.

i. **Experiment 1: sequential sample**

In this first experiment, a sample of 7200 seconds has been taken, that is, 2 hours per device, which in total adds up to 10 hours of samples. Figure 3 shows the deviation (offset) accumulated in each device during those 2 hours.

As it can be seen, there are 2 types of behavior. Devices 1, 2 and 3 remain monotonic at the beginning and then greatly increase their deviation. On the other hand, devices 4 and 5 remain without major changes in the entire sample. This can be seen more clearly in Fig. 4.

The final objective is to obtain a way to statistically distinguish the data, for this a graph is generated that shows between which values this data moves in each device,

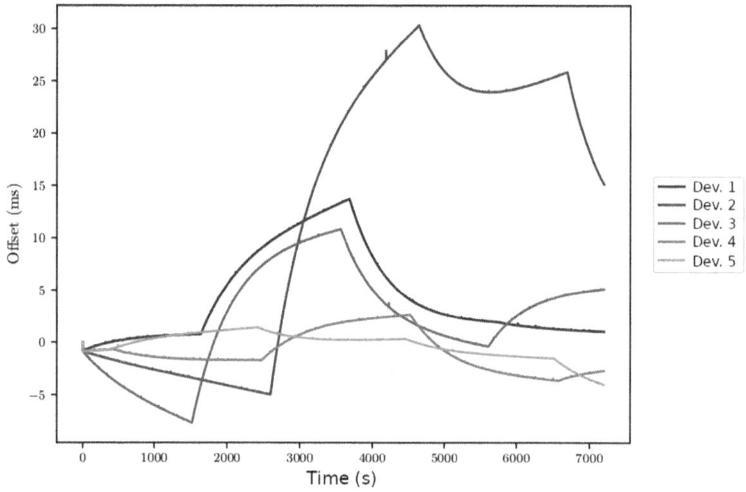

Fig. 3 Sequential offset

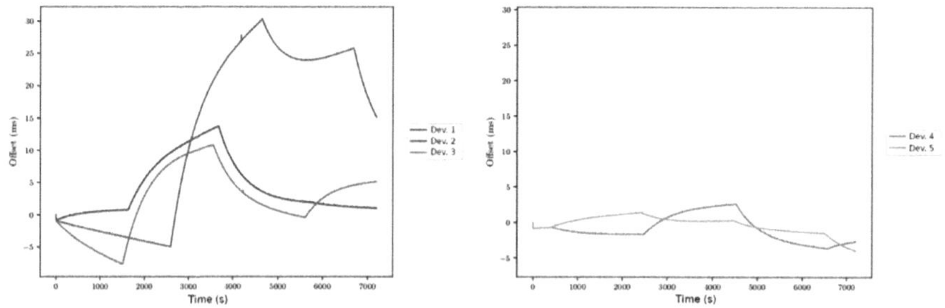

Fig. 4 Differences Between Device Offsets

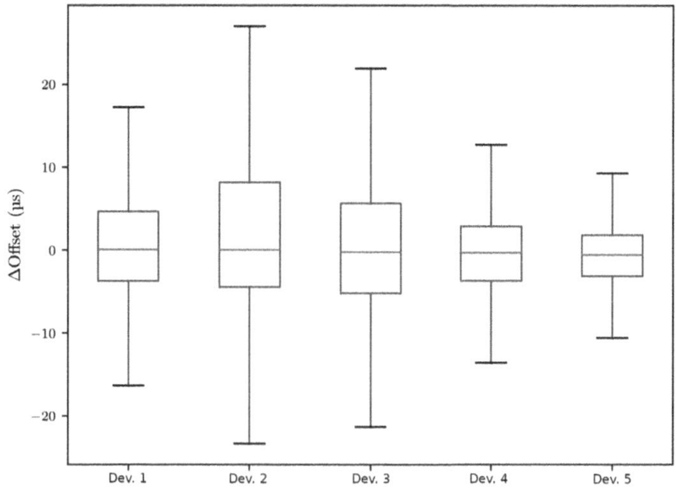

Fig. 5 Box diagram of the sequential sample

in short, it can be seen with a box plot (Fig. 5). Outliers have been removed since at such a small scale they do not reveal the true results.

Devices 4 and 5 are the ones that vary the least, this can be seen in that the boxes are small, and the median is centered between the quartiles (Q1 and Q3). In the other 3 devices, it is observed that there is more space between the median and the third quartile than between the median and the first quartile. This is because they have many large increments, which is why they grow so sharply. With this, you have an idea that the devices are statistically distinguishable, which confirms that you can train a model that identifies them.

ii. **Experiment 2: parallel sample**

For the second experiment, data was captured from all the devices simultaneously for a period of 43,200 seconds, that is, 12 hours.

Figure 6 shows the cumulative deviation in each device in this period. Certain points have also been marked on this graph where the devices seem to "sync". By this we mean that around these timestamps all devices change their trend abruptly and seem to repeat themselves with a certain periodicity. The timestamps on this graph correspond to points {4000, 16,500, 29,000, 41,000} that are approximately equidistant (approximately 12,000 seconds).

This "synchronization" between devices seems to be due to some kind of updating of the internal clock of the observing device (RockPro64) by some daemon in the operating system. The main daemon that handles this task is the NTP service, which was disabled for these tests.

Consulting another box diagram with the increase in the deviation at each instant (Fig. 7) it is observed how all the devices are more on par than in the sequential

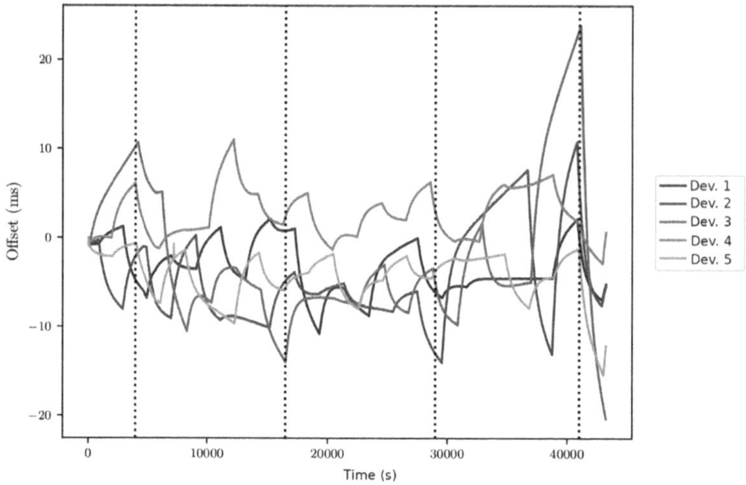

Fig. 6 Parallel offset

sample (Fig. 5). Even so, they still have certain differences, with which it can be concluded that they are also statistically differentiable.

iii. **Sample selection**

Analyzing the increments of the deviation in both samples (sequential and parallel) it can be seen that the parallel sample has values more similar to those expected. These deviation increment values are smaller and more similar among all the devices, which is why the algorithms will be trained with the data from the parallel sample.

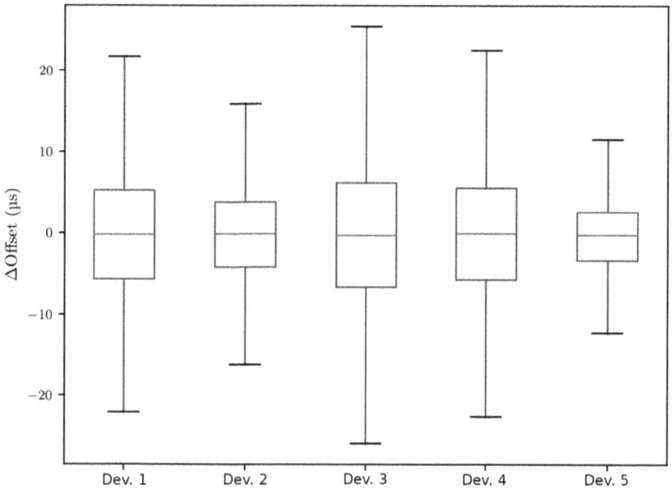

Fig. 7 Box diagram of the parallel sample

	Sum	Mean	Median	Mode	Std	IQR	Kurtosis	Skew	Max	Min	Device
1	-284.0	-4.733333333333333	-203.0	-10750.0	6531.321744049499	8739.5	-0.8026364427898236	0.266444555013173	12077.0	-10750.0	Disp. 1
2	-65895.0	-1098.25	106.5	-13344.0	3926.559099283938	2519.75	1.4605213340303709	-1.1040127142547507	7616.0	-13344.0	Disp. 2
3	96179.0	1602.9833333333333	815.0	-8136.0	5010.092595279735	6575.5	-0.39715065367509084	0.2484646585713819	12831.0	-8136.0	Disp. 3
4	109162.0	1819.3666666666666	2016.5	-10485.0	6159.084454763058	8290.5	-0.7264084617343212	-0.3858981208999922	11469.0	-10485.0	Disp. 4
5	-81317.0	-1355.2833333333333	-2127.0	-6378.0	3665.051390911538	2616.5	2.701193448053943	1.7089231615691112	10383.0	-6378.0	Disp. 5
6	19928.0	332.1333333333333	-147.0	-10750.0	6613.483928825726	10212.0	-0.8404647945245984	0.23473423365399895	12077.0	-10750.0	Disp. 1
⋮	⋮	⋮	⋮	⋮	⋮	⋮	⋮	⋮	⋮	⋮	⋮

Fig. 8 Feature sample

e. **Data preprocessing**

In this section, a pre-processing of the data will be carried out for its adequacy for the training of the Machine Learning models. Various statistical variables will be obtained from these data, and then the statistical variables that are correlated will be eliminated.

These variables will be obtained by using a 1-min sliding window on the increment of the device clock offset. The statistical variables that will be obtained from the parallel sample data will be:

- Sum
- Mean
- Median
- Mode
- Standard deviation
- Interquartile range (IQR)
- Kurtosis
- Asymmetry coefficient (skewness)
- Maximum
- Minimum

i. **Feature Extraction**

In this section, the statistical data of the parallel sample will be generated using the sliding window, the results that can be seen in Fig. 8 are obtained.

The decision to use statistical data as input is to obtain a larger amount of data for Machine Learning algorithms. Given the nature of this study, only one value is being taken into account, the clock deviation of each device. By obtaining these statistical variables, it is possible to generate much larger vectors that will provide more information to the different algorithms when learning.

ii. **Dimensionality reduction**

In this section we will look for statistical values correlated with each other. If there is a correlation greater than 0.9, these variables will be eliminated before training the models, since they do not provide information and will cause the algorithm to take even longer to train.

The correlation between the statistical variables can be seen in Fig. 9. It can be seen how sum, mean and median are the variables most correlated with each other,

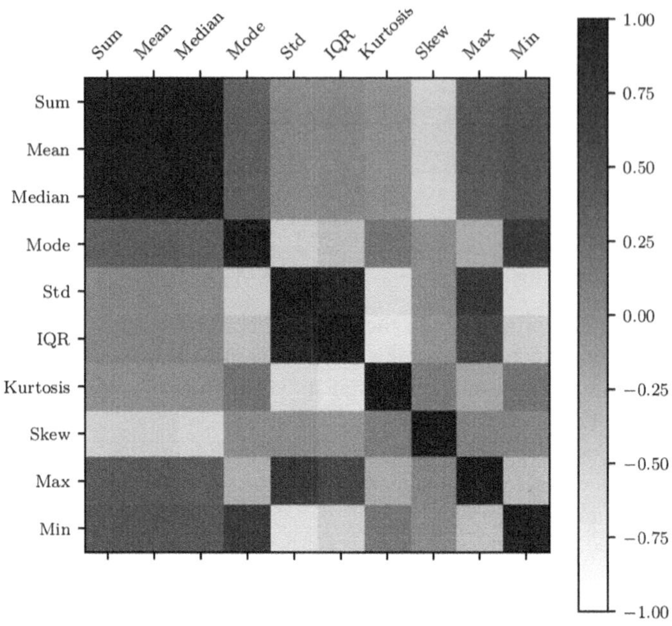

Fig. 9 Correlation between statistical variables

therefore 2 of them will be eliminated. In this case, mean and median are selected to be removed.

f. **Model training**

This section will show the training process of the different models. The scikit-learn library on the Python programming language has been used to perform these trainings. This library has classes that implement the main algorithms in the state of the art. The classes corresponding to each model are shown in Table 3.

i. **Classification Models**

When training Machine Learning algorithms, the first thing to do is to adjust their hyperparameters. The hyperparameters of a model are parameters of the model that can be modified to make it better adapt to the data it is working with or to change the way the model learns to adapt to these data.

To fit the hyperparameters, completeness of the data is not used, a training/validation/testing model is used. The data has been divided into a training and a test set, 70% for training and 30% for test, it is usual to have about one third of the data for training and the rest for test.

To partition the data, random but ordered samples will be taken to maintain the temporal correlation of the data. For example, given the data set × 1,...., × 20, a random sample could be × 15, × 8, × 3, × 17, × 5, × 12 but if these data present

Table 3 Algorithms tested

Algorithm	Implementation	Objective
Decision Tree	DecisionTreeClassifier [28]	Classification
Random Forest	RandomForestClassifier [29]	Classification
MultiLayer Perceptron (MLP)	MLPClassifier [30]	Classification
Naïve Bayes	GaussianNB [31]	Classification
k-NN	KNeighborsClassifier [32]	Classification
SVM	LinearSVC [33]	Classification
Isolation Forest	IsolationForest [34]	Anomaly Detection
Local Outlier Factor	LocalOutlierFactor [35]	Anomaly Detection
OneClass-SVM	OneClassSVM [36]	Anomaly Detection

a temporal correlation, as is the case in this work, this correlation is lost. Therefore, the data have to be returned to their natural order $\times$ 3, $\times$ 5, $\times$ 8, $\times$ 12, $\times$ 15, $\times$ 17.

The training set will be used only with the final model and with the hyperparameters already fitted, since it contains a large amount of data. To fit the hyperparameters of each model and to compare the models with each other a reduced set of this, 35% of the training data, will be used. This subset of the training data will also be divided into a training/test set, although in this case the test set will be called the validation set. The different algorithms will be trained with these sets in order to adjust the hyperparameters and decide the best among all of them.

To test which hyperparameters are the best fit to the data, the function GridSearchCV [37] will be used, which allows given an algorithm and a set of values for the hyperparameters to test all combinations and thus obtain an accuracy of each one, so that only the best ones can be compared and filtered out. This function allows the use of cross-validation.

ii. **Anomaly Detection Models**

When training unsupervised learning algorithms, a similar tuning process to the one seen in the previous section is performed. However, there is a difference and that is the input data of the algorithm.

While with the classification algorithms a division was performed only taking into account the size of the data set, in this case each algorithm will be trained with only 80% of the data from a device. Subsequently, 20% of the data from that device will be used to obtain a True Positive Rate (TPR) or Recall value and 20% of the data from the rest of the devices will be used to obtain a True Negative Rate (TNR) value. This process is performed in order to check if the model is able to detect as correct data those coming from its own device and to detect as anomalous data those coming from the rest of the devices.

5 Experimentation Results

Once all the algorithms have been trained with their best hyperparameters, the results of the classification models are obtained, which can be seen in Fig. 10; and also the results provided by the anomaly detection models, presented in Table 4.

To analyze all these models, we will make use of various metrics, such as those presented between formulas 5.1 and 5.4. To obtain these metrics it is necessary to make use of confusion matrices.

A confusion matrix is a table that evaluates the results of a classification-oriented Machine Learning model. In this type of table, well predicted and poorly predicted values are confronted by observing the actual labels and those predicted by the data. With this type of tool, it is easy to see the hits and misses of a model. This tool can also be generalized to multiclass classification, which has been the use in this work. With a confusion matrix it is possible to obtain the different metrics used to compare

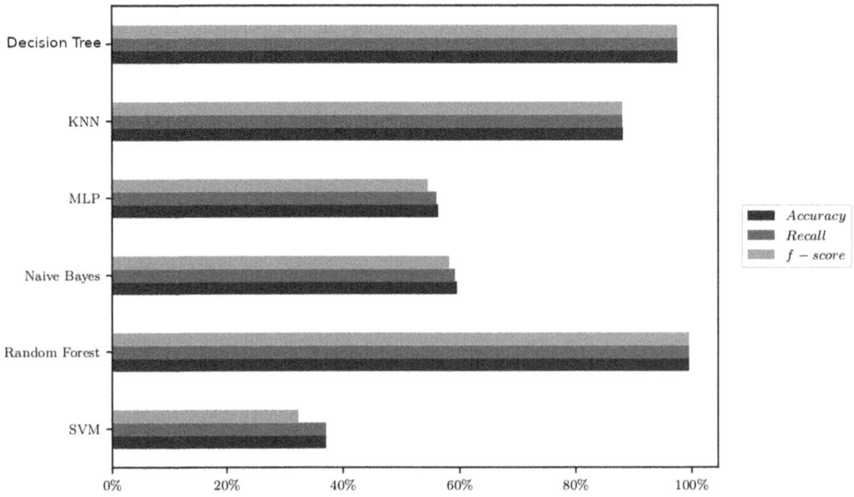

Fig. 10 Classification models comparison

Table 4 Anomaly detection model results

	Isolation Forest		Local Outlier Factor		OneClass-SVM	
	Recall (%)	TNR (%)	Recall (%)	TNR (%)	Recall (%)	TNR (%)
Device 1	95.12	10.36	98.99	10.68	49.48%	58.24
Device 2	94.98	28.87	98.45	57.74	50.90	70.36
Device 3	94.80	13.90	98.73	13.64	49.84	52.65
Device 4	95.80	16.61	98.88	8.74	50.54	64.66
Device 5	94.89	46.99	98.63	77.98	49.53	94.59

Machine Learning models such as Accuracy, Recall, True Negative Rate (TNR), f -score.

Analyzing first the classification models, it can be seen in the figure mentioned above, Fig. 10, that the models that show the best results are those based on decision trees, which are the decision tree algorithm itself and random forest. Another way to see these results is through the confusion matrices generated by each algorithm.

It can be seen in Fig. 10 that the tree-based models get it right practically in the totality of the occasions. In particular, the Random Forest algorithm is the best performer with approximately 99.5% in all the analyzed metrics, followed closely by Decision Trees, with approximately 97.6% in all the metrics.

From the point of view of the anomaly detection models, values of the metrics have not been obtained as high as in the classification algorithms seen above. The calculation of the Recall and TNR metrics that is done based on the two confusion matrices, Table 4, which generates the pair <Device, Algorithm> is given as an example. In the evaluation of detecting as correct values those of its own device, a Recall is obtained of 95.12%. On the other hand, in the task of detecting as anomalous values those coming from the rest of the devices, it does not obtain such a high value, only 10.36% TNR.

Considering the rest of the values of the anomaly detection models, it can be seen in Table 4 that none of the trained models is able to identify the different devices in a solvent way. The best among them is OneClass-SVM, but it is not able to obtain such high results compared to the classification models. However, it is observed that both Device 2 and Device 5 obtain, in general, better results.

Once all the models that have been created for this work have been compared, it is concluded that the best among them is the Random Forest classification algorithm. Therefore, a final model has been generated with the hyperparameters adjusted previously and with the totality of the training data. With these results we have a final Accuracy value of 99.38%, Recall of 99.39% and f -score of 99.38%.

6 Conclusion and Future Work

In this chapter, we have designed a model capable of classifying devices with which communication has been established based on small differences in the manufacture of the components, which alter the time it takes to execute certain tasks.

In the first part of the project, it has been seen that in order to obtain the desired timing accuracy, it has been necessary to use the TCP protocol and send time stamps in the body of the packet.

During the development of this part of the research it was found that the internal clock is susceptible to being altered by external processes to be always synchronized with the rest of the devices (NTP protocol). For this reason, this service had to be deactivated before any data collection, since the differences between times would not be those of each device.

Once the service was deactivated, data were still obtained that were not correct since a system clock that could be modified was being used. This clock was changed for a clock that could not be modified by external processes (steady_clock) and with that the data were more accurate.

Both sequential and parallel captures of the deviation of the device clocks were taken, from which their increments at each time point were obtained. With these increments and a sliding window of 1 min, the statistical variables that would be used to train the models were obtained. These variables were analyzed, and those that were correlated were eliminated.

Both classification and anomaly detection models were trained with the aforementioned variables and then compared with each other to decide which model best fit the data, with the Random Forest classification algorithm being chosen as the best.

Finally, the chosen algorithm, Random Forest, was trained with the data from the parallel sample, and the hyperparameters were considered the best when training with the training/validation sets. This model obtained 99.38% in the Accuracy value, 99.39% in the Recall value, and 99.38% in the f -score value.

Comparing this work with those that have been analyzed in the State of the Art chapter, it is concluded that a system capable of identifying devices with a similar accuracy to those reviewed has been created. However, these data have been obtained in a very controlled environment that is not representative of a real scenario. There have also been problems such as the one mentioned with the NTP service, which may have influenced the results.

One of the possible future avenues of this work would be the development of a real-time model of this system. For this purpose, a local copy of the footprints generated by certain devices should be available for comparison with the ones generated by other devices.

To realize this system in real time, it would also be necessary to create mechanisms that allow the system to update itself with new data, and thus generate new fingerprints for the devices. The Machine Learning models would also have to be modified because for the system to work in real time, they would have to be updated.

References

1. R. van der Meulen, 8.4 billion connected things will be in use 2017 | gartner, https://www.gartner.com/en/newsroom/press-releases/2017-02-07-gartner-says-8-billion-connected-things-will-be-in-use-in-2017-up-31-percent-from-2016 (2017)
2. K. Shafique, B.A. Khawaja, F. Sabir, S. Qazi, M. Mustaqim, Internet of things (iot) for next-generation smart systems: a review of current challenges, future trends and prospects for emerging 5g-iot scenarios. IEEE Access **8**, 23 022–23 040 (2020), ISSN: 2169–3536
3. B. Miller, D. Rowe, A survey scada of and critical infrastructure incidents, in *Proceedings of the 1st Annual Conference on Research in Information Technology*, ser. RIIT '12, Calgary, Alberta, Canada: Association for Computing Machinery, 51–56 (2012), ISBN: 9781450316439. [Online]. Available: https://doi.org/10.1145/2380790.2380805

4. A. Nicholson, S. Webber, S. Dyer, T. Patel, H. Janicke, Scada security in the light of cyber-warfare. Comput. Secur. **31**(4), 418–436 (2012), ISSN: 0167-4048. [Online]. Available: https://www.sciencedirect.com/science/article/pii/S0167404812000429
5. I. Bisio, C. Garibotto, F. Lavagetto, A. Sciarrone, S. Zappatore, Unauthorized amateur uav detection based on wifi statistical fingerprint analysis. IEEE Commun. Mag. **56**(4), 106–111 (2018), ISSN: 1558-1896
6. M.R. Nosouhi, K. Sood, M. Grobler, R. Doss, Towards spoofing resistant next generation Iot networks. IEEE Trans. Inform. Forensics Secur. ISSN: 1556–6021
7. N. Wang, L. Jiao, P. Wang, M. Dabaghchian, K. Zeng, Efficient identity spoofing attack detection for iot in mm-wave and massive mimo 5g communication, in *2018 IEEE Global Communications Conference (GLOBECOM)*, pp. 1–6 (2018)
8. N. Hoque, D.K. Bhattacharyya, J.K. Kalita, Botnet in ddos attacks: trends and challenges. IEEE Commun. Surv. Tutorials **17**(4), 2242–2270 (2015), ISSN: 1553-877X
9. P.M. Sánchez, J.M. Jorquera, A. Huertas, G. Bovet, M. Gil, G. Martinez, A survey on device behavior fingerprinting: Data sources, techniques, application scenarios, and datasets. IEEE Commun. Surv. Tutorials **23**(2), 1048–1077 (2021)
10. P.M. Sánchez, J.M. Jorquera, A. Huertas, G. Bovet, M. Gil, G. Martínez, A methodology to identify identical single-board computers based on hardware behavior fingerprinting. J. Netw. Comput. Appl. **212**, 103579 (2023)
11. N. Yousefnezhad, A. Malhi, K. Främling, Security in product lifecycle of it devices: a survey. J. Netw. Comput. Appl. **171**, 102 779 (2020), ISSN: 1084-8045. [Online]. Available: https://www.sciencedirect.com/science/article/pii/S1084804520302538.
12. A. Al-Omary, A. Othman, H.M. AlSabbagh, H. Al-Rizzo, Survey of hardware-based security support for iot/cps systems. KnE Eng., 52–70 (2018)
13. P. Oser, F. Kargl, S. Lüders, Identifying devices of the internet of things using machine learning on clock characteristics, in G. Wang, J. Chen, L.T. Yang (eds.) *Security, Privacy, and Anonymity in Computation, Communication, and Storage*. Cham: Springer International Publishing, pp. 417–427 (2018), ISBN: 978-3-030-05345-1
14. D.A. Borman, R.T. Braden, V. Jacobson, TCP Extensions for High Performance, Tech. Rep. 1323, May 1992, 37 pp. [Online]. Available: https://rfc-editor.org/rfc/rfc1323.txt
15. S.A. Hamad, W.E. Zhang, Q.Z. Sheng, S. Nepal, Iot device identification via network-flow based fingerprinting and learning, in *2019 18th IEEE International Conference On Trust, Security And Privacy In Computing And Communications/13th IEEE International Conference On Big Data Science And Engineering (TrustCom/Big-DataSE)*, IEEE, pp. 103–111 (2019)
16. A. Aksoy, M.H. Gunes, Automated iot device identification using network traffic, in *ICC 2019–2019 IEEE International Conference on Communications (ICC)*, pp. 1–7 (2019)
17. M. Miettinen, S. Marchal, I. Hafeez, N. Asokan, A.-R. Sadeghi, S. Tarkoma, Iot sentinel: Automated device-type identification for security enforcement in iot, *2017 IEEE 37th International Conference on Distributed Computing Systems (ICDCS)* (2017). [Online]. Available: https://doi.org/10.1109/ICDCS.2017.283
18. M. Hall, E. Frank, G. Holmes, B. Pfahringer, P. Reutemann, I.H. Witten, The weka data mining software: an update. SIGKDD Explor. Newsl. **11**(1), 10–18 (2009), ISSN: 1931-0145. [Online]. Available: https://doi.org/10.1145/1656274.1656278
19. H. Jafari, O. Omotere, D. Adesina, H.-H. Wu, L. Qian, Iot devices fingerprinting using deep learning, in *MILCOM 2018–2018 IEEE Military Communications Conference (MILCOM)*, pp. 1–9 (2018)
20. D. Gislason, Zigbee Wireless Networking, Pap/Onl. USA: Newnes, (2008), ISBN: 978-0-7506-8597-9
21. M. Abadi et al., TensorFlow: Large-scale machine learning on heterogeneous systems, Software available from tensorflow.org (2015). [Online]. Available: https://www.tensorflow.org/
22. F. Lanze, A. Panchenko, B. Braatz, A. Zinnen, Clock skew based remote device fingerprinting demystified, in *2012 IEEE Global Communications Conference (GLOBECOM)*, pp. 813–819 (2012)

23. Y. Meidan et al., Profiliot: a machine learning approach for iot device identification based on network traffic analysis, in *Proceedings of the Symposium on Applied Computing*, ser. SAC '17, Marrakech, Morocco: Association for Computing Machinery, pp. 506–509 (2017), ISBN: 9781450344869. [Online]. Available: https://doi.org/10.1145/3019612.3019878
24. L.C.C. Desmond, C.C. Yuan, T.C. Pheng, R.S. Lee, Identifying unique devices through wireless fingerprinting, in *Proceedings of the First ACM Conference on Wireless Network Security*, ser. WiSec '08, Alexandria, VA, USA: Association for Computing Machinery, pp. 46–55 (2008), ISBN: 9781595938145. [Online]. Available: https://doi.org/10.1145/1352533.1352542
25. C. Veenman, M. Reinders, E. Backer, A maximum variance cluster algorithm. IEEE Trans. Pattern Anal. Mach. Intell. **24**(9), 1273–1280 (2002)
26. "std::chrono::system_clock", [Online]. Available: https://en.cppreference.com/w/cpp/chrono/system_clock
27. "std::chrono::steady_clock", [Online]. Available: https://en.cppreference.com/w/cpp/chrono/steady_clock
28. "DecisionTreeClassifier", [Online]. Available: https://scikit-learn.org/stable/modules/generated/sklearn.tree.DecisionTreeClassifier.html#sklearn.tree.DecisionTreeClassifier
29. "RandomForestClassifier", [Online]. Available: https://scikit-learn.org/stable/modules/generated/sklearn.ensemble.RandomForestClassifier.html#sklearn.ensemble.RandomForestClassifier
30. "MLPClassifier", [Online]. Available: https://scikit-learn.org/stable/modules/generated/sklearn.neural_network.MLPClassifier.html#sklearn.neural_network.MLPClassifier
31. "GaussianNB", [Online]. Available: https://scikit-learn.org/stable/modules/generated/sklearn.naive_bayes.GaussianNB.html#sklearn.naive_bayes.GaussianNB
32. "KNeighborsClassifier", [Online]. Available: https://scikit-learn.org/stable/modules/generated/sklearn.neighbors.KNeighborsClassifier.html
33. "LinearSVC", [Online]. Available: https://scikit-learn.org/stable/modules/generated/sklearn.svm.LinearSVC.html
34. "IsolationForest", [Online]. Available: https://scikit-learn.org/stable/modules/generated/sklearn.ensemble.IsolationForest.html#sklearn.ensemble.IsolationForest
35. "LocalOutlierFactor", [Online]. Available: https://scikit-learn.org/stable/modules/generated/sklearn.neighbors.LocalOutlierFactor.html#sklearn.neighbors.LocalOutlierFactor
36. "OneClass-SVM", [Online]. Available: https://scikit-learn.org/stable/modules/generated/sklearn.svm.OneClassSVM.html#sklearn.svm.OneClassSVM
37. "GridSearchCV", [Online]. Available: https://scikit-learn.org/stable/modules/generated/sklearn.model_selection.GridSearchCV.html

Development of IoT Soil Sensor Node: Installation and Modeling

Waqas A. K. Afridi, Ignacio Vitoria, and Subhas C. Mukhopadhyay

Abstract Smart sensors, robots, wireless networks, big data analytics, machine learning, and more advanced technologies are now part of Agriculture 4.0. Experts in meteorology and agronomy can now make financially advantageous and ecologically responsible choices thanks to these cutting-edge technologies. To gather information about soil moisture, several sensing techniques have recently been developed. But when it came to efficacy and dependability, the studies showed serious shortcomings. This study describes the design, development, and field testing of a low-cost, multi-depth IoT soil sensor node that can communicate real-time field data with scalability and high spatial resolution. Before being used in a real pilot farm, the sensors are calibrated and tested in the laboratory. To improve the general comprehension of land and climate conditions, the proposed sensor node additionally used an environmental sensor and a tipping rain sensor. Weather forecasting and irrigation scheduling are two long-term applications that can benefit greatly from the initial data from the field sensors. Such a system with collaboration with industry can be seen as a cost-effective solution to expensive commercial tools.

Keywords Sensors · Soil moisture · IoT · System development · Data analytics · Agriculture · Modeling

W. A. K. Afridi
School of Engineering, Macquarie University, Sydney 2109, Australia
e-mail: Waqas.afridi@mq.edu.au

I. Vitoria
Department of Electrical, Electronic and Communication Engineering, Public University of Navarra, Arrosadía Campus, 31006 Pamplona, Spain
e-mail: ignacio.vitoria@unavarra.es

S. C. Mukhopadhyay (✉)
School of Engineering, Macquarie University, Sydney, NSW 2109, Australia
e-mail: subhas.mukhopadhyay@mq.edu.au

B. Pradhan and S. Mukhopadhyay (eds.), *IoT Sensors, ML, AI and XAI: Empowering A Smarter World*, Smart Sensors, Measurement and Instrumentation 50,
https://doi.org/10.1007/978-3-031-68602-3_5

1 Introduction

The Earth's ecosystem depends heavily on soil moisture, which has profound effects on environmental sciences, meteorology, and agriculture. For many purposes, especially for farmers and meteorologists, the measurement of soil moisture is crucial. To optimize crop yields and ensure sustainable agricultural practices, farmers must have a basic understanding of and ability to measure soil moisture [1]. Overall crop health, root development, and plant growth are all directly impacted by the soil moisture content. Stressed-out crops can have lower yields, crop failure, and insufficient or excessive soil moisture. Farmers are better able to plan irrigation schedules, apply fertilizer, and choose crops when they accurately measure the amount of moisture in the soil. These decisions are critical to productive and ecologically friendly farming [2]. Therefore, for weather forecasting and climate monitoring, meteorologists also heavily rely on soil moisture data. Influencing variables like temperature, humidity, and precipitation significantly affect the soil moisture levels which are critical in determining weather patterns. More precise predictions of floods, droughts, and other climatic events are generally driven by the use of accurate soil moisture data [3]. In this regard, a few recent studies have made significant advancements toward the development of a soil moisture monitoring system. However, the studies are constrained by several issues that affect their efficacy and dependability, including calibration and validation processes, the lifespans of the sensors, energy needs, data transmission and accessibility, temporal resolution, and cost [4–7]. This work proposes a low-cost multi-depth smart sensor node system that can obtain high spatial resolution and scalable real-time data. The proposed system can be expanded further after installation without requiring a sizable infrastructure investment. Essentially, to improve the system's overall comprehension of agricultural conditions and facilitate data-driven decision-making, an environmental monitoring sensor has also been incorporated [8]. Overall, the proposed system aims to fulfill the demand for a smart quantification method of soil moisture that is more easily obtainable, expandable, and comprehensive in terms of data. Due to its distinctive features, it is ideally adapted to data-driven techniques and sustainable resource management in modern agriculture.

2 Sensor Node Development

The design, testing, and optimization phases made up the development process of a multi-depth IoT soil sensor node. At first, the sensor nodes are designed to fulfill particular needs and limitations like price, depth, power usage, and scalability. Subsequently, the accuracy, dependability, and robustness of the nodes were verified under various weather conditions, including exposure to sun and rain. This required both controlled laboratory and field experiments [9]. Additionally, the system's performance has been examined through ongoing monitoring, evaluation, and optimization of data. Comprising of commercially available sensors, including digital soil temperature sensors, tipping bucket rain gauge sensors, capacitive analog soil moisture sensors, and an I2C protocol environmental sensor, the electronic system is housed in an electrical junction box. With compatible shields like LoRaWAN, a solar charger, and a Qwiic shield stacked on top for simple electronics integration, an Arduino Mega 2560 microcontroller serves as the brain that makes decisions about the node's operation. The embedded microcontroller system harvests energy from 6 V 6W solar panels and 3.7 V rechargeable lithium-ion battery. Additionally, to cut down power consumption the microcontroller incorporates a sleep mechanism [10]. With the addition of sleep mode, the power consumption has been reduced significantly throughout the embedded system resulting in a notable increase in battery life. The soil probe used sensors at three successive layers inserted at about 600 mm depth, which is unique to the previously suggested methods [11, 12]. The specifications for hardware components used in the development of IoT sensor node are compiled in Table 1.

2.1 Sensor Node Schematic

Figure 1 displays the sensor node's wiring schematic diagram. The central portion represents the Arduino Mega 2560 microcontroller stack made up of a LoRa shield, a Solar cell shield, and a Qwiic shield. While the I_2C bus protocol is used to integrate the environmental sensors with the microcontroller. The analog input pins A1, A2, and A3 on the MCU are connected to three analog soil moisture sensors. An n-channel transistor is used to connect three DS18B20 digital soil-temperature sensors to a single digital input pin 4 of MC. A 100 Ω resistor is used to attach the transistor's gate to pin 5. Additionally, a 1 MΩ pull-up resistor connects pin 3 of the MCU to a digital tipping bucket rain sensor. The Qwiic shield's I2C serial ports are used to integrate the BME280 environmental sensor. A rechargeable lithium-ion battery (3.0 ~ 5.0 V) using 6 V–6 W solar power cells powers the microcontroller.

2.2 Sensor Node Testing

The chosen v2.0 capacitive-type soil moisture sensors have undergone laboratory testing to optimize the standard output accuracy. Because these analog sensors don't have exposed electrodes, they won't corrode when reading the dielectric and salinity of soil. Thus, they offer superior accuracy in comparison to inexpensive resistive sensors [13, 14]. Moreover, these sensors work well with low-power microcontrollers (3 ~ 5 V), which is perfect for the testing conditions used in the study. The developed

Table 1 Component used in sensor node development

Component name	Specification
ATMega2560 microcontroller	Operating voltage: 5 V I/O pins: 54-digital, 16-analogue Flash memory: 256 KB Clock speed: 16 MHz
Capacitive-type analogue soil-moisture sensor	Operating voltage: 3.3 ~ 5.5 V Output voltage: 0 ~ 3.0 V Dimension (L × W): 175 × 30 mm/6.89 × 1.18 inches **Need calibration** of analog output
DS18B20 digital soil-temperature sensors	Operating voltage: 3.0 ~ 5.5 V ±0.5 °C Accuracy from −10 °C ~ +85 °C Multiple sensors can share one pin
BME280 environmental sensor	BME280–Humidity, pressure, temperature Operating voltage: 1.7 ~ 3.6 V Power consumption: 60 mW at IAQ measurement every second
Digital tipping-bucket rain sensor	Material: ABS Power type: AA battery Dimension: 4.0–6.9 Inches
3D-printed housings for sensors	Material: PLA Dimension (L × W): 112 × 50 mm Screwed with PVC conduits to form a probe
Polyvinyl chloride pressure conduits	Material: PVC Dimension (L × W): 90 × 40 mm Screwed with 3D housings for sensors
6 V 6W Solar panel	Cell type: Monocrystalline Cell efficiency: 19%+ 6.15 Watts Peak Power
Polymer lithium-ion battery	3.7 V at 6000 mAh 2C continuous discharge Standard 2-pin JST-PH connector
Dipole antenna-916-CW-HWR-SMA	Electrical type: Dipole Frequency range: 900–930 MHz Gain: 1.2 dBi, Impedance: 50 Ω

probes were then placed inside a specially designed soil sampling container as illustrated in Fig. 2a filled with loamy textured soil perfect for gardening and plant growth. The performance of the sensors is assessed by pouring measured water volume and subjecting probes to a 100 W lightbulb, respectively, to simulate subsequent rain events and longer solar hours. Capacitive sensors are observed to require a minimum of 1 to 5 min to settle analogous to the soil, whereas the dielectric output readings typically fall within the range of approximately 700 to 10, based on the soil texture and moisture content [15]. The sensor output voltage in principle, is dependent upon changes in soil dielectric constant and capacitance, both of which are proportionate

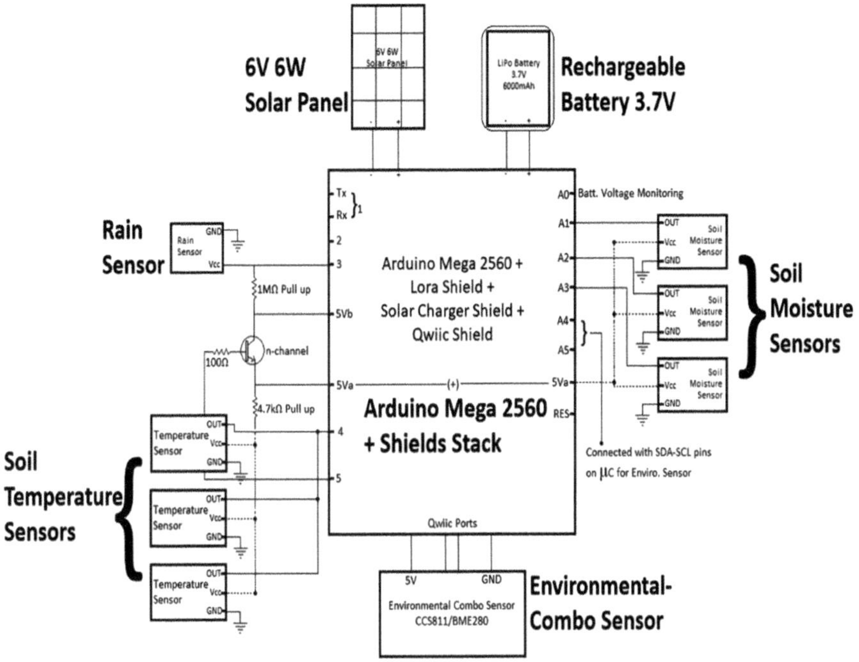

Fig. 1 Sensor node schematic

to the volume of water in the soil [16]. For example, when water percolates into the soil, its capacity to hold charge (capacitance) changes, resulting in a proportional voltage. The microcontroller's analog-to-digital pin is used to measure this change in voltage, which is calibrated to the soil moisture content. As a result, the obtained dielectric readings are mapped into a moisture percentage hierarchy from 0 to 100%, as shown in Fig. 2b corresponding to dry to wet soil conditions. An Arduino IDE serial monitor was used to record the observations and to confirm the accuracy of the model, a regression coefficient of $R^2 = 0.9912$ was obtained.

2.3 *Study Location*

In the northeast township of Narrabri, New South Wales, Australia (30°16′06.2"S 149°51′18.8"E), a grazing grassland site at L'lara farm has been piloted to trial the developed system as shown in Fig. 3. The entire land area of the farm is 1850 hectares, of which 1070 hectares are utilized for dryland crops. With mostly hot summers and cold winters, the location typically has a humid tropical climate. The average annual precipitation recorded on the site is 658 mm. The topography of the L'lara farm

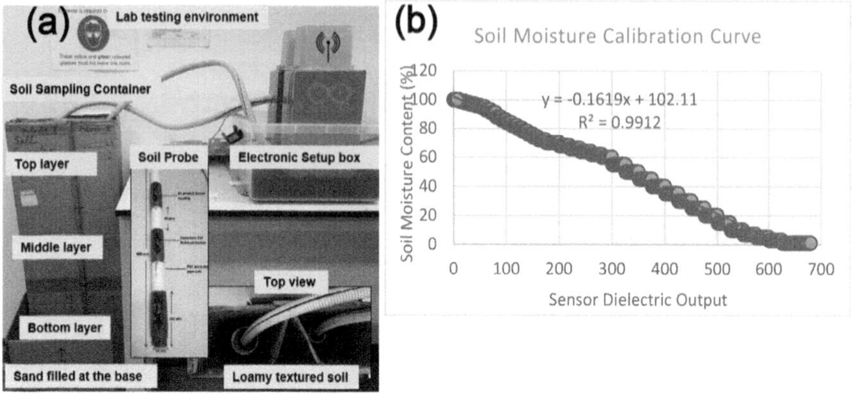

Fig. 2 **a** Laboratory testing environment **b** Soil moisture calibration curve

is primarily made up of floodplain regions and hills covered in sand. These sand-covered hills are primarily made up of moderately weathered soil groups like the Luvisol soil that are mainly used for grazing beef cattle. Although the areas within the floodplain have been fully allocated to dryland cropping on Vertisol soils [17].

Fig. 3 Pilot L'lara farm in Narrabri, NSW, Australia

2.4 Field Installation

A pilot network site has been chosen for the deployment of the sensor nodes to collect primary field data that will inform analytical estimations for weather forecasting and irrigation scheduling. Figure 4a shows the field installation of the sensor node. Through the use of an on-stack LoRa shield via a LoRaWAN gateway, the system is linked to the wireless network [18]. Furthermore, a live data visualization is done on an open-access IoT analytics server called ThingSpeak [19]. The Arduino IDE development software is used to program the microcontroller to communicate primary field data every hour and every ten minutes in the event of rain. Figure 4b illustrates the IoT framework of the proposed system.

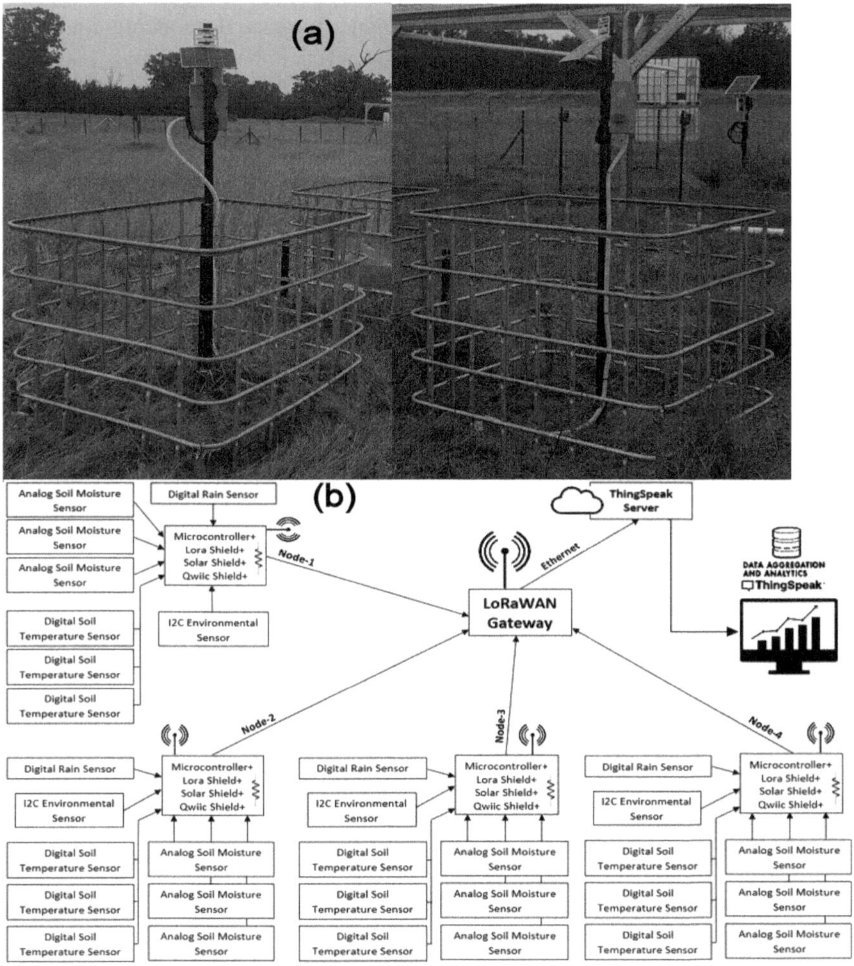

Fig. 4 **a** Installed nodes in the field **b** Sensor nodes IoT framework

3 Field Results

The data collected from the deployed sensor node is updated in real-time using the LoRaWAN communication protocol. The network protocol is low cost, and low power, which is suitable for long-range greenhouse applications [20, 21]. Figure 5 displays the live data analytics dashboard, which is set up to track and record field variables in real-time, including relative air humidity, ambient temperature, soil moisture content, and soil temperature.

Additionally, Table 2 shows how the percentage of soil moisture changed at three successive layers during the first week following installation. The dielectric readings from capacitive soil moisture sensors serve as the basis for the data. It is evident for both nodes that as rainwater percolates through succeeding layers, the soil moisture content has grown gravimetrically. The evapotranspiration during the next few days without rain may have contributed to the notable decrease in moisture content at each layer. Interestingly, at node 4, it was discovered that the bottom layer had more moisture than the middle layer, suggesting a high field capacity at a specific soil depth [22].

Additionally, Fig. 6 is plotted to demonstrate how capacitive soil moisture sensors respond to percolating rainwater at the top, middle, and bottom layers. It is observed that during the spring season, the multi-depth soil moisture sensors have shown proportional variance to the rain events. Such that, on significant rainy days, the capacitive sensors have shown a substantial increase in the soil moisture content. For instance, From 19–21 October a total of 113.4 mm of rain was measured bringing an exponential rise to moisture at all three layers, where topsoil (20 cm) increased from 35–38%, the middle layer reached 35–37%, and the bottom layer rose 36–39%. Interestingly, the bottom layer (60 cm) at the pilot site is relatively more moistured

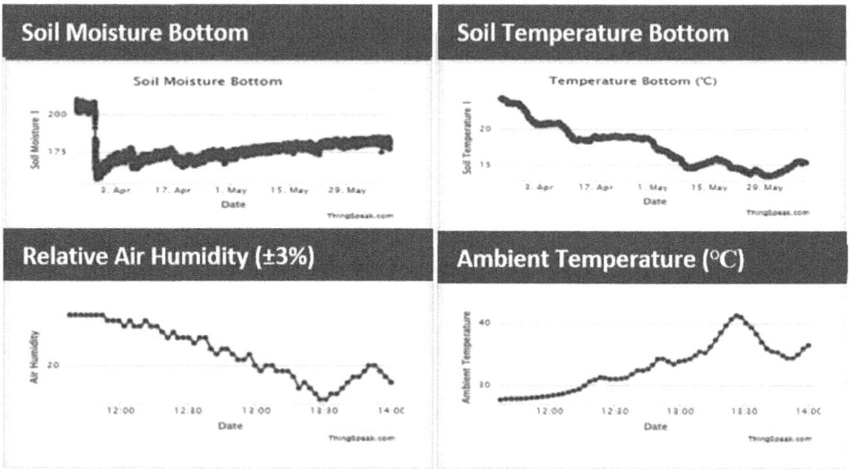

Fig. 5 ThingSpeak live data monitoring

Table 2 Change in soil moisture content at vertical depths

Date	20/10/2022	24/10/2022	28/10/2022
Precipitation	18.9 mm	0	0
Node 3			
Top	98%	96%	95%
Middle	84–100%	86–84%	80%
Bottom	60–64%	62%	60%
Node 4			
Top	95%	80–72%	70%
Middle	55–60%	50%	50%
Bottom	64–66%	62%	62%

than the above two layers indicating a high field capacity at a deeper layer. The findings support the US Department of Agriculture's finding that dry soils have higher infiltration rates [23].

Furthermore, Fig. 7 illustrates the performance evaluation of the DS18B20 digital soil temperature sensors against the BME280 environmental temperature sensor at three subsequent depths. Typically, December is taken as a pilot month when higher temperatures are observed in a year. It is found that the ambient temperature proportionally drives the temperatures at subsequent soil layers. Essentially the topsoil has shown higher variance in daily temperatures compared to the middle and bottom layers. For instance, on 15-12-22 minimum ambient temperature of 14.7 °C drives the dropping of soil temperatures to 17.9–17.2 °C, 18.4–17.7 °C, and 19.1–18.5 °C at the top, middle, and bottom layers, respectively. Similarly, on 26-12-22 maximum ambient temperature of 29.1 °C influenced a rise in soil temperatures to 18.7–19.8 °C,

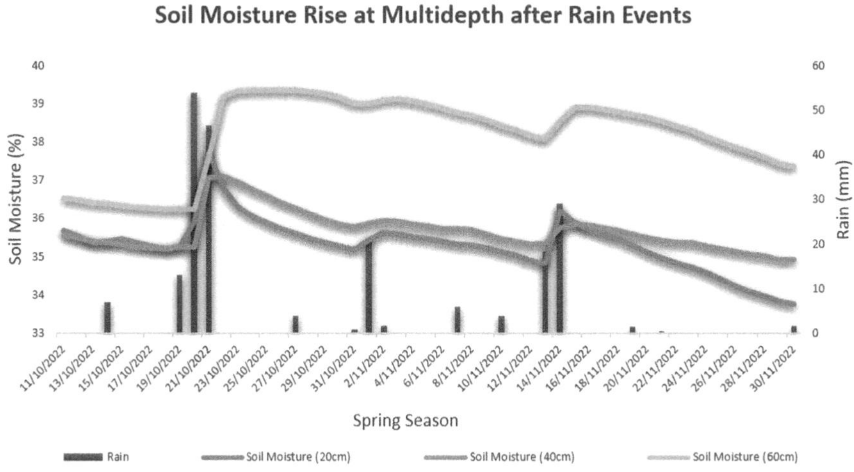

Fig. 6 Rise in soil moisture content at vertical depths during rain

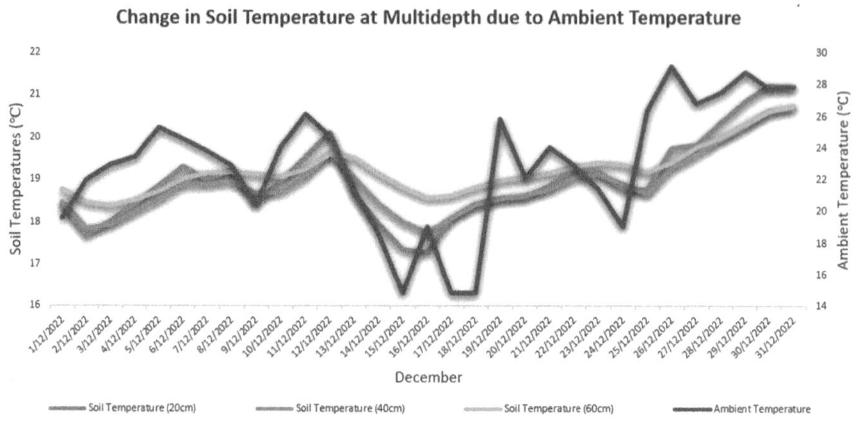

Fig. 7 Change in soil temperature at multidepth due to ambient temperature

18.6–19.5 °C, and 19.1–19.3 °C at the top, middle, and bottom layers, respectively. The findings are consistent with a case study that found a strong positive correlation between soil temperature and atmospheric temperature up to a depth of 20 cm in the soil using a regression model [24].

4 Machine Learning Modeling

The data obtained from the deployed sensor node is used to model evapotranspiration (ET) using a machine-learning modeling technique. The challenges of traditional approaches, like the need for precise, real-time predictions and the temporal and spatial variability of environmental factors, can potentially be resolved by integrating machine learning with real-time sensor node data [25]. In the process of developing the model, key input parameters such as air temperature, atmospheric pressure, wind speed, relative humidity, rain, sun exposure, soil temperature, and change in soil moisture are measured in real time and computed using mathematical Eq. 1 of the Penman-Monteith model [26]. The main dataset utilized in this exploration spans October 2022 to October 2023.

$$ET_o = \frac{0.408\Delta(R_s - G) + \gamma\left(\frac{900}{T+273}\right)u_2(e_s - e_a)}{\Delta + \gamma(1 + 0.34u_2)} \tag{1}$$

where,

Δ = Slope of Saturation Vapor Pressure Curve (kPa/°C)
R_s = Solar Radiation (MJ/m^2 day^{-1})
G = Soil Heat Flux ≈ 0
Υ = Psychrometric Constant

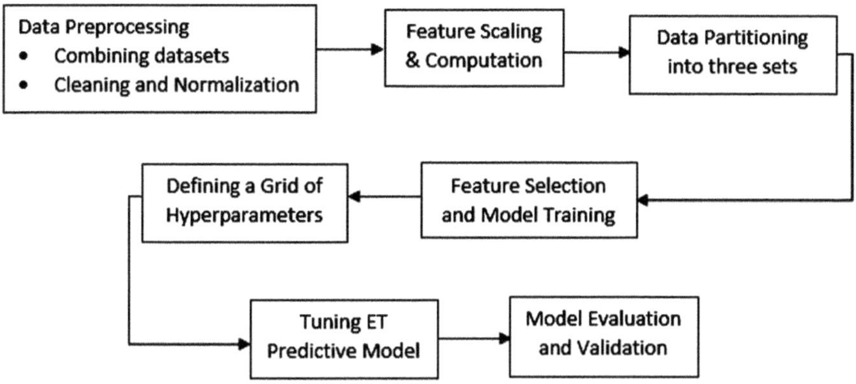

Fig. 8 Model implementation flowchart

T = Air Temperature (°C)
u_2 = Wind Speed (km/h)
e_s = Saturation Vapor Pressure (kPa)
e_a = Actual Vapor Pressure (kPa)

Furthermore, a Support Vector Machine (SVM) learning algorithm has been chosen to construct an ET predictive model owing to its superior robustness over other machine learning algorithms [27]. SVM is used because it can make accurate predictions from larger, more complicated datasets, especially when used with real-time sensor data [28]. The study used a combination of three data partitions for reiterative model learning, that is, sets for validation, testing, and training. Moreover, it is also hypothesized that the predictability of the ET model has been greatly improved by the addition of soil moisture measurement at subsequent soil depths [29]. It gives users quantitative insights into how evapotranspiration accounts for the majority of water loss following rain events and in hot summer months. Figure 8 displays the flowchart of the model implementation.

The evaluation metrics such as mean absolute error (MAE), coefficient of determination (R^2), and root mean squared error (RMSE) have been used to evaluate the predictive model performance. These metrics generally provide important insights into the model's generalization and accuracy capabilities [30]. In this regard, Fig. 9 shows an SVM prediction model for evapotranspiration that has the train, test, and validation sets divided into 80:10:10. The high R^2 value of 0.97 indicates that the model has excellent predictive accuracy. The minimal prediction errors indicated by the low MAE of 0.14 and RMSE of 0.19 further support the model's dependability in estimating evapotranspiration. The graph probably demonstrates a narrow range of deviations between the computed and predicted ETc values, indicating that the SVM model predicts the ETc values with good accuracy across a range of measurement levels.

Figure 10 illustrates a positive correlation between solar radiation and both actual and predicted ET_c values during the summer season. The predicted ET_c (using SVM)

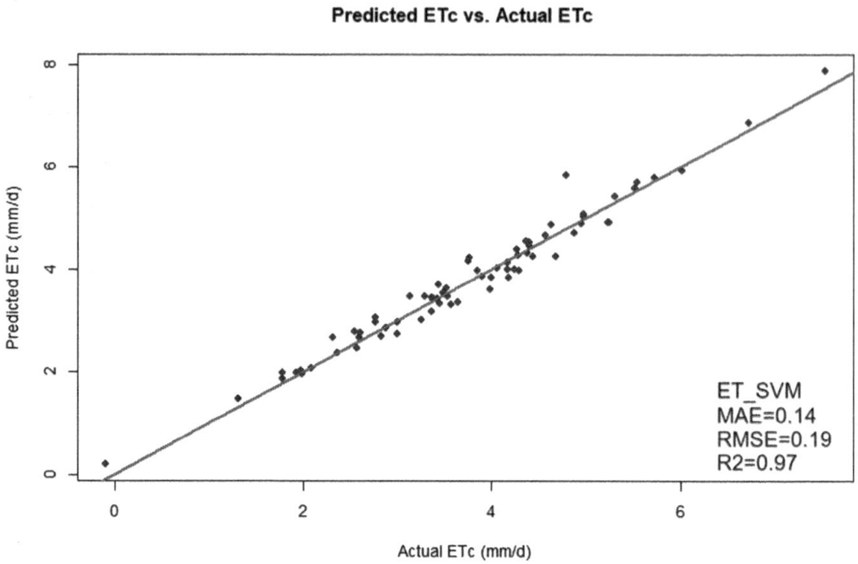

Fig. 9 Evapotranspiration model using support vector machine learning algorithm

closely follows the pattern of the actual ET_c (PM), which indicates that the SVM model is reasonably good at estimating ET_c based on the given inputs, which include solar radiation as a significant predictor. However, there are instances where the prediction deviates from the actual values as well as it does not always peak exactly when solar radiation is at its highest, indicating that other factors, such as temperature, wind, and humidity also play significant roles in evapotranspiration.

The close match between the predicted and actual ET_c values suggests that the SVM model could be used for forecasting ET_c, which is valuable for water resource planning, especially in agriculture. However, The discrepancies between the predicted and actual values could be a focus for further research to improve the SVM model's accuracy, perhaps by incorporating additional environmental parameters.

Figure 11 depicts a comparison between the different methods employed to determine water loss through Evapotranspiration (ET) across different climatic seasons, such as an empirical Penman-Monteith method applied for principle referencing, a support vector machine learning model constructed for driving ET predictions, and the daily average values reported by the Australian bureau of meteorology used for validation purposes. It is found that the water loss through evapotranspiration varies with each season potentially due to the driving climatic factors such as temperature, sunlight, and humidity, which also vary seasonally. It is observed that the BOM has throughout reported higher ET_a values compared to the other two methods possibly due to differing sensitivity to specific seasonal factors, or it might mean their method is overestimating ET_a due to some bias in its calculation. However, the predicted SVM values appear to be more consistent with actual PM values, which indicates that the machine learning model has been effectively learning from the patterns

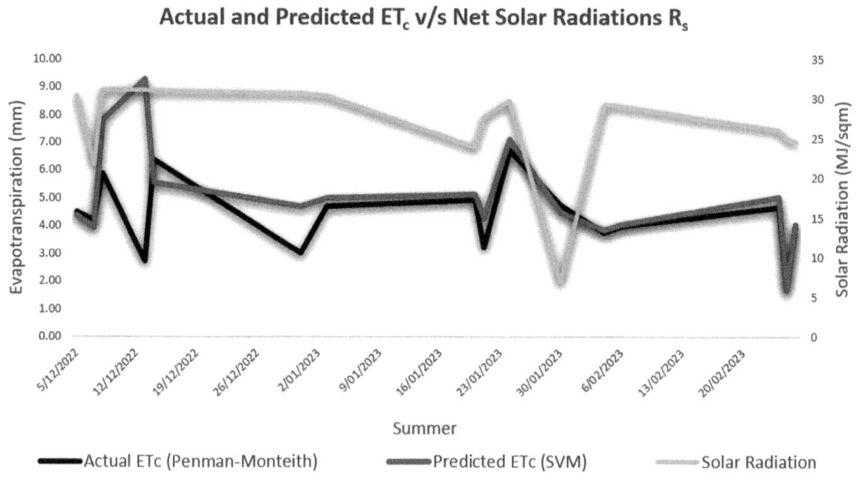

Fig. 10 Actual and predicted ET_c versus net solar radiations

presented in the data it was trained on. Interestingly, the Summer season shows the maximum evapotranspiration across all methods, which makes sense given that summer is typically the hottest and often the sunniest season. The BOM data stands out with significantly higher values ($ET_a = 7.66$ mm) than the other two methods, indicating a possible difference in computation methodology. On the contrary, during the Winter season, minimum evapotranspiration was recorded, which is consistent with lower temperatures and potentially less sunlight. When looking at the annual data, the actual PM values and the predicted SVM values are very close in their estimates, while the measured BOM shows a slightly higher value.

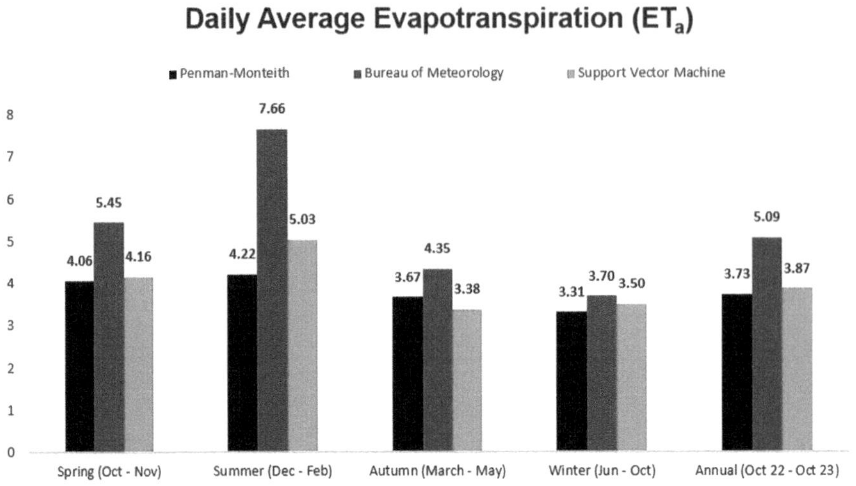

Fig. 11 Daily average evapotranspiration at Narrabri farm during the four pilot seasons of the year

5 Conclusion

The primary purpose of this experimental study is to bridge the gap between traditional soil monitoring methods and modern technological advancements, providing a more efficient and real-time solution to monitor soil conditions. The development section introduced the hardware and software intricacies of sensor node creation, emphasizing the importance of selecting appropriate components and ensuring seamless connectivity. Further, the installation segment shed light on the practical considerations of deploying these nodes in diverse soil environments highlighting best practices for optimal data collection and longevity. Perhaps most crucially, the modeling aspect showcased how the data harvested from the sensor node can be processed, analyzed, and modeled to yield actionable insights.

The implications of the IoT-enabled system with the integration of machine learning extend far beyond the experimental realms. In the real world, this technology stands poised to revolutionize agriculture and environmental management practices. Farmers and land managers can utilize real-time data to implement precision agriculture techniques, tailoring their interventions to specific soil conditions and minimizing resource wastage. Moreover, through rigorous data analysis and modeling, the support vector machine learning model has demonstrated promising capabilities in accurately estimating ET based on sensor-derived parameters, offering a robust alternative to traditional methods. The model's performance metrics, including high accuracy rates and low error margins, validate its efficacy in capturing intricate relationships between sensor data and ET values. Such precision not only enhances our understanding of water consumption dynamics in ecosystems but also empowers stakeholders with timely, actionable insights crucial for informed decision-making in agriculture, water resource management, and environmental conservation.

References

1. M. Zeri et al., Importance of including soil moisture in drought monitoring over the Brazilian semiarid region: an evaluation using the JULES model, in situ observations, and remote sensing. Clim. Resil. Sustain. **1**(1), 1–18 (2022). https://doi.org/10.1002/cli2.7
2. I. Haqiqi, D.S. Grogan, T.W. Hertel, W. Schlenker, Quantifying the impacts of compound extremes on agriculture. Hydrol. Earth Syst. Sci. **25**(2), 551–564 (2021). https://doi.org/10.5194/hess-25-551-2021
3. C. Forgotson et al., How satellite soil moisture data can help to monitor the impacts of climate change: SMAP case studies. IEEE J. Sel. Top. Appl. Earth Obs. Remote Sens. **13**, 1590–1596 (2020). https://doi.org/10.1109/JSTARS.2020.2982608
4. E.A.A.D. Nagahage, I.S.P. Nagahage, T. Fujino, Calibration and validation of a low-cost capacitive moisture sensor to integrate the automated soil moisture monitoring system. Agric. **9**(7) (2019). https://doi.org/10.3390/agriculture9070141
5. M.Q. Aringo, C.G. Martinez, O.G. Martinez, V.B. Ella, Development of low-cost soil moisture monitoring system for efficient irrigation water management of upland crops. IOP Conf. Ser. Earth Environ. Sci. **1038**(1) (2022). https://doi.org/10.1088/1755-1315/1038/1/012029
6. M. Hardie, Review of novel and emerging proximal soil moisture sensors for use in agriculture. Sensors (Switzerland) **20**(23), 1–23 (2020). https://doi.org/10.3390/s20236934
7. D. Schwamback, M. Persson, R. Berndtsson, L.E. Bertotto, A.N.A. Kobayashi, E.C. Wendland, Automated low-cost soil moisture sensors: trade-off between cost and accuracy. Sensors (Switzerland) **23**(2451) (2023). https://doi.org/10.3390/s23052451
8. T. Truong, A. Dinh, K. Wahid, An IoT environmental data collection system for fungal detection in crop fields, in *Canadian Conference on Electrical and Computer Engineering* (2017), pp. 1–4. https://doi.org/10.1109/CCECE.2017.7946787
9. W.A.K. Afridi, F. Akhter, I. Vitoria, S.C. Mukhopadhyay, A technology review and field testing of a soil water quality monitoring system, in *Lecture Notes in Electrical Engineering*, vol. 1035 (2023), pp. 460–475. https://doi.org/10.1007/978-3-031-29871-4_47
10. I. Kuzminykh, A. Carlsson, M. Yevdokymenko, A performance evaluation of sensor nodes in the home automation system based on Arduino, in *2019 IEEE International Scientific-Practical Conference Problems of Infocommunications, Science and Technology (PIC S&T)*, pp. 511–516 (2019). https://doi.org/10.1109/PICST47496.2019.9061442
11. C. Rusu et al., Miniaturized wireless water content and conductivity soil sensor system. Comput. Electron. Agric. **167**(November), 105076 (2019) https://doi.org/10.1016/j.compag.2019.105076
12. B. Kempegowda, Real-time soil monitoring system for the application of agriculture a precision farming for water optimization view project, November (2016). https://doi.org/10.4010/2016.1304
13. J.C. Songara, J.N. Patel, Calibration and comparison of various sensors for soil moisture measurement. Meas. J. Int. Meas. Confed. **197**(May), 111301 (2022). https://doi.org/10.1016/j.measurement.2022.111301
14. S. Adla, N.K. Rai, S.H. Karumanchi, S. Tripathi, M. Disse, S. Pande, Laboratory calibration and performance evaluation of low-cost capacitive and very low-cost resistive soil moisture sensors. Sensors (Switzerland) **20**(2) (2020). https://doi.org/10.3390/s20020363
15. W.A.K. Afridi, I. Vitoria, K. Jayasundera, S.C. Mukhopadhyay, Z. Liu, Development and field installation of smart sensor nodes for quantification of missing water in soil. IEEE Sens. J. **23**(21), 26495–26502 (2023). https://doi.org/10.1109/JSEN.2023.3317418
16. Radi, Murtiningrum, Ngadisih, F.S. Muzdrikah, M.S. Nuha, F.A. Rizqi, Calibration of capacitive soil moisture sensor (SKU:SEN0193), in *Proceedings—2018 4th International Conference on Science and Technology, ICST 2018*, vol. 1, no. August (2018), pp. 1–6. https://doi.org/10.1109/ICSTC.2018.8528624
17. E.J. Jones, P. Filippi, R. Wittig, M. Fajardo, V. Pino, A.B. Mcbratney, Mapping soil slaking index and assessing the impact of management in a mixed agricultural landscape. Soil **7**(1), 33–46 (2021). https://doi.org/10.5194/soil-7-33-2021

18. G. Gupta, R. Van Zyl, Energy harvested end nodes and performance improvement of LoRa networks. Int. J. Smart Sens. Intell. Syst. **14**(1), 1–15 (2021). https://doi.org/10.21307/IJSSIS-2021-002
19. N.R. Harris, J. Curry, Development and range testing of a LoRaWAN system in an urban environment. Int. J. Electron. Commun. Eng. **12**(1), 43–51 (2018)
20. A. James, A. Seth, S.C. Mukhopadhyay, IoT enabled sensor node: a tutorial paper. Int. J. Smart Sens. Intell. Syst. **13**(1), 1–18 (2020). https://doi.org/10.21307/ijssis-2020-022
21. S.J. Juneidi, Smart greenhouses using internet of things: case study on tomatoes. Int. J. Smart Sens. Intell. Syst. **15**(1) (2022). https://doi.org/10.2478/ijssis-2022-0019
22. W.A.K. Afridi, S.C. Mukhopadhyay, I. Vitoria, Multi-depth capacitive soil sensor node testing and deployment, in *16th International Conference on Sensing Technology (ICST)*, pp. 1–5 (2023). https://doi.org/10.1109/ICST59744.2023.10460836
23. USDA, Soil Quality Indicators. *USDA Natural Resources Conservation Service* (2008). https://www.nrcs.usda.gov/sites/default/files/2022-10/Infiltration.pdf
24. K.I. Islam, A. Khan, T. Islam, Correlation between atmospheric temperature and soil temperature: a case study for Dhaka, Bangladesh. Atmos. Clim. Sci. **05**(03), 200–208 (2015). https://doi.org/10.4236/acs.2015.53014
25. T.K. Anyachebelu, M. Conrad, T. Ajmal, Modeling and prediction of surface water contamination using on-line sensor data. Int. J. Smart Sens. Intell. Syst. **7**(5), 1–5 (2014). https://doi.org/10.21307/IJSSIS-2019-117
26. B. Benli, A. Bruggeman, T. Oweis, H. Üstün, Performance of Penman-Monteith FAO56 in a semiarid highland environment. J. Irrig. Drain. Eng. **136**(11), 757–765 (2010). https://doi.org/10.1061/(asce)ir.1943-4774.0000249
27. A. Van Messem, *Support Vector Machines: A Robust Prediction Method with Applications in Bioinformatics*, vol. 43, 1st edn. (Elsevier B.V., 2020). https://doi.org/10.1016/bs.host.2019.08.003
28. S. Amani, H. Shafizadeh-Moghadam, A review of machine learning models and influential factors for estimating evapotranspiration using remote sensing and ground-based data. Agric. Water Manag. **284**(May), 108324 (2023). https://doi.org/10.1016/j.agwat.2023.108324
29. J. Dong, R. Akbar, D.J. Short Gianotti, A.F. Feldman, W.T. Crow, D. Entekhabi, Can surface soil moisture information identify evapotranspiration regime transitions? Geophys. Res. Lett. **49**(7) (2022). https://doi.org/10.1029/2021GL097697
30. A. Raza et al., Performance evaluation of five machine learning algorithms for estimating reference evapotranspiration in an arid climate. Water (Switzerland) **15**(21) (2023). https://doi.org/10.3390/w15213822

A Novel Determination of Chemical Oxygen Demand (COD) Using Thermal Sensor: A Key Initiative in Wastewater Treatment

Nishant Chavhan, Sharda Chandak, Bhushan Chaware, Prasad Bongarde, Sanket Bodhe, and Gayatri Gawande

Abstract Water pollution is a pressing global concern, and the methodologies employed for water assessment play a pivotal role in gauging its severity. The quantification of organic compounds within water bodies assumes paramount significance in the meticulous monitoring and management of pollution levels. India, a nation grappling with the deteriorating health of its rivers, faces an enduring predicament. A staggering 40 million liters of wastewater are introduced into rivers and other aquatic reservoirs every day, with a mere 37% of it undergoing appropriate treatment. At the outset, Chemical oxygen demand (COD) is widely used as one of the most important an index for the appraisal of water quality. The paper describes the design and development of a novel portable Chemical oxygen demand (COD) detection method based on thermal sensors. The shortage of sustainable water resources and increasing awareness of environmental protection refers to the desideratum for an onsite, environmentally friendly, and rapid analytical method. The proposed work utilizes the principle of change in degree of enthalpy i.e., heat generated during the oxidation of organic compounds, and the heat is proportional to the amount of organic compounds contained in the water sample is a part of the thermal sensor's development. The presented work utilizes Sodium hypochlorite (NaClO) solution is employed as an oxidizer after examining several other oxidizers. Introduced method results to claim, further comprising with COD and were compared against the conventional method using Potassium dichromate. The COD values of water samples from various sources were correlated with those obtained using the standard dichromate method, yielding a linear regression equation of CODts = 1.0215 CODCr − 27.725 and a performing correlation coefficient of 0.9838. The paper compares the results of a few best-performing Machine learning (ML) algorithms like Linear regression, Regression tree, Random forest, and decision tree with grid search. At the outset, the

N. Chavhan (✉) · S. Chandak · B. Chaware · P. Bongarde · S. Bodhe · G. Gawande
Department of Chemical Engineering, Vishwakarma Institute of Technology, Pune–37, India
e-mail: chavhannishant19@gmail.com

B. Pradhan and S. Mukhopadhyay (eds.), *IoT Sensors, ML, AI and XAI: Empowering A Smarter World*, Smart Sensors, Measurement and Instrumentation 50,
https://doi.org/10.1007/978-3-031-68602-3_6

most dynamic model was observed to be the Decision tree with grid search achieving an overall accuracy i.e., the determination coefficient (R^2) being 97.85%. Furthermore, It's important to establish a new environmentally friendly chemical oxygen demand detection method with safe and high efficiency in the future.

Graphical Abstract

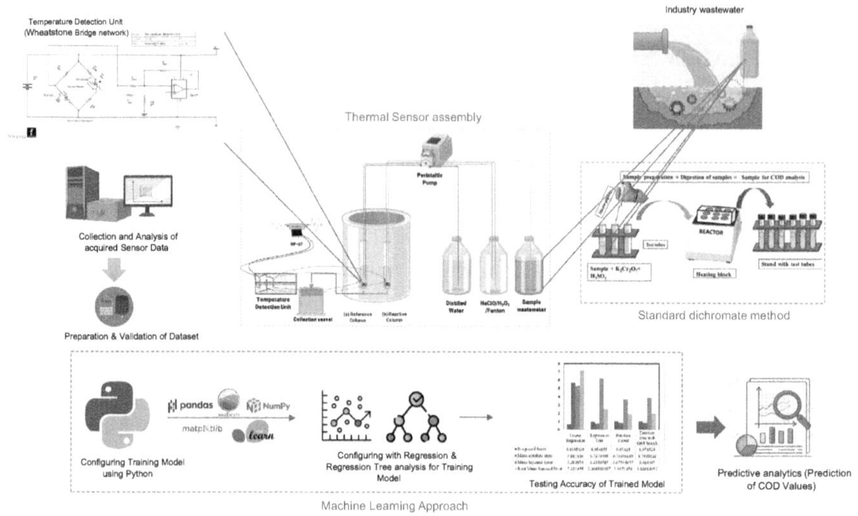

Keywords Wastewater · Thermal sensor · Chemical oxygen demand · OP07 · Wastewater treatment

1 Introduction

India faces a profound challenge of widespread water contamination, with a staggering 1,95,813 communities documented to have inadequate water quality. This dire situation poses significant health risks to the population. Alarmingly, pollution in the country has led to over 2.3 million premature deaths in 2019, according to a recent study [1]. Among these fatalities, approximately 1.6 million were attributed to air pollution, while water pollution was responsible for over half a million deaths. The 'Composite Water Management Index' report, released by NITI Aayog in June 2018, underscores India's most severe water crisis to date [2]. Around 600 million individuals are grappling with high to extreme levels of water stress. The report also reveals India's dismal ranking of 120th out of 122 countries in the water quality index, with a concerning 70% of its water sources contaminated. As a result, the necessity for water quality monitoring has intensified, with COD (Chemical Oxygen

Demand) emerging as a pivotal monitoring parameter for surface water, domestic sewage, and industrial wastewater [3].

Domestic and industrial wastewater exhibit a range of physical, chemical, and biological characteristics, as articulated in Fig. 1 [2]. Physically, wastewater can appear cloudy, colored, and have an unpleasant odor. Its temperature fluctuates depending on location, season, and industrial activities. While rare, radioactive contamination can also occur. Chemically, wastewater contains inorganic and organic impurities. Inorganic impurities can harm aquatic life and pollute water sources, while organic impurities deplete oxygen in water. The amount of organic matter is measured using Chemical Oxygen Demand (COD) and Biochemical Oxygen Demand (BOD). Biologically, wastewater teems with microorganisms, some beneficial for treatment processes and others harmful pathogens. Chemical oxygen demand (COD) serves as a critical parameter for assessing water quality. Effective management of pollution levels hinges on accurately quantifying the presence of organic compounds in water. Both COD and BOD assess the degree of Chemical and Biological oxidation, respectively, thereby offering insights into the organic matter content in polluted water [4]. COD is described as the amount of molecular oxygen necessary to completely oxidize and break down every organic and inorganic compound present in the water sample, resulting in water and carbon dioxide [5]. Portable, efficient, remote, and real-time monitoring of the determination of COD is essential for effective monitoring and control of wastewater treatment processes and environmental protection [6]. An essential use of the COD determination lies in quantifying soluble COD in wastewater, as characterization of the overall COD in wastewater is crucial for precise modeling of biotransformation during wastewater treatment procedures. The shortage of sustainable water resources and rising concern for environmental protection refers to the desideratum for an onsite, environmentally friendly, and quick analytical approach [7].

The discharge of large volumes of wastewater from industrial and agricultural activities, coupled with rapid urbanization, is a significant global water pollution issue. Inadequate wastewater treatment results in the direct release of pollutants into natural water bodies, exacerbating the already pressing problem of global water scarcity and impeding the achievement of Millennium Development Goals. The contaminants found in such wastewater can comprise a diverse array of organic and inorganic substances, creating a hazardous 'toxic molecular soup' that has detrimental effects on both aquatic and terrestrial ecosystems, as well as human health. Various classical and instrumental analytical methods, including titrimetry [8], spectrophotometry [9–14], electrochemistry [15–18], and spectroelectrochemistry [10], are employed to measure and analyze the organic and inorganic contaminants responsible for water pollution.

The conventional standard methods utilized for COD analysis have long been prevalent in environmental monitoring and wastewater treatment applications. However, these methods suffer from several drawbacks, including low sensitivity of detection, prolonged reflux time (typically 2–4 h), susceptibility to interferences such as chloride ions, and the use of consumptive and hazardous reagents like silver

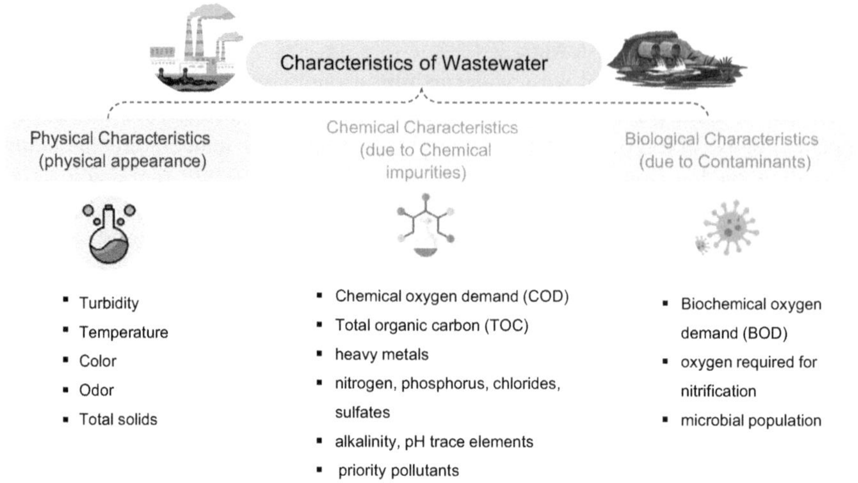

Fig. 1 Characteristics of wastewater

ions, potassium dichromate, and mercury sulfate. These limitations curtail their suitability for COD measurements and impede their utility for real-time environmental assessments. In recent years, considerable efforts have been directed toward devising reliable, precise, and nearly real-time methods for COD determination in water bodies. Innovative digestion techniques such as microwave-assisted or ultrasonic digestion methods [4, 8], along with analytical approaches like spectrophotometry (UV-Vis) [9], chemiluminescence (CL) methods [11, 13, 17], and Flow injection analysis (FIA) [11, 19–23], have emerged as effective alternatives for COD determination, as delineated in Figs. 2 and 3. These advanced methods have demonstrated enhanced detection sensitivity, affordability, high accuracy, and improved throughput. Nonetheless, they still necessitate the use of hazardous reagents such as dichromate ions ($Cr_2O_7^{2-}$) and mercury sulfate ($HgSO_4$) [4, 8, 24]. Overall analytical differentiation and characterization of an array of COD detection techniques are depicted in Fig. 2.

Conventionally, COD analysis has relied on complex and time-consuming laboratory methods, such as the closed reflux method or the dichromate method, which involve the use of strong chemical oxidizing agents viz., dichromate or permanganate. However, the National standard method utilizes the Standard dichromate-based COD method due to its higher oxidizing power and superior reproducibility. However, there is growing interest in developing alternative methods that are more convenient, rapid, and cost-effective, without compromising accuracy.

The propitious and novel approach is the use of thermal sensors for COD detection in wastewater. Thermal sensors are designed to measure the temperature changes that occur during a chemical reaction, which can be correlated to the amount of COD present in the sample. This method offers several advantages, including real-time monitoring, simplicity, portability, and reduced reliance on hazardous chemicals.

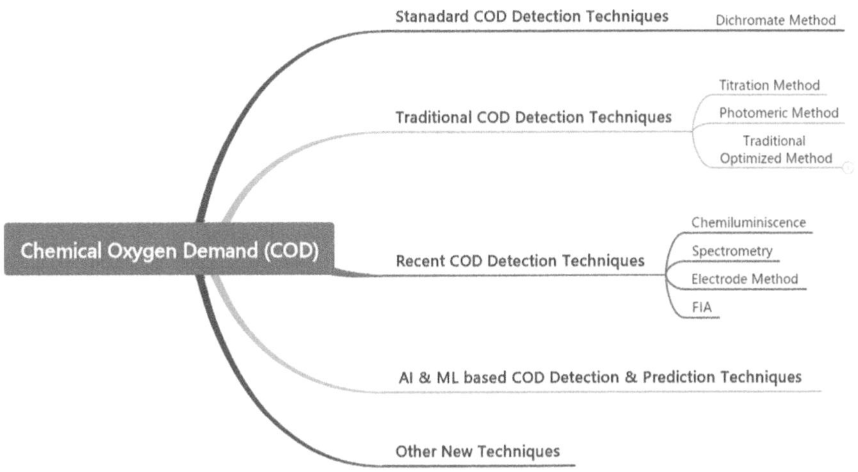

Fig. 2 Differentiation and characterization for an array of COD techniques

The principle behind thermal-based COD detection is based on the exothermic reactions that take place when organic matter in wastewater is oxidized. As the organic compounds react with an oxidizing agent, heat is released, causing a measurable change in temperature. By carefully monitoring this temperature change using a thermal sensor, the COD level can be determined.

Thermal sensors for COD detection come in various forms, such as thermistors, thermocouples, or infrared sensors. These sensors are highly sensitive to temperature changes and can be integrated into monitoring systems or portable devices for on-site analysis. The use of thermal sensors offers real-time measurements, allowing for prompt adjustments and decision-making in wastewater treatment processes. The adoption of thermal-based COD detection methods has gained attention due to their potential for automation, cost-effectiveness, and ease of use. Furthermore, they can be applied in a wide range of wastewater treatment scenarios, including industrial, municipal, and agricultural applications.

The proposed study aims to explore and validate the effectiveness of thermal sensor-based COD detection for wastewater analysis. The study investigates the correlation between the temperature changes recorded by the thermal sensor and the COD levels determined using traditional laboratory methods by virtue of a machine learning model associated with it. The results obtained will contribute to the development of a reliable and practical COD detection technique that can be implemented in wastewater treatment plants, enabling efficient monitoring and optimization of treatment processes.

Nonetheless, the utilization of thermal sensors for COD detection in wastewater represents a promising advancement in the field of water quality monitoring. By harnessing the advantages of thermal-based analysis, we can enhance our ability to assess and manage the pollution levels of wastewater, ensuring the preservation and protection of our precious water resources.

2 COD Analysis by Standard Methods

Water pollution presents a significant challenge to society, necessitating comprehensive water treatment processes that encompass analytical parameters to assess the extent of contamination at both the onset and culmination of treatment. Among the myriad pollutants, organic matter holds particular importance, leading to the development of several techniques for quantifying its presence in water. Chemical Oxygen Demand (COD) emerges as a swift assessment tool for monitoring water quality throughout treatment procedures. COD serves as a metric for gauging contamination levels in water bodies attributable to organic matter. It denotes the amount of oxygen equivalent to the organic content within a sample, which potent chemical oxidants like potassium permanganate or potassium dichromate can oxidize. These oxidants, tailored for specific water types, facilitate COD determination: COD_{Mn} (Chemical Oxygen Demand with Manganese) finds utility in slightly contaminated water bodies, whereas COD_{Cr} (Chemical Oxygen Demand with Chromium) is suitable for heavily contaminated water or wastewater, owing to the formidable oxidizing capabilities of the reagent.

Recently, there has been a significant emphasis on the development of novel techniques for COD determination in order to mitigate the adverse effects associated with dichromate and standard methods, such as toxicity, secondary pollution, and interference issues. Approaches based on techniques like chemiluminescence (CL), flow injection analysis (FIA), and advanced oxidation processes (AOP) have emerged as viable alternatives and have found widespread application in COD assays, illustrated in Fig. 3. These techniques offer various advantages, including ease of operation,

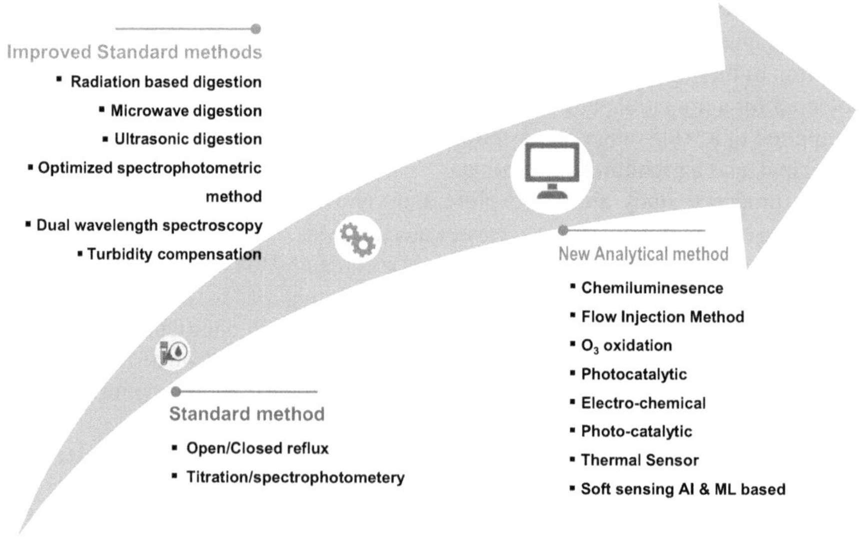

Fig. 3 An array of COD determination approaches

rapid analysis, reduced dependence on toxic reagents, automation capabilities, and minimal generation of secondary pollution. These advancements address the limitations of traditional methods and provide more sustainable and environmentally friendly options for COD analysis.

COD determination typically comprises two primary stages: sample oxidation or digestion, with the subsequently applied measurement of the oxidant consumed during the process. Digestion approaches can include either open reflux or closed reflux techniques and the determination of the use of the oxidant is mostly done by using titration and spectrophotometric methods [4, 8, 24]. These methodologies are delineated into two main categories in the standard procedures: (i) manual operations of poultices digestion co-paired with potassium dichromate ($K_2Cr_2O_7$) titration, and (ii) automated, closed digestion skied with spectro-photometric methods described in ISO 6060-1989 [25] and ISO 15705-2002 [26], respectively. ISO 6060-1989 is intended as a national standard that provides the principles for the determination of Chemical Oxygen Demand (COD) in wastewater samples. This protocol includes acid reflux digestion and the sample is heated with a potassium dichromate process for a specified time under reflux. Following digestion, any extra dichromates are titrated with ferrous ammonium sulfate to measure the amount of oxidant that reacted. The COD value is then calculated by the volume of the reducing agent. This standard guarantees the technically accurate and consistent COD analysis that is necessary for the reliable assessment of the organic pollutants in wastewater.

The outline of standard procedure in International Standards ISO 6060-1989 [25, 27] and ISO 15705-2002 [26], includes the oxidation in the presence of other substances like potassium biphthalate with potassium dichromate and sulfuric acid during a two hours reflux process at 150 °C in an open or closed system. $HgSO_4$ in the reaction mixture is the source of Hg^{2+} ions which combine with Cl^- ions to form the precipitate, but in a reaction sense, silver salts are catalysts.

Chromium(VI) ions consumed in the reaction can be further quantified using ferrous ammonium sulfate titrated against a $K_2Cr_2O_7$ solution, and ferroin as an indicator. Reduction-charged biphthalate will have first contact with dichromate. Then, in the subsequent stage, the dichromate ($K_2Cr_2O_7$) is changed to oxygen (O_2) Consequently, bronze metal dichromate forms after acid reflux and is oxidized to dichromate in surplus. Then there is iron solution titration (Fe^{2+}) with a specific chemical reaction. Instead of this, a closed reflux procedure that allows the detection of any residual dichromate via spectrophotometry techniques (either monitoring at the absorbance wavelength of 420 nm or recording the increase of the trivalent chromium ions (Cr(III)) at the wavelength of 600 nm) ensures the tracing and quantification of the mentioned chemical compound.

In the standardized procedure for determining Chemical Oxygen Demand (COD), the presence of organic material is simulated using potassium hydrogen phthalate (KHP). This method entails refluxing a sample portion with mercury(II) sulfate, a specified amount of potassium dichromate ($K_2Cr_2O_7$), and a silver sulfate (Ag_2SO_4) catalyst in concentrated sulfuric acid over a duration of two hours. During this process, a fraction of the dichromate is reduced owing to the presence of oxidizable substances. The reaction between KHP and $K_2Cr_2O_7$ is illustrated as follows:

$$2KC_8H_5O_4 + 10\,K_2Cr_2O_7 + 41H_2SO_4 \rightarrow 16CO_2 + 46H_2O + 10Cr_2(SO_4)_3 + 11K_2SO_4 \quad (1)$$

When $K_2Cr_2O_7$ is substituted by O_2, the equation becomes:

$$2KC_8H_5O_4 + 15O_2 + H_2SO_4 \rightarrow 16CO_2 + 6H_2O + K_2SO_4 \quad (2)$$

Following digestion, the dichromate that remains unreduced is quantified through potentiometric titration employing a solution of Fe^{2+}, based on the equation:

$$Cr_2O_7^{2-} + 6Fe^{2+} + 14H^+ \rightarrow 2Cr^{3+} + 6Fe^{3+} + 7H_2O \quad (3)$$

The quantity of oxidized organic matter is determined in terms of oxygen equivalents, where one equivalent weight of an electron corresponds to 8 g of oxygen. The following equation expresses this calculation:

$$\text{COD (mg } O_2L^{-1}) = (X - Y) * M * 8000/V \quad (4)$$

where, X and Y represent the volumes of Mohr's salt used for the blank and sample, respectively (in mL), M is the molarity of Mohr's salt, and V is the volume of the sample (in mL).

Table 1 summarizes an array of Techniques used for the determination of the COD Standardized method along with new analytical methods of Advanced oxidation processes (AOPs). The COD determination methods which are conventionally experienced are well accepted worldwide, however, they have got several constraints. They also entail deferral or prolonged digestion and titration again, expensive and wastage of corrosive and toxic chemicals such as silver ions (Ag^+), concentrated sulfuric acid (H_2SO_4), mercury sulfate ($HgSO_4$), and potassium dichromate ($K_2Cr_2O_7$), which are environmentally hazardous. Moreover, outside interference from the inorganic substances including chloride and nitrite, as well as the incomplete oxidation of the volatile compounds, are the disadvantages of these techniques. Open reflux-titration method offers advantages such as shorter reaction times, limited chemical consumption, lower costs and less production of secondary pollutants because of its direct comparison to the human language. Nevertheless, a number of challenges pose typical issue, for example, bubble formation, turbidity, color interference, and limited spectrophotometric device capabilities associated with COD assessment. In addition to that, the application of both methods has been regarded as deficient in online and real-time monitoring purposes.

UV spectroscopy harnesses the remarkable absorption characteristics exhibited by diverse organic compounds present in water samples across the ultraviolet spectrum. The methodology entails partitioning samples into two categories: one is subjected to evaluation through the conventional national standard COD method, while the other undergoes recording of its UV spectral absorbance curve. Utilizing methodologies

Table 1 An array of techniques used for the determination of COD [28]

Technique	Standardization	Linear range (mg L^{-1})	Detection range (mg L^{-1})
Standardized method	Standardized method Open reflux	NA	> 50
	Standardized method 5220D Closed reflux	30–600	NA
Standardized method: Oxidation processes	AOPs: photocatalysis	3.4–20	1.2
	AOPs: Fenton-H_2O_2 process	2.0–50	2
	Oxidation with permanganate	20–500	−8
Standardized method: New analytical methods	Flow injection analysis (FIA)		
	Fluorescence	1–100	0.9
	Chemiluminescence	0.16–19.24	0.1
	Electrochemical detection: nano-Ni sensor	10–1533	1.1
	Electrochemical sensor: Pt NPs	NA	1.83
	Photocatalytic and electro-photocatalytic detection	20–300	15
New techniques	Thermal biosensors	5.0–3000	1.84

like the least squares method or advanced algorithms such as neural networks facilitates the construction of a COD prediction model, elucidating the correlation between UV spectrum and COD. This approach facilitates rapid COD estimation once the UV absorbance pattern is identified, drastically reducing detection time from prolonged potassium dichromate digestion to a mere 2-min optical scan, thereby significantly enhancing detection efficiency. By spectrophotometric method for COD determination the absorbance of a colored complex, formulated in the interaction of sample and oxidizing agent is measured. Regression models based on the adsorption-desorption process in the UV-vis spectroscopy provide a COD-measurement alternative method [9]. This indirect procedure is based on registering absorbance at single UV-vis wavelengths. COD concentrations are determined by employing calibration standards obtained using chemical procedures, which correlate the absorbance readings to the concentrations. COD value can be deduced by comparing the absorbance of the sample to the calibration curve. In another instance, Zhao Youquan et al. [20] designed a COD monitoring system utilizing UV-Vis spectrometry. The system of the apparatus uses a pulsed xenon lamp, concave holographic grating, and spectral scanning structure to capture the spectral data in the range of 200–720 nm, which correlates to the COD concentrations from 30 to 1000 mg/l by computing UV absorbance. UV-vis spectroscopy presents an energy- and economy-efficient alternative for COD analysis that bypasses the elaborate chemical digestion step. Nevertheless, reliable reference regression models are needed to obtain accurate and confirmed results.

Chemiluminescence (CL) is an advanced technique for determining COD that operates based on the detection of light emitted during a chemical reaction [11]. In this method, the organic compounds within the sample undergo oxidation by a suitable oxidant, resulting in the emission of light. The intensity of this emitted light directly correlates with the COD concentration. The abbreviation CL, referring to Chemiluminescence, has become synonymous with a reliable analytical tool boasting notable advantages, including instantaneous response time, high sensitivity, a wide dynamic range, portability, process simplicity, and cost-effectiveness. Recent applications have showcased its effectiveness in determining Chemical Oxygen Demand (COD) across various water samples from the environment. The catalytic process involves luminol oxidation with hydrogen peroxide in a neutral environment with the addition of chromium. CL stands out as a potent detection technique, offering several benefits such as high sensitivity, rapid response time, a broad linear range, straightforward instrumentation, and portability. Furthermore, the energy required for CL emission originates from a chemical reaction rather than an external light source. This method relies on chromium, which catalyzes the luminol oxidation reaction with hydrogen peroxide in a basic medium. The production of Cr^{3+} occurs during the chemical oxidation of COD substances in the samples and is measured with the assistance of the luminol-H_2O_2 system.

$$\text{Luminol} + H_2O_2 + Cr^{+3} \rightarrow h\gamma + \text{production} \tag{5}$$

Photocatalytic methods for COD determination exploit the principle of photocatalysis to oxidize organic compounds present in the sample. A photocatalyst, typically titanium dioxide (TiO_2) [29], is activated by ultraviolet (UV) light, leading to the generation of reactive species that facilitate the oxidation of organic compounds. The reduction in COD can then be quantified using diverse analytical techniques. Photocatalytic methods present several benefits such as high sensitivity, swift analysis, and the potential for real-time monitoring. In 2005, Junshui Chen et al. [21] developed a TiO_2 photocatalytic sensor employed in Flow Injection Analysis (FIA) for determining chemical oxygen demand (COD). This sensor, subjected to a positive bias potential of 0.4 V (vs. Ag/AgCl) and illuminated by a 12-W quartz UV lamp, recorded photocurrent resulting from charge transfer at the interface of the TiO_2 sensor and the passing solution. The change in photocurrent (ΔI Photo) caused by the detected sample was then calculated to characterize the COD value of the sample [16].

In 2022, Wenchao Duan et al. [6] presented a study on the development and evaluation of a straightforward electrochemical sensor designed for precise measurements of dissolved Chemical Oxygen Demand (COD) in water samples collected from urban wastewater treatment plants (UWWTP). This sensor facilitates discrete and accurate COD assessments directly from the sample to the result. Additionally, Wang and del Valle [24] introduced the electronic tongue technique, which incorporates electrochemical electrodes to qualitatively and quantitatively evaluate wastewater. This innovative approach offers a comprehensive analysis of water quality parameters using electrochemical sensors. Electrocatalytic methods employed for COD determination utilize electrodes to catalyze the oxidation of organic compounds within the sample. The resulting oxidation reaction generates an electric current that is proportionate to the COD concentration. Various electrode materials, including boron-doped diamond (BDD) and platinum, can be utilized in these methods. Electrocatalytic techniques offer notable advantages such as high sensitivity, selectivity, and the ability to provide real-time monitoring of COD levels in water samples. Photoelectrocatalytic methods for COD determination combine the principles of photocatalysis and electrocatalysis. They involve the use of a photocatalyst, such as TiO_2, in conjunction with an electrode to oxidize organic compounds in the sample. The photocatalyst absorbs light, generating reactive species that promote the oxidation reaction, which is detected as an electric current that offers high sensitivity, selectivity, and potential for real-time analysis.

In 2018, Giriprasad et al. [10] unfolded, presenting an innovative real-time water quality monitoring system. This system, meticulously crafted, aimed at not only being cost-effective but also profoundly efficient. Its capabilities extend to the measurement of vital parameters such as water temperature, dissolved oxygen (DO), turbidity, conductivity, and pH. Within this technological marvel, sensor readings find their sanctuary, meticulously processed within the core controller. Through a series of meticulous experiments, the feasibility and effectiveness of this method came to light, heralding a new era in water quality monitoring. The application of soft measurement modeling within the wastewater treatment industry stood as a testament to the system's promising potential. In the annals of 2017, Nguyen Thi Binh Phuong [15] embarked on, harnessing the prowess of artificial neural networks to discern spatial nuances in chemical oxygen demand (COD) concentrations through optical remote sensing images from Landsat 8. In a grand revelation, the study unveiled a substantial correlation ($R^2 = 0.89$) between Landsat 8's spectral reflectance values and COD concentrations, all thanks to the application of the ingenious artificial neural network methodology. Through this visionary approach, the spatial tapestry of COD concentrations emerged, subtly surpassing national irrigation standards as delineated in column B1 of the QCVN.

The thermal sensor demonstrated robustness and tolerance to interferences, allowing for the analysis of over 90 samples within a 30-day period. However, the accuracy of the method was found to be relatively low, as the measured values were only one-third of the values obtained using the standard dichromate method for approximately half of the samples. Additionally, the oxidation efficiency of the sensor was limited for a wide range of organic compounds, which restricted its potential implementation.

3 Methodology

In contemporary research endeavors, the predominant focus lies on the eradication of organic contaminants from wastewater, which pose a significant threat to water quality. These organic residues represent a substantial source of water pollution. Within this context, the parameter of Chemical Oxygen Demand (COD) emerges as a pivotal metric for gauging the extent of organic pollutants present in wastewater. The process involves the oxidation of organic compounds within the wastewater by potent oxidizing agents such as dichromates and permanganates. The resultant consumption of oxygen equivalents during this oxidation process quantifies the COD, thereby providing a measure of the organic pollutant load within the wastewater sample. While alternative methodologies have been explored and suggested, they have yet to match the efficiency and efficacy offered by the COD assessment technique.

A thermal sensor has been developed for the determination of COD (Chemical Oxygen Demand) in wastewater samples. This sensor exploits the principle that heat is released during the oxidation of organic pollutants present in the wastewater. By measuring the heat produced during this oxidation process, the sensor quantifies the COD levels. The assay is coupled with flow injection analysis (FIA) to enable continuous analysis, allowing for precise control over the amount of test water sample and carrier solution used. In all conceptual framework for the determination of Chemical oxygen demand is illustrated in Fig. 4.

The thermal sensor setup includes a peristaltic pump with two channels for tubing, a port injection valve, an amplifier, and two thermistors, all connected to a computer for data collection and analysis. The thermistors are sensitive to changes in temperature corresponding to the oxidation process, and the change in enthalpy of the organic contents is proportional to the thermometric peak height observed. Unlike other sensors, the thermistor used in this setup is specifically designed to be insensitive to the electrochemical, optical, and other properties of the water sample, ensuring accurate COD measurements. One of the significant advantages of this thermal sensor is its high detection efficiency, allowing for precise and accurate determination of COD values [9]. The low degree of interference from other characteristics of the test water sample further enhances the accuracy of the results obtained.

Chemical Oxygen Demand (COD) detection using thermal sensors can be done by monitoring the change in temperature of a sample as it is oxidized by a strong chemical oxidizing agent, typically Sodium hypochlorite (NaClO) [30]. The amount of heat generated during the oxidation reaction is proportional to the amount of organic

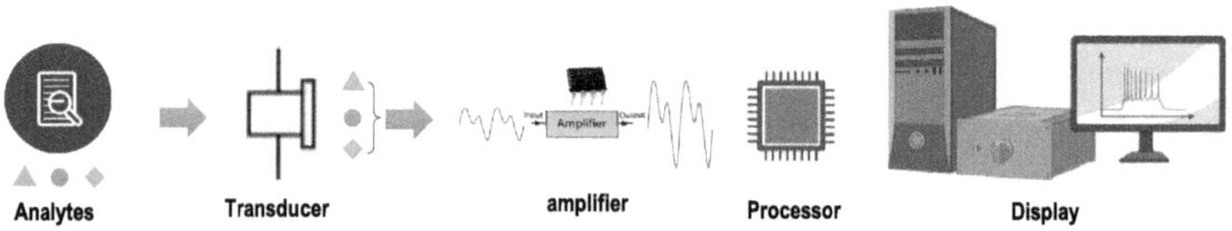

Fig. 4 Conceptual framework for the determination of COD via thermal sensor

matter in the sample, and thus to its COD. The change in temperature can be measured using a thermometer or a thermal sensor, and the COD can be calculated from the change in temperature using a calibration equation. This method is commonly used in industrial and environmental applications for the rapid determination of the organic content of wastewater and other aqueous sample. The overall flow of the system can be demonstrated through the following Fig. 5 which characterizes from Hardware assembly to Predicting COD values through a machine learning model.

3.1 Principle of COD Calibration

The principle of calibration for a thermal sensor used in COD (Chemical Oxygen Demand) determination involves establishing a relationship between the heat generated during the oxidation of organic compounds and the corresponding COD concentrations in a sample [9]. The calibration relationship equation used to relate the change in temperature to the COD of the sample typically takes the form:

$$\mathrm{COD} = \mathrm{k} * \Delta \mathrm{T} + \mathrm{b} \tag{7}$$

where; COD is the Chemical Oxygen Demand in mg/L, ΔT is the change in temperature in degrees Celsius, k is the calibration factor, which relates the change in temperature to the COD of the sample, b is the y-intercept, which represents the residual COD that cannot be measured by the method.

The calibration factor (k) and the y-intercept (b) are typically determined experimentally using a set of standard samples of known COD. These parameters can then be used to calculate the COD of unknown samples based on their change in temperature. Assuming all the organic compositions present in a given sample are completely decomposed to carbon dioxide and water in the reaction and the reaction is given by-

$$C_6H_{12}O + 12NaClO \rightarrow 12NaCl + 6CO_2 + 6H_2O + \Delta H \tag{6}$$

The circuit design and calibration procedure will depend on the specific requirements of the application, such as the temperature range, the desired accuracy, and the availability of components. A comprehensive understanding of electronics and programming is necessary for the successful design and implementation of a COD detection system using a thermistor. It is important to note that the calibration equation is specific to the conditions used for the COD determination (e.g. reaction time, oxidizing agent, temperature, etc.), and may need to be redetermined if these conditions change.

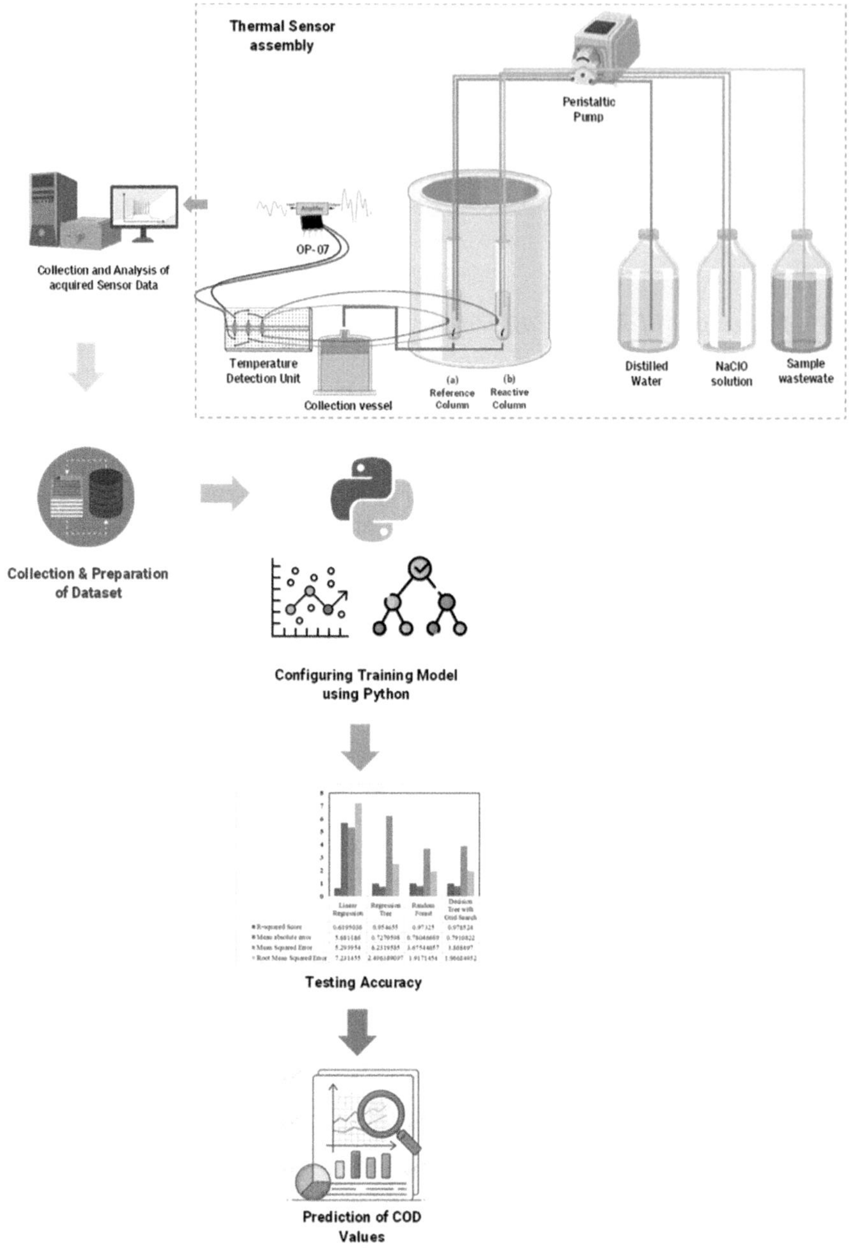

Fig. 5 Block flow diagram for determination of COD through thermal sensor

3.2 Chemicals and Material

It is critical to identify a type of universally applicable oxidizer. In the proposed study, Sodium Hypochlorite (NaClO) (available chlorine > 10%), is utilized as the oxidizer because it is widely applicable, and high activity, and can completely oxidize organic compounds present in water, along with Hydrogen Peroxide (H_2O_2), Fenton, Potassium dichromate ($K_2Cr_2O_7$), Silver sulfate (Ag_2SO_4), Mercury(II) sulfate ($HgSO_4$), Sulphuric acid (H_2SO_4), Ferrous ammonium sulfate (FAS) (($(NH_4)2Fe(SO_4)_2$), and Sodium Hydroxide (NaOH) were purchased from *D. Haridas & Co (A Unit of Advance Labs Scientific Pvt. Ltd.)*, Pune, India. Electronics components i.e., Thermister and resistor (10 kΩ), OP07 (Operational amplifier) used in order to design the network utilized as Temperature Detection Unit in Thermal sensor were purchased from *Pioneer Electronics, (Electronic Parts supplier)* Pune, India. The wastewater used as the water sample to optimize the analytical parameters was collected from an industry nearby in Pune.

3.3 Instrument Design

The designed novel Thermal sensor can be segregated into four main parts (shown in Fig. 6).

The first part of the sensor consists of storage units for Reagents viz., Oxidizing agent (NaClO/H_2O_2/Fenton), Distilled water utilized in the reference column, and the wastewater Sample whose COD is to be calibrated used for the detection process. Prior to incorporating NaClO solution into this investigation, we explored

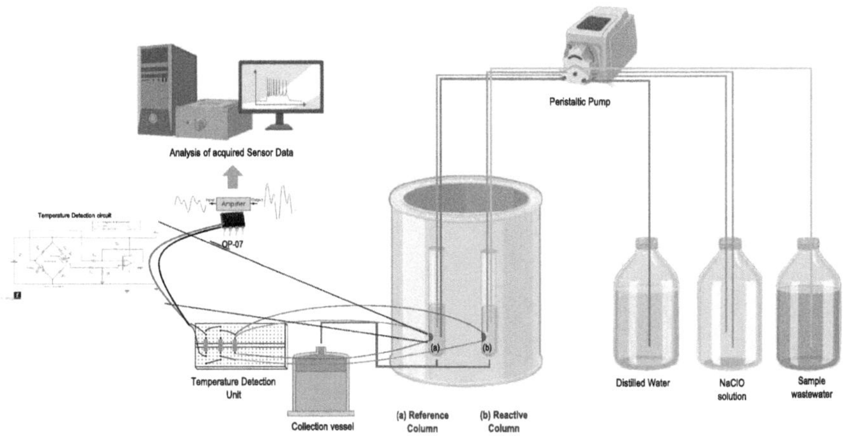

Fig. 6 Design and schematic of designed thermal sensor

the viability of employing Fenton's reagent, sulfate solution, and hydrogen peroxide as oxidants for COD determination using the thermal sensor.

The first part i.e. Storage units followed by the core component of the portable COD detection sensors. It facilitates the mixing of the sample and reagent, allows for the necessary reactions to occur, and enables the temperature to be detected by the virtue of Thermister which changes its resistance in accordance with heat released during the oxidation of the sample [31].

The third part is the temperature detection unit, which converts the heat variation generated during the reaction into a voltage signal which is further amplified by the OP-07 operational amplifier.

At the outset, the Fourth part is data processing and prediction of COD value using a machine learning model with resulted in voltage data from the third part of the thermal sensor as input dataset for the programmed model.

3.4 *Temperature Detection Unit*

The temperature detection unit plays a crucial role in measuring heat changes and converting them into voltage signals. Due to the small volume of reaction and heat release, precise and sensitive thermistors and specialized measuring circuits are required to detect weak signals. The Wheatstone bridge configuration is employed to measure the voltage signal from the thermistor, enabling indirect measurement of temperature changes. Figure 7 illustrates the composition of the Wheatstone Bridge. This setup allows for accurate and reliable detection of temperature variations in the system.

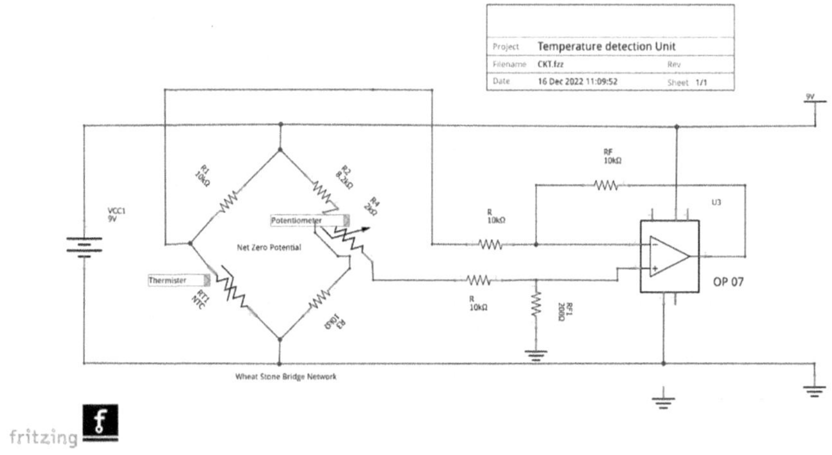

Fig. 7 Designed Wheatstone Bridge network utilized as temperature detection unit in thermal sensor

In COD detection using a thermal sensor, the temperature detection unit comprises a thermistor, a resistor, and an operational amplifier (OP07) as an amplifier. The thermistor is a temperature-sensitive resistor that changes its resistance with variations in temperature. It acts as the sensing element and is connected in a Wheatstone bridge configuration with the resistor. The Wheatstone bridge is a circuit arrangement that provides a balanced output voltage when the bridge is in equilibrium. Any change in temperature causes an imbalance in the bridge, resulting in a voltage difference across the bridge. This voltage signal is amplified by the OP07 operational amplifier to increase its strength and make it suitable for further processing and analysis. The amplified signal can be used to determine the change in temperature, which is then correlated with the COD concentration in the sample.

The core purpose of the temperature detection unit is to detect variations in heat and convert them into voltage signals. Achieving this requires a thermistor with high accuracy and sensitivity, along with a corresponding measuring circuit capable of accurately capturing subtle signals. This necessity arises because the heat release and reaction volumes are typically minuscule [31]. To accomplish this, a high precision thermistor is essential, coupled with a matching measuring circuit to ensure accurate detection of these weak signals [32]. The voltage signal generated by the thermistor is assessed using a Wheatstone bridge configuration, which facilitates the indirect measurement of temperature fluctuations. Through this method, even minor changes in temperature can be effectively detected and quantified.

In the Fig. 7 resistors R1, R2, and R3 are specified as standard precision metal film resistors with a resistance value of 10 kΩ, following the 1/1000 standard. The thermistor MF51 serves as RX, characterized by a B value of 3950 (25/850 °C) and a temperature coefficient of 4.45% [25]. At room temperature (25 °C), the resistance of the thermistor is 10 kΩ, and at 25 °C, the thermistor resistance is 10 kΩ. The supply voltage (Vcc) is set to 10 V. The operational amplifier employed is the OP07, renowned for its high precision, ultralow offset voltage and drift, low bias current, high common-mode rejection ratio, and excellent power supply rejection characteristics. The amplification circuit is configured for a gain of 1 to amplify the output of the amplifying circuit.

The OP07 [32–34] is widely recognized for its remarkable features of extremely low offset and long-term stability. This outstanding performance is achieved through an advanced low-noise, bipolar-input-transistor amplifier circuit. In most scenarios, external components are unnecessary for offset nulling and frequency compensation. The OP07 boasts an impressively low input offset voltage (with a maximum of 75 μV for the OP07E), achieved through precision trimming during wafer production. This minimal offset voltage typically eliminates the need for external nulling. Additionally, the OP07 features low input bias current (±4 nA for the OP07E) and high open-loop gain (200 V/mV for the OP07E), making it particularly well-suited for high-gain instrumentation applications [35].

The OP07 offers a wide input voltage range (±13 V minimum) and exceptional common-mode rejection ratio (106 dB for the OP07E) [36], combined with high input impedance, ensuring high accuracy in noninverting circuit configurations. Even at high closed-loop gains, it maintains excellent linearity and gain accuracy. Furthermore, the OP07 demonstrates outstanding stability of offsets and gain over time and temperature variations.

Available in two standard performance grades, the OP07E operates within the 0 °C to 70 °C temperature range, while the OP07C is specified for temperatures ranging from −40 °C to +85 °C [36]. With true differential input, a wide voltage range, and excellent common-mode rejection, the OP07 offers maximum flexibility and performance in high-noise environments and noninverting applications. Its accuracy, stability, and freedom from external nulling have established the OP07 as an industry standard for instrumentation applications.

The OP-07 is an operational amplifier known for its high gain, low offset voltage, and low input bias current. These features are crucial in accurately amplifying the small voltage signals produced by the Wheatstone Bridge in response to temperature changes. As OP-07 amplifier exhibits excellent precision and stability, making it suitable for sensitive temperature measurements. Its low offset voltage minimizes any voltage errors in the output signal, ensuring accurate temperature detection. Additionally, the low input bias current helps maintain the balance and integrity of the Wheatstone Bridge. The OP-07 amplifier also has a wide bandwidth, allowing it to handle fast temperature variations effectivelyThis characteristic is beneficial for capturing rapid temperature changes in real time.

3.5 Machine Learning Model

The machine learning model for the prediction of COD (Chemical Oxygen Demand) using a thermal sensor involves training a model to establish a relationship between the thermal sensor readings and the corresponding COD concentrations in water samples can be examined in Fig. 8.

The model is trained using a dataset that includes a range of water samples with known COD values and their corresponding thermal sensor readings. The dataset is divided into a training set and a validation set. The features used for training the model include the temperature changes detected by the thermal sensor during the oxidation process in the form of outputted voltage signal through the amplifier. These temperature readings are pre-processed and normalized to ensure consistency in the input data.

Various machine learning algorithms can be employed, such as linear regression, support vector machines (SVM), or neural networks using several Python libraries viz., Seaborn, pandas, numpy, Scikitlearn, and matplotlib. The training data is used to optimize the model parameters, adjusting the algorithm's weights and biases to minimize the prediction error. During the training process, the model learns the underlying patterns and correlations between the thermal sensor readings and the COD concentrations. It captures the complex relationships in the data, allowing it to make predictions on unseen samples.

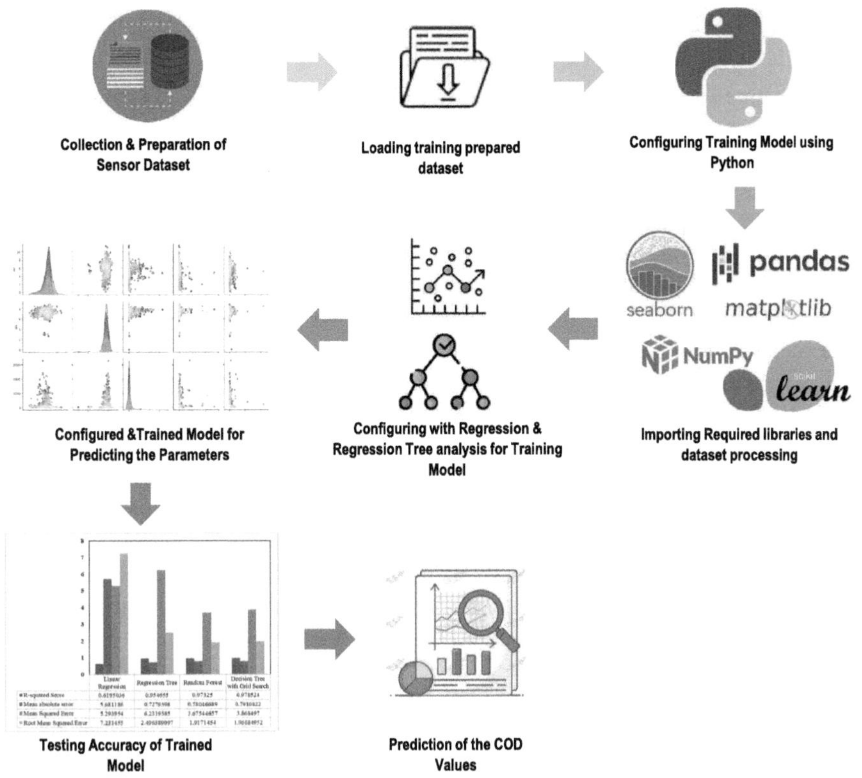

Fig. 8 Machine learning approach for the determination of COD

Once the model is trained and validated, it can be used to predict the COD concentrations of new water samples based on their thermal sensor readings. The model takes the temperature changes as input and produces an estimate of the corresponding COD concentration as output. The accuracy and performance of the model can be evaluated using various metrics, such as mean squared error (MSE) or coefficient of determination (R-squared). If the model performs well on the validation set and meets the desired criteria, it can be deployed for real-time COD prediction using the thermal sensor.

It is important to periodically update and retrain the model using new data to account for any changes or variations in the measurement process or the characteristics of the water samples. This ensures that the model maintains its accuracy and reliability over time.

4 Results and Discussion

The machine learning model developed for predicting COD concentrations using a thermal sensor, specifically employing a thermistor with an OP-07 amplifier, has demonstrated promising outcomes in accurately estimating COD levels in water samples. Leveraging a diverse and dynamic dataset, the model was trained to discern the intricate correlation between thermal sensor readings and corresponding COD concentrations. This dataset encompassed a wide spectrum of water samples with known COD values, alongside their associated thermal sensor readings. The results of the tests conducted indicate a strong correlation between the mixing behavior and heat release and the concentration of NaClO, with other parameters held constant. Building upon these findings, we propose the development of a portable COD detection instrument based on a thermistor and the OP-07 amplifier. To validate the feasibility of this instrument, a test experiment is designed. The robust performance observed in both simulation and experimental testing serves as compelling evidence of the practical viability of the proposed portable COD detection method and instrument. Furthermore, the data generated from the simulation and experimental testing offer valuable insights for future testing and application endeavors. This amalgamation of simulation and experimental data establishes a solid foundation for further research and utilization of the proposed COD detection approach.

The trained model demonstrated high accuracy in predicting COD concentrations when tested on unseen data. The model was able to capture the non-linear relationships between the temperature changes detected by the thermal sensor and the COD concentrations, providing reliable and precise predictions. The utmost important consideration of using a thermal sensor, such as a thermistor with an OP-07 amplifier, is its high sensitivity to temperature changes. This allows for precise measurement of the heat generated during the oxidation process, which is directly related to the COD concentrations in the water samples.

The regression plot demonstrates the model's accuracy in predicting COD concentrations by comparing predicted versus actual values, highlighting the non-linear correlation. The confusion matrix underscores the model's effectiveness in classifying COD levels, showcasing its ability to accurately distinguish between different concentration ranges. Additionally, the scatterplot visualizes the direct correlation between the voltage signal measured by the thermal sensor and the actual COD concentrations, emphasizing the relationship between these two parameters, as visualized in Fig. 9.

The use of machine learning techniques further enhances the predictive capabilities of the thermal sensor-based COD detection system. By utilizing algorithms like regression, Random forest, and Decision tree using Grid search, support vector machines (SVM), or neural networks, the model can effectively learn the underlying patterns and correlations in the data, resulting in accurate and robust predictions [19]. Table 2 summarises the Error and accuracy of the Proposed Prediction machine learning model for COD through the design and development of thermal sensors.

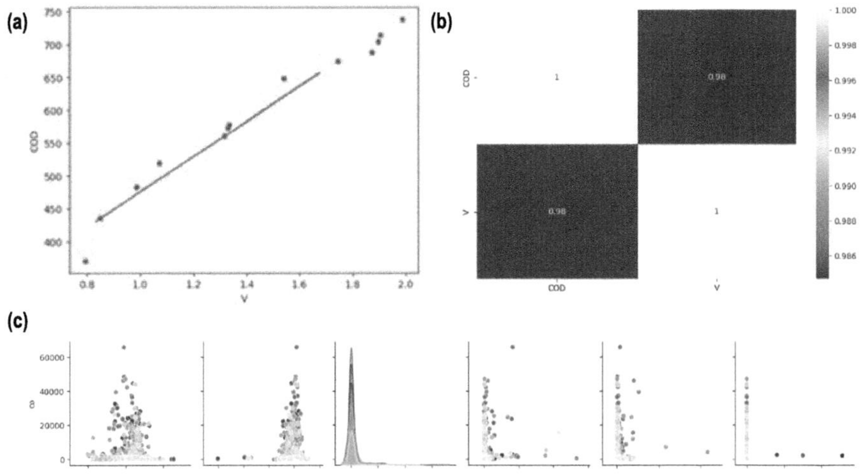

Fig. 9 (**a**) Regression- (**b**) Confusion matrix- (**c**) Scatterplot- Voltage signal vs CODactual for proposed ML model

Table 2 Interpreting error and accuracy for the proposed prediction machine learning model for COD

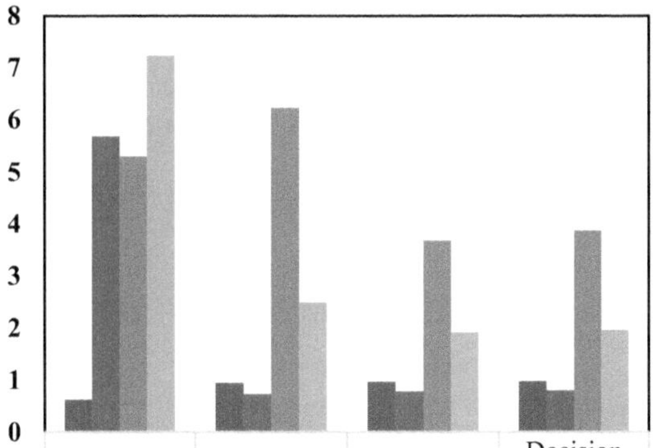

	Linear Regression	Regression Tree	Random Forest	Decision Tree with Grid Search
R-squared Score	0.6195036	0.954655	0.97325	0.978524
Mean absolute error	5.681186	0.7279598	0.78046689	0.7910822
Mean Squared Error	5.293954	6.2319585	3.67544657	3.868497
Root Mean Squared Error	7.231455	2.496389097	1.9171454	1.96684952

However, it is important to note that the performance of the machine learning model may depend on the quality and representativeness of the training dataset. The availability of diverse samples with varying COD concentrations and corresponding thermal sensor readings is crucial for achieving reliable and generalizable predictions. Additionally, the model's performance may be influenced by external factors, such as the quality and calibration of the thermal sensor, the accuracy of the amplification circuit (OP-07), and the overall system stability. Regular maintenance and calibration of the hardware components are necessary to ensure accurate and consistent results. Nevertheless, the combination of a thermal sensor, such as a thermistor with an OP-07 amplifier, and a well-trained machine learning model provides a reliable and efficient approach for COD prediction. This system offers the potential for real-time monitoring of COD concentrations in water samples, contributing to improved water quality assessment and environmental management.

4.1 Effect of pH

Figure 10 illustrates how pH affects the voltage response in samples where NaClO is used as the oxidizer. Initial experiments indicated that the pH of the NaClO solution influenced the assay outcomes. The graph shows that the signal response increases as the pH rises, reaching its peak at a pH of 12. However, significant deviation from the baseline was observed, making it impossible to determine COD values when the pH exceeded 12. Consequently, subsequent experiments were conducted at a basic pH of 12.

4.2 Effect of Concentration of NaClO Solution

Experiments were conducted to assess the impact of NaClO solution concentration on assay results illustrated in Fig. 11. Prior to incorporating NaClO solution into this investigation, we explored the viability of employing Fenton's reagent, sulfate solution, and hydrogen peroxide as oxidants for COD determination using the thermal sensor. Various concentrations of NaClO solutions (all with a pH of 7) were introduced into the thermal sensor. High concentrations of NaClO intensified degradation. Additionally, highly concentrated NaClO can cause damage to the instrument due to its strong corrosive nature. Therefore, an experimental concentration of NaClO solution containing 0.1% available chlorine was selected.

The correlation between COD values determined by both the thermal sensor (CODTs) and conventional standard dichromate method (CODCr)was investigated, illustrated in graphically in Fig. 12. Through the calibration equation derived from the standard sample method, the CODET values were determined. The equation CODts $=$ 1.0215 CODCr $-$ 27.725 was employed, yielding a correlation coefficient (R) of 0.9838. The dataset consisted of the average values obtained from three

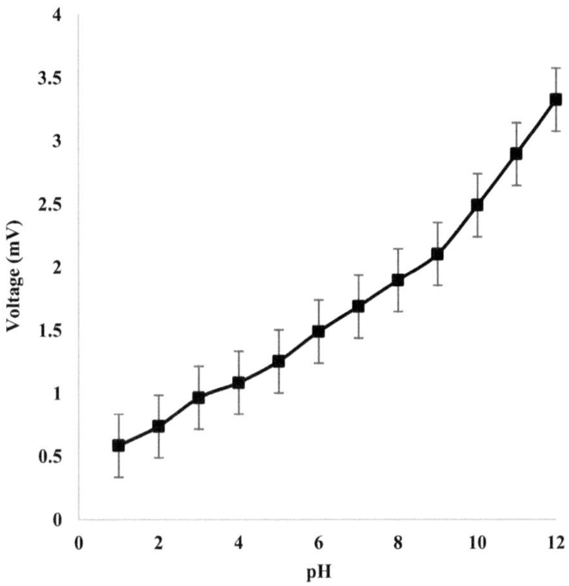

Fig. 10 Effect of pH on the voltage response (in mV)

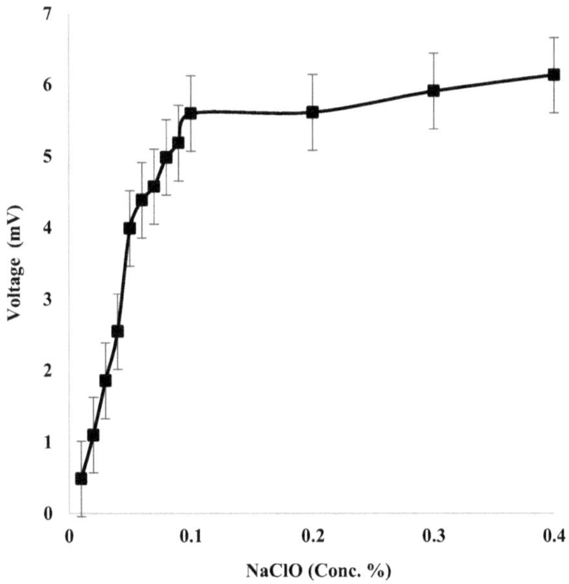

Fig. 11 Effect of NaClO concentration (%) on the voltage response (in mV)

measurements, with a total sample size (n) of 50. The high correlation coefficient (R = 0.9707) between the COD values determined by the thermal sensor (CODTs) and the conventional method (CODCr) suggests a strong agreement between the two

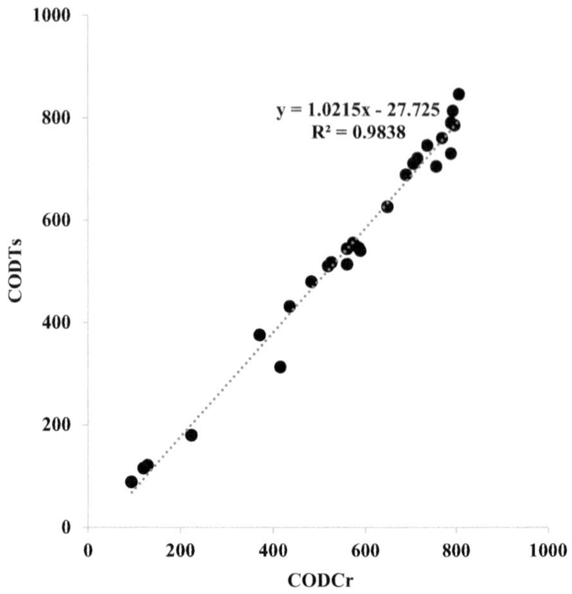

Fig. 12 The relationship between COD values determined by both the thermal sensor (CODTs) and conventional standard dichromate method (CODCr)

measurement techniques. This indicates that the thermal sensor method is reliable for accurately assessing COD levels in real water samples for industry wastewater.

The utilization of the calibration equation allowed for the accurate conversion of thermal sensor readings into COD values, enhancing the applicability of the thermal sensor in environmental monitoring and water quality assessment. The consistency observed across multiple measurements further strengthens the reliability of the thermal sensor approach. These findings underscore the potential of thermal sensors as efficient tools for rapid and accurate COD measurement in environmental monitoring applications. Their ability to provide reliable data while offering advantages such as real-time monitoring and cost-effectiveness positions thermal sensors as valuable assets in efforts aimed at safeguarding water quality and environmental health.

5 Conclusion

The proposed machine learning model for predicting COD along with the design and development of a thermal sensor, with a low-cost, highly robust thermistor and OP-07 amplifier, has shown promising results in accurately estimating COD concentrations in water samples. The predicted results demonstrated that the Decision tree with grid search has achieved an accuracy of 97.85% resulting into a reliable technique for the determination of COD. The combination of the thermal sensor and machine learning techniques offers a reliable and efficient approach for real-time monitoring of COD. The robust results from the test experiments demonstrate the viability of the portable COD detection method and equipment that are suggested in the research. Additionally, the development of autonomous and potable COD diagnostic instruments is a prospective application area for this COD determination approach. Further refinement of the model can be done by incorporating more diverse and representative datasets, including samples from different geographical locations and varying water quality conditions. This would enhance the model's robustness and generalizability. The study revealed a confined correlation between the COD values obtained via the conventional standard dichromate method (referred to as CODCr) and those determined using the thermal sensor approach (referred to as CODTs). The linear regression equation derived from this correlation was CODts = 1.0215 CODCr − 27.725, yielding a high correlation coefficient of 0.9838. This coefficient signifies a substantial agreement between the two methodologies. Consequently, the findings suggest that the novel thermal sensor method is both feasible and accurate for COD detection.

The proposed model can be extended to handle other parameters and contaminants in addition to COD. By integrating additional sensors or incorporating data from other analytical techniques, such as pH, turbidity, or specific pollutant measurements, a comprehensive water quality monitoring system can be developed. Furthermore, advancements in sensor technology can contribute to the improvement of the thermal sensor-based COD detection system. The development of more sensitive and reliable thermal sensors, as well as optimization of the amplification circuitry, can enhance the accuracy and precision of the measurements. By utilizing IoT (Internet of Things) technologies, the machine learning model can be deployed in a network of connected sensors, allowing for continuous monitoring and automated data processing. This would enable timely response and decision-making in water treatment processes and environmental management.

Acknowledgements We extend our heartfelt gratitude to our mentor, whose timely guidance and unwavering support were invaluable throughout the course of this project. We are also deeply appreciative of our Institute, the Department of Chemical Engineering, and Bajrangdas Lohiya Central Library of Vishwakarma Institute of Technology for entrusting us with this remarkable opportunity and the freedom to embark on this project.

Declaration of Competing Interest The authors state that they have no conflicts of interest to declare. They also confirm that they do not have any financial or personal connections that could have influenced the content of this paper.

References

1. Parched or polluted: on India's water crisis – The Hindu. Retrieved from https://www.thehindu.com/opinion/editorial/parched-or-polluted/article59780601.ece. Accessed 03 Sep 2024 [Online].
2. Aayog, N. (2019). COMPOSITE WATER MANAGEMENT INDEX In association with Ministry of Jal Shakti and Ministry of Rural Development.

3. India's rivers | Editorial: Poisoned water – Telegraph India. Retrieved from https://www.telegraphindia.com/opinion/editorial-poisoned-water/cid/1854819. Accessed 03 Sep 2024 [Online].
4. Li, X., Shang, Y., Fernandez, C., Pei, T., & Wang, L. (2021). Electrochemical determination of chemical oxygen demand in mixed organic solution by Al/SnO_2-TiO_2 electrode. *International Journal of Electrochemical Science, 16*. https://doi.org/10.20964/2021.11.45
5. Liu, Y. et al. (2020) A current sensing biosensor for BOD rapid measurement. *Archaea, 2020*. https://doi.org/10.1155/2020/8894925
6. Duan, W., del Campo, F. J., Gich, M., & Fernández-Sánchez, C. (2022) In-field one-step measurement of dissolved chemical oxygen demand with an integrated screen-printed electrochemical sensor. *Sensors & Actuators, B: Chemical, 369*. https://doi.org/10.1016/j.snb.2022.132304
7. POLLUTED RIVER STRETCHES FOR RESTORATION OF WATER QUALITY-2022.
8. Qi, M., Han, Y., Zhao, Z., & Li, Y. (2021). Integrated determination of chemical oxygen demand and biochemical oxygen demand. *Polish Journal of Environmental Studies, 30*(2). https://doi.org/10.15244/pjoes/122439
9. Yao, N., Liu, Z., Chen, Y., Zhou, Y., & Xie, B. (2015). A novel thermal sensor for the sensitive measurement of chemical oxygen demand. *Sensors (Switzerland), 15*(8). https://doi.org/10.3390/s150820501
10. Kumaraguru, K., Giriprasad, J., Mohammad Shifath Ali, L., Beena, A., & Rajinigirinath, D. (2018). Monitoring of the chemical oxygen demand and identify the presence of gas. *International Journal of Trend in Scientific Research and Development, 2*(3), 1740–1746. https://doi.org/10.31142/IJTSRD11533
11. Almeida, C. A., González, P., Mallea, M., Martinez, L. D., & Gil, R. A. (2012). Determination of chemical oxygen demand by a flow injection method based on microwave digestion and chromium speciation coupled to inductively coupled plasma optical emission spectrometry. *Talanta, 97*. https://doi.org/10.1016/j.talanta.2012.04.030
12. Rohyami, Y., Aprianto, T., & Marjono. (2021) Validation method on determination of chemical oxygen demand using indirect UV-Vis spectrometry. *Advanced Materials Research, 1162*. https://doi.org/10.4028/www.scientific.net/amr.1162.101
13. Cao, X. et al. (2017) An on-line microfluidic analysis system for seawater chemical oxygen demand using ozone chemiluminescence. *Guang Pu Xue Yu Guang Pu Fen Xi/Spectroscopy and Spectral Analysis, 37*(12). https://doi.org/10.3964/j.issn.1000-0593(2017)12-3698-05
14. Zhong, Y., Xia, F., Lian, J. (2017) Determination of chemical oxygen demand in wastewater by ultraviolet and near-infrared spectroscopy. *Chinese Journal of Environmental Engineering, 11*(2). https://doi.org/10.12030/j.cjee.201510021
15. Phuong, N. T. B., Tri, V. P. D., Duy, N. B., & Nghiem, N. C. (2017) Remote sensing for monitoring surface water quality in the Vietnamese Mekong Delta: The application for estimating chemical oxygen demand in river reaches in Binh Dai, Ben Tre. *Vietnam Journal of Earth Sciences, 39*(3). https://doi.org/10.15625/0866-7187/39/3/10270
16. Kim, Y. C., Sasaki, S., Yano, K., Ikebukuro, K., Hashimoto, K., & Karube, I. (2002) A flow method with photocatalytic oxidation of dissolved organic matter using a solid-phase (Tio_2) reactor followed by amperometric detection of consumed oxygen. *Analytical Chemistry, 74*(15). https://doi.org/10.1021/ac015678r
17. Wang, J. et al. (2015). Electrochemical tuning of the activity and structure of a copper-cobalt micro-nano film on a gold electrode, and its application to the determination of glucose and of Chemical Oxygen Demand. *Microchimica Acta, 182*(3–4). https://doi.org/10.1007/s00604-014-1353-z
18. Wang, J., Wu, C., Wu, K., Cheng, Q., & Zhou, Y. (2012). Electrochemical sensing chemical oxygen demand based on the catalytic activity of cobalt oxide film. *Analytica Chimica Acta, 736*. https://doi.org/10.1016/j.aca.2012.05.046
19. Zhang, X., Zhu, Y., Chen, X., Shen, W., & Lutes, R. (2018). Soft-sensing modeling of chemical oxygen demand in photo-electro-catalytic oxidation treatment of papermaking wastewater. *Journal of Bioresources and Bioproducts, 3*(2). https://doi.org/10.21967/jbb.v3i2.112

20. Zhao, Y., Li, Y., Zhen, Y., & Fang, Y. (2011). A novel monitoring system for COD using optical ultraviolet absorption method. In *Procedia Environmental Sciences*. https://doi.org/10.1016/j.proenv.2011.09.366
21. Chen, J., Zhang, J., Xian, Y., Ying, X., Liu, M., & Jin, L. (2005). Preparation and application of TiO_2 photocatalytic sensor for chemical oxygen demand determination in water research. *Water Research, 39*(7). https://doi.org/10.1016/j.watres.2004.12.045
22. Li, J., Li, L., Zheng, L., Xian, Y., Ai, S., & Jin, L. (2005). Amperometric determination of chemical oxygen demand with flow injection analysis using F-PbO_2 modified electrode. *Analytica Chimica Acta, 548*(1–2). https://doi.org/10.1016/j.aca.2005.05.068
23. Yu, H., Ma, C., Quan, X., Chen, S., & Zhao, H. (2009). Flow injection analysis of chemical oxygen demand (COD) by using a boron-doped diamond (BDD) electrode. *Environmental Science & Technology, 43*(6). https://doi.org/10.1021/es8033878
24. Wang, Q., & Del Valle, M. (2021). Determination of chemical oxygen demand (Cod) using nanoparticle-modified voltammetric sensors and electronic tongue principles. *Chemosensors, 9*(3). https://doi.org/10.3390/chemosensors9030046
25. ISO 6060:1989 – Water quality—Determination of the chemical oxygen demand. Retrieved from https://www.iso.org/standard/12260.html. Accessed 03 Sep 2024 [Online].
26. ISO 15705:2002 – Water quality—Determination of the chemical oxygen demand index (ST-COD)—Small-scale sealed-tube method. Retrieved from https://www.iso.org/standard/28778.html. Accessed 03 Sep 2024 [Online].
27. Dan, D., Sandford, R. C., & Worsfold, P. J. (2005). Determination of chemical oxygen demand in fresh waters using flow injection with on-line UV-photocatalytic oxidation and spectrophotometric detection. *Analyst, 130*(2). https://doi.org/10.1039/b409810b
28. Ma, J. (2017). Determination of chemical oxygen demand in aqueous samples with non-electrochemical methods. https://doi.org/10.1016/j.teac.2017.05.002
29. Qiu, J., Zhang, S., Zhao, H. (2011). Recent applications of TiO2 nanomaterials in chemical sensing in aqueous media. *Sensors & Actuators, B: Chemical, 160*(1). https://doi.org/10.1016/j.snb.2011.08.077
30. Yao, N., Wang, J., & Zhou, Y. (2014). Rapid determination of the chemical oxygen demand of water using a thermal biosensor. *Sensors (Switzerland), 14*(6). https://doi.org/10.3390/s140609949
31. Shi, J., Fu, P., Zheng, W., Yin, H., Feng, L., & Qiao, J. (2018). Chemical oxygen demand detection method research and portable instrument development based on microfluidic chip. In *I2MTC 2018—2018 IEEE International Instrumentation and Measurement Technology Conference: Discovering New Horizons in Instrumentation and Measurement, Proceedings*. https://doi.org/10.1109/I2MTC.2018.8409608
32. Bai, Y., He, S., & Wei, K. (2021). Design of a signal regulating circuit for wearable ECG sensor applications. In *2021 3rd International Conference on Intelligent Control, Measurement and Signal Processing and Intelligent Oil Field, ICMSP 2021*. https://doi.org/10.1109/ICMSP53480.2021.9513370
33. Sanjaya, W. S. M., Anggraeni, D., Multajam, R., Subkhi, M. N., & Muttaqien, I. (2017). Design and experiment of electrooculogram (EOG) system and its application to control mobile robot. *IOP Conference Series: Materials Science and Engineering, 180*(1), 012072. https://doi.org/10.1088/1757-899X/180/1/012072
34. Precision Operational Amplifier OP07 APPLICATIONS • Wireless Base Station Control Circuits • Optical Network Control Circuits • Instrumentation • Sensors and Controls • Precision Filters DESCRIPTION
35. Devices, A. Ultralow offset voltage operational amplifier. Retrieved from www.analog.com. Accessed 03 Sep 2024 [Online]
36. OP07x Precision Operational Amplifiers, 2023. Retrieved from www.ti.com. Accessed 03 Sep 2024 [Online]

Developing a Cloud-Based Weapon Detection System Using Transfer Learning and Generative Adversarial Networks

Om Nankar, Shambhavi Patil, Sahil Gupte, Shilpa Gite, and Biswajeet Pradhan

Abstract Public safety is a massive concern for governments and civilians alike. Measures taken to ensure public safety range from urban and safe city planning at the government level to self-defence programs for the public. Among these measures, baggage checking and surveillance systems are already in place. The motivation for this paper is to enhance these existing systems. In this paper, the authors propose an object detection system for baggage security that will significantly automate the surveillance process. The proposed system uses deep learning techniques, such as Transfer Learning (YOLO and Detectron2 pre-trained models achieved loss values of 0.01345 and 0.04498, respectively), Generative Adversarial Networks-GAN (for synthetic image creation with a batch size of up to 32), and cloud computing. It aims to identify threatening objects and isolate them during the X-ray scanning process in baggage security. A large dataset is generated using the GAN technique and data augmentation to train the model. Threats are then detected using popular models such as You Only Look Once (YOLO) and Detectron2. Additionally, this paper proposes a cloud-based system to facilitate seamless conversion between multiple object detection annotation formats, such as PascalVOC, YOLO, and Common Objects in Context (COCO). The cloud-based system uses client-side calculations and compression techniques to convert annotation data interchangeably between the mentioned formats, providing users this service free of charge.

O. Nankar · S. Patil · S. Gupte · S. Gite (✉)
Artificial Intelligence and Machine Learning Department, Symbiosis Institute of Technology, Symbiosis International (Deemed) University, Pune 412115, India
e-mail: shilpa.gite@sitpune.edu.in

S. Gite
Symbiosis Centre of Applied AI (SCAAI), Symbiosis International (Deemed) University, Pune 412115, India

B. Pradhan (✉)
Centre for Advanced Modelling and Geospatial Information Systems (CAMGIS), School of Civil and Environmental Engineering, Faculty of Engineering & IT, University of Technology Sydney, Sydney, NSW 2007, Australia
e-mail: biswajeet.pradhan@uts.edu.au

B. Pradhan and S. Mukhopadhyay (eds.), *IoT Sensors, ML, AI and XAI: Empowering A Smarter World*, Smart Sensors, Measurement and Instrumentation 50,
https://doi.org/10.1007/978-3-031-68602-3_7

Keywords Object detection · Generative adversarial networks · Transfer learning · You only look once · Detectron · Annotation conversion

1 Introduction

The main motivation behind this paper was to address the issue of public safety and to regulate the presence of weapons such as guns, knives, and other sharp objects in public settings [1]. Security in all public venues, including shopping malls, hostels, airports, and train stations, must be of utmost importance. Having a system [2, 3] that can identify potentially dangerous items carried openly in public is highly advantageous for preventing attacks. The use of such object detection technology can provide a sense of security in public spaces for survivors of past major incidents. This paper proposes a method that can be especially helpful in areas with large crowds, such as airports (for baggage scanning), shopping malls, supermarkets, and more.

The scope of this research is to grab an image from a video, detect specific objects, and identify a predefined set of harmful objects prohibited in public places. It aims to alert users upon the successful detection of a prohibited object. Thus, the main objective of this work is to provide real-time object detection for prohibited baggage items in public places. The system is designed to identify objects belonging to five classes: forks, handguns, knives, screwdrivers, and wrenches. In existing baggage security systems, human monitoring is required, and the scanning process only covers the baggage being passed through, necessitating continuous monitoring. Moreover, these existing systems for surveillance and scanning are costly and require heavy machinery [4]. The research work implements various techniques, including cloud-based annotation conversion, GAN-based dataset creation, transfer learning, and real-time object detection.

The paper is divided into six sections, each laying a foundation for the next while defining the system architecture; Sect. 2: Literature Review (where the authors define the problem and objectives). Section 3 describes the system design and other model architecture incorporated in the methodology and implementation, Sect. 4: Implementation, Sect. 5: Results and Discussion, and the proposed work is concluded in Sect. 6: Conclusion and Future Scope.

2 Literature Review

In a recent work, Tahir et al. [5] discussed the concerns raised in this paper regarding public safety. The authors of this paper consider the use of CCTV surveillance and deep learning for real-time monitoring of suspicious activities in public places. They construct a custom dataset using various sources and implement algorithms like VGG16 [6], SSD-MobileNet [7], Inception-ResNet [8], Faster-RCNN [9], and YOLO [10]. The authors concluded that YOLOv4 outperforms the others in all aspects. In a separate work, Dhiraj et al. [11] recommended the use of a deep neural network-based framework for threat object identification. In this article, they propose an imaging model and create a dataset due to the lack of availability. Models such as Yolo and Faster Region-based Convolutional Neural Network (FRCNN) are employed. The study explored the performance of transfer learning on the dataset across four categories: Gun, Shuriken, Razor-blade, and Knife. To achieve better results, FRCNN incorporates regional ideas in the first step. The FRCNN model is trained with the pre-trained ResNET [12] on the ImageNet [13] dataset, achieving a 98.4% accuracy rate and processing images in 0.16 s per picture.

Priscilla et al. [14] introduced an object area recommendation approach employing a region proposal network that provides the detector module with an object score and allows precise localisation of potential human body hazards. This system utilises a modified loss function to reduce classification errors. The suggested model's results indicate a 15% improvement in object localisation. Consequently, the enhanced Faster R-CNN achieves an overall detection accuracy of 27% while reducing processing time by half. This improved, faster R-CNN outperforms the state-of-the-art system in terms of detection accuracy, making it applicable for risk detection and security audits. Xingyuan et al. [15] introduced GAIA, a transfer learning system that automatically and efficiently provides customised solutions based on various downstream requests. GAIA can provide robust pre-trained weights, select models that align with downstream requirements, such as latency constraints and specific data domains, and generate valuable results with minimal data points. GAIA performs well across a range of datasets, rapidly producing models with latencies ranging from 16 to 53 ms and average precision (AP) scores ranging from 38.2 to 46.5 ms on the COCO dataset [16]. While the AP score is relatively lower, the results are promising.

Zou et al. [17] conducted an extensive review of the history of object detection spanning the past two decades. This review encompassed the evolution of models, bounding boxes, and datasets. In this comprehensive article, the authors assessed over 400 works on object detection, tracing technological advancements from the 1990s to 2019. They delved into historical detectors, detection datasets, metrics, fundamental components of detection systems, acceleration techniques, and the majority of existing popular detection methods deemed 'state-of-the-art.' The study also explored critical applications of detection, such as pedestrian detection, face recognition, text detection, and more, assessing their challenges and recent technological advancements. The future outlook presented in this paper includes video object detection, small object detection, and the utilisation of GANs for research purposes, all of which are implemented in the work. Ruhin et al. [18] introduced a real-time object detection security system tailored for enhancing security in banks. This approach can also be advantageous for other security applications with limited human access requirements. The proposed method finds utility across a broad spectrum of security-related applications, achieving an accuracy rate of 84% using single-shot detectors (SSD), particularly SSD-Mobilenet-v2, on the COCO dataset. Sanam et al. [19] presented a system designed for the recognition of weapons such as guns and rifles. The authors employed the YOLO algorithm and transfer learning for object detection on a custom dataset, with the aim of using the system for surveillance purposes. Marks et al. [20] trained a weapon detection system using YOLOv5 on various data sources, focusing on video surveillance images. This system is capable of real-time inferences, operating at 33 frames per second on Nvidia's Jetson AGX Xavier. It achieved an accuracy rate of 98.56%, a notable result in comparison to other available research.

According to Jesus Salido et al. [21], their research employed three convolutional neural network (CNN) models to identify guns in video surveillance images. Their research aimed to see whether the inclusion of postural information, specifically related to handgun holding positions in images from the training dataset, could help reduce false positives. When trained on the dataset with posture information, they used ResNET-50 as the backbone, fine-tuning the model using RetinaNET, which produced the highest average precision (96.36%) and recall (97.23%). On the other hand, YOLOv3 had the highest accuracy (96.23%) and F1 scores (93.36%). In a separate study, Jain et al. [22] conducted experiments with different datasets and models. They utilised two types of models, the Single Shot Detector (SSD) algorithm and the Faster R-CNN algorithm, on two datasets, one with pre-labeled images and another with manually labeled images, to detect various types of guns, including AK-47 and Colt M191, among others. Their results showed that the pre-labeled images gave better accuracy, and between the two algorithms, SSD showed faster processing speed, while FRCNN exhibited better accuracy.

After a thorough review of previous work, as summarized in Table 1, it was revealed that model implementations were relatively unrelated. Results were primarily obtained and compared based on performance metrics, with insufficient emphasis on security aspects in the results and discussion sections. Regarding the dataset, previous research primarily focused on detecting threats, which were predominantly firearms. However, as highlighted by the United States Secret Service National Threat Assessment Center [23], mass attacks in public spaces also involve weapons such as knives and glass bottles. This research takes a broader approach, considering readily available weapons like wrenches and knives, in addition to firearms, for offensive purposes. Moreover, for model selection, this work aims to integrate object detection models into a framework to enhance object detection. Object detection in real-life scenarios is a formidable challenge. Weapons are often small in size and can be challenging to detect in the presence of other objects is challenging, particularly as non-harmful objects may be mistaken for threats.

The primary aim of this paper is to introduce new technologies, including Pro-GANs for dataset creation in weapon detection, a custom-built and cloud-hosted annotations converter, a novel method to create synthetic images using Pro-GANs, and the integration of open-source datasets. Additionally, data augmentation techniques are employed to build a custom dataset for multi-class weapon detection, encompassing guns, knives, wrenches, screwdrivers, and forks. The main contribution of the paper lies in the comparative analysis of YOLOv5 and Detectron2 models' performance on the custom dataset, coupled with the identification of their use cases.

Table 1 Summary of previous work

Reference	Field of work	Algorithms/ Models used	Pros	Cons
[5]	Weapon detection	VGG16, SSD-MobileNet, Inception-ResNet, Faster-RCNN, and YOLO	Wide range of models custom dataset	No evidence on real-time footage
[11]	Baggage screening	YOLO, FRCNN	Wide range of objects	Only X-ray images included
[14]	Threat detection	Faster R-CNN	Recognizes the threat and need for better monitoring	Only FRCNN used for testing dataset
[18]	Security system	SSD-Mobilnet	Practical application	Compared to SSD, YOLO is a better choice
[19]	Weapon detection	YOLOv3	System developed in view of practical world	Model limited to guns/rifles
[20]	Weapon detection	YOLOv5	Solution centric approach	Nvidia jetson xavier is expensive to install
[21]	Handgun detection	CNN, YOLOv3	Promising F1 scores and precision values	Dataset limited to gun holding positions
[22]	Weapon detection	SSD, Faster RCNN	Addresses the dataset availability issue	Doesn't address the security aspect of weapon detection

2.1 Models Used

In this paper, we implemented the dataset with two popular object detection algorithms: You Only Look Once (YOLOv5), and Detectron2. We chose these models to evaluate their performance and because they originate from different architectures, which may offer valuable insights regarding performance and data compatibility. The proposed approach employs a dataset consisting of both manually labeled images and machine-generated images created using Progressive Generative Adversarial Networks (Pro-GANs) [24]. Image generation alone is insufficient for object detection; images must first be labeled and then used for training [25]. These annotated photos are then used to train models developed by Facebook AI Research, such as You Only Look Once (YOLO) [26] and Detectron2 [27]. (FAIR). We employ transfer learning with pre-trained weights on this dataset to achieve our results. In addition, this work describes a cloud-based system for converting numerous object annotation formats, including PascalVOC, YOLO, and COCO.

2.2 *Detectron2*

Detectron2 is a state-of-the-art object detection software system developed by Facebook AI Research, built on the PyTorch framework [27]. It excels in object detection tasks, offering capabilities such as instance segmentation, keypoint detection, and panoptic segmentation. Detectron2 represents a complete overhaul of the original Detectron, initially based on the mask rcnn-benchmark. Thanks to its new, modular design, Detectron2 is highly versatile and extensible, facilitating swift training on single or multiple GPU servers.

2.3 *Architecture of Detectron2*

Backbone Network: Feature extraction maps from the picture are provided as inputs at various scales, specifically at 1/4, 1/8, 1/16, 1/32, and 1/64 scales. However, the output is only taken from a 1/16 scale. The Region Proposal Network (RPN) uses multi-scale characteristics to recognize object regions. The default setting generates 1000 boxes, each with associated confidence scores. The Box Head segment cuts and warps the feature maps at different scales using predefined boxes of fixed scale. A fully connected layer refines these boxes, resulting in the displayed outcome boxes. Following all refinement and selection processes, 100 boxes are typically chosen as the default value, but this can be adjusted. The architecture can be viewed in Fig. 1, which is customized to fit the dataset and desired output.

2.4 *YOLOv5*

One of the most popular object detection models, YOLO, which stands for "You Only Look Once," is known for its speed, precision, and incredible efficiency for its size [28]. It has consistently improved with the introduction of new versions. YOLO is a quick and accurate method for detecting objects. The system merely looks at a picture once and identifies all of the items and their locations. [29] YOLO operates by dividing photos into grid cells, each of which recognizes an item on its own. These grid cells predict bounding boxes and assign a confidence score to each prediction to assess its accuracy. In this paper, the research focuses on Glenn Jocher's YOLOv5, the newest model of the YOLOv5 series [30], which was introduced in 2020.

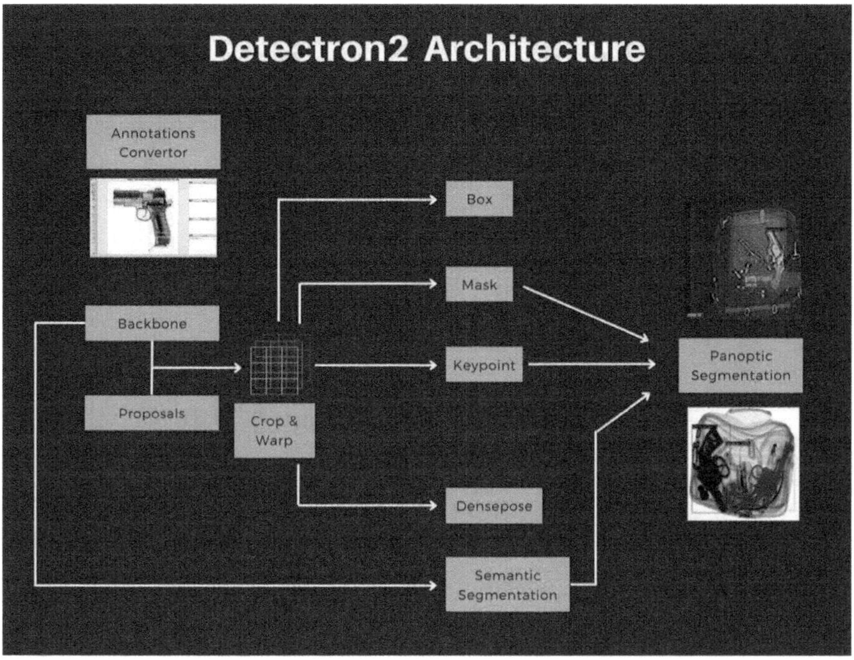

Fig. 1 Detectron2 architecture

2.5 *Architecture of YOLOv5*

YOLOv5 differs from its predecessor models in a number of ways. To begin, PyTorch is used instead of Darknet [30]. The backbone CSPDarknet53 reduces parameters and enhances inference speed, accuracy, and model size. The PANet is used as a bottleneck for improving the information flow. By enhancing localization at lower levels, PANet improves the object's localization accuracy. Furthermore, the head remains unchanged, resulting in three distinct outputs of the feature map for multi-scale prediction. It also helps improve the model's capacity to predict small to large things. The journey of an image is that it is sent for feature extraction via CSPDarknet53 and subsequently for feature fusion through PANet. Lastly, the head layer generates the output. The architecture can be viewed in Fig. 2, which has been customized to fit the dataset and the desired output.

3 Methodology

Figure 3 illustrates the complete model methodology, encompassing all the steps from the dataset to the final output.

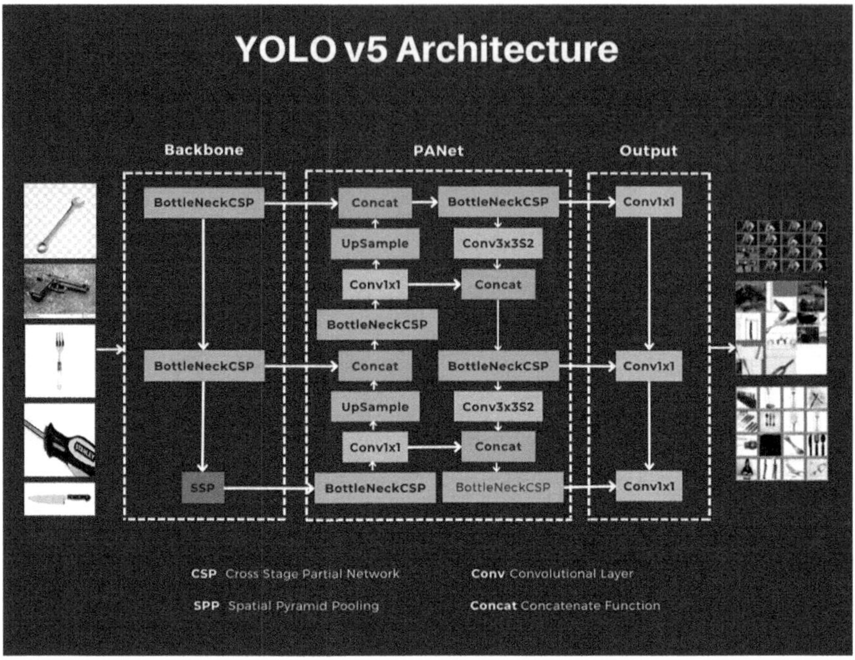

Fig. 2 YOLOv5 architecture

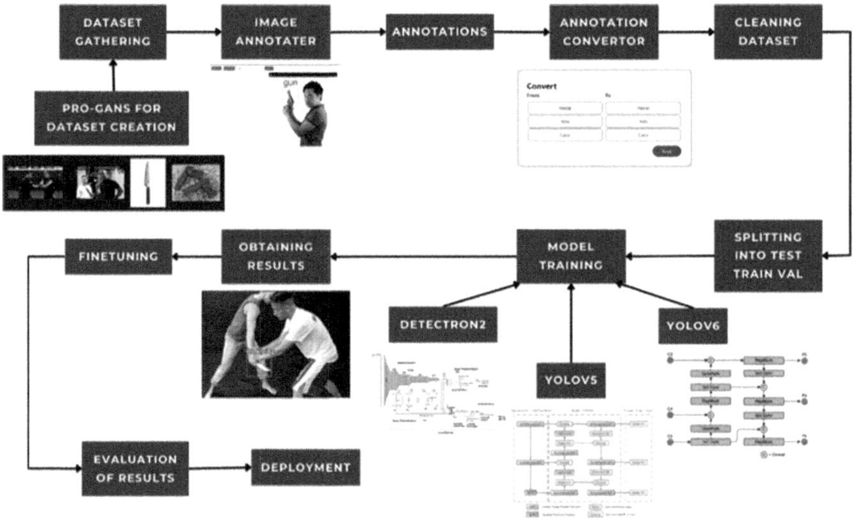

Fig. 3 YOLOv5 architecture

3.1 Pro-GANs

GANs, or Generative Adversarial Networks [31], are relatively recent contributions to the deep learning discipline. They find applications across various data types, including text, graphics, and audio, in numerous domains. Progressive GANs (Pro-GANs) [26], a sort of GAN that allows for the steady training of generator models capable of creating big, high-quality pictures, have garnered much appeal for projects, prompting researchers to create new GANs. It requires starting with a tiny picture and batch size and progressively adding layers. With the addition of each layer, two things happen: first, increase the generator's output size and the discriminator's input size [26]. This goes on till the defined required image size is seen in the output. This technology is effective in the creation of high-quality synthetic images [32].

For every step, the batch size increments are calculated using Eq. 1.

$$batch\,size = config.BATCH\_SIZES[int(\log\_2(\frac{imagesize}{4}))] \tag{1}$$

3.2 Dataset

GANs, or Generative Adversarial Networks [31], are relatively recent. The dataset being used is a compilation from various sources. These sources included:

- Open-Source datasets [33, 34]
- Data Augmentation techniques (rotation, cropping, angles)
- Pro-GANs.

These images were divided into five classes, and the total size of the dataset was 6641 images. The image data distribution of each class can be seen in Fig. 4.

The combination of all these images is displayed in Fig. 5. However, all of these images were not trained in the model implementation due to computational complexities. The images alone were not enough for object detection; the images had to be annotated. For annotation purposes, the tools LabelIMG [35] and LabelME [36] were utilized, but different models required different types of annotations to tackle this issue, so the authors propose the annotation converter. Regarding the type of images, the dataset included not only images of the objects defined in the classes but also situations of threat and attack to give the model an idea of identifying these objects in such situations. Progress and output observation were mentioned using a Tensorboard.

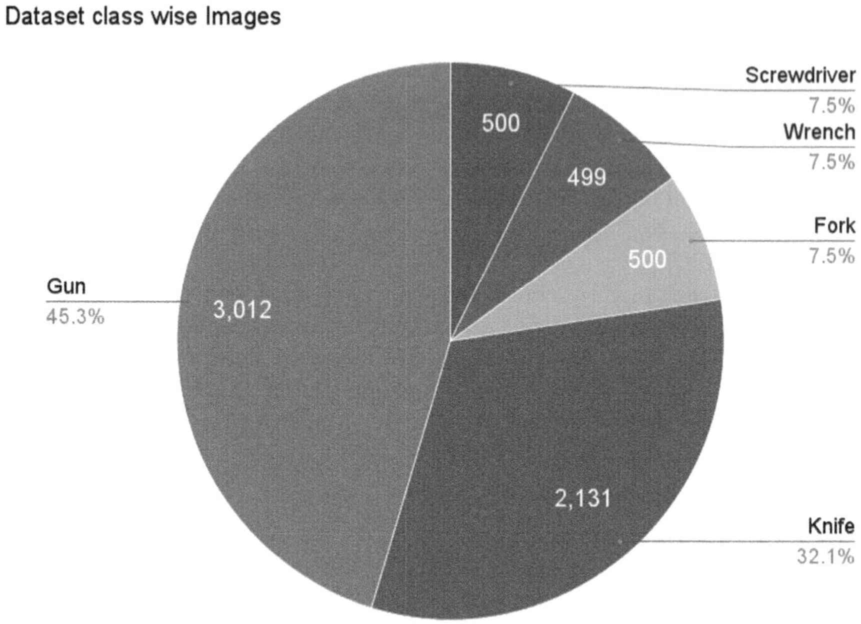

Fig. 4 Class-wise image distribution in the dataset

Fig. 5 Sample images from the dataset

3.3 Annotation Converter

While training the dataset on different models, an issue arose, and the authors realized that each model required annotations in a different format. These annotations were generated using different software for each model. Initially, 'labelIMG' was used to generate annotations for the dataset in PascalVOC format [37]. However, while training the dataset using the Detectron2 model, annotations were required in COCO format [16]. It would not accept the PascalVOC format. For that, another software had to be used called 'labelme.' Usually, annotations can be written simply in 4 coordinate pairs [38]. PascalVOC simplifies this and uses four numbers: xmin, xmax, ymin, and ymax. It is a well-known fact that YOLO uses relative geometry to do its processing

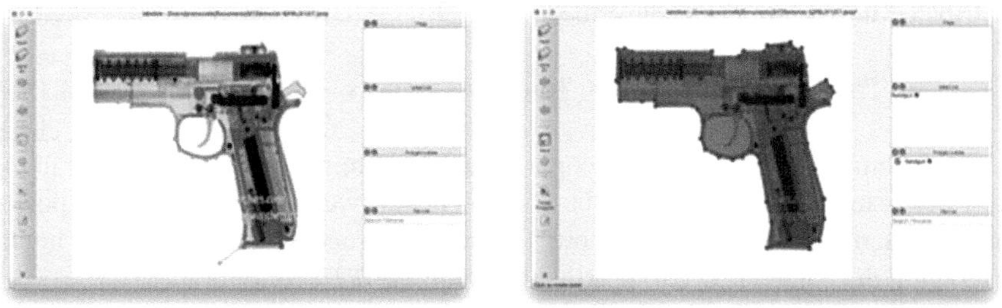

Fig. 6 **a** Image annotated using Labelme; and **b** complete poly annotation

[39]. The annotations are represented in floating-point numbers <1 & >0. COCO uses a coordinate system similar to PascalVOC, but its coordinates differ. To tackle this inconsistency, we developed our own solution-an Annotation Converter. This cloud-based web application seamlessly converts annotations between PascalVOC, COCO, and YOLO formats, offering flexibility and interchangeability.

3.4 Detectron2

Dataset labelling (using Labelme): To use a custom dataset in Detectron2, we used poly segmentation, which is the best use case of Detectron2. Custom Poly Segmentation requires approximately traced dataset images in JSON format (COCO). During implementation, we focused more on the handgun class because Poly segmentation worked best. The handgun dataset included 44 training images and 7 testing images with dimensions 416 px × 416 px. The process is shown in Fig. 6.

Implementation of Detectron2 was carried out in four phases:

- Basic setup—Installing Detectron2 based on PyTorch 1.8 from the model zoo.
- Configuration update—Registering the Training and Testing Dataset and updating dataset/poly parameters.
- Instance segmentation—Training the Detectron2 segmentation model.
- Visualization—Inferring the data using the trained model.

During the implementation of Detectron2, observation was that fewer well-labeled images produced better results during model training than a greater number of generic images of a specific class did.

3.5 YOLOv5

From the novel dataset, 1450 images of similar sizes (image size: 416) were selected to undergo the training process using the YOLOv5 model. The images were randomly distributed into three sets as described in Fig. 7:

$$\rightarrow \text{ Train : 1014 images } \rightarrow \text{ Test : } \ 147 \text{ images } \rightarrow \text{ Val : } \ 289 \text{ images}$$

The training took place in the Google Colab environment, where multiple training runs were executed to test the results. The best results were obtained after 40 epochs, and the outcomes are presented below. The training process took almost three hours, leveraging the colab GPU equipped with NVidia K80/T4 configuration.

Fig. 7 **a** Dataset distribution for YOLOv5; and **b** Class-wise instances for YOLOv5

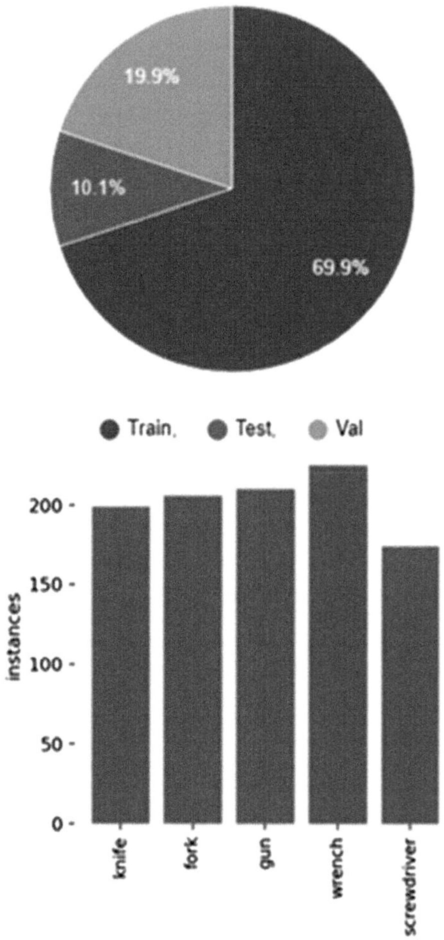

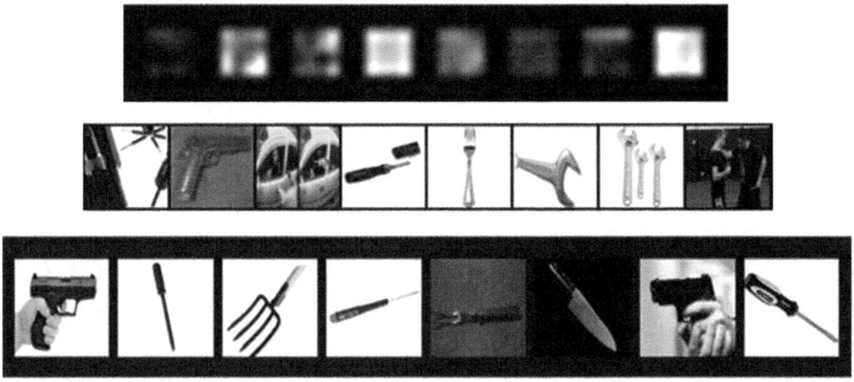

Fig. 8 Pro-GAN results: **a** first step; **b** step 40; and **c** step 80

Table 2 Annotation type and format

Annotation type	Format
Pascal VOC	XML
YOLO	txt
COCO	JSON

4 Results

4.1 *Pro-GANs*

Figure 8 displays the results obtained after training the Pro-GAN on 1500 training images.

4.2 *Annotation Converter*

The annotation converter successfully converts the vastly different Pascal, YOLO, and COCO annotation formats. As a part of this research, all three formats described in Table 2 were used.

4.3 *Detectron2*

After training on 44 handgun class dataset images for 10,000 iterations, the following results were obtained. Our training model exhibits 100% detection and over 90% accuracy in X-Ray test images. The obtained results, as displayed in Table 3, were

Table 3 Loss summary during Detectron2 training

Total loss	0.04498
Class loss	0.04486
Box loss	0.0082
Mask loss	0.0307

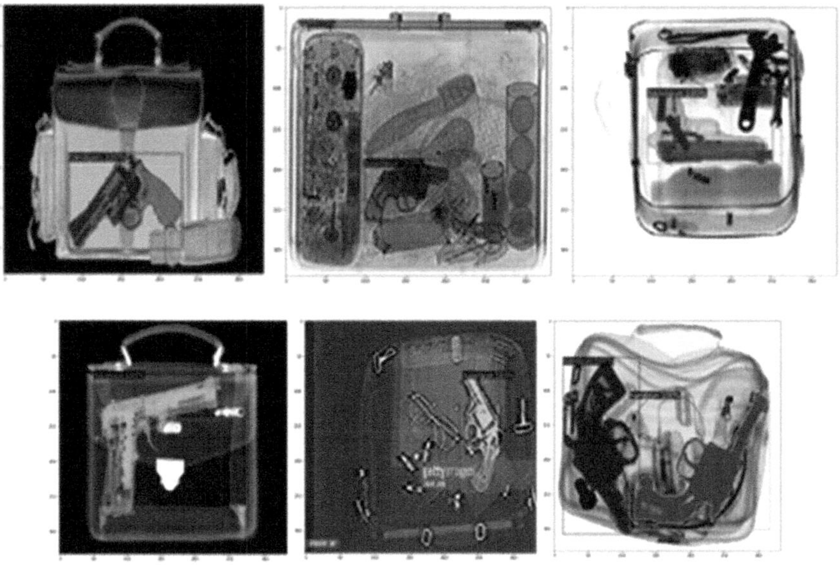

Fig. 9 Visualizing the trained model on the test dataset

tested and observed in extreme conditions of brightness and contrast, with lots of disturbances in the images characterized by low-quality static (Fig. 9).

The baggage scanning system serves as a use case for the application of Detectron2, the results for which can be seen in Fig. 10.

For models like Detectron2, merely increasing the quantity of data does not necessarily enhance the quality of object detection. The dataset plays a huge role in the performance of the model. Increasing the number of instances labeled in each image in the training dataset for multi-class classification models improves the chances of detection in testing and validation data.

4.4 D. Detectron2

After training for 40 epochs, Fig. 11 shows the metrics, which are discussed in the following subsection:

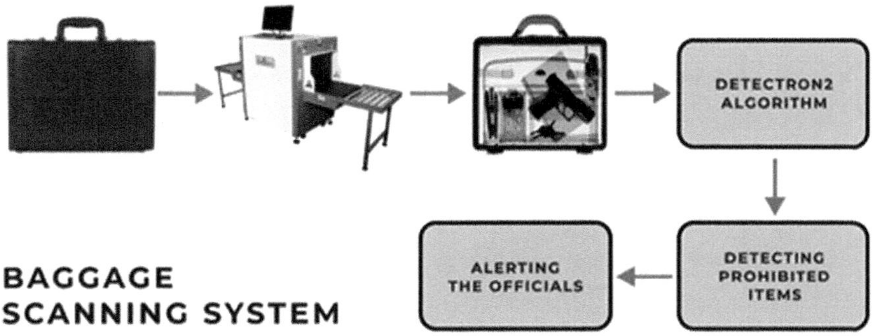

Fig. 10 Baggage scanning system design

Class	Images	Labels	P	R	mAP@.5	mAP@.5:.95:
all	289	290	0.848	0.787	0.847	0.554
knife	289	75	0.881	0.787	0.833	0.463
fork	289	63	0.841	0.746	0.847	0.501
gun	289	58	0.77	0.637	0.714	0.387
wrench	289	45	0.909	0.886	0.96	0.755
screwdriver	289	49	0.838	0.878	0.881	0.662

Fig. 11 Training results for YOLOv5

The overall mAP was 84.7%, considered state-of-the-art (SOTA). The highest mAP (mean average precision) was achieved for the wrench class, while the lowest was observed for the gun class. A reasonable explanation for this is that the gun images in the dataset either need more training or have insufficient bounding box annotations, potentially affecting the Intersection over Union (IoU) calculations.

The box loss was 0.03, the object loss 0.01345, and the classification loss was 0.019019, as viewed in Table 4. These are standard metrics produced by YOLO models, which can be viewed in Fig. 12.

To further discuss the results, the following three equations were used to have a meaningful understanding of it.

$$F_1 = 2 * \frac{precision * recall}{precision + recall} \tag{2}$$

$$precision = \frac{TP}{TP + FP} \tag{3}$$

Table 4 Loss summary during YOLOv5 training

Box loss	0.03074
Object loss	0.01345
Classification loss	0.01019

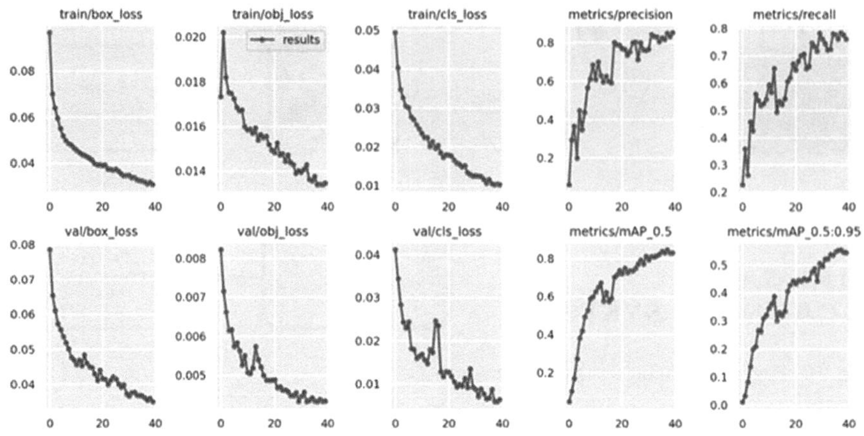

Fig. 12 Losses plot for YOLOv5 training

$$recall = \frac{TP}{TP + FN} \tag{4}$$

From the F1_curve (Fig. 13), the optimized confidence value for both precision and recall can be identified. Here, the F1 score reaches a maximum of 0.81, and the confidence level can be selected as 0.75, which is not far from 0.81, as observed. The authors assume any confidence above 0.7 to yield optimal results, which aligns with the precision and recall curves shown in Fig. 14.

Using the values from the Precision curve and Recall curve, we plotted the PR curve (Fig. 14) for the mAP at 0.5, resulting in a score of 0.847. The Confusion Matrix is considered one of the best metrics for evaluating a model's performance, particularly for a multi-class training model. In the confusion matrix (Fig. 15), we can assess the model's performance. The wrench and screwdriver classes exhibit the

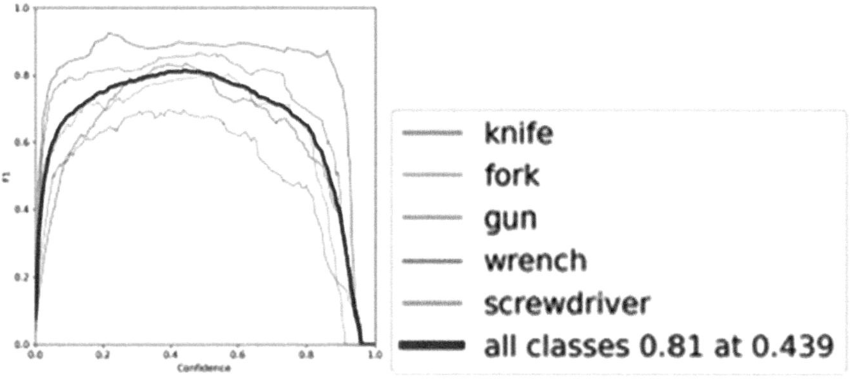

Fig. 13 F1_curve

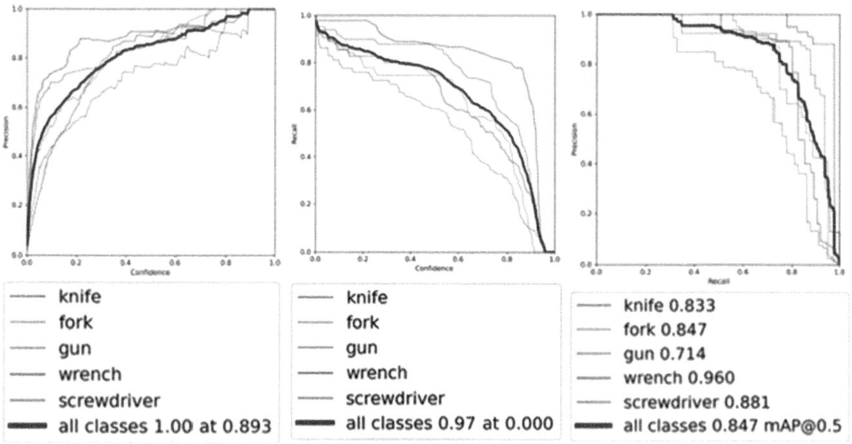

Fig. 14 **a** P_curve; **b** R_curve; and **c** PR curve

highest accuracy, leading to high True Positives and mAP scores of 0.91 and 0.86, respectively. The gun and fork classes have high false background negatives, which explains their precision and recall scores.

Figure 16 presents examples of the results produced on validation images, with the confidence levels mentioned for each detection.

In order to check and validate the performance of the trained model, it was also used to detect objects in video footage. The images below are snippets from the results.

After implementing YOLOv5, the algorithm can be applied to a prohibited items scanning system based on the results. Samples of how it will function as a weapon detection system are shown in Fig. 17 and the system design is displayed in Fig. 18.

Here are some of the challenges encountered during the object detection work in the paper: In real-time scenarios, the unavailability of a standard dataset tailored to the specific classes and scenarios required for the study posed a significant challenge. Creating a new dataset manually proved to be a time-consuming endeavor, further exacerbated by the manual labeling process necessary to annotate the dataset appropriately. Compounding this issue was the use of multiple object detection models, each employing its own labeling format, which was not always compatible with others. Consequently, each model necessitated different annotations to be adhered to during the dataset's pre-processing steps, adding complexity and further extending the time required for data preparation.

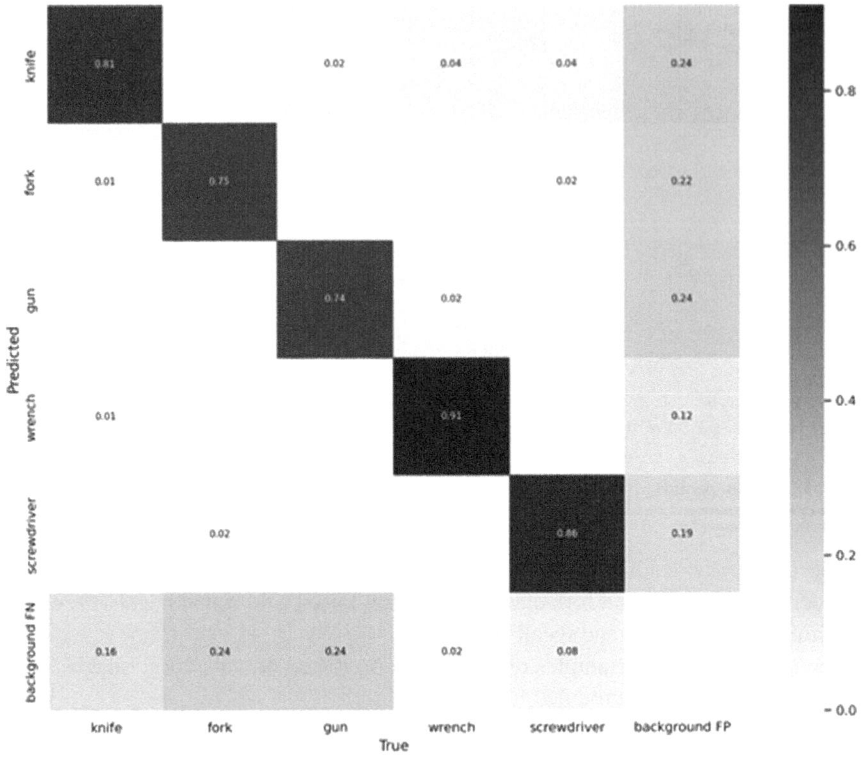

Fig. 15 Confusion matrix

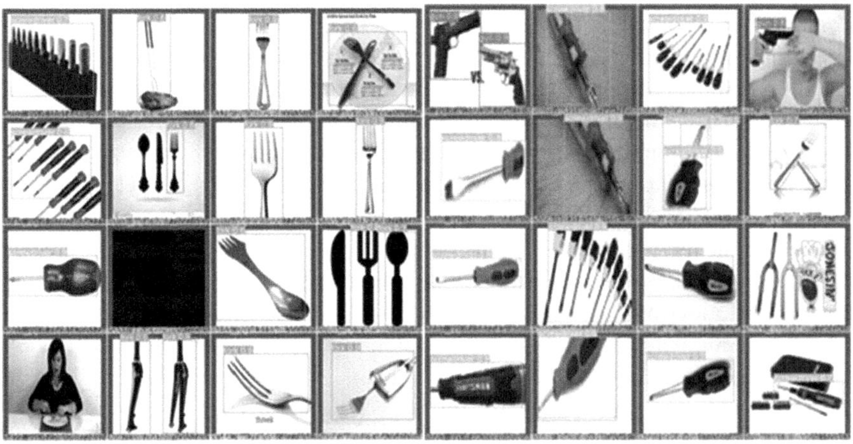

Fig. 16 Training results on validation images

Fig. 17 Snips from object detection on video

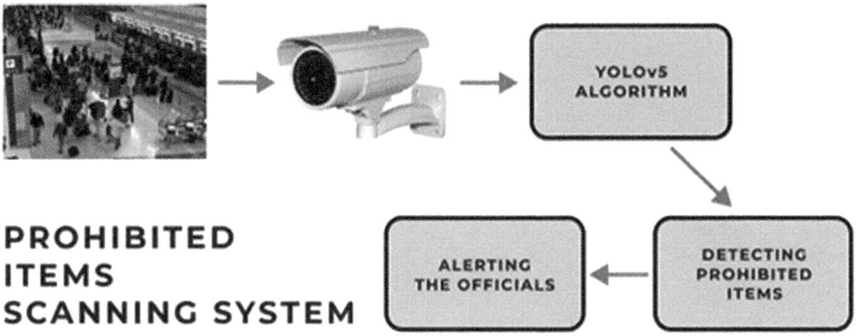

Fig. 18 Prohibited items scanning system design

5 Discussions

Figure 11 discusses the different metric values achieved for the model training using YOLOv5. The metrics used include precision, recall, and mean average precision. The highest overall values were obtained for the wrench class, while the gun class had the lowest values. The architectures of both models are significantly different. Faster R-CNN is a two-stage object identification model utilized by Detectron2, which includes the RPN sub-network for evaluating object proposal samples. However, this paper employed more than one technique for object detection on a small dataset.

On the other hand, one-stage detector models aim to locate relevant objects without the need for a region proposal filtering phase. These one-stage detectors typically have simpler structures and are often faster but slightly less accurate. An example is the YOLOv5 architecture, with some of the lighter-configured models achieving high frame rates (FPS), albeit at the cost of detection quality, while Faster R-CNN operates at lower FPS. From the obtained results and after implementation, three crucial factors on which the two models-Detectron2 vs. YOLOv5, can be evaluated and differentiated are summarized in Table 5.

In terms of use cases, the Detectron2 model would be more suitable for baggage scanning, given the smaller number of samples. Deploying a heavy model like Detectron2 is more viable when detection needs to be performed on fewer objects within the frame. On the other hand, the YOLOv5 model is much lighter, making it more practical for security surveillance in parking lots and hostels to monitor activity and reduce threats. These observations are summarized in Table 6.

Table 5 Detectron2 versus YOLOv5

Factor	Detectron2	YOLOv5	Explanation
Training data size and accuracy	✔ (for small datasets)		In the event that images are less than 50, Detectron2 would be our best bet in most cases. If one does not want to compromise on the speed, YOLOv5 can be used only if all the classes have unique features that are easily distinguishable (a handgun and a fork, for example)
Model size		✔	YOLOv5 is a considerably smaller model as compared to Detectron2. So, if both models perform similarly on the said dataset, YOLOv5 would be a better choice
Resource usage and training time		✔	YOLOv5 uses fewer resources compared to Detectron2, partly because of its small size. When both these models give decent results, the model that uses fewer resources can be chosen

Table 6 YOLOv5 vs Detectron2 observations

Detectron2	YOLOv5
Suitable for extensive datasets	Suitable for small datasets
Smaller model size	Larger model size
Requires lesser resources	Requires more resources
Takes longer time to train	Takes lesser time to train

6 Conclusion

Detectron2 (for luggage object detection only) and YOLOv5 have been effectively trained on the five predefined types of objects, expanding the scope to general object detection. The multi-class object detection achieved precision, recall, and mean average precision (mAP) values of 0.848, 0.787, and 0.847, respectively. Each model produced satisfactory results based on its specific training parameters. Initially, the project aimed to develop a luggage object detection system. YOLOv5 and Detectron2 were chosen for training after a thorough analysis of prior work in the literature and the evaluation of available resources. The preceding sections clearly demonstrate results for various use cases. This study can be further extended by expanding baggage object detection into a comprehensive security object detection system, encompassing both baggage detection and surveillance scanning in video footage. The YOLO model architecture provides higher accuracy in such scenarios, and the test cases align better with this objective.

References

1. D.J. Carr, K. Godhania, P.F. Mahoney, Edged weapons awareness. Int. J. Legal Med. **133**(4), 1217–1224 (2019)
2. B. Madhushree, K.N. Sowmya, H.R. Chennamma, Automatic weapon detection in video using deep learning, in *Data Engineering and Intelligent Computing* (Springer, Singapore, 2021), pp. 503–510
3. A. Warsi, M. Abdullah, M.N. Husen, M. Yahya, S. Khan, N. Jawaid, Gun detection system using YOLOv3. In *2019 IEEE International Conference on Smart Instrumentation, Measurement and Application (ICSIMA)* (IEEE, 2019), pp. 1–4
4. O. Elharrouss, N. Almaadeed, S. Al-Maadeed, A review of video surveillance systems. J. Vis. Commun. Image Represent. **77**, 103116 (2021)
5. M.T. Bhatti, M.G. Khan, M. Aslam, M.J. Fiaz, Weapon detection in real-time cctv videos using deep learning. IEEE Access **9**, 34366–34382 (2021)
6. K. Simonyan, A. Zisserman, Very deep convolutional networks for large-scale image recognition (2014). arXiv preprint arXiv:1409.1556
7. A.G. Howard, M. Zhu, B. Chen, D. Kalenichenko, W. Wang, T. Weyand, H. Adam, Mobilenets: Efficient convolutional neural networks for mobile vision applications (2017). arXiv preprint arXiv:1704.04861
8. C. Szegedy, S. Ioffe, V. Vanhoucke, A.A. Alemi, Inception-v4, inception-resnet and the impact of residual connections on learning, in *Thirty-first AAAI Conference on Artificial Intelligence* (2017)
9. S. Ren, K. He, R. Girshick, J. Sun, Faster r-cnn: Towards real-time object detection with region proposal networks. Adv. Neural Inf. Process. Syst. **28** (2015)
10. J. Redmon, S. Divvala, R. Girshick, A. Farhadi, You only look once: unified, real-time object detection, in *Proceedings of the IEEE Conference on Computer Vision and Pattern Recognition* (2016), pp. 779–788
11. D.K. Jain, An evaluation of deep learning based object detection strategies for threat object detection in baggage security imagery. Pattern Recogn. Lett. **120**, 112–119 (2019)
12. K. He, X. Zhang, S. Ren, J. Sun, Deep residual learning for image recognition, in *Proceedings of the IEEE Conference on Computer Vision and Pattern Recognition* (2016), pp. 770–778

13. J. Deng, W. Dong, R. Socher, L.J. Li, K. Li, L. Fei-Fei, Imagenet: A large-scale hierarchical image database, in *2009 IEEE Conference on Computer Vision and Pattern Recognition* (IEEE, 2009), pp. 248–255
14. P. Steno, A. Alsadoon, P.W.C. Prasad, T. Al-Dala'in, O.H. Alsadoon, A novel enhanced region proposal network and modified loss function: threat object detection in secure screening using deep learning. J. Supercomput. **77**(4), 3840–3869 (2021)
15. X. Bu, J. Peng, J. Yan, T. Tan, Zhang, Z, GAIA: A transfer learning system of object detection that fits your needs, in *Proceedings of the IEEE/CVF Conference on Computer Vision and Pattern Recognition* (2021), pp. 274–283
16. T.Y. Lin, M. Maire, S. Belongie, J. Hays, P. Perona, D. Ramanan, C.L. Zitnick, Microsoft coco: Common objects in context, in *European Conference on Computer Vision* (Springer, Cham, 2014), pp. 740–755
17. Z. Zou, Z. Shi, Y. Guo, J. Ye, Object detection in 20 years: a survey (2019). arXiv preprint arXiv:1905.05055
18. R.M. Saji, N.V. Sobhana, Real time object detection using SSD for bank security, in *IOP Conference Series: Materials Science and Engineering*, vol. 1070, no. 1 (IOP Publishing, 2021), p. 012060.
19. S. Narejo, B. Pandey, C. Rodriguez, M.R. Anjum, Weapon detection using YOLO V3 for smart surveillance systems. Math. Probl. Eng. (2021)
20. M. Dextre, O. Rosas, J. Lazo, J.C. Gutiérrez, Gun detection in real-time, using YOLOv5 on Jetson AGX Xavier, in *2021 XLVII Latin American Computing Conference (CLEI)*. (IEEE, 2021), pp. 1–7
21. J. Salido, V. Lomas, J. Ruiz-Santaquiteria, O. Deniz, Automatic handgun detection with deep learning in video surveillance images. Appl. Sci. **11**(13), 6085 (2021)
22. H. Jain, A. Vikram, A. Kashyap, A. Jain, Weapon detection using artificial intelligence and deep learning for security applications, in *2020 International Conference on Electronics and Sustainable Communication Systems (ICESC)*. (IEEE, 2020), pp. 193–198
23. L. Alathari, A. Blair, A. Carlock, S. Driscoll, D. Drysdale, J. McGarry, Mass attacks in public spaces-2018 (2019)
24. T. Karras, T. Aila, S. Laine, J. Lehtinen, Progressive growing of GANs for improved quality, stability, and variation (2017). arXiv preprint arXiv:1710.10196
25. O. Ciga, A.L. Martel, Learning to segment images with classification labels. Med. Image Anal. **68**, 101912 (2021)
26. G. Jocher, K. Nishimura, T. Mineeva, R. Vilariño, YOLOv5. *Code repository* (2020)
27. Y. Wu, A. Kirillov, F. Massa, W.Y. Lo, R. Girshick, Detectron2 (2019)
28. G. Jocher, K. Nishimura, T. Mineeva, R. Vilariño, *YOLOv5. Code repository* (2020). https://github.com/ultralytics/yolov5
29. K. Liu, H. Tang, S. He, Q. Yu, Y. Xiong, N. Wang, Performance validation of YOLO variants for object detection, in *Proceedings of the 2021 International Conference on Bioinformatics and Intelligent Computing* (2021), pp. 239–243
30. D. Thuan, Evolution of YOLO algorithm and YOLOv5: the state-of-the-art object detection algorithm (2021)
31. I. Goodfellow, J. Pouget-Abadie, M. Mirza, B. Xu, D. Warde-Farley, S. Ozair, Y. Bengio, Generative adversarial nets. Adv. Neural Inf. Process. Syst. **27** (2014)
32. H. Navidan, P.F. Moshiri, M. Nabati, R. Shahbazian, S.A. Ghorashi, V. Shah-Mansouri, D. Windridge, Generative adversarial networks (GANs) in networking: a comprehensive survey & evaluation. Comput. Netw. **194**, 108149 (2021)
33. https://public.roboflow.com/object-detection/pistols/1
34. https://www.kaggle.com/code/mpwolke/bionic-wrench/data
35. D. Tzutalin, LabelImg. *GitHub Repository, 6* (2015)
36. B.C. Russell, A. Torralba, K.P. Murphy, W.T. Freeman, LabelMe: a database and web-based tool for image annotation. Int. J. Comput. Vision **77**(1), 157–173 (2008)
37. S. Vicente, J. Carreira, L. Agapito, J. Batista, Reconstructing pascal voc, in *Proceedings of the IEEE Conference on Computer Vision and Pattern Recognition* (2014), pp. 41–48

38. Y. He, C. Zhu, J. Wang, M. Savvides, X. Zhang, Bounding box regression with uncertainty for accurate object detection, in *Proceedings of the IEEE/cvf Conference on Computer Vision and Pattern Recognition* (2019), pp. 2888–2897
39. P. Jiang, D. Ergu, F. Liu, Y. Cai, B. Ma, A review of YOLO algorithm developments. Procedia Comput. Sci. **199**, 1066–1073 (2022)

Implementation of IoT and Machine Learning Techniques in Smart Irrigation Systems

Abhirup Paria and Ruma Das

Abstract The lack of water supplies and the need for improved farm management are serious issues that is required to be fixed immediately. This study proposes an automated irrigation system created on the Internet of Things (IoT) for farms. This cost-effective automated irrigation monitors over-irrigation, soil erosion, and crop-specific watering needs. This new strategy promoted a sustainable practice by reducing the wastage of the excess water used in farming. The approach involves establishing a distributed wireless sensor network (WSN) across the farm, with various sensors transmitting data to a central server. Machine learning algorithms is implemented here to analyze the server data for the prediction of the optimal irrigation designs built on crop types and weather conditions, ensuring an appropriate water management.

Keywords Irrigation automation · Internet of Things (IoT) · Distributed wireless sensor network (WSN) · Machine learning (ML) · Sustainable irrigation

1 Introduction

Agriculture is a significant contributor to the gross domestic product (GDP) of both developing and developed countries [1]. In the current date, the development of smart farming technologies is crucial. Affordable and accessible technologies are essential for extensive adoption of global challenges like climate change [2], food scarcity [3], and malnutrition [4].

A. Paria (✉)
Department of Computer Science & Engineering (Artificial Intelligence & Machine Learning), Haldia Institute of Technology, Haldia, India
e-mail: abhirup.paria@gmail.com

R. Das
School of Applied Science & Humanities, Haldia Institute of Technology, Haldia, India
e-mail: rumadas1993@gmail.com

B. Pradhan and S. Mukhopadhyay (eds.), *IoT Sensors, ML, AI and XAI: Empowering A Smarter World*, Smart Sensors, Measurement and Instrumentation 50,
https://doi.org/10.1007/978-3-031-68602-3_8

Now a days, smart home systems is observed to change quickly, but smart farming is not growing up. This research proposal connecting this gap between conventional farming and smart farming by automating irrigation systems using low cost, easy accessible Internet of Things (IoT) [5] and Machine Learning techniques.

A wireless sensor network (WSN) installed across the farm gathers real-time data on many parameters. Machine learning algorithms analysis this data and open-source files to figure out the type of irrigation each crop needs based on the time of year and the weather. This data-driven approach optimizes water usage [6], promotes soil health, and restocks groundwater.

The proposed system leverages readily available microcontrollers like Raspberry Pi and Arduino for cost-effectiveness and ease of implementation. Sensors monitor soil temperature, moisture, and fertility, enabling control over these crucial factors.

This study adds to what has already been written and studied by using high-quality study papers and machine learning algorithms. We analyze the advantages and disadvantages of various sensors and microcontrollers for effective farm management and irrigation. This work unlocks the vast potential of technology to revolutionize agriculture.

2 Literature Review

This research leverages a comprehensive literature review to identify effective knowledges and algorithms for developing a Smart agriculture Monitoring System. The review explores various approaches, including microcontroller-based irrigation systems for cost-effectiveness [7–10], farm pest detection using ultrasonic sensors [11], and complete IoT-enabled irrigation system architectures [12]. Additionally, the review considers machine learning for soil classification [13], farm vehicle transmitting [14], plant detection with deep learning for irrigation [15], and deep learning with altitude-based irrigation [12]. Furthermore, it explores WSN for precision farming [16], cloud and edge computing to intelligent agriculture [16], and a comprehensive analysis of intelligent farming challenges, enablers, and opportunities [17]. Combining these insights, this paper proposes a precise solution for farm needs, including a prototype for a distributed sensor network.

3 Proposed Framework

The proposed system utilizes two microcontrollers: Raspberry Pi 3 and ESP32. These choices balance computational power, cost-effectiveness, and ease of access. Various sensors will continuously monitor critical environmental factors, enabling crop-specific irrigation based on real-time data. Figure 1 illustrates the key components of this system.

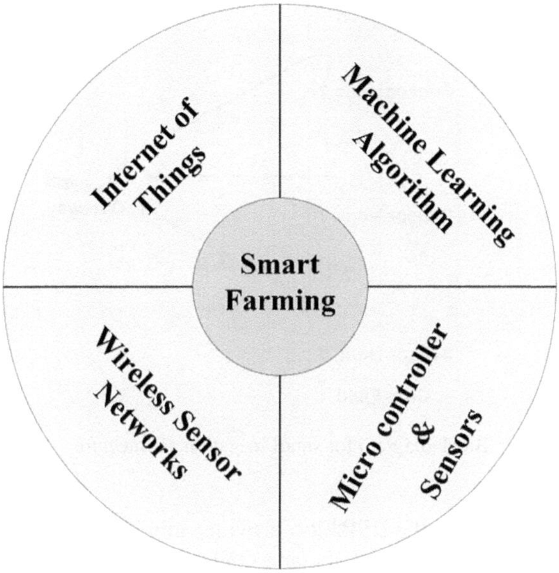

Fig. 1 Schematic for technology enabled smart irrigation

3.1 *Smart Irrigation Framework*

The initial step for the innovative irrigation system involves establishing a wireless sensor network (WSN). This network consists of interconnected sensor nodes that transmit data to a central server. Each node utilizes a Wi-Fi module for communication, ensuring efficient data transfer. An automated Python script running on the server regularly polls this data. The script can trigger alerts or initiate irrigation actions as needed based on the analysis. While Fig. 2 depicts a general network topology, the layout may vary depending on the specific field characteristics.

Data collection is crucial for an automated irrigation system. Many sensors are strategically placed throughout the ground to monitor vital environmental conditions. A Raspberry Pi 3 is the central gateway node, facilitating communication with all sensor nodes (ESP32) [17]. Each ESP8266 unit is equipped with an array of sensors, including advanced soil moisture sensors, Wi-Fi modules (for data transmission), temperature and humidity sensors, an MQ-135 gas sensor for air quality monitoring, a water level indicator, an alarm system, a clock unit, a battery, and a relay component for controlling irrigation valves. Additionally, an ultrasonic sensor is integrated to detect potential animal intrusions. Sensor data is relayed to the Raspberry Pi gateway, where a Python script processes and stores it on a server/cloud platform. The user can then access this data through a dedicated mobile application. Internet connectivity is optional for data broadcast from Raspberry Pi to the server. An intranet can be established for efficient system operation in areas with limited or no internet access.

The connection between sensors and the microcontroller follows fundamental electronic principles—receiver, spreader, ground, and voltage. Most networks are

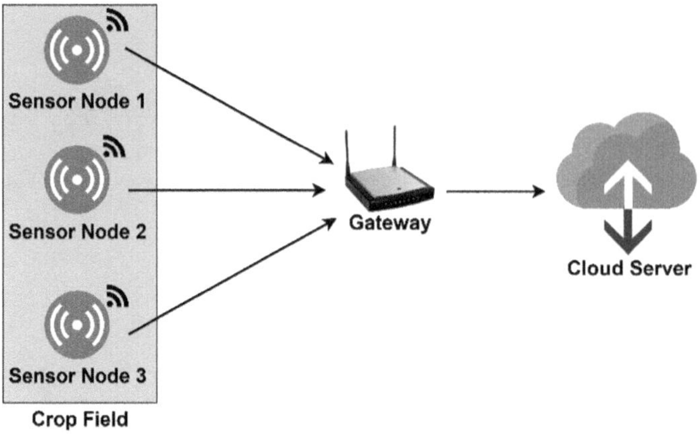

Fig. 2 Block diagram for smart irrigation architecture

similar, and the ESP8266 provides ample pins (RXD, TXD, digital, analogue) and power supply pins (Vcc and GND) to facilitate these connections. Establishing this network topology and data collection process represents the initial phase of the proposed application.

3.2 Node Structure

This research proposes the ESP8266 microcontroller as the primary choice for each sensor node due to its capabilities aligning well with the solution's requirements. However, alternative microcontrollers like Node MCU can be viable for smaller-scale implementations. Each node will have a Wi-Fi module for data transmission, a DHT22 sensor for measuring temperature and humidity, an MQ-135 gas sensor for air quality monitoring, and a battery module for power supply. Figure 3 illustrates the block diagram of an ideal sensor node.

3.3 Data Analysis and Decision Making

The sensor-collected data undergoes comprehensive analysis to trigger appropriate actions and alerts within the irrigation system. A Python script on the Raspberry Pi gateway examines various conditions derived from sensor data and online information. It seamlessly integrates real-time weather data from open-source APIs to make informed decisions. If a high rainfall probability (exceeding 98%) is forecast, irrigation is temporarily withheld to conserve water. However, a safety mechanism

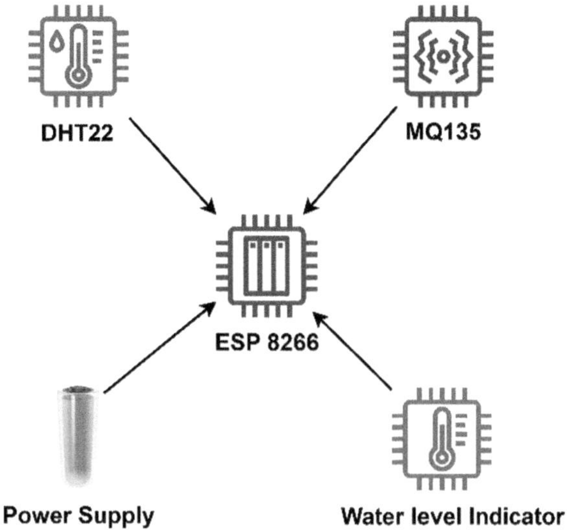

Fig. 3 Combination of sensors for each sensor node

is in place to initiate irrigation if soil humidity falls below a crop-specific threshold, ensuring optimal moisture levels.

The system upholds a crop-specific approach, adjusting irrigation based on soil type and plant requirements. Machine learning algorithms classify soil types to automate this process. As highlighted in the literature review [17], Support Vector Classification (SVC) is the most suitable algorithm for this task.

Following a thorough analysis of parameters, the Raspberry Pi gateway broadcasts a control signal to the actuators, primarily solenoid valves, for water pump activation. The system continuously monitors soil humidity at defined intervals. Once the desired humidity level is reached, irrigation ceases, preventing water wastage.

3.4 Soil Moisture Measurement and Machine Learning Integration

The system continuously monitors soil humidity using DHT22 sensors measuring moisture and temperature. Relative humidity (RH) readings range from 15 to 85%, providing insights into the current irrigation state and crop water requirements.

Machine learning algorithms were implemented to optimize irrigation based on historical data. A comparative study evaluated the performance of Support Vector Regression [18], and Random Forest Regression [19] for predicting future soil moisture levels. The results of this evaluation are shown in Fig. 4.

For more precise irrigation for a specific crop, the system can integrate a well-established formula as proposed by Brouwer et al. group [20]. Equation 1 calculates the daily water requirement ($Water_{need}$) based on evapotranspiration (ET_o) and

Fig. 4 Actual and predicted graph for relative humidity with days

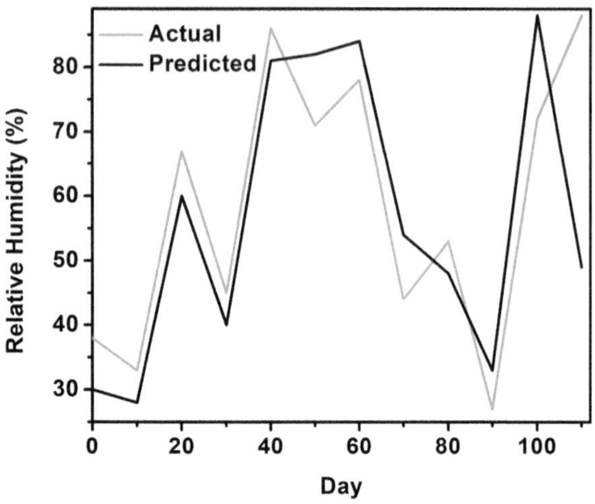

specific crop factor (SC_f).

$$\text{Water}_{\text{need}} = ET_o * SC_f \quad (1)$$

Traditionally, one-size-fits-all methods of irrigation often fail because they do not consider the type of food, weather, or the stage of growth. This formula confirms high irrigation accuracy by considering these crucial factors, promoting efficient and sustainable water management in this present automated farm monitoring system.

4 Result Analysis

Here, an automated farm monitoring system is presented to address food security challenges and contribute to sustainable agriculture. This system is developed with IoT and Machine learning algorithms, including Support Vector Regression (SVR) with a radial basis function kernel. These algorithms assist to classify the type of soil and identify crop types. Along with that, these algorithms predict the accurate irrigation required for specific crops. For the better performance of the automated irrigation system, a comprehensive evaluation of algorithms is carried out. On that context, Random Forest Regression (RFR) delivers the most promising results, bagging an accuracy of ~91.4% and the maximum F-score for predicting relative humidity (RH%). The RFR model also shows an insignificant p-value (<0.08) for the link between independent and dependent variables suggesting the implementation of the proposed method in the real field application.

This innovative farming approach offers a scalable solution, benefiting large-scale agribusinesses, small-scale farmers, and even household gardening enthusiasts. Interconnected devices facilitate seamless data collection and automated processes,

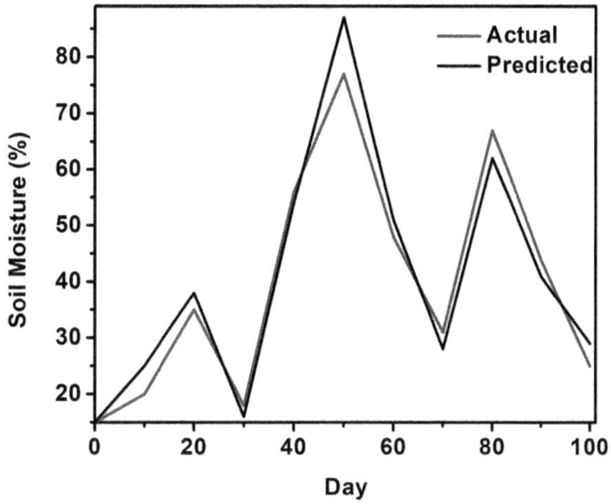

Fig. 5 The variation of actual and predicted soil moisture with days

enhancing farm management efficiency. Figure 5 illustrates the system's soil moisture prediction results.

5 Challenges

While the projected system offers significant benefits, certain factors can influence its performance. Firstly, proper installation of the sensor network is decisive for accurate data collection. Secondly, extensive training datasets, especially incorporating regional variations in soil composition and weather patterns are necessary for the accurate prediction in Machine learning. Though it has animal recognition features, it needs to be fixed by hand to avoid the damage of hardware by wildlife. Data visualization demands a dedicated server or network storage solution. A significant challenge is the need for more readily available, structured data on crop water requirements. This highlights the importance of data quality, as the accuracy of the model's predictions hinges on the data it's trained on. In conclusion, SVR performance is sensitive to hyperparameter tuning, requiring careful optimization to achieve optimum results.

6 Conclusion and Future Scope

This research proposes a sustainable and computationally effective method for smart farming built upon the IoT framework. Creating a well distributed sensor network is vital for accurate predictions from machine learning algorithms SVR and RF. Interconnected sensor nodes ensure comprehensive field monitoring, capturing critical data points across the entire area.

In conclusion, we envision a practical mobile system which will support farmers throughout the entire agricultural cycle, from initial seeding to crop harvesting. This method will not only make their work easier, but also it will assist them for correct decisions. Moreover, we propose exploring tools for automating hyperparameter tuning and leveraging ensemble learning techniques to meaningfully improve prediction accuracy. This portable, dynamic, and strong automation approach represents the exciting future scope of in the smart farming.

References

1. G.M. Idroes et al., Economic growth and environmental impact: assessing the role of geothermal energy in developing and developed countries. Inn. Green Dev. **3**(3), 100144 (2024)
2. P. Ritala, Grand challenges and platform ecosystems: scaling solutions for wicked ecological and societal problems. J. Prod. Innov. Manag. **41**(2), 168–183 (2024)
3. S. Akter et al., Tackling the global challenges using data-driven innovations. Ann. Oper. Res., 1–16 (2024)
4. E. Rothenberg et al., Global Leadership Initiative on Malnutrition (GLIM) for the diagnosis of malnutrition-a framework for consistent dietetic practice. Clin. Nutr. ESPEN (2024)
5. R. Molaei, K.R. Fard, A. Bouyer, Time and cost-effective online advertising in social Internet of Things using influence maximization problem. Wireless Netw. **30**(2), 695–710 (2024)
6. A. Mahule et al., AquaBamboo data-driven suggested system for water management and sustainable growth of bamboo: a review. Adv. Bamboo Sci., 100072 (2024)
7. Ç. Ersin, R. Gürbüz, A.K. Yakut, Application of an automatic plant irrigation system based Arduino microcontroller using solar energy. Solid State Phenom. **251**, 237–241 (2016)
8. Y. Liu et al., The research of precision irrigation decision support system based on genetic algorithm, in *2006 International Conference on Machine Learning and Cybernetics* (IEEE, 2006)
9. N. Agrawal, S. Singhal, Smart drip irrigation system using Raspberry pi and Arduino, in *International Conference on Computing, Communication & Automation* (IEEE, 2015)
10. S. Koprda et al., Proposal of the irrigation system using low-cost Arduino system as part of a smart home, in *2015 IEEE 13th International Symposium on Intelligent Systems and Informatics (SISY)* (IEEE, 2015)
11. A.S. Ahouandjinou, P.M. Kiki, K. Assogba, Smart environment monitoring system by using sensors ultrasonic detection of farm pests, in *2017 2nd International Conference on Bio-engineering for Smart Technologies (BioSMART)* (IEEE, 2017)
12. R. Kumar, A.M. Tripathi, M. Tyagi, Programmed dribble water system framework (ADIS) to identify soil moisture by using a dirt adhesive sensor, in *2022 3rd International Conference on Intelligent Engineering and Management (ICIEM)* (IEEE, 2022)
13. D. Smith, W. Peng, Machine learning approaches for soil classification in a multi-agent deficit irrigation control system, in *2009 IEEE International Conference on Industrial Technology* (IEEE, 2009)

14. C.-C. Wu et al., Design and development of farm vehicle monitoring and intelligent dispatching system, in *Proceedings of 2004 International Conference on Machine Learning and Cybernetics* (IEEE Cat. No. 04EX826) (IEEE, 2004)
15. J. Kwok, Y. Sun, A smart IoT-based irrigation system with automated plant recognition using deep learning, in *Proceedings of the 10th international conference on computer modeling and simulation* (2018)
16. M. Martinelli et al., A WSN-based solution for precision farm purposes, in *2009 IEEE International Geoscience and Remote Sensing Symposium* (IEEE, 2009)
17. M. Bacco et al., Smart farming: opportunities, challenges and technology enablers, in *2018 IoT Vertical and Topical Summit on Agriculture-Tuscany (IOT Tuscany)* (2018), pp. 1–6
18. S. Bandyopadhyay et al., Deep learning approaches for analysis and detection of tomato leaf diseases, in *International Symposium on Artificial Intelligence* (Springer, 2022)
19. A. Paria, S. Jana, Prediction of crops production using random forest regression, in *Proceedings of International Conference on Computational Intelligence, Data Science and Cloud Computing: IEM-ICDC 2021* (Springer, 2022)
20. C. Brouwer, M. Heibloem, Irrigation water management: irrigation water needs. Training Manual 3, pp. 1–5 (1986)

Autonomous Driving from an Architectural Perspective

Tripti Singh, Samarth Srivastava, Nikhil Kale, Nakul Saha, Shilpa Gite, and Biswajeet Pradhan

Abstract As the sensor technology and reliability of the obstacle detection techniques advances automated driving is going to become the most pivotal technology that will be the birth of the next revolution in mobility industry. The impact of autonomous systems will no longer be limited to automobile industry, but to the mankind. Controversies, discussions, and ethical laws often correlated with the study of autonomous driving. This paper brings out several facets of autonomous driving and integrates them to give an overview of technical architecture of a self-driving car in both software and hardware domain. Case studies on current challenges, fatalities, high-end architectural design, and collaborative safety frameworks are comprehensively reviewed. We conclude by highlighting some challenges in the sensor fusion field, and propose possible future research revolving around perception, planning and control of the autonomous vehicles.

T. Singh · S. Srivastava · N. Kale · N. Saha · S. Gite (✉)
Artificial Intelligence and Machine Learning Department, Symbiosis Institute of Technology, Symbiosis International (Deemed) University, Pune 412115, India
e-mail: shilpa.gite@sitpune.edu.in

T. Singh
e-mail: tripti.singh.btech2018@sitpune.edu.in

S. Srivastava
e-mail: srivastava.samarth.btech2018@sitpune.edu.in

N. Kale
e-mail: kale.nikhil.btech2018@sitpune.edu.in

N. Saha
e-mail: saha.nakul.btech2018@sitpune.edu.in

S. Gite
Symbiosis Centre of Applied AI (SCAAI), Symbiosis International (Deemed) University, Pune 412115, India

B. Pradhan (✉)
Centre for Advanced Modelling and Geospatial Information Systems (CAMGIS), University of Technology Sydney, Ultimo, NSW 2007, Australia
e-mail: Biswajeet.Pradhan@uts.edu.au

B. Pradhan and S. Mukhopadhyay (eds.), *IoT Sensors, ML, AI and XAI: Empowering A Smarter World*, Smart Sensors, Measurement and Instrumentation 50,
https://doi.org/10.1007/978-3-031-68602-3_9

Keywords Object detection · Generative adversarial networks · Transfer learning · You only look once · Detectron · Annotation conversion

1 Introduction

According to the World Health Organization (W.H.O) Global Status Report on Road Safety of December 2018, it was highlighted that the number of annual road traffic deaths has reached to a massive 1.35 million. That directly states that each day, almost 3700 people die in car crashes involving cars, trucks, buses, motorcycles, bicycles, or pedestrians [1]. India tops the world in road crash deaths with 11 percent, despite accounting for just 1 percent of the world's vehicles [2] and apart from deaths, around 45–50 million people suffer from just the road accident injuries and fatality.

Prime objective of this paper is to discuss architectural development of autonomous driving. Designing of an autonomous car is integration of several modules and consideration of thousands of rules and regulations.

In 2014, SAE International introduced J3016 "Levels of Driving Automation" standard for consumers. Standard chart defines 6 distinct levels of driving automation, starting from Level 0 where driver is in full control of vehicle, to Level 5 where vehicles themselves can control all aspects of dynamic driving tasks without human intervention (Fig. 1).

There are three main domains of autonomous driving: Perception to evaluate the environment, motion planning to ensure the route planning and the controller

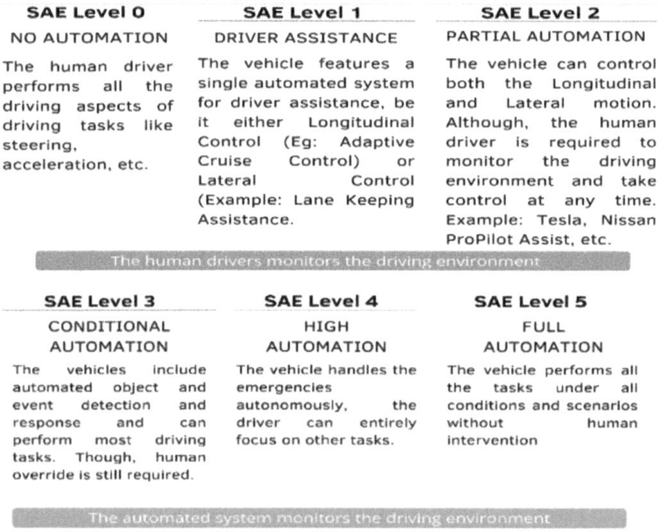

Fig. 1 SAE's levels of driving automation [3]

which manages the break, steering angle, and gear predictions. We also consider the operational design domain (ODD) which comprises of the ideal conditions for the successful working of the automobile.

Other tasks involve lateral control (steering and navigating laterally on the road), longitudinal control (for controlling the position or velocity of the car along the roadway through braking and acceleration), OEDR (Object and Event Detection and Response) for object detection, event prediction and immediate response actions.

2 Architectural Overview

The entire autonomous driving [4, 5] architecture can be broken down into perception, planning and control. The paper follows the same sequence to discuss architectural model of autonomous driving (Fig. 2).

Sensors hold the foundation to the perception of vehicle surroundings in automated driving systems. Secondly, the software comes into integration which takes the input from all the sensors, analyses them, puts the information of different categories of sensors together, and performs sensor fusion to obtain a collective data of a particular time. This data is then analyzed collectively by the CPU and a certain action is performed by the vehicle based on the situation around the car. The CPU/GPU/ Processors used are extremely powerful and can compute billions of operations per millisecond. Finally, we will be discussing about the safety frameworks and rules that are taken into consideration before putting an automated vehicle to the test on road. The automated car should have the high end, fully working sensors that are able to detect each object or the obstacle that the vehicle might encounter on the road from the predictable traffic and the traffic signals [6] to the unpredictable pedestrian or even a construction site.

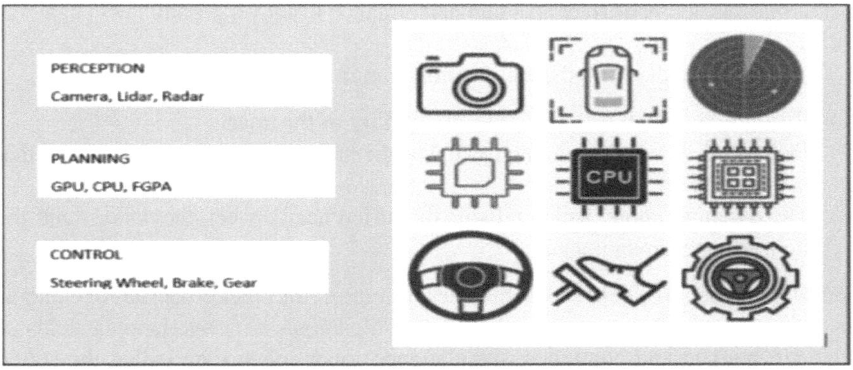

Fig. 2 The hardware and the components that are integrated with processor and CPU/GPU

3 Perception Task

Perception comprises of perceiving the environment the static and dynamic objects and making decisions accordingly. It is a subcategory of OEDR i.e., recognition of events and response through certain actions. We need to understand the different objects in the environment whether they are static or dynamic, prediction of their movement to plan our movement around them.

There are many limitations like abruption and ambiguity of sensor data due to weather conditions, occultation, and reflections. Perception overcome sensor data loss by processing large amounts of sensor data.

A. Sensors and it's Types:

Sensors are the driving forces for detection of movement data, the selection of sensors for various autonomous driving is a crucial task.

Sensors detect and sense movement of any property of the environment, or changes to that property over time. Depending on what property they record, sensors are generally categorized into two types:

(1) External movement of the environment Is recorded by exteroceptive sensors.
(2) If the sensors record a property of the ego vehicle, it is proprioceptive.

Exteroceptive Sensors: Any changes in the external environment are captured by the exteroceptive sensors such as camera, Lidar, Radar, Sonar. A brief discussion surrounds the choice of sensors in prioritized action implementation.

(a) CAMERAS:

First in the queue of all the exteroceptive sensors, we have the most used and the most important sensor for any automated vehicle, the camera. Cameras are a passive, most accurate way to create a visual representation of the world. Although, some groups believing that the camera should be the only sensor truly required for self-driving, is just wrong. A picture-perfect accurate performance is not yet possible with cameras alone.

Cameras generally have their three most important metrics:

(1) Resolution, which is the measure for quality of the image.
(2) Field of view, which is defined as the horizontal and vertical angular extent that is visible to the camera.
(3) The dynamic range, which tells us the difference between the darkest and the lightest tones in an image.

Focal length, depth of field and frame rate are some of the other properties of cameras that affect perception exist as well. When two cameras with overlapping fields of view are merged and combined, their aligned image planes are called the stereo camera. These cameras have the job of doing an in-depth estimation from synchronized image pairs. Pixel values from image is matched/mapped to the other image producing a disparity map of the scene. This disparity can then estimate depth at each differentiating pixel.

Fig. 3 LiDAR image of the surrounding [7]

(b) LiDAR:

Next in the sensors line-up, there is Light Detection and Ranging sensor, also referred to as the LiDAR sensor. It works by sending out beams of light in all the directions and then calculating how long it takes for the light to hit an object/obstacle and reflect to the LiDAR scanner. The distance to the object can then be calculated using the speed of light that hits back at the spinning LiDAR. Typically placed on the top of the car, they fire thousands of light beams per second. A 3D representation called a point cloud can be created based on the collected data, to represent the environment around the car. The big advantage of LiDAR sensors is their accuracy as they can identify details of only a few centimeters from objects that are 100 m away. The downside is the cost, which is currently even 10 times more than cameras and radar (Fig. 3).

LIDAR are not affected by the environments lighting. Therefore LIDAR, as opposed to cameras, do not face the same challenges when operating in poor or variable lighting conditions [8]. Some of the characteristics of LiDAR are listed below:

1. The more detailed the 3D point cloud can be, the faster the point collection will be.
2. The higher the rotation rate, the faster the 3D point clouds are updated.
3. Detection range is also important and is fully dependent on the power output of the light source.
4. Lastly, the field of view, which is the extent of the angle visible to the LIDAR sensor.

(c) RADAR:

RADAR which stands for Radio Detection and Ranging, is an electromagnetic sensor used to locate and robustly detect large objects in the environment. Radars are mostly unaffected by precipitation; they are particularly useful in bad weather conditions. RADAR is selected based on its detection range (which ranges up to long distance of about 250m). A field of view, of and the position and speed measurement accuracy. RADAR particularly have a wide angular field of view but short range. Or have a narrow field of view but a longer range (Fig. 4).

(d) Sonar:

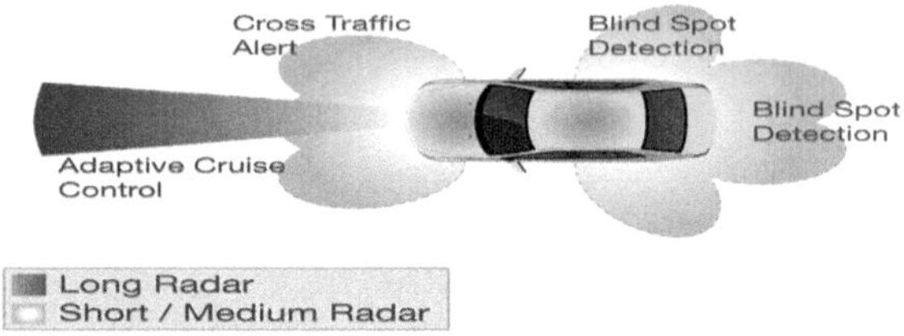

Fig. 4 RADAR sensors and their positioning wrt different scenarios

Sound Navigation and Ranging sensor, or better known as the SONAR's or the ultrasonic sensors which uses sound waves and sound propagation to navigate, measure the range, and to detect objects. These sensors are useful for a short-range observation and detection of any obstacle around the vehicle's environment. This makes them the most suitable device for the ego-vehicles in parking scenarios, where they need to make movements very close to other cars that are parked in front of it and behind it (Figs. 5 and 6).

Fig. 5 RADAR sensors and their positioning wrt different scenarios [9]

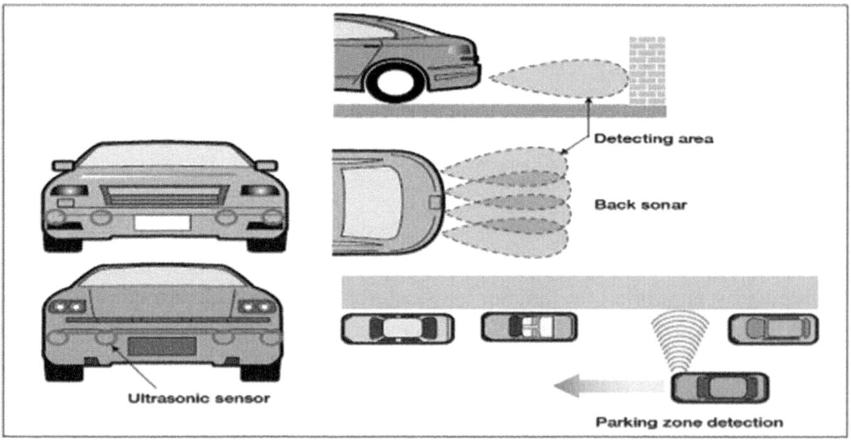

Fig. 6 RADAR sensors and their positioning wrt different scenarios during parking [9]

SONAR sensors are available at a comparatively cheaper cost. And since they detect the object properties using sound waves, they do not get affected adverse weather conditions (like RADAR) or by the heavy lighting.

Proprioceptive Sensors: Proprioceptive sensors are the sensors that sense ego properties. The most common ones here are Global Navigation Satellite Systems (GNSS), such as GPS (Global Positioning System), and an IMU (Inertial Measurement Unit). GNSS receivers are mostly used for measuring the ego vehicle position, its velocity, and most importantly, its heading, which is the most important for vehicle control. The accuracy depends substantially on the actual positioning methods and therefore the corrections used. Apart from these, the IMU also measures the angular rotation rate, accelerations [10] of the ego vehicle, and hence the 3D orientation of the vehicle can be estimated by the combined measurements.

And finally in list of sensors, we have wheel odometry sensors. This sensor is used to keep track of rates of rotation of wheel and uses these to estimate the speed and heading rate of change of the ego car. This equivalent sensor that tracks the mileage of vehicle.

So, to summarize, main sensors used nowadays for autonomous driving perception include cameras, RADAR, LIDAR, sonar, GNSS, IMUs, and wheel odometry modules (Fig. 7).

The most crucial part of the Hardware architecture [12] used in recent autonomous driving is the computing brain, the decision-making unit of the car. It takes altogether sensor data and outputs the commands needed to drive the vehicle.

There are sensors that would detect the traffic coming from the sides in a roundabout, or simply when we want to overtake a parked car in a lane. The cameras should be integrated to each other and give precise readings about where the car is going, whether it is inside the lane or not, detect the traffic coming from behind while overtaking, and many other factors.

The stereo camera being used for extracting the depth information, LIDAR for all weather 3D input, RADAR for object detection, SONAR for short-range 3D input

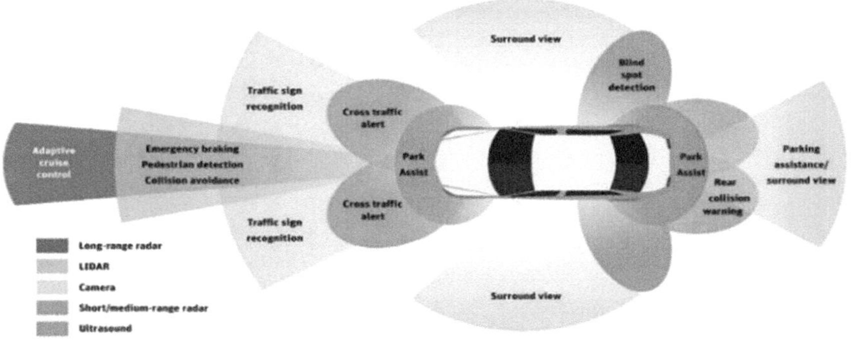

Fig. 7 RADAR sensors and their positioning with respect to different scenarios [11]

and GNSS/IMU data and wheel odometry for ego state estimation. All sensors have different range of views and different configurations.

The cars can have some type of sensors which deal with the requirement of a particular condition or situation the best. We want our sensors to capture the Operational Design Domain (ODD) [description of conditions in which the autonomous vehicle is designed to drive and operate safely we have in mind. There can be many possible scenarios in driving such as driving in the rainy season, driving in rural areas where it becomes hard to locate the lanes [13].

Lastly, for roundabouts we need a wide-range, short distance sensor laterally since the traffic is slow but we also need a wide-range short distance sensor longitudinally because of how movement around the roundabout occurs [14].

4 Hardware Architecture

We can imagine an Autonomous Vehicle (AV) as a human body. Just like how human body contains eyes, ears, brain, and other organs to differentiate between different kinds of things, AV's need a brain and a vision too, to visualize and process different things and act upon it. The CPU works as a brain of the car and does processing of the information and triggers the car and its mechanical sensors to take actions (Fig. 8).

A. **Depth Perception Using Vision**:

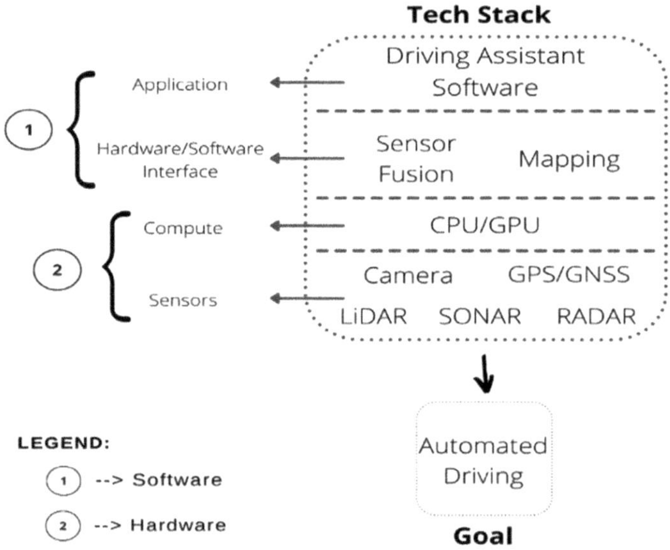

Fig. 8 Stack overview of how the hardware and software are lined up in an autonomous vehicle [15]

There are at least 3 important sensors that are needed in an automated car, the cameras for the vision and capturing the pixels, the ultrasonic sensors/SONARs for detection of the close by environment around the car and important in situations of traffic and parking, and the third one can be either LiDAR or RADAR or both (Fig. 9).

LiDAR, in coordination with the vision of the camera, generates an overview of the entire environment around the car; and by tracking these points and their coordinates, the CPU/GPU can map the entire environment around the car and have the knowledge about the position of each object [17, 18].

To penetrate occlusions and bridge the weather gap, companies like TESLA uses and promotes the use of more RADAR's built into their cars and no LiDAR's. When RADAR, combined with the GNSS and the GPS Navigation System and the vision from the Camera is much more efficient than the LiDAR alone itself [19] (Figs. 10 and 11).

Fig. 9 When vision and LiDAR combines, it gives the image of the surroundings like so [16]

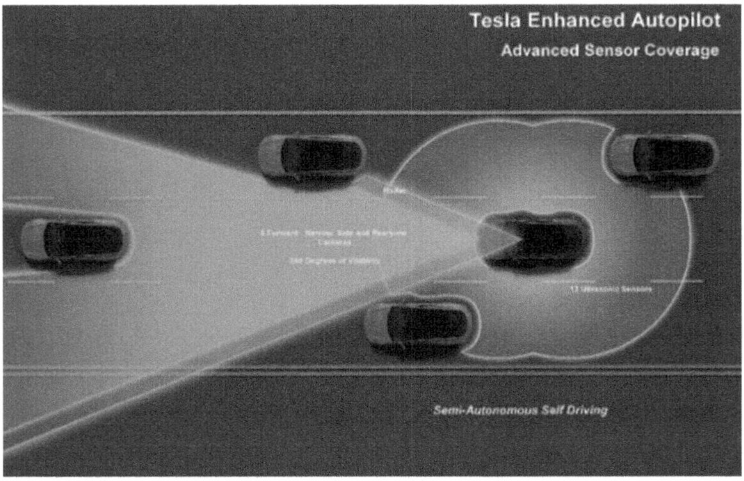

Fig. 10 How TESLA uses the power of RADAR and SONAR [20]

Fig. 11 Object detection and 3D virtualization image

The image shown above is how the CPU/GPU processes it after receiving the inputs just from the front camera, combined with RADAR and GPS [21]. The left-hand side image shows the bird eye view of the positioning of the car with respect to the lanes and the other vehicles it is being surrounded by, and the minimal distance from the car to the other vehicle is being shown by the yellow circles. Similarly, on the right-hand side image, the Camera through its vision feeds the CPU the images and the processor classify everything as a particular object/vehicle/person and what are their 3D dimensions and how big or small things are around the Autonomous Vehicle's (AV) environment.

5 Software Architecture

The basic software architecture [22] for an autonomous driving is broadly categorized as environment perception, environment mapping, motion planning, controller and system supervisor (Fig. 12).

5.1 Overview of Software Architecture

Localization of the autonomous vehicle and classification of the key elements [24] present in the environment are key tasks of the environment perception module. The environment mapping module is designed to create maps around the above obtained key points and the autonomous vehicle. Motion planning decides about the actions required supported by all the knowledge from the perception and mapping module. Controller takes charge to execute the planned route. Supervisor is the authoritative part of the entire structure. It is the main body which is responsible for alerting driver in case of maneuvers.

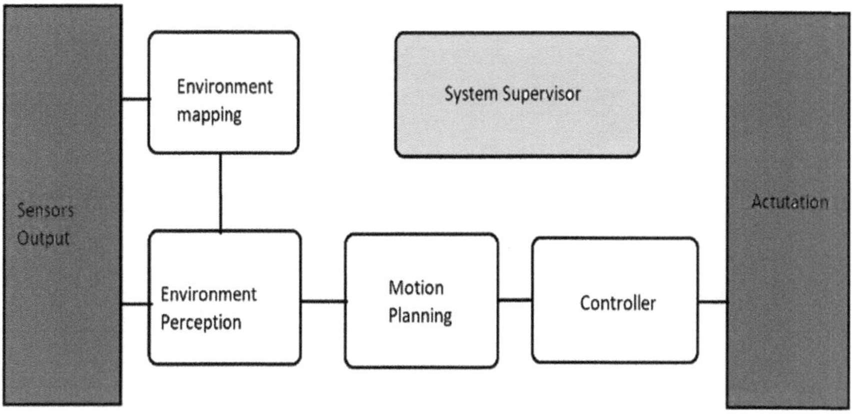

Fig. 12 System software architectural block diagram [23]

Environment Perception: Perception module is responsible for localization of ego vehicle and classification of environmental elements. The localization module incorporates GPS, IMU, odometry and Lidar and camera data [25]. History of path is maintained in coordination with road maps. Specific scenes map static objects with the help of camera input details and LiDAR (Fig. 13).

Environment Mapping: The environment maps create varied representation of current environment around the autonomous vehicle. Localization map improves estimation of the object, using lidar. Signs and signals [26] are stored for future motion planning. The map is a combination of historical pre-recorded data and new incoming data from the static elements of the environment (Fig. 14).

Motion Planning: Motion planning is a multi-integration process, involving a combination of many modules. It has numerous layers of abstraction.

Controller: A common self-driving controller is responsible for separating the control issue into two segments: longitudinal and lateral. The task of the lateral controller is to maintain the required steering angle, whilst the longitudinal controller's task is

Fig. 13 3-D bounding boxes around the objects

Fig. 14 Environmental mapping in self-driving cars [27]

to adjust the throttle, the gear, and the braking systems, so as to obtain the correct speed.

System Supervisor: System Supervisor is the supervising body of the software architecture. It gives warning signal in case of subsystem failure. Hardware supervisors keeps tabs on the hardware components to detect any malfunction or failure and software supervisor does the same for its respective sub systems.

6 Safety Frameworks for Autonomous Driving

We shall have a look at different definitions for safety, sources of hazards, what are the industry perspective for safety, and common frameworks for a safe system design. But first we should look at past crash reports by autonomous vehicles to see what exactly we are dealing with here (Fig. 15).

Fig. 15 Examples of automated vehicle crashes [28]

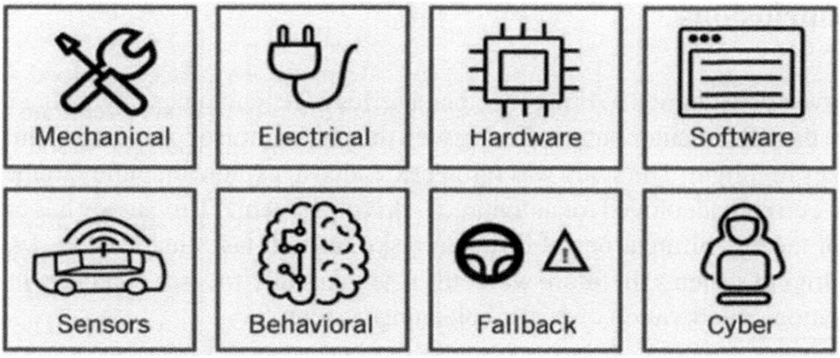

Fig. 16 Major hazard sources

An autonomous car while making predictions of its surroundings, cannot anticipate if a vehicle would try to squeeze in a smaller gap near it and if this does occur an accident is bound to happen (Fig. 16).

After seeing the various hazards possible for the project, ensuring a safety assessment framework is necessary. The National Highway Transportation Safety Administration (NHTSA) in the United States has established a twelve-part safety framework to structure autonomous driving safety assessments.

A. Industry Perspective:

WAYMO: All 12 NHTSA concepts are covered by Waymo, and they are organized into a five-level safety strategy.

Secondly, Waymo ensures that backups and redundancies are in place. When a fault arises, the vehicle can go to a backup or process in which the problem can be reduced and the vehicle can be returned to its safe position and if feasible, continue to drive.

GM MOTORS: GM's safety strategy concentrates on its implementation tactics for completing the mandatory safety assessments. They identify and address hazards with the goal of wholly removing them instead of reducing them. Finally, all of their systems adhere to the company's internal safety, dependability, and security criteria.

B. Frameworks:

The safety frameworks are:

(1) Generic Safety Frameworks include fault trees, failure modes and effects analyses or FMEA, and hazard and operability analysis or HAZOP
(2) FMEA is a "bottom-up" method, which examines each cause and determines all its effects.

7 Conclusions

The survey paper aimed to bring together a series of technologies now being used to explore the field of autonomous driving with the proposition of proposed architecture as the key highlight. The study was further extended to explore the safety frameworks that are currently deployed for autonomous driving systems. This survey has outlined some of the key bifurcations of different tasks and sub-tasks involved in designing of the driving systems. In future work, the research study focuses exploring in-depth information collaboration of control planning systems.

Acknowledgements This research is supported by Symbiosis Institute of Technology (SIT), Pune.

References

1. World Health Organization, *Global status report on road safety 2018: summary*. No. WHO/NMH/NVI/18.20. World Health Organization (2018)
2. S. Karnouskos, The role of utilitarianism, self-safety, and technology in the acceptance of self-driving cars. Cogn. Technol. (2020) https://doi.org/10.1007/s10111-020-00649-6
3. R. Ni, J. Leung, Safety and liability of autonomous vehicle technologies. Groups. Csail. Mit. Edu 1–49 (2014) [Online]. Available: https://groups.csail.mit.edu/mac/classes/6.805/student-papers/fall14-papers/Autonomous_Vehicle_Technologies.pdf
4. Y. Huang, Y. Chen, Survey of state-of-art autonomous driving technologies with deep learning, in *Proc. - Companion 2020 IEEE 20th Int. Conf. Softw. Qual. Reliab. Secur. QRS-C 2020*, (2020), pp. 221–228. https://doi.org/10.1109/QRS-C51114.2020.00045
5. W. Zong, C. Zhang, Z. Wang, J. Zhu, Q. Chen, Architecture design and implementation of an autonomous vehicle. IEEE Access **6**, 21956–21970 (2018). https://doi.org/10.1109/ACCESS.2018.2828260
6. J. Zhang, M. Huang, X. Jin, X. Li, A real-time Chinese traffic sign detection algorithm based on modified YOLOv2. Algorithms **10**(4), 1–13 (2017). https://doi.org/10.3390/a10040127
7. S. Royo, M. Ballesta-Garcia, An overview of lidar imaging systems for autonomous vehicles. Appl. Sci. **9**(19) (2019). https://doi.org/10.3390/app9194093
8. D. Gohring, M. Wang, M. Schnurmacher, T. Ganjineh, Radar/Lidar sensor fusion for car-following on highways, in *ICARA 2011 - Proc. 5th Int. Conf. Autom. Robot. Appl.* (2011), pp. 407–412. https://doi.org/10.1109/ICARA.2011.6144918
9. I. Bilik, O. Longman, S. Villeval, J. Tabrikian, The rise of radar for autonomous vehicles: signal processing solutions and future research directions. IEEE Signal Process. Mag. **36**(5), 20–31 (2019). https://doi.org/10.1109/MSP.2019.2926573
10. S.C. Lin et al., The architectural implications of autonomous driving: constraints and acceleration. ACM SIGPLAN Not. **53**(2), 751–766 (2018). https://doi.org/10.1145/3173162.3173191
11. R. Merco, Robust estimation of connected automated vehicles while performing cooperative tasks in presence of malicious agents, in *Proc. Annu. Conf. Progn. Heal. Manag. Soc. PHM* (2018). https://doi.org/10.36001/phmconf.2018.v10i1.714
12. A. Collin, A. Siddiqi, Y. Imanishi, E. Rebentisch, T. Tanimichi, O.L. de Weck, Autonomous driving systems hardware and software architecture exploration: optimizing latency and cost under safety constraints. Syst. Eng. **23**(3), 327–337 (2020). https://doi.org/10.1002/sys.21528
13. Z. Wang, W. Ren, Q. Qiu, LaneNet: Real-Time Lane Detection Networks for Autonomous Driving (2018). [Online]. Available: http://arxiv.org/abs/1807.01726

14. G. Villalón-Sepúlveda, M. Torres-Torriti, M. Flores-Calero, Traffic sign detection system for locating road intersections and roundabouts: the Chilean case. Sensors (Switzerland) **17**(6), 1–21 (2017). https://doi.org/10.3390/s17061207
15. S. Campbell et al., Sensor technology in autonomous vehicles : a review, in *29th Irish Signals Syst. Conf. ISSC 2018* (2018) pp. 1–4. https://doi.org/10.1109/ISSC.2018.8585340
16. Institute of Electrical and Electronics Engineers and Vehicular Technology Society, 2020 IEEE 91st Vehicular Technology Conference (VTC Spring) proceedings: Antwerp, Belgium, 25–28 May 2020 (2020), pp. 3–8
17. S. Virmani, S. Gite, Performance of convolutional neural network and recurrent neural network for anticipation of driver's conduct, in *8th Int. Conf. Comput. Commun. Netw. Technol. ICCCNT 2017* (2017). https://doi.org/10.1109/ICCCNT.2017.8204039
18. Virmani: Developing a novel Algorithm for identifying... - Google Scholar. https://scholar.google.co.in/scholar?hl=en&as_sdt=0,5&cluster=1796821290354728397. Accessed 20 Sept 2021
19. T.G. Phillips, N. Guenther, P.R. McAree, When the dust settles: the four behaviors of LiDAR in the presence of fine airborne particulates. J. F. Robot. **34**(5), 985–1009 (2017). https://doi.org/10.1002/rob.21701
20. S. Ingle, M. Phute, Tesla autopilot: semi autonomous driving, an uptick for future autonomy. Int. Res. J. Eng. Technol. **3**(9), 369–372 (2016). [Online]. Available: www.irjet.net
21. G. Reina, D. Johnson, J. Underwood, Radar sensing for intelligent vehicles in urban environments. Sensors (Switzerland) **15**(6), 14661–14678 (2015). https://doi.org/10.3390/s150614661
22. A.C. Serban, E. Poll, J. Visser, A standard driven software architecture for fully autonomous vehicles, in *Proc.—2018 IEEE 15th Int. Conf. Softw. Archit. Companion, ICSA-C 2018* (2018), pp. 120–127. https://doi.org/10.1109/ICSA-C.2018.00040
23. B. Leigh, R. Duwe, Software architecture of autonomous vehicles. ATZelectronics Worldw. **14**(9), 48–51 (2019). https://doi.org/10.1007/s38314-019-0104-7
24. J. Fayyad, M.A. Jaradat, D. Gruyer, H. Najjaran, Deep learning sensor fusion for autonomous vehicle perception and localization: a review. Sensors (Switzerland) **20**(15), 1–34 (2020). https://doi.org/10.3390/s20154220
25. E. Martí, M.Á. De Miguel, F. García, J. Pérez, A review of sensor technologies for perception in automated driving. IEEE Intell. Transp. Syst. Mag. **11**(4), 94–108 (2019). https://doi.org/10.1109/MITS.2019.2907630
26. S. Reddy, M. Mathew, L. Gomez, M. Rusinol, D. Karatzas, C.V. Jawahar, RoadText-1K: text detection recognition dataset for driving videos, in *Proc.—IEEE Int. Conf. Robot. Autom.*, (2020), pp. 11074–11080. https://doi.org/10.1109/ICRA40945.2020.9196577
27. S.D. Pendleton et al., Perception, planning, control, and coordination for autonomous vehicles. Machines **5**(1), 1–54 (2017). https://doi.org/10.3390/machines5010006

A Prototype Design for Effective Traffic Navigation

M. R. Pooja, N. Bindu, Chandana V. Kumar, D. S. Chinmayee, and R. Ganavi

Abstract Potholes are a common road hazard that can cause damage to vehicles and pose a safety risk to drivers. Over the years, numerous methods have been developed to detect and navigate around potholes. This review paper aims to summarize the various pothole detection and navigation techniques that have been proposed in the literature. We begin by discussing traditional approaches that rely on human inspection or visual sensors. We then review more recent techniques that utilize various sensors and machine learning algorithms to detect potholes in real-time. Finally, we discuss techniques for navigating around potholes, including the use of map data and vehicle control systems. The work offers a promising approach to identifying and mapping potholes in real-time, which could help to improve road safety and reduce vehicle damage. Overall, this review provides an overview of the state-of-the-art in pothole detection and navigation and identifies areas for future research.

Keywords Pothole detection · Sensor · Image processing · Deepbus · 3D reconstruction

1 Introduction

Potholes are a fairly common phenomenon on Indian streets and roads. Potholes are formed due to poor road construction and maintenance, heavy rains and movement of heavy vehicles. Because of their high visibility, potholes are among the most common local concerns of the general population and the biggest cause of accidents. It also affects the tourism industry of the country [2].

People lose control of their vehicle while driving over potholes which leads to many traffic accidents and loss of lives. According to government statistics, over the five years from 2016 to 2020, potholes claimed an average of 2,300 lives a year across

M. R. Pooja (✉) · N. Bindu · C. V. Kumar · D. S. Chinmayee · R. Ganavi
Vidyavardhaka College of Engineering, Mysuru, India
e-mail: pooja.mr@vvce.ac.in

B. Pradhan and S. Mukhopadhyay (eds.), *IoT Sensors, ML, AI and XAI: Empowering A Smarter World*, Smart Sensors, Measurement and Instrumentation 50,
https://doi.org/10.1007/978-3-031-68602-3_10

the country. Now the situation could only get worse as the pandemic affected repairs and maintenance. The epidemic affected repairs and maintenance, so the situation could only get worse [3]. The National Crime Records Bureau (NCRB) estimates that potholes account for roughly 5% of road deaths.

An effective depth-based pothole detection method that does not require high computational and processing effort is proposed. In the proposed method, we use the depth calculated by the ultrasonic sensor as the basis for pothole detection, and then its position is sent to the database.

Location coordinates are retrieved from the database using the app and pothole locations are updated on the map in real time. The results show that this method can be used to identify road potholes and support efficient navigation [4].

2 Types of Methods to Detect Potholes

2.1 Image Based Pothole Detection

Using photos or videos as input data, the vision-based method uses image processing and machine learning algorithms or deep learning to recognize potholes on the road surface. It requires less processing. Although vision-based approaches are cheaper than 3D laser scanning methods, it is difficult to accurately detect a pothole using these methods due to noise during image and video data collection [5].

2.2 Sensor Based Pothole Detection

In sensor based methods, the pothole is detected using the accelerometer sensor readings obtained from the vehicle. It is also used to calculate the depth of the pothole. The sensor-based approach is the most economical of the three because of its low cost and very fast processing capabilities. It can be implemented in real time easily. The data collection also requires less storage space. However, sensor-based techniques are subject to noise. In addition, only wheel track areas on the road directly adjacent to the lane boundary can be evaluated [6]. The sensors and vehicles used to collect the data also play a role in influencing the readings [1, 11–13].

2.3 3D Reconstruction-Based Pothole Detection

Of the three techniques, the method based on 3D reconstruction most accurately predicts the volume of the pothole and its dimensions also. 3D construction can be based on laser or stereo imaging. This model depends on how the cameras are aligned and what algorithms are used. The advantage of using this method is that it gives an accurate shape of the pothole along with its dimensions. The drawback is that its more expensive than other methods and it is not easy to detect when the potholes are filled with water or debris [7]. It will also be difficult to detect if the lighting is bad.

Of the three techniques, the method based on 3D reconstruction most accurately predicts the volume of the pothole and its dimensions also. 3D construction can be based on laser or stereo imaging. This model depends on how the cameras are aligned.

3 Literature Survey

The following sections explain the different approaches to detect potholes.

3.1 DeepBus

Deepbus is a methodology that aims to detect potholes and irregular surfaces on roads using the Internet of Things in real time. This methodology based on machine learning, was proposed in a paper titled "DeepBus: A Machine Learning-Based Real-Time Pothole Detection System for Smart Transportation Using IoT" published in 2020 by Kashish Bansal, Kashish Mittal, Gautam Ahuja, Ashima Singh, and Sukhpal Singh Gill. The goal of the solution was to (a) reduce pothole-related mortality (b) identify the location of the pothole to the public. The proposed solution consists of hardware and software components. This methodology is able to classify the pothole and record the location coordinates whenever the vehicle moves over it The hardware components consist of sensors, i.e. accelerometer and gyroscope together with a GPS module and Raspberry Pi. The software component consists of a map that calculates a route and displays potholes every time the user requests it [16–18]. A test rig is built using a Raspberry Pi, an accelerometer and a gyro sensor to create a dataset that will be used to train different machine learning models. 8 The experimental setup also includes a quad-core A53 (ARMv8) Broadcom BCM2837B0 clocked at 1.4 GHz, a Broadcom Videocore-IV GPU, and 1 GB of LPDDR2 SDRAM. Bumps and irregularities are detected using the accelerometer and gyroscope. The collected raw data consists of six elements (Ax, Ay, Az, Gx, Gy, Gz) [8]. Accelerometer data in three directions and gyroscope data in three directions. Various algorithms were trained, among which Random Forest proved to be the most effective with 86.8% precision, 83.7% precision, 84.4% recall, and 83.7% F1-score. A good number of technologies including computer vision and convolution Neural Network have been used to recognize the human computer interface [9, 20]. The advantages of this system are that Random Forest avoids overfitting problems and (b) with a single training pass it can be applied to both regression and classification problems. The disadvantage of the solution is that it does not indicate the severity of the pothole.

3.2 Image-Processing to Detect Potholes

The pothole detection solution proposed by Mae M. Garcillanosa, Jian Mikee L. Pacheco, Rowie E. Reyes, and Junelle Joy P. San Juan in their article "Smart Detection and Reporting of Potholes via Image-Processing using Raspberry-Pi Microcontroller" was published in 2018 is the use of image processing techniques. The goal of this solution is to provide a solid design for a low-cost, reliable system that deals with pothole detection using image analysis and pothole reporting without the use of smartphones. The system consists of software and hardware modules. The entire system sends the image of the pothole and its location to the main server. The hardware components include a camera, a gps module, a Raspberry Pi microcontroller and a Wifi module. In addition, a web server is used to store and display pothole information. The camera was placed at the top of the car, behind the rear view mirror, in order to obtain a wide field of view and provide reliable and accurate results. The system captures eight images per second in front of the vehicle's windshield and uses image processing techniques to detect potholes in real time. Image processing involves removing other objects from the image, extracting the road surface, and then using Canny's edge detection and contour detection to identify potholes based on their color and surface area. When a pothole is detected, the microcomputer sends an image of the pothole and its location to a master database over the Internet. The system also includes an LED indicator to show when a message has been sent. The GPS module is used to locate the location where the pothole was found and to obtain the coordinates of the location. Both ideal and less ideal conditions were used for data collection. The less than ideal dataset yielded the following results: Accuracy was 87.45% and specificity – (TP/(TP + FP)) * 100, was recorded as 85.71% [10]. The advantage of this solution is that smartphones are not needed, making it cheap and affordable. The disadvantage of the system is relatively low accuracy.

3.3 Vibration-Based Method

The article "Abnormal Road Surface Detection Based on Smartphone Acceleration Sensor" was authored by Ronghua Du, Gang Qiu, Kai Gao, Lin Hu, and Li Liu and was published on January 13, 2020. The authors propose a technique for detecting potholes and other irregularities on road surfaces using smartphone acceleration sensors. The process involves several stages, including data gathering, preprocessing, locating the abnormal surface, classifying the abnormal surface, and identification. The authors gathered data on the vehicle's position, speed, and acceleration using the global positioning and navigation system and an integrated accelerometer. They preprocessed the raw data using a Butterworth filter and used a Gaussian model with a z-axis acceleration threshold condition to identify irregular road surfaces. The authors used the K-Nearest Neighbors algorithm to categorize potholes, bumps, and uneven road surfaces. To evaluate the accuracy of their system, they compared field measurements of abnormal road surfaces with the system's predictions.

The used data set included the following:

- 118 bumps, 174 flat areas, and 103 pothole areas are seen in the training model.
- 151 bumps, 583 flat regions, and 68 pothole spots can all be seen in the testing model.

The experimental requirements for testing the model include the following:

- The experimental vehicle must be a class A vehicle such as a Cavalier or SUV (Qoros 5).
- A smartphone should be mounted preferably on the driver's seat handrail, and an android application should be operating in the background. For example, the Redmi Note 8 Pro smartphone.
- The sampling frequency required is 400 Hz.
- Based on the test results, potholes and bumps are accurately detected at 0.9603 and 0.9412 respectively.

3.4 3D Reconstruction-Based Method

On May 15, 2019, M. Uzair Ul Haq, Moeez Ashfaque, Senthan Mathavan, Member, IEEE, Khurram Kamal, and Adeel Ahmed published a paper on a novel automated system for detecting potholes on road surfaces using stereo imaging. The proposed system comprises six phases: data acquisition, data preprocessing, stereo matching, 3D data generation, 3D data post-processing, and pothole metrology. To collect the data, the system used static and dynamic images captured by the left and right cameras mounted on a vehicle. The dynamic pictures were filtered during the preprocessing step before applying stereo matching, which uses a hybrid matching scheme to combine keypoint matching with block matching techniques. The stereo triangulation technique was then used to generate 3D data from the stereo-matched results. After processing the 3D data, the system detected potholes and estimated their volume. The authors evaluated the system's performance using various methods, including accuracy qualification, stereo reconstruction, and volume calculations. This system offers a promising solution for detecting and measuring potholes on road surfaces, which can help improve road safety and maintenance.

The experimental requirements for testing the model include the following:

- It calls for an A4Tech PKS-732 G webcam with two stereo pairs.
- For static display, the web camera must be mounted on a tripod, and for dynamic display, it must be mounted on the back of the vehicle.
- The processing of raw images is done on a Windows computer.

In the static situation, the 3D measurements were accurate to within 3 mm, and in the dynamic situation, they were accurate to within 5 mm. When compared with static photos, potholes are depicted with an accuracy of −20% for volume, −15% for area, and −4% for depth at 10 km/h.

3.5 *Ultrasonic Sensor Method*

Shambhu Hegde, Harish V. Mekali and Golla Varaprasad published a paper in 2014 titled "Pothole Detection and Vehicle-to-Vehicle Communication". This study aims to identify potholes with a minimum depth of 1 inch and communicate this information to moving vehicles within a 100 m radius. The proposed system uses ultrasonic sensors to measure temperature and light intensity. The ultrasonic sensor used is LV-MaxSonar-EZ0. It also uses a Zigbee module and an MBED microcontroller based on NXP LPC1768 for communication [14]. Ultrasonic sensors work on the sonar principle. Pothole dimensions can be identified based on the time it takes for the ultrasound wave to travel back after bouncing back from the object. A vehicle with a connected sensor forms a transmission system. The microcontroller receives the sensor output and uses the motor drivers to control the motors. In addition, it sends data to the receiving vehicle. The system was tested in real time on Bangalore roads. If a pothole is detected, it will be displayed on the microcontroller screen [15, 19]. The disadvantage of the system is that the communication system is limited to only two vehicles.

4 Implementation

The system for pothole detection involves a hardware component, a server component which is the database and a mobile application. The hardware component consists of a Microcontroller namely, the ESP32-WROOM-32D which has a Wi-Fi and Bluetooth module. It is also equipped with an Ultrasonic sensor, HCSR04 and a GPS module, Neo-6M as shown in the Fig. 1. The connections for these components is as depicted in Fig. 2. The data picked up by the sensor and the GPS module when a vehicle is driven close to a pothole, is sent to the Firebase Database which will store the location coordinates as well as the dimensions of the pothole. This information is updated real-time and is reflected in the Mobile application. Users of this app can view the routes of various destinations on the map along with an animated imagery of potholes which informs the users about the routes which are not safe or are heavily damaged.

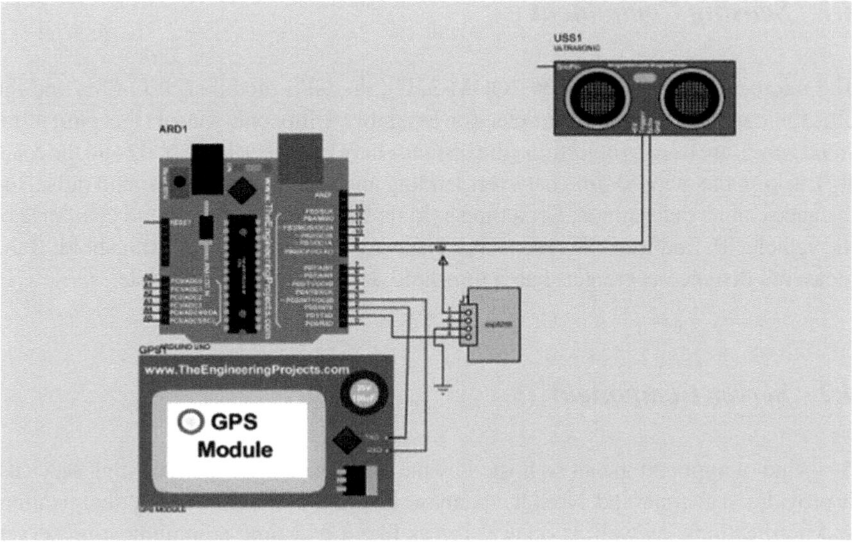

Fig. 1 Pin diagram of the hardware

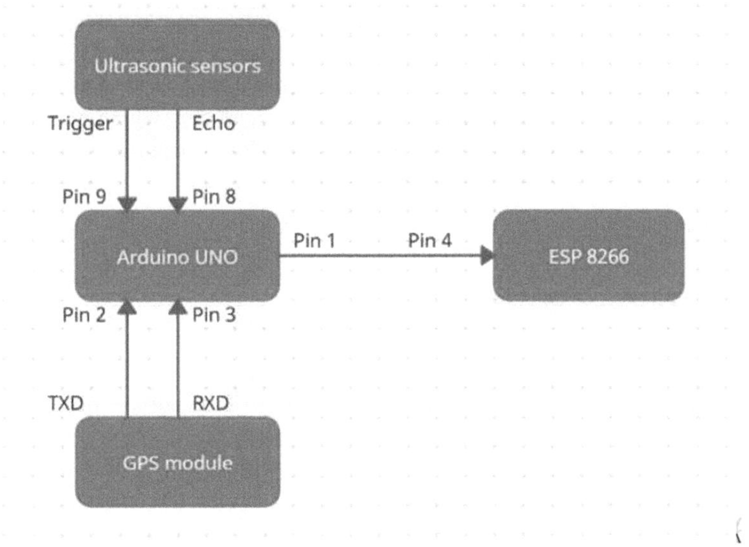

Fig. 2 Block diagram of the hardware

4.1 Sensing Component

The microcontroller (ESP32-WROOM-32D), the GPS module (NEO-6M) and the ultrasonic sensor constitute the detection assembly. Ultrasonic sensors that emit ultrasonic waves are used to determine the distance between the vehicle body and the road. By tracking the elapsed time between sending and receiving an ultrasound pulse, the distance can be determined. Set a threshold that depends on the ground clearance of the vehicle. To find traps, the measured distances are compared to a threshold. If the measured distance is greater than a threshold, it is considered a pothole.

4.2 Server Component

Any kind of application can be hosted by the Firebase collection of hosting services. It provides real-time and NoSQL database hosting, content, social authentication, and notifications, as well as other services like a real-time communication server. Using the WIFI module, the pothole's dimensions from the ultrasonic sensor and its GPS coordinates from the GPS module are sent to the firebase. The GPS receiver is constantly receiving information regarding its location coordinates, using this information the database is checked for any data around the given location coordinates. Any discovered information is obtained by the processor from the database via the Wi-Fi module and displayed on the user vehicle's smartphone. The same are reflected in the map app which. The same is reflected in the map app, which warns the user ahead of a pothole with a visual alert. Municipal employees can use this information as well for urgent maintenance or repair.

4.3 Software Component

The model detects the potholes using ultrasonic sensors, where a sensor will be attached to the vehicle which collects the dimensions of the pothole detected during the time of travel. The GPS module will then record the location coordinates of the pothole and the data from both these modules will then be sent to the database. This solution provides real time updated information on the whereabouts of potholes which increases accuracy. The app has been designed in the aspect of user needs and requirements, it consists of a Map of all the states, district and locality it is similar to the format of popular e-maps. The entire process of the system is shown in Fig. 3. The interface design is extremely user friendly because of the familiar format of the map and in order to indicate the potholes along the way, animated images of potholes will be shown on map for easy readability. The sensing components with the connections are shown in Fig. 4.

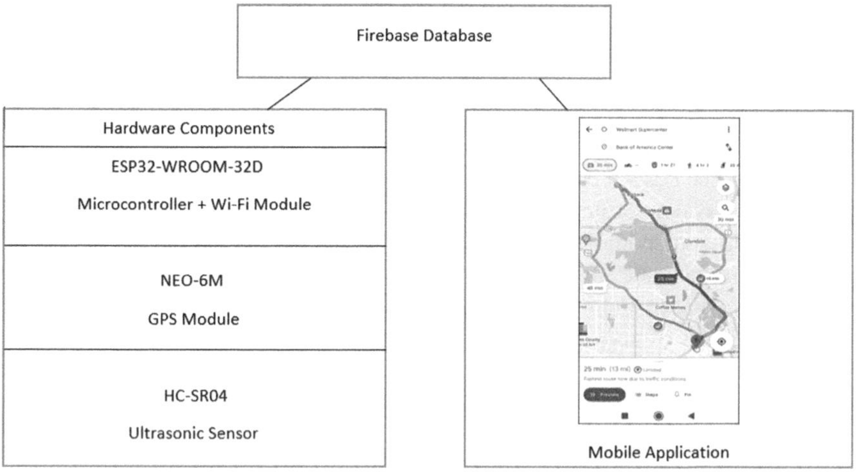

Fig. 3 Block diagram of the process

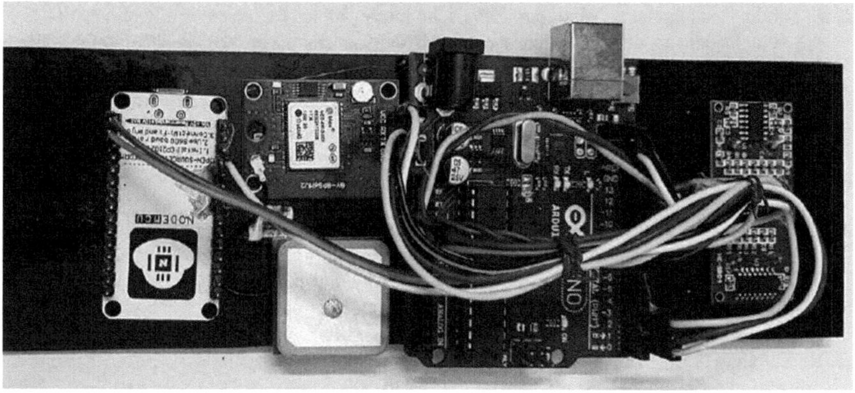

Fig. 4 Sensing component

5 Conclusion

Potholes are a significant road hazard that can cause damage to vehicles and pose a safety risk to drivers. A wide range of methods and automations have been invented to detect and navigate around potholes, including traditional approaches that rely on human inspection or visual sensors, as well as more recent techniques that use various sensors and machine learning algorithms. The existing systems included the use of Image based pothole detection, Sensor based pothole detection and 3D reconstruction-based method.

The system's hardware components, including ultrasonic sensors, microcontrollers, and a database, were successfully integrated and tested in a laboratory setting. The results showed that the system accurately detected and mapped potholes in real-time, providing location coordinates that were updated in the database and made available to users via an app. The system's potential benefits are significant, including improved road safety for drivers, cyclists, and pedestrians, reduced vehicle damage and maintenance costs, and increased efficiency of maintenance and repair efforts. The proposed system offers a practical solution to address these issues by providing realtime updates on pothole locations through an app, enabling drivers, cyclists, and pedestrians to navigate safely and efficiently on roads and highways. Moreover, it enables maintenance and repair crews to identify and repair potholes more quickly, reducing the risk of accidents and injuries caused by potholes. The need for the proposed system is further supported by the increasing demand for infrastructure monitoring and maintenance technologies in the transportation sector.

References

1. S. Sabde, P.S. Pagala, K. Mudaliar, K.S. Ashwini, Scalable machine learning and analytics of the vehicle data to derive vehicle health and driving characteristics, in *2021 International Conference on Disruptive Technologies for Multi-Disciplinary Research and Applications (CENTCON)* (2021)
2. Study of causes of potholes on bituminous roads—a case study by Dr. Umesh Sharma, Er. Abhishek Kanoungob
3. Table 1A.9 state-ut-city.pdf https://ncrb.gov.in/sites/default/files/adsi_reports_previous_year/Table%201A.9%20state-ut-city.pdf
4. E.J. Reddy, P.N. Reddy, G. Maithreyi, M.B. Chandra Balaji et al., Development and Analysis of Pothole detection and Alert based on NodeMCU, in *2020 International Conference on Emerging Trends in Information Technology and Engineering (ic-ETITE)* (2020)
5. S.-K. Ryu, T. Kim, Y.-R. Kim, Image-based pothole detection system for ITS service and road management system. Article ID 96836 (2015)
6. R. Suhail, H. Boparai, F. Ahmed, Automated sensor based pothole detection system for preventing unfortunate causality. IJERT **9**(07) (2020)
7. A. Ahmed, M. Ashfaque, M.U. Ulhaq, S. Mathavan, K. Kamal, M. Rahman, Pothole 3D reconstruction with a novel ımaging system and structure from motion techniques
8. K. Bansal, K. Mittal, G. Ahuja, A. Singh, S.S. Gill, DeepBus: machine learning based real time pothole detection system for smart transportation using IoT. Internet Technol. Lett. **3**, e156 (2020). https://doi.org/10.1002/itl2.156
9. R. Chandana, S. Jilani, S.J. Hussain, Smart surveillance system using thing speak and Raspberry Pi. Int. J. Adv. Res. Comp. Commun. Eng. **4**(7), 214–218 (2015)
10. M.M. Garcillanosa, J.M.L. Pacheco, R.E. Reyes, J.J.P. San Juan, *Smart Detection and Reporting of Potholes via Image-Processing using Raspberry-Pi Microcontroller* (IEEE, 2018)
11. R. Du, G. Qiu, K. Gao, L. Hu, L. Liu, Abnormal road surface recognition based on smartphone acceleration sensor. Sensors **20**, 451 (2020). https://www.mdpi.com/1424-8220/20/2/451
12. M.U.U. Haq, M. Ashfaque, S. Mathayan, S. Mathavan, K. Kamal, A, Ahmed, Stereo-based 3D reconstruction of potholes by a hybrid, dense matching scheme. IEEE Sens. J. **19**, 3807–3817 (2019). https://doi.org/10.1109/JSEN.2019.2898375
13. R. Tsai, A versatile camera calibration technique for high-accuracy 3D machine vision metrology using off-the-shelf TV cameras and lenses. IEEE J. Robot. Autom. **3**(4), 323–344 (1987). https://doi.org/10.1109/JRA.1987.1087109

14. R. Eaton, R. Joubert, E. Wright, *Pothole Primer: A Public Administrator's Guide to Understanding and Managing the Pothole Problem.* http://www.crrel.usace.army.mil/library/specialreports/SR81-21.pdf
15. S. Hegde, H.V. Mekali, G. Varaprasad, Pothole detection and ınter vehicular communication, in *2014 IEEE International Conference on Vehicular Electronics and Safety (ICVES)*
16. S.Das, M.R. Pooja, K.S. Anusha, *Implementation of Women's Self-Security System Using IoT-Based Device. Sustainable Computing: Transforming Industry 4.0 to Society 5.0* (Springer International Publishing, Cham, 2023), pp. 87–98
17. C. Sinchana, et al., Artificial ıntelligence applications in defense, in *Inventive Communication and Computational Technologies: Proceedings of ICICCT 2019* (Springer Singapore, 2020)
18. S. Das, M.R. Pooja, K.S. Anusha, IOT based case study: applications and challenges, in *2021 5th International Conference on Computing Methodologies and Communication (ICCMC)* (IEEE, 2021)
19. R.K. Pattanayak, et al., Implementation of E-commerce application with analytics check for updates, in *Applications of Computational Intelligence in Management & Mathematics: 8th ICCM, Nirjuli, AP, India,* vol. 417 (2023), p. 51
20. N. Soujanya, et al., Application prototypes for human to computer ınteractions, in *Sentiment Analysis and Deep Learning: Proceedings of ICSADL 2022* (Springer Nature Singapore, Singapore, 2023), pp. 27–36

Integrating Topology and Geospatial Knowledge for Mapping Road Network Layers from High-Resolution Remote Sensing Images

Madhumita Dey and Bharath H. Aithal

Abstract The active research in artificial intelligence has inspired a radical transformation in problem-solving capabilities across various domains. Among all, it has proved its effectiveness in the remote sensing research community wherein researchers are utilizing different deep learning techniques for remote sensing scene understanding. Meanwhile, with high-resolution Earth observation data availability, automatically extracting spatial information has become a vital challenge. Additionally, mapping spatial entities such as road features is crucial for decision-making and improving transportation systems, urban planning, disaster rescue systems, and many more. This study proposes a novel deep learning-based convolutional neural network pipeline for road extraction and vectorization. The novelty of the proposed work lies in eradicating the occlusion problem faced due to the shadow of trees, tall buildings, or vehicles for road extraction tasks. The proposed model achieves 85.14% mIoU and 83.56% MCC on unseen test datasets compared to other state-of-the-art architectures.

Keywords Road extraction · Road vectorization · Segmentation · Deep learning · Remote sensing · Orthophoto images

1 Background

Advancements in artificial intelligence (AI) and the Internet of Things (IoT) have drawn new challenges and opportunities in the domain of geospatial and remote sensing. IoT devices, for instance, sensors and satellites, provide vast amounts of data at various resolution scales that are impossible to handle manually [34]. Incorporating these techniques has enabled researchers to comprehend complex real-world challenges with time optimization. The advancements, specifically in attaining high

M. Dey · B. H. Aithal (✉)
Energy and Urban Research Group, RCG School of Infrastructure Design and Management, Indian Institute of Technology Kharagpur, Kharagpur, West Bengal 721302, India
e-mail: bharath@infra.iitkgp.ac.in

B. Pradhan and S. Mukhopadhyay (eds.), *IoT Sensors, ML, AI and XAI: Empowering A Smarter World*, Smart Sensors, Measurement and Instrumentation 50,
https://doi.org/10.1007/978-3-031-68602-3_11

spatial and temporal resolution, are overwhelming [28]. Subsequently, the applications of AI in geospatial and remote sensing go beyond data collection and analysis. AI reduces manual intervention and enables users to create intelligent tools or techniques with high accuracy and speed. They can be further used to analyze intricate spatial patterns and relationships for precise mapping, land-use classification, urban planning, transport management, network routing, disaster resilience, etc. [27]. Additionally, IoT provides high-resolution and real-time remotely sensed images that can be processed to extract features, identify patterns, and detect anomalies or valuable insights, further facilitating informed decision-making at different scales. Among all others, extracting and mapping road features encompasses infrastructure maintenance and planning. The up-to-date road maps ensure informed decisions about road networks, traffic management, parking optimization, etc., thereby providing a holistic approach to transportation and urban planning [7].

1.1 Overview of Road Networks as a Spatial Entity

Road network feature extraction and mapping have received worldwide attention in recent decades. Road network maps are used to plan and manage urban and transportation systems [29]. Essentially, road networks are linear geographical features with regular shapes. They possess uniform grey distribution and have interconnected network topology (6). Roads have always been a significant determinant of urban expansion. Since the spatial development of urban spaces is directly influenced by road networks. Therefore, road networks are called the "skeleton of urban expansion" [19]. It can be characterized in high-resolution remotely sensed images based on the following:

1. Geometric features: The road is a line feature whose width does not abruptly change. Its length is always longer than its width. The junctions are represented with the signs "T", "Y", or "+". These features are directly related to the road shapes.
2. Photometric features: Roads are photometric and radiation features as they emit white light. The grey values of roads are consistent and change very slowly. However, they vary differently from neighbouring features such as trees, buildings, and other non-road features.
3. Topological features: Roads have intersections and are not suddenly broken but continuous features.
4. Texture features: The textual features are somewhat homogeneous compared to the neighbourhood's spatial distribution of pixel grey levels.

Road networks integrate accessibility, connectivity, and transport suitability within the cities. Road research is paramount for extracting geospatial features for updating or creating databases or geoportals essential for urban growth pattern analysis, city planning, navigation, map updating, vehicle routing, road monitoring, etc. [3]. It involves an intelligent understanding of images and road characteristics.

Road extraction utilizing remote sensing images has always been categorized as semantic segmentation in image processing and feature extraction [10]. Here, in road extraction, semantics refers to allocating class labels to individual pixels, and segmenting means translating pixel-to-pixel knowledge to an image. Specifically for roads, semantic information must include the properties of road structures such as topology, texture, etc. Existing methodologies consider road mapping into two categories- road extraction and vectorization. Road extraction is further classified into two aspects. First is road centerline extraction [32], here, the goal is to generate single-pixel-based strip information from the road centerline. They might have a particular buffer to provide width information based on a given threshold value. The latter is road area extraction [13], which aims to provide pixel-to-pixel road information. Both methods are crucial when it comes to delivering road extraction. The former can be obtained by using conventional morphological operations. However, three individual-level tasks are required for road area extraction, i.e., road surface segmentation, centerline extraction, and edge detection [5]. The obtained binary maps of road and non-road values or probability maps are further vectorized. The final road vector maps are essential to various applications in several domains for decision-making. Different morphological algorithms, skeletonization, etc., were used in the literature to carry out road vectorization.

1.2 Challenges in Mapping Road Networks Using High-Resolution Images

In the era of smart cities and autonomous driving systems, road networks are integral. Additionally, with technological advancement, specifically in IoT devices, such as unmanned aerial vehicles (UAVs), we have overcome the shortcomings of high-resolution remote sensing data. UAV images facilitate high-resolution images with high spatial and spectral information at a better cost but with several challenges. As mentioned, roads appear elongated and slender in high-resolution remote-sensing images and often exhibit similar spectral reflectance to building features. Besides, they get obscured due to tall trees and high-rise buildings, which hinders the automatic extraction of feature information. To solve this issue, it is essential to consider the geometric and topological information of the road networks. Along with the feature information, it is fundamental to understand that the road may be straight, curved, or diverging; therefore, edge information plays a significant role.

Meanwhile, output vector road maps must include geographical coordinates for precision mapping. Consequently, with the localization information, several pieces of data can be generated, such as the type of neighborhood location, categorization of road type, and many more. These data are critical regarding urban land use applications and transport management. Therefore, they need special attention while performing road vectorization.

1.3 Literature Review

This section presents an exhaustive review of the methods and algorithms developed for road extraction and vectorization tasks. We have explicitly focused on two categories: conventional methods and AI-based methods. These procedures have significantly contributed to the road extraction and vectorization tasks. Here, we aim to analyze and compare these methods at both quantitative and qualitative scales and provide valuable insights with refined prospects for the proposed task.

1.3.1 Conventional Methods

The conventional road extraction and vectorization techniques were semi-automatic and were based on morphological algorithms. Karimi and Liu [14] have explored region-growing algorithms followed by edge detection techniques to extract road features from high-resolution satellite images. Laplacian Gaussian edge detector has been adopted to identify road features from high-resolution satellite images. Findings showed that the adopted method couldn't eliminate the noisy edges, leading to incoherent output.

Christophe and Inglada [6] utilized spectral angles, which are rich in color information for road pixels. In this spectral angle, line detection with a gradient filter was applied to enhance the pixel knowledge of the algorithm. The algorithm could extract roads from radar images but failed for complex road networks.

An integrated approach has been adopted by Poullis and You [23] that merges perceptual growing theory with the segmentation algorithm to address road detection and classification. The authors refined the classification of road features by combining the Gabor filter with tensor voting. They have encoded the data that eliminates manual intervention for thresholding. Further, gaussian filters are utilized for automatic road information extraction to create final road segments. However, the visual findings showed that the results could have been more accurate for the satellite images due to the spectral reflectance distribution of roads and background pixels.

Mirnalinee et al. [22] exploited road feature properties in a neural network architecture using a probabilistic SVM and DSM approach. The segmentation approach has also been adopted as a post-processing technique to refine the output results. The authors have utilized a constraint satisfaction neural network with a dynamic window that adds complimentary feature information such as edges and regions. The findings signify that the model could efficiently extract urban roads with larger pixel widths.

Yang et al. [31] adopted a 3D road detection model based on mobile laser scanning point cloud data. Essentially, the model segments point cloud data into road cross sections, which is further refined using a moving window. The model ensures that the obtained road cross-sections are globally consistent and locally similar. However, fixing the length of the moving window is complicated based on the variations of local and global data points for certain road areas.

Senthilnath et al. [25] have explored the computational complexity of GPU using a normalized cuts algorithm for road extraction. The author found that using an image size of 900 × 900 pixels significantly reduces the time complexity of the model performance. Furthermore, the normalized cut algorithm considers inter-group and intra-group characteristics of the road features, resulting in better extraction capabilities.

Maboudi et al. [18] employed a linearity index approach that eliminates road segments from other objects concerning structural, spectral, textural, and contextual information. They considered an edge-preserving guided filter that smoothens the road segments to reduce heterogeneity. The tensor voting approach filled the missing parts of the output road binary map and was finally vectorized.

Hong et al. [12] proposed a convolutional features-based road extraction technique that utilizes side-output image fusion to obtain road probability maps. Multi-scale and multi-level information was fed to the network to improve road extraction abilities. Furthermore, road centerlines were vectorized to integrate spatial attribute information to facilitate various application purposes. However, they couldn't generate accurate road width due to mixed pixels. Therefore, past literature on conventional methods failed to develop precise road information. They were primarily based on manual processing and customized tensor-voting or graph-cut methods. Subsequently, conventional methods are limited to coarse-resolution satellite images and simpler road networks.

1.3.2 AI-Based Methods

With high-resolution remote sensing images and advancements in AI-based methods, road extraction research has accelerated in the current era. AI technologies have revolutionized the remote sensing domain with developments in machine learning, IoT devices, and especially deep learning, proving their efficiency in a wide variety of applications. These deployments are turning semi-automatic techniques into automatic techniques with zero human interventions. Manandhar et al. [20] suggested an automated road extraction technique by updating the Volunteered Geographic Information (VGI) and deep neural networks. The initial output from VGI was reduced to an 8 × 8 pixel size and further fed to a convolutional neural network architecture for road classification. They performed post-processing to maintain a continuous road network consisting of spectral, geometric, and topological information.

Xin et al. [30] proposed the DenseUNet model with minimal parameters and robust characteristics compared to the existing Unet model. The model consists of dense blocks that accept information from the previous input layers, transfer characteristics of feature maps, and accelerate computation. Gao et al. [9] modified the residual network and Unet and proposed a deep residual convolutional neural network integrating post-processing steps. The architecture's novelty lies in using a blind tensor voting algorithm based on math morphology that joins misclassification regions that occurred due to visual occlusion. The line road features are integrated with the output to connect the broken parts or spurs. However, the visual findings ascertained that advanced post-processing techniques are required to accurately extract roads, especially in urban regions.

Senthilnath et al. [26] applied three deep learning algorithms, namely, conditional generative adversarial networks (cGAN), Cycle GAN, and fully convolutional network (FCN) with an ensemble classifier. They used a transfer learning approach to increase the model's generalizability in the road extraction task. The ensemble classifier combines the outputs and applies the tensor voting method to refine the final road regions.

Abdollahi et al. [1] presented a simultaneous automatic road extraction and vectorization approach using improved deep learning methods. The architecture consists of interlinked Unet networks, wherein the first improves the extraction capability to produce coherent predictions. The latter maps the final road regions, acquiring knowledge from previous feature maps. They used dense dilated spatial pyramid pooling that provides an enlarged field of view with dense scale flexibility. They employed median frequency balancing based focal loss weighted function that investigates complex samples provocatively. However, the visual results address the need to incorporate topological knowledge into the model architecture. Chen et al. [4] implemented an end-to-end framework consisting of two modules: node proposal network and node connectivity. The node proposal network comprises road nodes derived from road masks essentially to improve road continuity. The latter shares the backbone parameters with the former and refines the final road vector map by projecting the road nodes onto the edges and inserting projection points. They further demonstrate the failure case wherein the proposed model couldn't predict complete road segments due to obstruction by trees and high-rise buildings.

Yang et al. [32] proposed a convolutional neural network based on spatially enhanced and densely connected Unet architecture. The network combines global and multi-level spatial information of road networks fetched from the densely connected CNN blocks. The authors used a joint loss function that ruled the hard-to-demarcate road pixels. A semi-automatic pipeline for 3D road boundary extraction using an MLS point cloud has been demonstrated by Mi et al. [21]. Firstly, pre-processing methods are applied to localize the road areas by eliminating the off-ground points. Curb points were obtained using the super voxel method to refine the road boundaries. Further, they were clustered and represented as a series of cubic Bezier curves, and finally, the missing points were filled using the Kalman filtering approach.

He et al. [11] proposed an automatic method to improve the topology of predicted road vectors by integrating object detection and node iterative search. They located the stacked and average road intersections through the object detection method. The stacked road points were matched with the ground truth nodes to refine the results further. Finally, they used the iterative correction method to eliminate the incorrect topology information via road nodes. The authors mention using the Libra R-CNN approach to match the areas with significant discrepancies. Hu et al. [13] explored the potential of transformers integrated with CNN. The authors proposed a network based on multiscale deformable vision transformer knowledge capable of exploring global feature map dependencies. They designed an offset component that acquires information about narrow and elongated road shapes and their distribution in selected regions. They further utilized residual blocks with depth convolution blocks to investigate local feature information and Gaussian error linear units with a multilayer perceptron, resulting in more computational resources. Liu et al. [16] addressed the issue by investigating the power of Swin transformers with CNN architectures by combining spatial and channel-separable modules that capture integrated road features. However, the model provides scope for further improvement and tweaking. Khan and Singh [15] recently proposed the Unet model with the BRISQUE method and adjusted feature spaces. Findings suggested that the model performed well on complex road structures but failed when roads were obstructed due to shadowing.

Comparing the various techniques developed for road extraction and vectorization from the literature survey, it can be inferred that the existing models could surpass the accuracy for extracting road features; however, discrepancies persist due to occlusion. Additionally, integrating transformers with convolutional layers resulted in higher computational resources with slight accuracy increment. Therefore, accurate road extraction and vectorization with spatial coordinate reference remains challenging in this context. Gaining insights from the comprehensive literature review, this study leverages the power of CNNs to gain intricate details in the form of feature maps. We propose a cost-effective and automatic pipeline for road extraction and vectorization. The output vector maps consist of spatial information that could be utilized in several decision-making.

2 Method

The proposed pipeline has been categorized into three: (i) dataset preparation, (ii) road feature extraction, and (iii) road vectorization based on the principles of remote sensing and deep learning. The individual entities are significant in achieving the overall objective of the work. All three steps have been demonstrated in Fig. 1 with the necessary details pertaining to the separate entities. The workflow has been extensively tested on unseen test 3-channel orthophoto images. Computationally, this workflow is efficient compared to others regarding model configuration and output generation. This pipeline could be further adapted for any slender-like feature to obtain accurate feature information in the context of geographical details.

2.1 Dataset

In this study, we acquired drone-captured aerial images from various regions of Karnataka, India, as shown in Fig. 2. The acquired 884 orthophoto images are of varied sizes, comprising rural and urban road networks. The datasets are captured as RGB images and have a very high resolution, with a 3 cm spatial resolution, and each tile is 3200*3200 pixels in size. The dataset is acquired in the context of the occlusion problem.

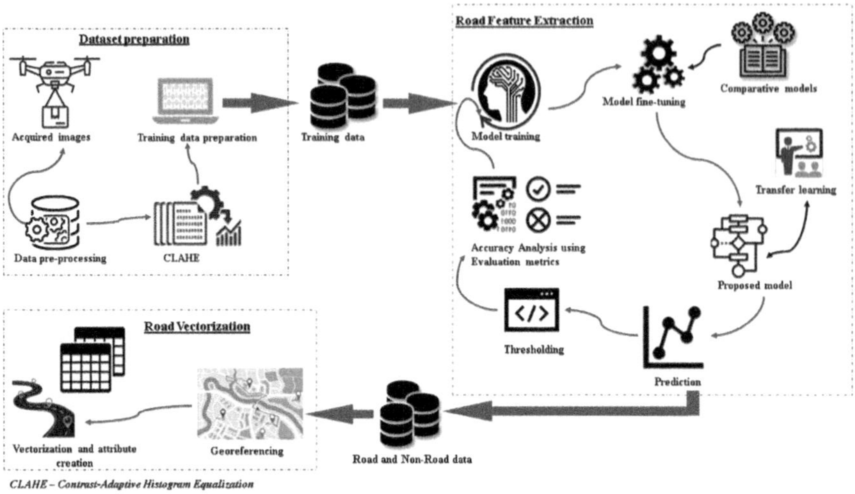

Fig. 1 Overall workflow adapted in this study

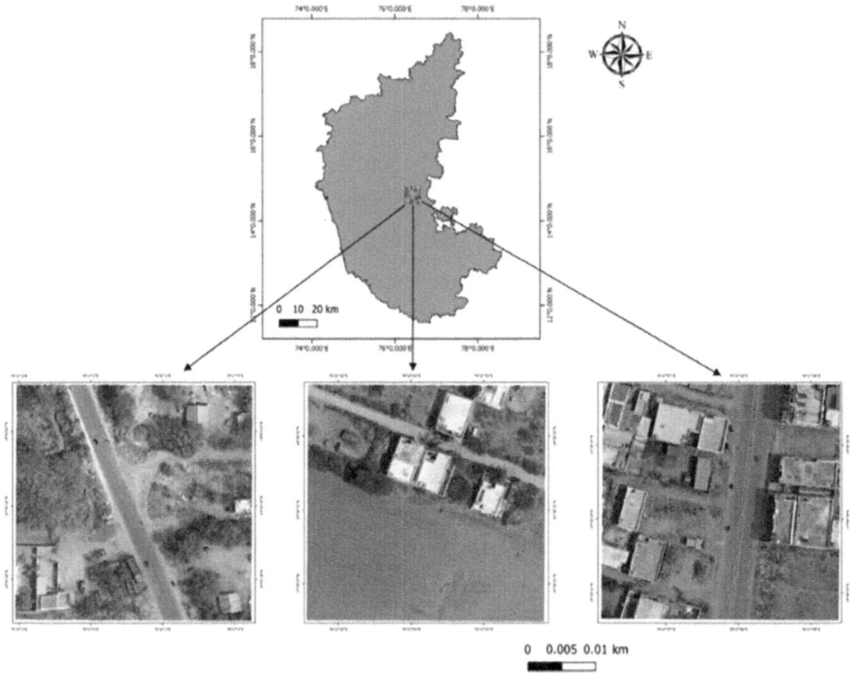

Fig. 2 Orthophoto images of this study

The dataset is further pre-processed based on remote sensing and image processing principles. We have used CLAHE [33] as an image pre-processing algorithm to improve the contrast of the images that sometimes deteriorate because of several reasons, such as weather conditions, shadows, etc. CLAHE is an acronym for contrast-limited adaptive histogram equalization as a substitute for adaptive histogram equalization. It reduces the errors caused by local contrast and addresses the limitations of noise amplification. The reason for using CLAHE as a pre-processing algorithm is that it works specifically on the feature without considering other elements present in the given image. The algorithm has two parameters, block size and clip limit, that are adjustable according to our requirements. This study used a clip limit of 40 obtained on a trial-and-error process.

After performing the image pre-processing step, the dataset was divided into training, test, and validation sets to feed the architecture further. The images are resized to 512*512 to speed up the training process and reduce the computational efficiency. We have used a split ratio of 75:15 on 871 images, and the remaining 13 images were used for the testing set, resulting in 740 training images and 131 validation images.

2.2 Network Architecture

Over the last decades, extensive work has been done involving road extraction and segmentation. However, through vast literature, it has been studied that none of the methods has directly fused topology knowledge, such as road structure information, into the model architecture. In this context, we have designed convolutional and deconvolutional blocks that feed the edge information related to various road structures and pixel information directly to the feature map, as shown in Fig. 3.

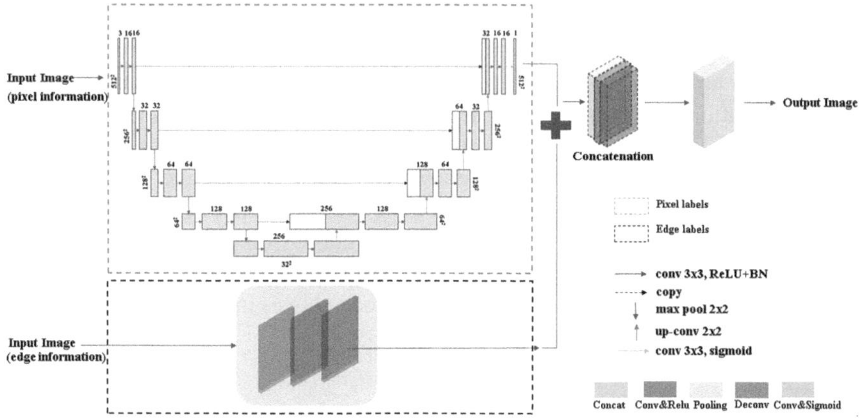

Fig. 3 Proposed architecture adapted for road extraction

The architecture accepts the image's pixel and edge information simultaneously and passes through convolutional and deconvolutional blocks to process the feature maps and generate output feature maps. We have adopted the Unet architecture developed by Olaf [24] for its outstanding performance in semantic segmentation. Unet has already established itself as a superior architecture in terms of its model performance and computational resource utilization. Meanwhile, [5] mentioned that three individual-level tasks are required to extract road information and perform road vectorization further, i.e., road surface segmentation, centerline extraction, and edge detection. The architecture comprises encoder blocks of filter sizes 16, 32, 64, 128, and 256, followed by max-pooling operations, activation functions, bottom layer, and decoder blocks. The encoder-decoder blocks essentially decode feature information, reconstruct with spatial intricacy, and generate final segmentation maps. The flow between encoder-decoder layers is established by skip connections, which maintain the flow of information between the respective channels by preserving spatial information simultaneously. The convolutional layers with 3*3 filter size are expressed as Eq. 1.

$$S(i,j) = \sum_{p=1}^{P}\{\sum_{m=0}^{2}\sum_{n=0}^{2} I(i+m, j+n) * F(m,n)\} + b_k \tag{1}$$

where b_k is the bias of k-th filter shared among all values of (m, n), $S(i, j)$ is the value at each position (i, j) in the resultant feature map. $I(i + m, j + n)$ is the pixel value at a location $(i + m, j + n)$ in the given input image. $F(m, n)$ is the filter coefficient at location (m, n).

However, since our case is limited data, it has potential drawbacks, such as leakage of spatial information through skip connections. Additionally, overfitting has also been observed to affect the learning capability in diverse scenarios. Therefore, to rule out the same, we have fused convolutional and deconvolutional blocks comprising edge information to the model architecture. The output from the Unet architecture and the output from the fused layer have been merged using concatenation followed by one convolution block. In the final layer, we used a 1*1 convolutional block that translated 16 layers into a single layer and sigmoid activation function for binary output generation. The predicted output map generated consists of probability values in the range of 0 to 1 that have been translated using a binary thresholding algorithm.

Further, we have used a binary weighted cross-entropy loss function to regulate the model's performance. This loss function imperatively reduces biases in the case of binary segmentation, here, road and non-road class. Compared to other existing loss functions such as binary cross-entropy, Jaccard coefficient, and others, weighted cross-entropy considers the hard-to-demarcate road classes into consideration rather than focusing on simpler ones.

2.3 Georeferencing

In general, the generated segmentation maps from any AI-based model lack explicit geographical information, which may be because the models are trained on pixel-level information only. They do not operate on geographical coordinates. However, for the outputs to be utilized in a real-world spatial context, they must contain geographic coordinates. In our work involving road feature extraction, the output road features must have geographic coordinate information. This enhances the spatial accuracy of the components that could be further used for multiple applications. Therefore, this work proposes an automatic procedure that directly converts the generated road features with geographic coordinates details. The pseudo-code for the same is described below.

```
Reference: 3-channel RGB image with reference geocoordinate infor-
mation
Input: 3-channel RGB image without geocoordinate information
Output: 3- channel Georeferenced RGB image
1: Begin
2: Use GDAL to read image info.
3. Check geocoordinates for the input images
4. Read geocoordinates from the reference image using Rasterio
5. Create GCP points
6. Translate GCPs to the input images
7. Update src_profile for input images
8. Write the output image with a geocoordinate system
9. End
```

2.4 Vectorization

The overall method flow chart depicts the flow of vectorization once the output road segmentation output map has been generated. The entire procedure is automatic without manual involvement. First, the output maps are georeferenced as described above. The georeferenced maps are then vectorized in context to their spatial reference, followed by attribute table creation. Each road stretch has been given a unique ID so that further applications can be easily performed. This manner of vectorization is beneficial as it is cost-effective and accurate with time optimization. As vector maps are significantly smaller in size and are fast operated, that makes the entire procedure less cumbersome and precise. The pseudo-code for the same is depicted below.

```
Input: 3-channel RGB image with geocoordinate information
Output: Vector shapefile with its attribute table
1: Begin
2: Use GDAL to read image info.
3. Create a new vector layer
```

```
4. Define a field that stores class information (here, road class)
5. Create a band mask to exclude background pixels with 0 values
6. Perform vectorization using GDAL
7. Add attribute data to the created vector feature
8. Write the output vector file
9. End
```

2.5 Metrics

The model's qualitative performance was assessed using evaluation metrics such as mIoU and MCC (Matthew correlation coefficient). These metrics are extensively studied in various domains to measure the effectiveness of solving similar problems at a large scale. mIoU is a standard segmentation metric that gives the ratio of intersection and concatenation of the two sets, i.e., true labels and predicted labels [19]. The Matthew correlation coefficient has also been considered a significant metric to validate the segmentation task in an unbalanced class distribution scenario [1]. Essentially, MCC provides a correlation coefficient between the predicted and the encountered binary labels, and its value ranges between 1 to -1. The metric parameters are defined as:

$$\text{mIoU} = \frac{1}{2} * (\frac{TP}{TP + FN + FP} + \frac{FN}{TF + FN + FP}) \tag{2}$$

$$\text{MCC} = \frac{TP * TN - FP.FN}{\sqrt{(TP + FP)(TP + FN)(TN + FP)(TN + FN)}} \tag{3}$$

where true positive (TP) specifies the number of road pixels predicted as the actual class, true negative (TN) defines the number of non-road pixels or background pixels indicated as a false class, false negative (FN) specifies the number of road pixels wrongly classified as a non-road class or background class and false positive (FP) specifies the number of non-road pixels or background pixels predicted as road class.

3 Experimental Results

Mapping road networks through high-resolution orthophoto images plays a significant role in various applications such as autonomous vehicles, urban planning, updating road maps, etc. Hundreds of literatures have already showcased this type of work. However, they lacked accurate road maps and spatial information. Therefore, this study proposes an approach for road extraction and vectorization that requires much less computational resources and time. The proposed architecture has been performed on an NVIDIA V100 GPU with 16 GB onboard memory. The model has been implemented on a Keras and TensorFlow environment using a Python scripting language. We have used Adam optimizer for its converging power with a batch size of 2 and default learning rate for 30 epochs to avoid overfitting further.

3.1 Experimental Performance of Road Area Extraction

We tested the road area extraction on our proposed model for thirteen unseen typical test images. Here, we have presented three of the best cases that correctly highlight the occlusion scenario for the road extraction task. Further, the performance of the proposed model was quantitively verified utilizing the selected performance metric, as reported in Table 1. Additionally, to validate the model's performance, we have compared it with three other state-of-the-art architectures for semantic segmentation, namely, ResUnet [8], FCN [17], and SegNet [2]. From the comparison results, it can be inferred that ResUnet performed worst in road area extraction, followed by FCN. The reason is that additional residual units in the ResUnet architecture led to the vanishing gradient problem during backpropagation. Meanwhile, FCN has been designed to convolute the pixel information using upsampling and downsampling and directly generate the segmentation output. However, SegNet performed better than the other two methods regarding both mIoU and MCC. The architecture of SegNet is similar to Unet, which follows an encoder-decoder structure. While SegNet uses pooling layers to extract feature maps and spatial details, Unet uses an encoder and decoder to obtain intricate hierarchical information. Additionally, Unet uses skip connections to transfer high-resolution feature details, aiding flow between encoder and decoder layers and not leading to missing delicate spatial information that plays a significant role in the segmentation task. On the other hand, benefiting from the decoder layer fused with edge information and contextual details has led to the outstanding performance of our proposed model. Figure 4 presents the predicted segmentation output from our proposed model. From the figure, it can be inferred that the proposed model accurately extracts roads from complex orthophoto images. Since ours is a binary segmentation problem that considers only road and non-road classes, from the literature, MCC is one of the crucial metrics for this case. MCC gives a higher value if only a high correlation exists between actual and predicted values. Our proposed model attains 83.56% MCC, which further validates the model's prediction capability. Additionally, the findings confirmed the application of the model when roads are obscured by shadow.

Even though the proposed model performed excellently on the dataset, there are some failure cases where the model could have been predicted accurately. Figure 5 presents a few examples of failure cases in complex scenarios. Image 1 is an example wherein the road got obscured by vehicles; though it predicted the other car, it failed to maintain the road connectivity for the first car. The second image demonstrates a scenario where tall trees completely obscure a rural road. It is already mentioned that our dataset comprises a smaller number of rural roads. Meanwhile, if we have trained the model with more such images, there is a high chance that the model will

Fig. 4 Visualization of the road extraction by the proposed method on orthophoto images

Table 1 Quantitative comparison of the proposed model with other state-of-the-art architectures (in percentage)

Models	mIoU (%)	MCC (%)
ResUnet	53.3	39.82
FCN	75.25	70.28
SegNet	81.82	79.05
Proposed model	85.14	83.56

perform effectively. The third case is an interesting one, where it can be observed that the model failed to maintain the connectivity for the bituminous road; however, it has predicted the rural or kaccha road that needed to be marked in the ground truth. There are a few samples that we have kept unmarked to showcase the effectiveness of our proposed model. Since the model is fed with topology information and contextual knowledge, it has gained insights into the structure and spectral behaviors of various road categories, such as rural and urban roads.

In general, after obtaining the single channel output from a segmentation model, we transform that map using binary thresholding to extract features of interest. Practically we use a value of 0.5 that is values greater than 50% were considered and rest are discarded. In this study, to check the efficacy of the proposed model, we have increased the threshold value at set checkpoints at 50, 60, 80, and 90. Figure 6 depicts the performance of the model at different checkpoints. From the findings, it can be derived that at a threshold value of 90, the model gives an overall accuracy of more than 90% with about 58% of the IoU value. IoU score inferred the percent of overlap between ground truth labels and predicted labels. Meanwhile, other scores have their usual meaning and show positive changes similar to theoretical knowledge.

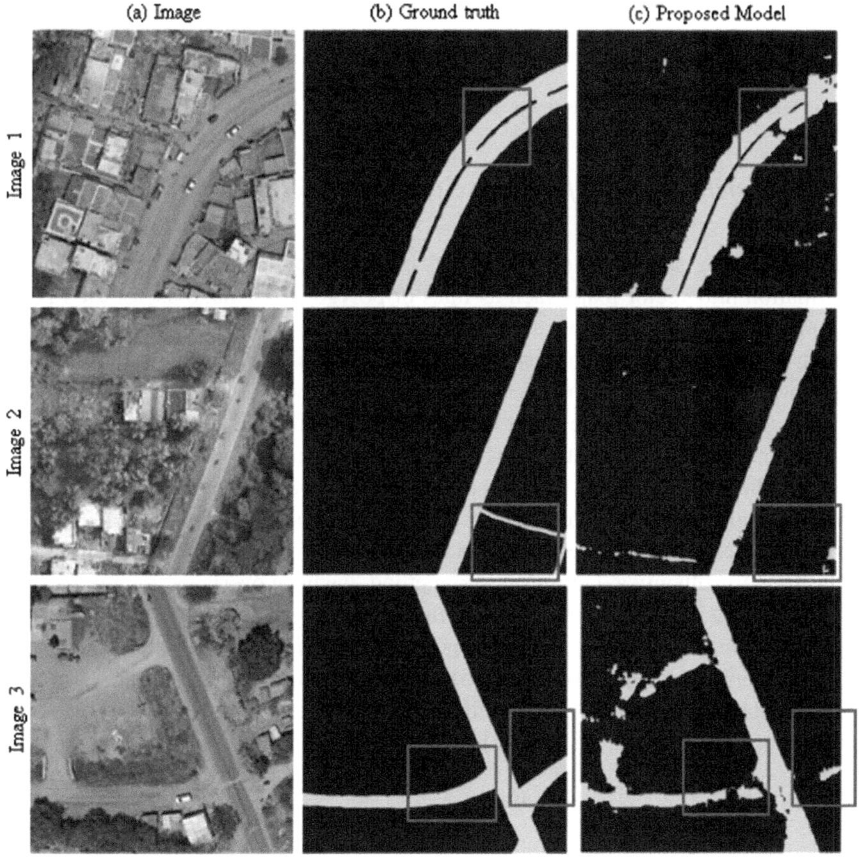

Fig. 5 Visualization of the failure cases for road extraction by the proposed method on orthophoto images

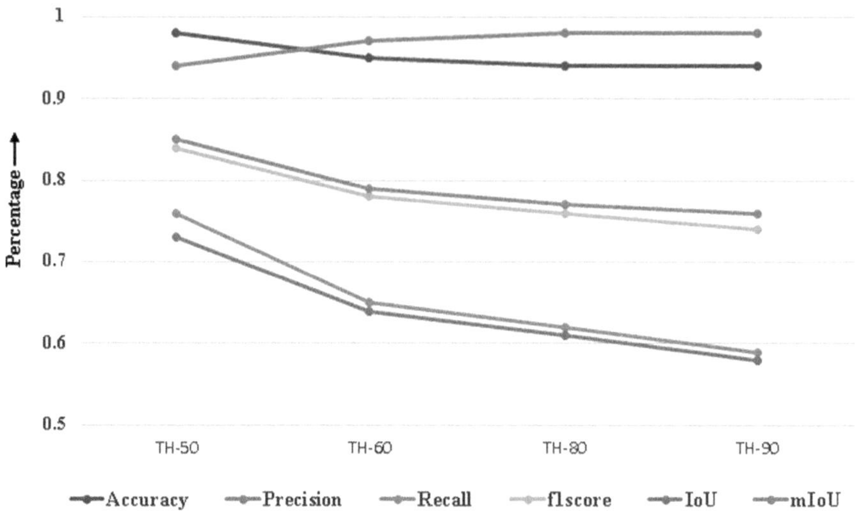

Fig. 6 Variation in evaluation metrics at different checkpoints of the threshold

3.2 *Experimental Performance of Road Vectorization*

After generating segmentation maps, it is essential to convert the rasterized road information into vector maps such as lines or polygons. To make use of the rasterized maps in a variety of applications, this procedure is significant. However, maintaining spatial information in this step is crucial. In this study, we have extracted the segmented maps in geotiff format and georeferenced them. After integrating geocoordinate information into the raster maps, we have vectorized the output maps. This procedure is entirely automatic, which is one of our concerns for further use in real-world applications. Figure 7 presents the vector forms of the segmented road maps. Once the segmented maps are vectorized, noise, such as erroneous segments, can be easily removed, and accurate vector road layers can be generated.

Further, to authenticate our proposed method, we have overlayed the final vector maps for some stretch of area on the orthophoto images. The final map shown in Fig. 8 validates the accuracy of our proposed approach. To overlay the vector maps on the orthophoto images, we used QGIS, which is open-source software. We have demonstrated our approach on a smaller stretch of the region. However, this work can also be performed at the regional or city level.

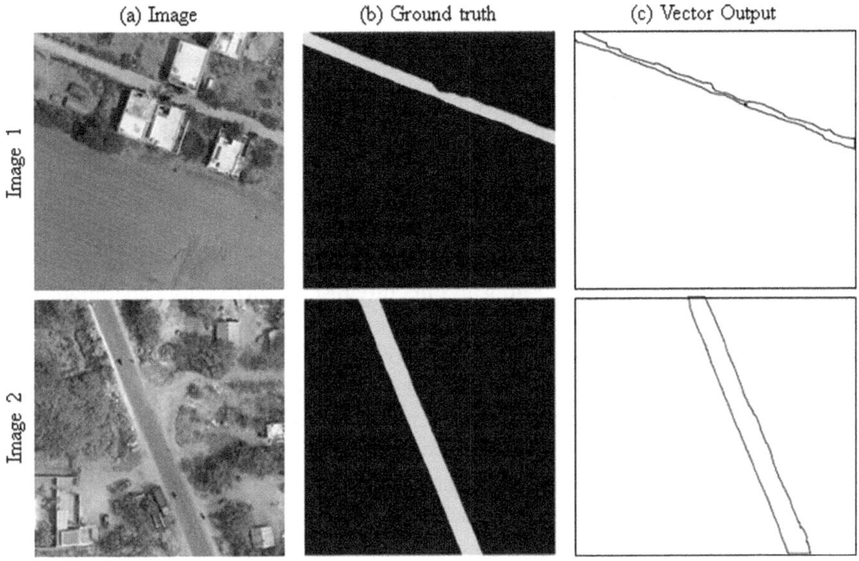

Fig. 7 Visualization of the road vectorization by the proposed method on orthophoto images

Fig. 8 Map showing a stretch of vector road map overlayed on the orthophoto image

4 Conclusions

This study is unique in presenting an end-to-end approach from scratch. Utilizing the advancements in IoT sensors and deep learning techniques, this study showcases a cost-effective and optimistic method for mapping road networks using high-resolution images. We have acquired high-resolution drone images with 3 cm spatial resolution to demonstrate the same. The proposed approach comprises two steps: road extraction and road vectorization. To perform road extraction tasks, we have proposed an end-to-end deep learning architecture fused with topological knowledge. The model has been implemented on thirteen unseen test images with complex road structures and occlusion scenarios. The demonstrated qualitative results performed on test images show that our proposed model obtained 85.14% mIoU and 83.56% MCC. Additionally, we have extended the model performance at different threshold checkpoints. Therefore, these values confirm the superiority of our model in the road extraction task. Subsequently, the obtained road segmentation maps are vectorized with geospatial coordinates integrated with them. We have overlayed the final vector road polygons on the orthophoto images and mapped them further with geo coordinates. The final output vector road layers are well-suited for various geospatial analyses. They can also be integrated with other geospatial datasets, including land use data, demographic data, etc. In other words, the proposed approach could accurately extract road features with their spatial details. Additionally, there are some advantages to the proposed method, which makes it practical for real-world applications. The proposed deep learning model utilizes a minimum number of parameters for road extraction tasks. The model could extract accurate roads in the scenarios when they are occluded by shadows. Since the vectorization has been performed using Python scripting, erroneous segments can be easily removed. Information related to any specific road pertaining to its road ID can also be extracted. Despite the advantages, there are a few shortcomings. Since the model has been trained on unbalanced data, it failed to generalize for rural roads accurately. In the future, we will introduce the model for a larger dataset and can utilize transfer learning to extricate the occlusion cases.

Acknowledgements We thank the Indian Institute of Technology Kharagpur and Ranbir and Chitra Gupta School of Infrastructure Design and Management for financial and Infrastructure support.

References

1. A. Abdollahi, B. Pradhan, A. Alamri, RoadVecNet: a new approach for simultaneous road network segmentation and vectorization from aerial and google earth imagery in a complex urban set-up. GIScience Remote Sens **58**(7), 1151–1174 (2021)
2. V. Badrinarayanan, A. Kendall, R. Cipolla, Segnet: A deep convolutional encoder-decoder architecture for image segmentation. IEEE Trans. Pattern Anal. Mach. Intell. **39**(12), 2481–2495 (2017)

3. D. Chen, Y. Zhong, Z. Zheng, A. Ma, X. Lu, Urban road mapping based on an end-to-end road vectorization mapping network framework. ISPRS J. Photogram Remote Sens. **178**, 345–365 (2021)
4. D. Chen, Y. Zhong, Z. Zheng, A. Ma, X. Lu, Urban road mapping based on an end-to-end road vectorization mapping network framework. ISPRS J. Photogram Remote Sens **178**, 345–365 (2021). Yang et al. (2022)
5. G. Cheng, Y. Wang, S. Xu, H. Wang, S. Xiang, C. Pan, Automatic road detection and centerline extraction via cascaded end-to-end convolutional neural network. IEEE Trans. Geos Remote Sens. **55**(6), 3322–3337 (2017)
6. E. Christophe, J. Inglada, Robust Road extraction for high-resolution satellite images. 2007 IEEE Int. Conf. Image Process. (5), 437 (2007)
7. D. Datta, N.I. Sarkar, Deep learning frameworks for internet of things. artificial intelligence-based. Internet Things Syst., 137–161 (2022)
8. F.I. Diakogiannis, F. Waldner, P. Caccetta, C. Wu, ResUNet-a: a deep learning framework for semantic segmentation of remotely sensed data. ISPRS J. Photogram Remote Sens. **162**, 94–114 (2020)
9. L. Gao, W. Song, J. Dai, Y. Chen, Road extraction from high-resolution remote sensing imagery using refined deep residual convolutional neural network. Remote Sens. **11**(5), 552 (2019)
10. L. Han, L. Hou, X. Zheng, Z. Ding, H. Yang, K . Zheng, Segmentation is not the end of road extraction: an all-visible denoising auto-encoder for connected and smooth road reconstruction. IEEE Trans. Geos Remote Sens. (2023)
11. J. He, J. Li, W. Li, Improving the road topological relation based on a road intersection detection method. In *IGARSS IEEE International Geoscience and Remote Sensing Symposium*, 5850–5853 (2023)
12. Z. Hong, D. Ming, K. Zhou, Y. Guo, T. Lu, Road extraction from a high spatial resolution remote sensing image based on richer convolutional features. IEEE Access **6**, 46988–47000 (2018)
13. P.C. Hu, S.B. Chen, L.L. Huang, G.Z. Wang, J. Tang, B. Luo, Road extraction by multi-scale deformable transformer from remote sensing images. IEEE Geos Remote Sens. Lett. (2023)
14. H.A. Karimi, S. Liu, Developing an automated procedure for extraction of road data from high-resolution satellite images for geospatial information systems. J. Transport Eng. **130**(5), 621–631 (2004)
15. M.J. Khan, P.P. Singh, Advanced road extraction using CNN-based U-Net model and satellite imagery. e-Prime- Adv. Electr. Eng. Electron. Energy **5**, 100244 (2023)
16. Y. Liu, T. Yuan, Y. Wang, Y. Wang, H. Zhao, VectorMapNet: end-to-end vectorized HD map learning. In *Inter Conference on Machine Learning*, pp. 22352–22369 (2023)
17. J. Long, E. Shelhamer, T. Darrell,. Fully convolutional networks for semantic segmentation. In *Proceedings of the IEEE Conference on Computer Vision and Pattern Recognition*, pp. 3431–3440 (2015)
18. M. Maboudi, J. Amini, M. Hahn, M. Saati, Road network extraction from VHR satellite images using context aware object feature integration and tensor voting. Remote Sens. **8**(8), 637 (2016)
19. D. Madhumita, H.A. Bharath, V.P. Devendra, B. Shivam, Road segmentation: exploiting the efficiency of skip connections for efficient semantic segmentation. J. South Asian Logist. Trans. **3**(1) (2023)
20. P. Manandhar, P.R. Marpu, Z. Aung, F. Melgani, Towards automatic extraction and updating of VGI-based road networks using deep learning. Remote Sens. **11**(9), 1012 (2019)
21. X. Mi, B. Yang, Z. Dong, C. Chen, J. Gu, Automated 3D road boundary extraction and vectorization using MLS point clouds. IEEE Trans. Intell. Transp. Syst. **23**(6), 5287–5297 (2021)
22. T.T. Mirnalinee, S. Das, K. Varghese, An integrated multistage framework for automatic road extraction from high-resolution satellite imagery. J. Indian Soc. Remote Sens. **39**, 1–25 (2011)
23. C. Poullis, S. You, Delineation and geometric modeling of road networks. ISPRS J. Photogram Remote Sens. **65**(2), 165–181 (2010)

24. O. Ronneberger, P. Fischer, T. Brox, U-Net: convolutional networks for biomedical image segmentation. In *Proceedings of the Medical Image Computing and Computer-Assisted Intervention*, Munich, Germany, pp. 234–241 (2015)
25. J. Senthilnath, S. Sindhu, S.N. Omkar, GPU-based normalized cuts for road extraction using satellite imagery. J. Earth Syst. Sci. **123**, 1759–1769 (2014)
26. J. Senthilnath, N. Varia, A. Dokania, G. Anand, J.A. Benediktsson, Deep TEC: deep transfer learning with ensemble classifier for road extraction from UAV imagery. Remote Sens. **12**(2), 245 (2020)
27. S.K. Singh, Y.S. Jeong, J.H. Park, A deep learning-based IoT-oriented infrastructure for secure smart city. Sustain. Cities Soc. **60**, 102252 (2020)
28. Y. Song, M. Kalacska, M. Gašparović, J. Yao, N. Najibi, Advances in geocomputation and geospatial artificial intelligence (GeoAI) for mapping. Inter J. Appl. Earth Obser. Geoinform, 103300 (2023)
29. J. Tao, Z. Chen, Z. Sun, H. Guo, B. Leng, Z. Yu et al., Seg-road: a segmentation network for road extraction based on transformer and CNN with connectivity structures. Remote Sens. **15**(6), 1602 (2023)
30. J. Xin, X. Zhang, Z. Zhang, W. Fang, Road extraction of high-resolution remote sensing images derived from DenseUNet. Remote Sens. **11**(21), 2499 (2019)
31. B. Yang, L. Fang, J. Li, Semi-automated extraction and delineation of 3D roads of street scene from mobile laser scanning point clouds. ISPRS J. Photogram Remote Sens. **79**, 80–93 (2013)
32. X. Yang, X. Li, Y. Ye, R.Y. Lau, X. Zhang, X. Huang, Road detection and centerline extraction via deep recurrent convolutional neural network U-Net. IEEE Trans. Geos Remote Sens. **57**(9), 7209–7220 (2019)
33. W.N. Yussof, M.S. Hitam, E.A. Awalludin, Z. Bachok, Performing contrast limited adaptive histogram equalization technique on combined color models for underwater image enhancement. Inter J. Interactive Digital Media. **1**, 1–6 (2013)
34. Y.B. Zikria, M.K. Afzal, S.W. Kim, A. Marin, M. Guizani, Deep learning for intelligent IoT: opportunities, challenges and solutions. Comput. Commun. **164**, 50–53 (2020)

Unlocking the Potential: Machine Learning and Deep Learning in Leukemia Diagnosis with Explainable AI

Nilkanth Mukund Deshpande, Shilpa Gite, and Biswajeet Pradhan

Abstract The human body can undergo changes in blood cell structure and characteristics when affected by contaminants. Examining the tiny images of blood cells helps identify potential infections or irregularities within the body, aiming to detect diseases. Accurately segmenting these cells significantly enhances disease detection, making it more precise and robust. Microscopic analysis of blood cells plays a pivotal role in pathological examinations, facilitating the identification of specific diseases and abnormalities, crucial for diagnosing various disorders, planning treatments, and assessing their outcomes. Leukemia is a disease that produces the infections within the bone marrow of a human being. Bone marrow generates the leukocytes (White Blood Cells), erythrocytes (Red Blood Cells), and other components of blood. Due to leukemia infection, this blood-forming capacity of bone marrow is hampered. This chapter reviews the diverse applications of microscopic blood cell imaging in disease detection. It provides a brief overview of blood composition, followed by a generalized approach to analyzing microscopic blood cell images for specific medical imaging purposes. Leukemia systematics, symptoms, and different types are explored in the survey. Additionally, the chapter discusses and compares different methodologies proposed by researchers for disease detection through the analysis of microscopic blood cell images. Different machine learning, deep learning methods are explored in relation to leukemia diagnosis. Prominent phases of leukemia diagnosis, including segmentation and classification, are reviewed, and different algorithms

N. M. Deshpande
School of Electrical and Communication Sciences, JSPM University, Pune, Maharashtra, India

S. Gite (✉)
Department of Computer Science, Symbiosis Institute of Technology, Symbiosis International (Deemed University), Lavale, Pune, Maharashtra 412115, India
e-mail: shilpa.gite@sitpune.edu.in

Symbiosis Center for Applied Artificial Intelligence (SCAAI), Symbiosis International (Deemed University), Lavale, Pune, Maharashtra 412115, India

B. Pradhan (✉)
Centre for Advanced Modelling and Geospatial Information Systems (CAMGIS), School of Civil and Environmental Engineering, University of Technology Sydney, Ultimo, NSW 2007, Australia
e-mail: Biswajeet.Pradhan@uts.edu.au

B. Pradhan and S. Mukhopadhyay (eds.), *IoT Sensors, ML, AI and XAI: Empowering A Smarter World*, Smart Sensors, Measurement and Instrumentation 50,
https://doi.org/10.1007/978-3-031-68602-3_12

employed for these phases are explored. The generalized methodology of disease detection along with a survey of state-of-the-art methods. As the explanability and interpretability issues make the deep learning frameworks as black boxes, revealing these frameworks' black box nature is necessary to instill trust in the diagnosis decisions. Various explainable AI frameworks are explored in this chapter.

Keywords Leukemia · Machine learning · Deep learning · Explainable AI · Disease diagnosis

1 Introduction

Leukemia is related to blood abnormalities. Exploring different blood components provides deep insights into leukemia infection and its effects on blood components. Blood, the most vital component of the body, made up of leukocytes and erythrocytes which are also termed as white blood cells (WBC), and red blood cells (RBC) respectively, as well as disc-shaped cell fragments called platelets, and plasma. Cells and platelets collectively account for approximately 45% of human blood, with plasma accounting for the remaining 55% [1, 2]. Various blood components undergo changes in their physical properties due to infections in the body, including alterations in shape, color, size, and cell count. There are various pathological procedures for disease detection. Microscopic imaging is often used to predict and detect abnormalities and the infection of diseases within the body. Various components of blood are explored in Fig. 1. Erythrocytes, leukocytes, and platelets [3] are the main components of blood. The plasma membrane encompasses all these components of blood. Leukocytes are subdivided into granulocytes and agranulocytes, where the former have sub-components such as neutrophils, eosinophils, and basophils, and the latter include lymphocytes and monocytes. Blood plasma comprises different gases, proteins, enzymes, nutrients, wastes, and hormones. Platelets are bone marrow cell fragments. Red blood cells function in the transport of oxygen to various parts of the body organs from the lungs. Lungs get back the carbon dioxide, which is exhaled in the lungs. RBC count is used to diagnose the abnormalities in the human body due to the attack of diseases. Less count of RBC indicates anemia, while a high count indicates poly-Cynthia.

The work of White blood cells is to defend the body from infection. The various blood components are identified to learn about a person's health. Microscopic images of blood smears are analyzed to detect various diseases [4].

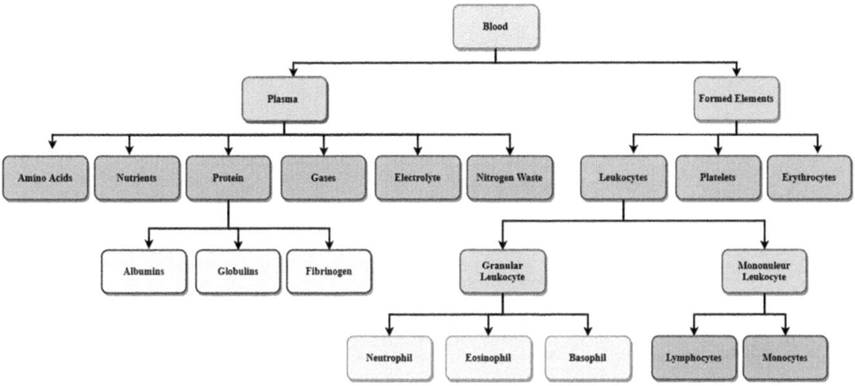

Fig. 1 Different blood components

2 Leukemia Broad Types

This is a type of blood and bone marrow cancer that originates in the bone marrow, affecting the blood-forming cells and eventually spreading through the bloodstream over time (Fig. 2). The disease typically progresses to other parts of the body. Being classified as chronic usually indicates that the leukemia spreads and grows slowly. However, Chronic Myeloid Leukemia (CML) can transition from a slow progression to a rapidly growing, acute form of leukemia that can affect almost any organ in the body. Leukemia generally affects the blood formation capacity of the bone-marrow. Leukemia begins in a bone marrow cell. The cell changes and becomes a type of leukemia cell. Leukemia cells may grow and survive better than normal cells after the marrow cell undergoes a leukemic change. Leukemia cells crowd out or suppress the development of normal cells over time. Each type of leukemia has a different rate of progression and how the cells replace normal blood and marrow cells. Leukemia is a blood infection that increases the number of blast cells [5]. These are young WBC blasts. There is very little space for RBC to flow through the blood due to the higher percentage of blasts. As a result, the percentage of RBC decreases, resulting in anemia and other abnormalities [6]. Leukemia is divided into chronic and acute, with lymphoblastic and myeloid leukemia subdivided [7].

Figure 3 illustrates various blood cells, both infected with and without leukemia. Leukemia is broadly classified as depicted in Fig. 4.

2.1 Acute Lymphocytic Leukemia (ALL)

It is also known as acute lymphoblastic leukemia and acute lymphoid leukemia, a type of blood cancer caused by the accumulation of abnormal WBCs (leukemia cells) in bone marrow. ALL spreads quickly, and replaces the functional lymphocytes with

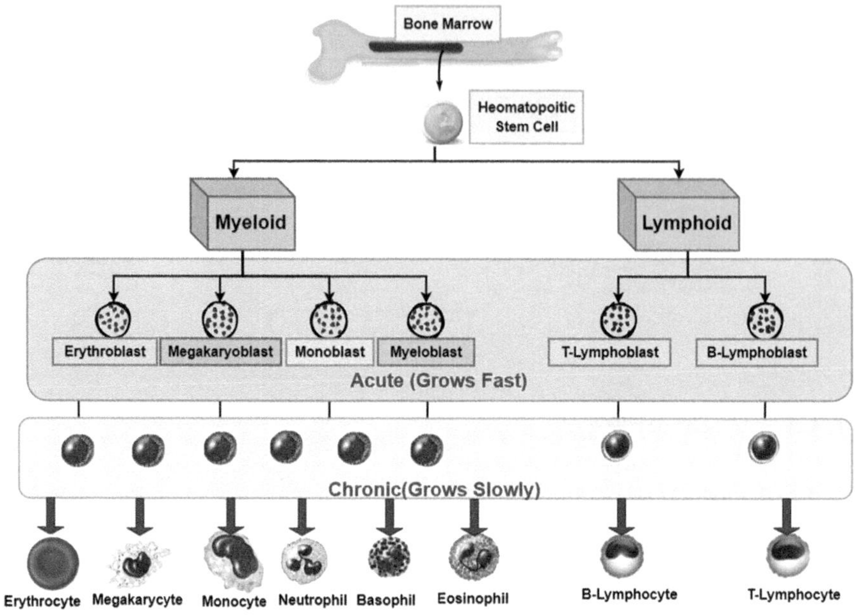

Fig. 2 Formation of chronic and acute leukemia

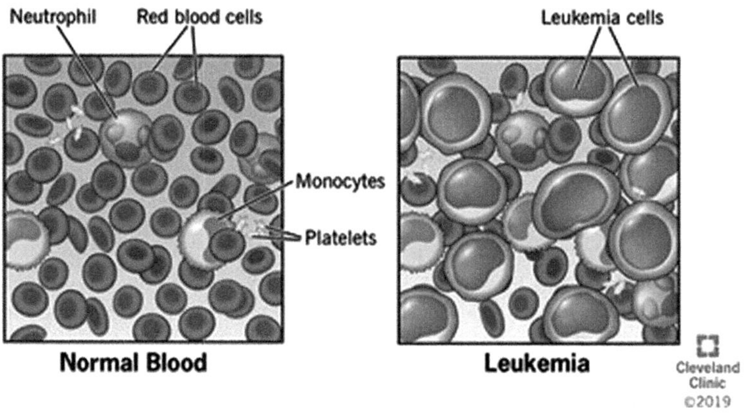

Fig. 3 Normal and leukemia cells

abnormal cell which are leukemia cells which cannot be matured in a proper way [8]. Leukemia cells approaches to other organs through the bloodstream and can get accumulated in the brain, liver, lymph nodes, and testes. In these organs the cells grow and divide. These leukemia cells' growing, dividing, and spreading can cause various symptoms [9]. ALL is typically associated with more B lymphocytes than T lymphocytes. B and T cells play active roles in protecting the body from infections

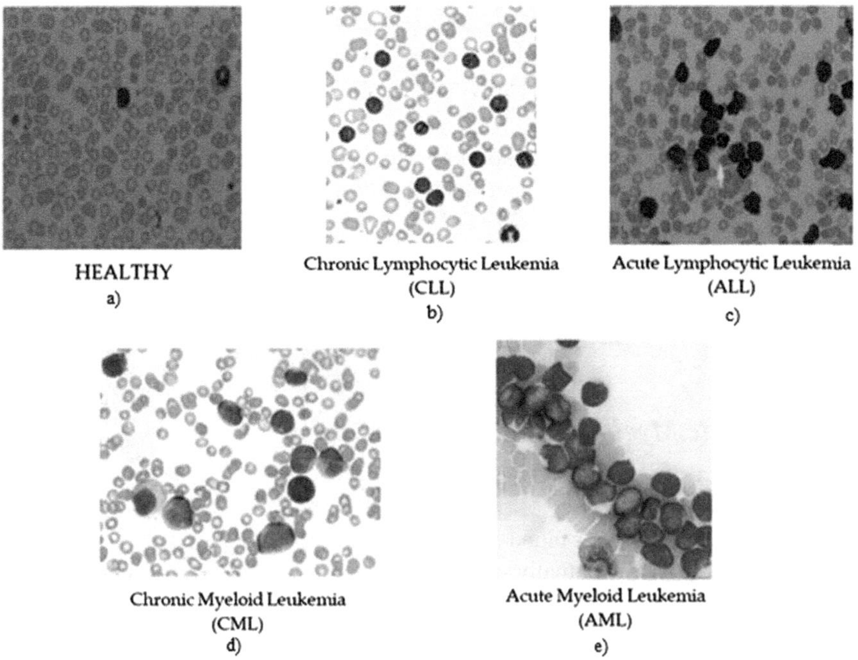

Fig. 4 Different types of leukemia

and germs and destroying infected cells. B cells, in particular, aid in preventing germ infection, whereas T cells destroy infected cells. ALL can strike at any age, but acute lymphocytic leukemia is most common in people under 15 or over 45. Although ALL accounts for most leukemia diagnoses in children under 15 (specifically, between the ages of two and four), adults rarely develop this disease [10].

2.2 *Acute Myeloid Leukemia (AML)*

It is known as acute myelogenous leukemia, acute myeloblastic leukemia, acute granulocytic leukemia, or acute nonlymphocytic leukemia, a type of blood and bone marrow cancer that proliferates. The most common type of acute leukemia is AML [11]. It happens when the bone marrow starts producing blasts, which are immature cells. Normally, these blasts develop into white blood cells. In AML, however, these cells do not develop and cannot fight infections. The bone marrow in AML may also produce abnormal red blood cells and platelets. These abnormal cells rapidly increase, and the abnormal (leukemia) cells crowd out the body's normal white blood cells, red blood cells, and platelets.

2.3 Chronic Lymphocytic Leukemia (CLL)

This slow-growing cancer begins in the bone marrow and spreads to the blood [12]. It can also infiltrate lymph nodes and organs like the liver and spleen. Chronic lymphocytic leukemia occurs when an abnormally large number of lymphocytes proliferate, crowding out normal blood cells and making it difficult for the body to fight infection. The term "chronic" refers to the disease's slow progression. The abnormal lymphocytes develop and multiply more slowly. As a result, it may take several years for a disease like CLL to become serious. Acute lymphocytic leukemia (ALL) progresses quickly in comparison. ALL may progress in a much shorter time frame.

2.4 Chronic Myeloid Leukemia (CML)

Chronic myeloid leukemia (CML) is a type of blood and bone marrow cancer. It begins in the bone marrow's blood-forming cells and then spreads to the bloodstream over time. The disease eventually spreads to other parts of the body. Being classified as chronic usually means that leukemia spreads and grows slowly. Conversely, CML can progress slowly to a rapidly growing, acute form of leukemia that can spread to almost any organ in the body.

3 Symptoms of Infection

Due to the infection of leukemia, the blood-forming capability of bone marrow is disturbed. This will cause the production of abnormal cells and shortens the number of normal WBCs, and RBCs [13]. There are different symptoms of the disease, depending upon the type of leukemia. Figure 5 shows a detailed description of the same. However, the general symptoms are fever, fatigue, easy bleeding, anemia, body aches, headaches, etc.

Although leukemia is a rare disease, there is quite a large number of new cases and deaths observed in the year 2022. As shown in Fig. 6 Worldwide Leukemia cases are 482,500 and the death cases were 310,000 in the year 2022.

4 Leukemia Diagnosis

Leukemia can be diagnosed with the help of pathological procedures by observing the morphology of stained blood smears.

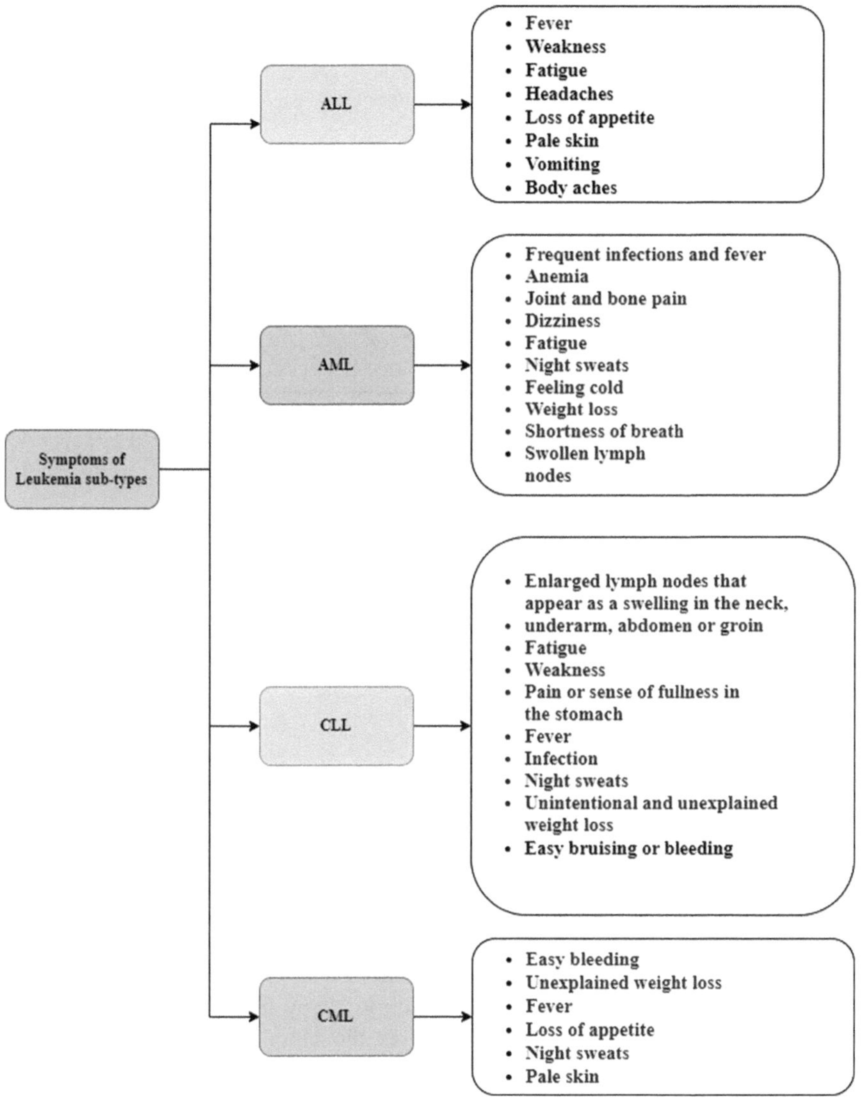

Fig. 5 Symptoms of leukemia infection

4.1 Diagnosis via the Pathological Procedure

Diagnosis of leukemia is very crucial and critical, as it involves an extremely trained pathologist [15]. Figure 7 shows the procedure. The diagnosis procedure includes collecting blood samples, and staining the same to form a blood smear. After this process, a blood slide is produced, which will be observed under a microscope for morphological changes, if any. These changes give rise to the decision about the

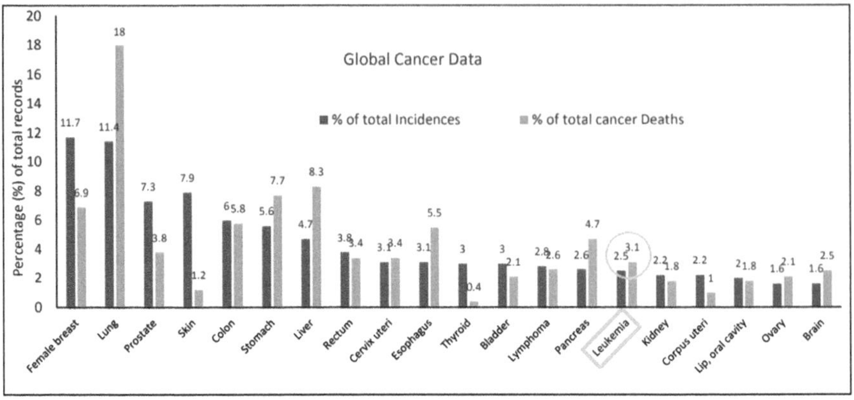

Fig. 6 Relative representation in the percentage of total incidences (19.3 million) and total deaths (10 million) due to cancer in 2022 [14]

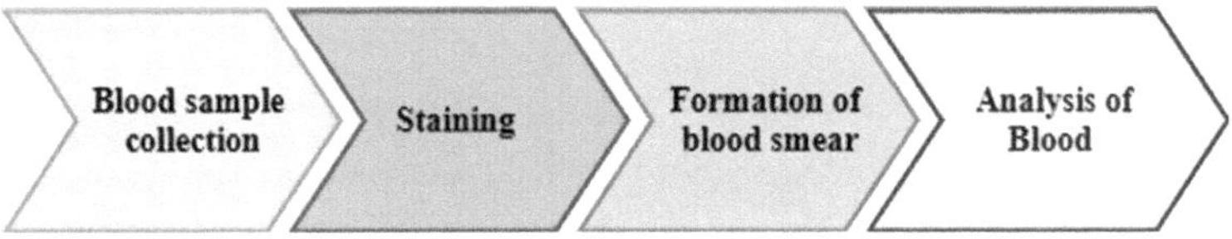

Fig. 7 Pathological procedure of leukemia detection

presence of leukemia and its sub-type [16]. A pathologist collects a blood sample from a patient. It is crucial to collect the sample carefully to ensure accurate results and maintain proper hygiene. The process of analyzing microscopic blood images requires the preparation of a blood film, which involves the use of a glass slide. Prior to examination under a microscope, the blood film needs to be stained. This preparation typically involves a slide, a tube, and a blood spreader, commonly employing the wedge method [1].

A drop of blood is placed on a base slide to create the blood film. The blood spreader slide is then moved backward over the blood drop, ensuring uniform spreading across the slide. The spreader slide should be angled at approximately 30° to 45° for optimal results relative to the blood base slide. Once the smear is prepared, it is necessary to dry it using an air dryer. To fix the dried smear, either absolute methanol or ethyl alcohol can be employed [17].

Following the fixation step, the prepared blood smear is stained using various liquids depending on the desired staining method. These may include rewmanosky stain, Leishman stain, may-crawled giema stain, or wright-Giemsa stain. Each staining technique offers unique advantages and facilitates different aspects of analysis. After that, the stained slides are examined under a microscope [18]. A trained pathologist does diagnosis of leukemia. However, researchers are motivated to ex-explore this area due to the crucial and critical diagnosis procedure the requirement of a very experienced and trained pathologist. The preceding section explores and discusses a generalized system for leukemia diagnosis via microscopic analysis.

4.2 Leukemia Diagnosis via Computer-Aided-Diagnosis System

In Fig. 8, a methodology for the analysis of microscopic blood cells is presented. The process consists of several stages: image acquisition, image segmentation, feature extraction, and disease detection. To begin, a skilled pathologist collects a blood sample from the patient and prepares a blood smear on a slide. The prepared slide is observed afterwards under a microscope, generating an image. This image can be captured directly using a camera or microscope-connected adaptor. It serves as the basis for further examination.

During image acquisition, it is possible for unwanted regions and overlapping of different blood components to be present in the acquired images. In order to address these issues, a suitable image enhancement technique is employed to improve the quality and clarity of the image [19]. This enhanced image is then ready for subsequent analysis and evaluation. After pre-processing, the enhanced image obtained allows for high-quality analysis. The next step involves separating different blood components, including red blood cells (RBCs), white blood cells (WBCs), plasma, and platelets. Segmentation is performed by considering the general characteristics of these blood components, allowing for the division of the regions of interest and enabling further classification [20].

Subsequently, the classified RBCs, WBCs, and other components are categorized into sub-classes. This categorization facilitates the extraction of specific sub-class images for feature extraction, which plays a crucial role in the analysis of blood cells and disease detection. Once the segmentation is completed, various features are extracted, considering the characteristics of different blood components. These include color, size, shape and the count of blood components [21].

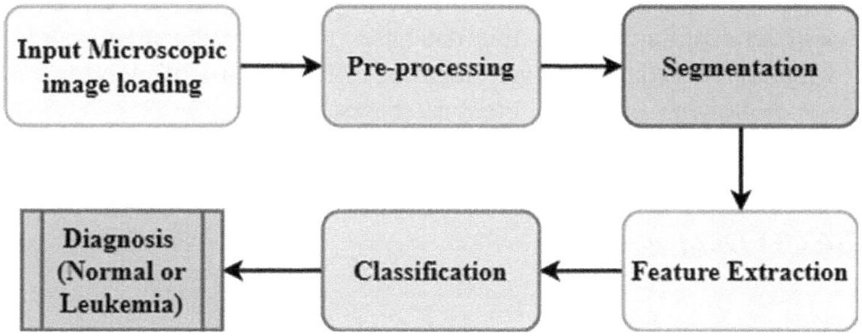

Fig. 8 Computer-aided-system of leukemia detection

Diseases can be detected by analyzing these extracted features, or cell counts can be determined. The decision regarding the presence of a particular disease is made based on the different features extracted. To facilitate these decisions, various classifiers can be designed, considering the specific features and patterns associated with different diseases. These classifiers aid in accurately detecting and diagnosing diseases based on the analyzed features [22].

Microscopic imaging analysis of blood cells is done in many diagnostic procedures [23]. There are different application areas where blood cell analysis is preferred for the diagnosis of diseases. The significance of blood cell analysis via morphological and other characteristics is to be explored, which will lead us to focus closer to the problem [24].

5 Application Areas of Microscopic Imaging in Disease Diagnosis

The literature is reviewed, considering the application areas of microscopic imaging in relation to the diagnosis of different diseases. Table 1 shows this application area with the potential for research in the particular area.

Out of the different applications explored in the above sections, leukemia is considered to be one of the most dangerous life-threatening diseases. The diagnosis of leukemia is done with microscopic imaging by observing the morphology of different cells. Although morphological diagnosis plays a major role in its diagnosis, other symptoms and blood parameters are also got affected due to the infection of leukemia. The next section discusses different blood parameters and other abnormalities after the leukemia infection. The prominent categories of leukemia ALL, AML, CLL, and CML are also explored in view of these parameters and abnormalities.

6 Datasets Available for Microscopic Imaging Diagnosis

There are several publicly accessible databases for analyzing microscopic blood images. Majorly these dataset have images of different parts of blood taken under microscope. Following are the different databases.

6.1 BCCD Dataset

This data contains 12,500 images in total. Each subtype has approximately 3,000 images: Eosinophil, Lymphocyte, Monocyte, and Neutrophil. This dataset also

Table 1 Analysis of different applications of blood cells analysis

Application	WBC, RBC segmentation
References	[7, 25–36]
Potential of further research	Blood cell analysis focuses on the crucial task of segmenting white blood cells (WBC) and red blood cells (RBC). The accurate identification of these cells' morphology holds significant value in diagnosing various diseases. While segmentation techniques continue to advance, there exists considerable untapped potential for further research in the field of blood cell segmentation
Application	RBC/WBC counting
References	[6, 37, 38]
Potential of further research	The quantification of red blood cells (RBC) and white blood cells (WBC) serves as an indicator for various infections present in the body. Typically, this analysis is carried out using equipment-based methods, which are cost-effective and part of routine procedures. However, researchers have also made advancements in developing computer-aided frameworks for microscopic analysis. Despite these developments, there is untapped potential in this field, particularly in enhancing the segmentation of distinct components of blood, including RBCs, WBCs, and platelets, which is an ongoing area of improvement
Application	Anemia or sickle cell detection
References	[39–41]
Potential of further research	The detection of anemia primarily relies on counting red blood cells (RBCs). Additionally, certain shape alterations in RBCs can also be indicative of anemia
Application	Malaria/dengue and other viral diseases detection
References	[42–44]
Potential of further research	In viral diseases such as malaria and dengue, the evaluation of platelet count becomes crucial. Furthermore, the presence of parasites resulting from these infections can be detected through morphological analysis using microscopic imaging. Exploring this area holds considerable promise, as the quantity and variety of parasites play a significant role in determining the disease severity and subsequent treatment guidelines
Application	Thalassemia detection
References	[45, 46]
Potential of further research	Thalassemia encompasses a spectrum of 12 distinct types, characterized by variations in the size and shape of red blood cells (RBCs). Typically, the detection of thalassemia involves identifying infected and non-infected cells. However, there remains untapped potential for further advancements in the detection methods of this disease
Application	Leukemia detection

(continued)

Table 1 (continued)

Application	WBC, RBC segmentation
References	[22, 47–58]
Potential of further research	Leukemia, a condition characterized by abnormal white blood cells produced by the bone marrow, encompasses two significant subtypes: acute leukemia and chronic leukemia. Additionally, leukemia can be further classified into specific types, including acute lymphocytic (ALL), acute myelogenous (AML), chronic lymphocytic (CLL), and chronic myelogenous (CML). The detection of leukemia primarily relies on morphological analysis, with each subtype guiding distinct treatment guidelines. Extensive research efforts have been dedicated to the detection and diagnosis of leukemia. However, there remains ample room for improvement and increased confidence in existing frameworks through the development of hybrid, optimized algorithms combining machine learning and artificial intelligence

includes 410 additional WBC and RBC images in JPEG and xml metadata formats. This database is licensed under the MIT license [59].

6.2 ALL-IDB (Acute Lymphoblastic Leukemia Database)

This specific dataset serves as a valuable resource for the analysis and research of acute leukemia. It is organized into two distinct subtypes: ALL-IDB1 and ALL-IDB2. Captured using a Canon Powershot G5 camera, the images were observed under a microscope with magnifications ranging from 300 to 500. The dataset employs JPEG images with a color depth of 24 bits. The ALL-IDB1 dataset comprises 109 images with a resolution of 2,592 $\times$ 1,944 pixels, totaling 3,900 elements. Within this dataset, a total of 510 lymphoblasts are included. On the other hand, the ALL-IDB2 dataset consists of 260 images with a resolution of 257 $\times$ 257 pixels, encompassing a total of 257 elements. Among these images, there are 130 lymphoblasts, with 118 specifically identified as such [60].

6.3 Nivaldo Mediros' Atlas of Hematology

The database comprises a diverse collection of microscopic blood cell images encompassing various types. These image types encompass the maturation sequence, erythrocyte alterations, images related to anemia, leukemia, parasites, fungus, as well as other miscellaneous images (totaling 119 images). The photographs were taken using Zeiss and Nikon photomicroscopes, with magnifications ranging from 200 to 1,000 [61].

6.4 The ASH Image Bank

The creation of this database can be attributed to the American Society of Hematology (ASH). Functioning as an online image bank, it serves as a comprehensive resource for images of leukemia cells. Accessible through the internet, this image library offers a wide range of categories within the field of hematology. Moreover, the database provides images with varying resolutions, catering to different requirements and preferences [56].

6.5 Leukocyte Images for Segmentation and Classification (LISC)

The dataset comprises images of individuals in a healthy state. It consists of 400 images derived from 100 distinct blood slides representing eight healthy subjects. The image dimensions are 720 × 576 pixels. These images were obtained from the Hematology-Oncology and BMT Research Center at Imam Khomeini Hospital in Tehran, Iran. For experimentation purposes, a total of 250 images are available. These images showcase different leukocyte elements, including eosinophils, basophils, lymphocytes, monocytes, and neutrophils [62].

6.6 The C-NMC Dataset

The ownership of this database belongs to the Cancer Imaging Archive (TCIA) public access. It comprises a total of 15,135 images involving 118 participants. All images in this database are in the bitmap image format (BMP). The data is divided into distinct sets, namely the train sets, preliminary test sets, and final test sets. Each set contains images of both cancerous and normal tissue, amounting to a total of 122 images.

Out of these datasets, LISC and BCCD databases have normal blood cell images with an emphasis on different blood components. The researchers utilize these to segment the different blood components and identify them. Atlas of Hematology is employed for morphologic similarities in images. ASH image bank have different types of cancerous and non-cancerous images, but the images need to be clearly sorted in different folders, such as cancerous and non-cancerous. The C-NMC dataset has a large number of images but has only single cells in each image. This dataset is much like a segmented dataset. It could give a fair amount of accuracy due to the single cell per image [63].

Major researchers utilize ALL-IDB1 and ALL-IDB2 for their experimentation. This dataset has clearly categorized as cancerous and normal cells. This database will prove the perfect one for binary classification. Also, to justify the proposed work, a literature comparison is necessary; in that regard, the same database is utilized by major of the researchers. Although, for the leukemia diagnosis, multiple leukemia sub-types need to be detected, and using real images is also essential to make the system robust and generalize. Hence in addition to ALL-IDB, a private dataset is prepared in consultation with a pathologist, and the framework is explored over the same.

7 Diagnostics of Different Leukemia Sub-Types

There are various types of blood tests that can be conducted to diagnose and monitor leukemia, and the specific results would depend on the type and stage of leukemia, as well as the individual's overall health [64]. However, some general information about blood tests is commonly used in the diagnosis and monitoring of leukemia (Fig. 9).

7.1 *Family History of Leukemia*

A family history of leukemia can play a role in assessing an individual's risk for developing the disease. In some cases, there may be a genetic predisposition or familial

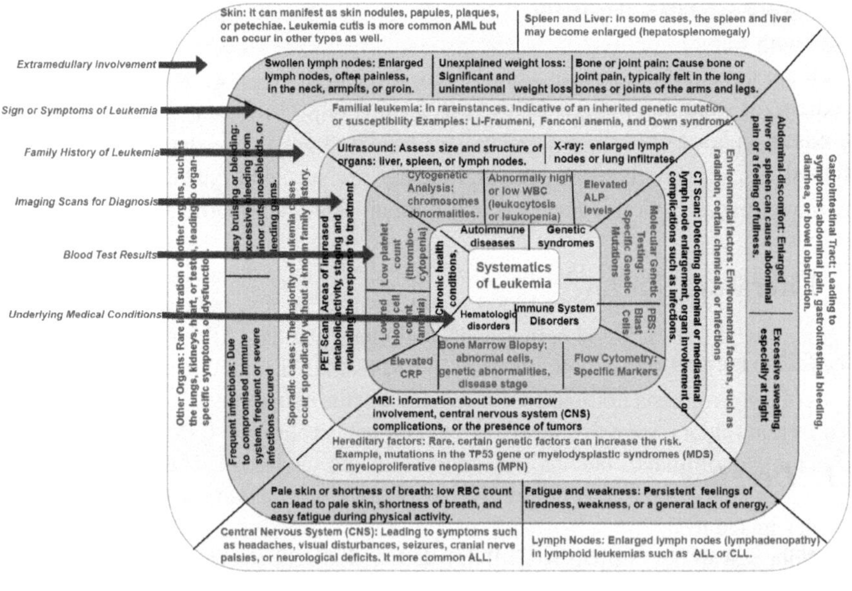

Fig. 9 Systematics of leukemia

clustering of leukemia, suggesting an increased likelihood of the disease running in families. However, it's important to note that the majority of leukemia cases occur sporadically without a clear familial link. Here are some points to consider regarding family history and leukemia: Familial leukemia: In rare instances, there are families where multiple members have been diagnosed with leukemia. This can be indicative of an inherited genetic mutation or susceptibility that increases the risk of developing the disease. Examples of genetic syndromes associated with familial leukemia include Li-Fraumeni syndrome, Fanconi anemia, and Down syndrome [65]. Hereditary factors, while most cases of leukemia are not directly inherited, certain genetic factors can increase the risk. For example, certain gene mutations, such as mutations in the TP53 gene or specific changes in genes related to myelodysplastic syndromes (MDS) or myeloproliferative neoplasms (MPN), can increase the likelihood of developing leukemia. It is also important to consider that family members may be exposed to similar environmental factors, such as radiation, certain chemicals, or infections, which could contribute to the development of leukemia. The majority of leukemia cases occur sporadically without a known family history. This means that even if no other family members have been diagnosed with leukemia, an individual can still develop the disease. If there is a family history of leukemia or other hematologic malignancies, it's advisable for individuals to inform their healthcare provider. In such cases, genetic counseling may be recommended to evaluate the risk and determine if further genetic testing or monitoring is necessary for the individual or other family members.

7.2 *Imaging Scans*

In the diagnosis and management of leukemia, imaging scans may be used to assess the extent of the disease, identify potential complications, and monitor treatment response. The specific imaging scans performed can vary depending on the type and stage of leukemia, as well as individual patient factors. Here are some imaging scans that may be utilized in leukemia. X-rays are commonly used to evaluate the chest and can help identify any abnormalities, such as enlarged lymph nodes or lung infiltrates. Computed Tomography scans provide detailed cross-sectional images of the body. They can help assess the size and location of tumors, lymph nodes, and organs affected by leukemia. CT scans are particularly useful for detecting abdominal or mediastinal lymph node enlargement, organ involvement, or complications such as infections [66]. Magnetic Resonance Imaging (MRI) uses magnetic fields and radio waves to generate detailed images of the body. It can provide valuable information about bone marrow involvement, central nervous system (CNS) complications, or the presence of tumors in specific locations [67]. Positron Emission Tomography (PET) Scan involve injecting a small amount of radioactive material into the body. The scan detects areas of increased metabolic activity, which can be indicative of cancerous cells. PET scans are useful for staging and evaluating the response to treatment in some types of leukemia [68]. Ultrasound uses sound waves to create images of internal organs and tissues. It can be used to assess the size and structure of organs, such as the liver, spleen, or lymph nodes. It's important to note that not all leukemia patients will undergo all of these imaging scans. The choice of imaging depends on the specific clinical situation, symptoms, and suspected complications. The healthcare team will determine the most appropriate imaging modality for each individual case.

7.3 *Common Symptoms*

There are certain signs and symptoms that can suggest the possibility of leukemia. However, it's important to note that these symptoms are not exclusive to leukemia and can be caused by other conditions as well. If you or someone you know is experiencing any of these symptoms, it is important to consult a healthcare professional for an accurate diagnosis. The common signs and symptoms associated with leukemia are persistent feelings of tiredness, weakness, or a general lack of energy, anemia resulting from a low red blood cell count can lead to pale skin, shortness of breath, and easy fatigue during physical activity, due to a compromised immune system, individuals with leukemia may experience frequent or severe infections, such as respiratory infections, urinary tract infections, or skin infections. Leukemia can affect the production of platelets, leading to easy bruising, excessive bleeding from minor cuts, nosebleeds, or bleeding gums, enlarged lymph nodes, often painless, in the neck, armpits, or groin. Significant and unintentional weight loss without an apparent cause, bone or joint pain, typically felt in the long bones or joints of the arms and legs enlarged liver or spleen can cause abdominal pain or a feeling of fullness. Excessive sweating, especially at night that are unrelated to an external temperature change [69]. It's important to remember that these symptoms can vary depending on the type and stage of leukemia, and not all individuals with leukemia will experience the same symptoms. Additionally, some individuals may have no symptoms at the time of diagnosis, and leukemia may be incidentally detected during routine blood tests or imaging scans. The persistence of symptoms has to be viewed seriously and it is best to consult a healthcare professional for the accurate diagnosis and proper management of treatment (Fig. 10).

7.3.1 Complete Blood Count (CBC)

A CBC provides information about the number and characteristics of different blood cells. In leukemia, the CBC may show abnormalities such as: Low red blood cell count (anemia) Low platelet count (thrombocytopenia), abnormally high or low white blood cell count (leukocytosis or leukopenia) [70].

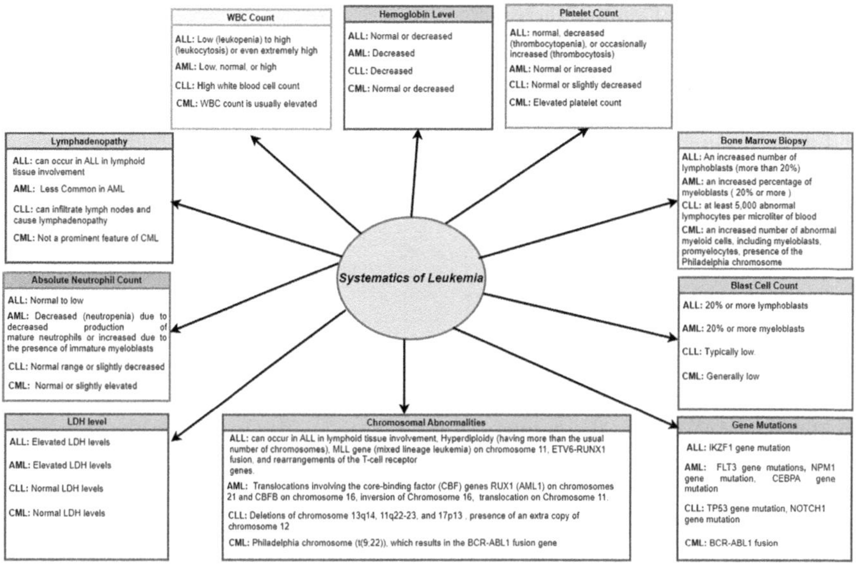

Fig. 10 Leukemia systematics considering different blood tests

7.3.2 Peripheral Blood Smear

A blood smear involves examining a sample of blood under a microscope to evaluate the size, shape, and appearance of blood cells. In leukemia, abnormal or immature cells, such as blast cells, may be observed. Bone Marrow Aspiration and Biopsy: These procedures involve obtaining a sample of bone marrow to examine the cells and determine the type and extent of leukemia. The results can provide information about the percentage of abnormal cells, the presence of genetic abnormalities, and the stage of the disease [71].

7.3.3 Flow Cytometry

Flow cytometry is a technique that analyzes the characteristics of cells, including their size, shape, and surface markers. It is commonly used to identify and classify different types of leukemia cells based on their specific markers [72].

7.3.4 Cytogenetic Analysis

Cytogenetic analysis involves examining the chromosomes in leukemia cells to identify any structural or numerical abnormalities. This can help determine the subtype of leukemia and provide prognostic information.

7.3.5 Molecular Genetic Testing [73]

Molecular tests can detect specific genetic mutations or rearrangements associated with leukemia. These tests can help refine the diagnosis, guide treatment decisions, and monitor disease progression. It's important to consult with a healthcare professional for accurate interpretation of blood test results and to discuss the specific findings in the context of an individual patient's situation.

7.3.6 White Blood Cell Count [74]

In leukemia, the white blood cell count can vary depending on the specific type and stage of the disease, as well as individual factors. Leukemia is characterized by an abnormal increase in the number of white blood cells (WBCs) in the blood, but the degree of elevation can differ. Here are some general considerations regarding white blood cell counts in leukemia. In Acute Lymphoblastic Leukemia (ALL), the white blood cell count can vary widely. It may range from low (leukopenia) to high (leukocytosis) or even extremely high. In some cases, the WBC count can be in the normal range. The presence of blast cells, which are immature white blood cells, is an important factor in the diagnosis of ALL. Acute Myeloid Leukemia (AML) is characterized by the rapid growth of abnormal myeloid cells. The white blood cell count can also vary, and it may be low, normal, or high. In some cases of AML, a high WBC count (leukocytosis) is observed, often due to the accumulation of immature myeloid cells. Chronic Lymphocytic Leukemia (CLL) is typically associated with a high white blood cell count, particularly in the lymphocytes. However, the white blood cell count can vary, and some individuals with CLL may have normal or low WBC counts. In chronic Myeloid Leukemia (CML), the white blood cell count is usually elevated and can be significantly high. The presence of an abnormal chromosome called the Philadelphia chromosome is a defining characteristic of CML. It's important to note that the white blood cell count alone cannot definitively diagnose leukemia, as other conditions can also cause abnormalities in WBC count. Diagnosis and classification of leukemia require a combination of clinical evaluation, blood tests (including complete blood count and peripheral blood smear), bone marrow examination, and potentially other genetic or molecular tests.

7.3.7 Hemoglobin Level

The hemoglobin level in leukemia can vary depending on the specific type and stage of the disease, as well as individual factors. Hemoglobin is the protein in red blood cells that carries oxygen throughout the body. In leukemia, the hemoglobin level can be affected due to various factors, including bone marrow involvement, anemia, and other related complications. In ALL, the hemoglobin level can be normal or decreased. Anemia, indicated by a low hemoglobin level, can occur due to several factors, such as replacement of normal bone marrow cells with leukemic cells or suppression of red blood cell production. AML also shows the variation in the hemoglobin level. Anemia is common in AML and can occur due to the displacement of normal bone marrow cells by leukemic cells or the suppression of red blood cell production. CLL may or may not be associated with a low hemoglobin level. Anemia can occur in CLL due to various factors, including bone marrow involvement by leukemia cells or autoimmune hemolytic anemia, which is a condition in which the immune system attacks and destroys red blood cells. CML has hemoglobin level can be normal or decreased [75]. Anemia can occur in CML due to factors such as bone marrow involvement by leukemic cells or suppression of red blood cell production. It's important to note that the hemoglobin level alone cannot definitively diagnose leukemia, as other conditions can also cause abnormalities in hemoglobin levels. The diagnosis and management of leukemia require a comprehensive evaluation that includes clinical assessment, blood tests (such as a complete blood count), bone marrow examination, and potentially other diagnostic tests.

7.3.8 Platelets Count

In leukemia, the platelet count can vary depending on the specific type and stage of the disease, as well as individual factors. Platelets are blood cells responsible for blood clotting, and their count can be affected in leukemia due to various factors, including bone marrow involvement, treatment-related effects, and other related complications. In ALL, the platelet count can be normal, decreased (thrombocytopenia), or occasionally increased (thrombocytosis). Thrombocytopenia is more commonly observed and can occur due to various factors, including bone marrow involvement by leukemic cells, chemotherapy treatment, or autoimmune destruction of platelets. Thrombocytopenia is common in AML and can occur due to factors such as bone marrow involvement by leukemic cells or suppression of platelet production. However, in some cases, AML can present with a normal or increased platelet count. CLL is typically associated with a normal or slightly decreased platelet count. However, in some cases, CLL can also cause thrombocytopenia. CML has platelet count in CML can be normal, decreased, or increased. Thrombocytosis, characterized by an elevated platelet count, is more commonly observed in CML [76]. It's important to note that the platelet count alone cannot definitively diagnose leukemia, as other conditions can also cause abnormalities in platelet counts. The diagnosis and management of leukemia require a comprehensive evaluation that includes clinical as-assessment, blood tests (such as a complete blood count), bone marrow examination, and potentially other diagnostic tests.

7.3.9 Bone Marrow Biopsy

The presence of abnormal cells in a bone marrow biopsy is a significant finding in the diagnosis and classification of leukemia. In leukemia, the bone marrow biopsy is an essential procedure to evaluate the cellular composition of the bone marrow and determine the presence of abnormal cells. The specific characteristics of these abnormal cells can provide valuable information about the type and stage of leukemia. In case of the ALL, the bone marrow biopsy typically reveals an increased number of lymphoblasts, which are immature white blood cells. The presence of more than 20% lymphoblasts in the bone marrow is a major diagnostic criterion for ALL. AML is characterized by an excess of immature myeloid cells in the bone marrow. The bone marrow biopsy in AML shows an increased percentage of myeloblasts, which are abnormal immature white blood cells. AML diagnosis typically requires the presence of 20% or more myeloblasts in the bone marrow. In the infection of CLL, the bone marrow biopsy shows an increased number of abnormal lymphocytes. These lymphocytes may have a characteristic appearance, such as small round cells with condensed chromatin. The presence of at least 5,000 abnormal lymphocytes per microliter of blood is one of the diagnostic criteria for CLL. CML infection has an increased number of abnormal myeloid cells, including myeloblasts, promyelocytes, and other immature cells. The presence of the Philadelphia chromosome, a genetic abnormality, is a hallmark of CML and can be detected through specialized testing on the bone marrow sample [77]. It's important to note that the specific findings of a bone marrow biopsy in leukemia can vary depending on the type and stage of the disease, as well as individual factors. A bone marrow biopsy is a crucial tool for diagnosing leukemia and providing important information for treatment planning and monitoring. The interpretation of bone marrow biopsy results requires the expertise of a hematopathologist or oncologist who will consider various factors to make an accurate diagnosis.

7.3.10 Blast Cells Counts

The blast cell count, also known as the percentage of blasts, is an important parameter in the evaluation of leukemia. Blast cells are immature or abnormal cells that are characteristic of leukemia. The blast cell count indicates the proportion of these cells present in the bone marrow or blood sample and can help determine the type and stage of leukemia. In ALL, the blast cell count refers to the percentage of lymphoblasts, which are immature lymphoid cells. A blast cell count of 20 can indicate the presence of leukemia. In AML, the blast cell count represents the percentage of myeloblasts, which are abnormal immature myeloid cells. AML diagnosis also requires the presence of 20. CLL is characterized by the accumulation of abnormal mature lymphocytes, and blast cells are generally not a prominent feature. The blast cell count in CLL is typically low. In CML, the blast cell count is generally low during the chronic phase of the disease [78]. However, as the disease progresses to the accelerated phase or blast crisis, the blast cell count increases. It's important to note that the blast cell count alone cannot definitively diagnose leukemia, as other factors and diagnostic criteria are considered. Additionally, blast cell counts can vary within different subtypes and stages of leukemia. The specific blast cell count and other diagnostic criteria would be determined through individual laboratory tests, bone marrow examination, and evaluation by a healthcare professional. Interpreting blast cell counts and making a diagnosis requires the expertise of a hematopathologist or oncologist who specializes in hematological malignancies.

7.3.11 Absolute Neutrophil Count

The absolute neutrophil count (ANC) is a measure of the number of neutrophils, a type of white blood cell, in the blood. ANC is an important parameter in assessing a patient's immune function, particularly their ability to fight bacterial infections. In leukemia, the ANC can be affected depending on the specific type and stage of the disease, as well as the presence of treatment-related effects. The normal range for ANC can vary slightly depending on the laboratory and the population being tested. However, a generally accepted normal range for ANC is between 1,500 and 8,000 neutrophils per microliter of blood. In leukemia, the ANC can be altered in different ways. The ANC in ALL can vary depending on the extent of bone marrow involvement and the impact on normal blood cell production. It can range from normal to low, depending on factors such as the stage of the disease and treatment received. AML can affect the production of normal blood cells, including neutrophils. As a result, the ANC in AML can be variable. It may be decreased (neutropenia) due to decreased production of mature neutrophils or increased due to the presence of immature myeloblasts in the blood. In CLL, the ANC may be within the normal range or slightly decreased. CLL is characterized by the accumulation of abnormal lymphocytes rather than affecting the production of neutrophils directly. In the chronic phase of CML, the ANC is usually normal or slightly elevated [79]. However, in advanced stages or during blast crisis, the ANC can be variable and may be decreased due to the effects of the disease or treatment. It's important to note that the ANC alone does not provide a complete picture of a patient's immune function or the status of their leukemia. The ANC is typically evaluated along with other blood cell counts, clinical symptoms, and diagnostic findings to assess the overall condition of the patient and guide treatment decisions.

7.3.12 C-Reactive Protein

C-reactive protein (CRP) is an acute-phase reactant protein produced by the liver in response to inflammation or infection. CRP levels can increase in various conditions, including infections, autoimmune diseases, and certain cancers. In the context of leukemia, elevated CRP levels may be observed due to the inflammatory response associated with the disease or secondary infections. However, it's important to note that CRP levels alone are not specific to leukemia and can be influenced by other factors as well. The specific CRP level in a leukemia patient would be determined through laboratory tests and interpretation by a healthcare professional. The reference range for CRP can vary depending on the laboratory and the specific assay used. Elevated CRP levels would indicate the presence of inflammation, but they do not provide a specific diagnosis or indicate the severity of the leukemia itself.

CRP levels are often used as one of many indicators to assess the inflammatory response in the body and monitor treatment response. Rising CRP levels may suggest an ongoing infection or disease progression, while declining levels may indicate a positive response to treatment or resolution of inflammation [80].

7.3.13 Serum Alkaline Phosphatase (ALP)

(ALP) is an enzyme found in various tissues, including the liver, bones, intestines, and placenta. Elevated ALP levels can occur in several conditions, including liver or bone disorders. In the context of leukemia, ALP levels may be influenced by factors such as liver involvement, bone marrow infiltration, or treatment-related effects. However, it's important to note that ALP levels alone are not specific to leukemia and can be influenced by other factors as well. The specific serum ALP level in a leukemia patient would be determined through laboratory tests and interpretation by a healthcare professional. The reference range for ALP can vary depending on the laboratory and the specific assay used. In leukemia, elevated ALP levels may be observed in the following circumstances: Liver Involvement: Leukemia can affect liver function, leading to increased ALP levels. Liver involvement may occur due to infiltration of leukemia cells, hepatic congestion, or treatment-related effects. Bone Marrow Infiltration: Leukemia cells can infiltrate the bone marrow, potentially impacting bone metabolism and leading to increased ALP levels. Certain leukemia treatments, such as chemotherapy or corticosteroids, can cause temporary elevations in ALP levels [81]. It's important to note that the interpretation of ALP levels in a leukemia patient should be done in the context of the patient's overall clinical presentation, other laboratory findings, and medical history.

7.3.14 Chroosonal Abnormalities

Chromosomal abnormalities are commonly observed in leukemia patients and play a significant role in the diagnosis, classification, and prognosis of the disease. These chromosomal abnormalities can be detected through specialized tests, such as cytogenetic analysis, fluorescence in situ hybridization (FISH), or molecular genetic techniques. In ALL T (9; 22), known as the Philadelphia chromosome, is found in a subset of ALL cases. It results in the fusion of the BCR (breakpoint cluster region) gene on chromosome 22 with the ABL1 (Abelson tyrosine kinase 1) gene on chromosome 9. Hyperdiploidy (having more than the usual number of chromosomes) is another common abnormality observed in pediatric ALL cases. Other recurrent chromosomal abnormalities in ALL include rearrangements involving MLL gene (mixed lineage leukemia) on chromosome 11, ETV6-RUNX1 fusion, and rearrangements of the T-cell receptor genes. AML has translocations involving the core-binding factor (CBF) genes RUNX1 (AML1) on chromosome 21 and CBFB on chromosome 16, such as t(8;21)(q22;q22) and inv(16)(p13;q22), are characteristic abnormalities in AML. In the version of chromosome 16, inv(16)(p13;q22), which involves the CBFB gene, is associated with the formation of the fusion gene CBFB-MYH11. Translocations involving the mixed lineage leukemia (MLL) gene on chromosome 11 (e.g., t(9;11)(p22;q23), t(6;11)(q27;q23)) are observed in some AML cases. CLL is characterized by various chromosomal abnormalities, including deletions of chromosome 13q14 (del(13q)), deletions of chromosome 11q22-23 (del(11q)), and deletions of chromosome 17p13 (del(17p)). Trisomy 12 (presence of an extra copy of chromosome 12) is another common chromosomal abnormality in CLL. CML is associated with the presence of the Philadelphia chromosome (t(9;22)), which results in the BCR-ABL1 fusion gene. This genetic abnormality is central to the pathogenesis of CML [82]. These are just a few examples of the chromosomal abnormalities that can be observed in leukemia. The specific chromosomal abnormalities and their prognostic significance can vary among individual patients and different subtypes of leukemia. The detection and analysis of chromosomal abnormalities are important for accurate diagnosis, risk stratification, and guiding treatment decisions in leukemia patients. The interpretation of these abnormalities requires the expertise of a hematopathologist or oncologist specialized in cytogenetics or molecular genetics.

7.3.15 Lactate Dehydrogenase Level

Lactate dehydrogenase (LDH) is an enzyme found in various body tissues, including blood cells. In the context of leukemia, LDH levels can be measured as part of routine blood tests to assess disease activity and monitor treatment response. Elevated LDH levels are often associated with increased cell turnover and tissue damage, which can occur in leukemia. However, it's important to note that LDH levels alone are not specific to leukemia and can be influenced by other factors as well. In ALL, elevated LDH levels can occur, particularly in cases with high tumor burden or extensive involvement of organs or tissues. LDH levels may be used as a marker to monitor disease response to treatment. AML cases with a high percentage of blasts or extensive tissue involvement may show elevated LDH levels. LDH levels can be useful in assessing treatment response and disease progression. CLL is generally associated with normal LDH levels [83]. However, elevated LDH levels can occur in cases with extensive lymph node involvement or complications such as Richter's transformation. In CML, LDH levels are typically within the normal range during the chronic phase. However, in cases of disease progression to the accelerated phase or blast crisis, LDH levels may increase due to increased cell turnover. It's important to remember that LDH levels should be interpreted in conjunction with other clinical and laboratory findings. The specific LDH level and its significance in a leukemia patient would be determined through individual laboratory tests and evaluation by a healthcare professional. Monitoring LDH levels over time can provide valuable information about disease progression and treatment response in conjunction with other clinical parameters.

7.3.16 Genetic Mutations

Genetic mutations play a crucial role in the development and progression of leukemia. These mutations can occur in specific genes and pathways involved in the regulation of cell growth, differentiation, and survival. The presence of certain genetic mutations can help in the diagnosis, classification, and treatment selection for leukemia patients. ALL shows the mutations such as philadelphia chromosome (BCR-ABL1 fusion) in which mutation is present in a subset of ALL cases and results from the fusion of the BCR (breakpoint cluster region) gene on chromosome 22 with the ABL1 (Abelson tyrosine kinase 1) gene on chromosome 9. IKZF1 gene mutation: This mutation is observed in a subset of B-cell ALL cases and is associated with a higher risk of relapse. AML cases have certain mutations as shown. Internal tandem duplications (ITD) or point mutations in the FLT3 gene are found in a significant proportion of AML cases and are associated with a poorer prognosis. NPM1 mutations are frequently observed in AML and are associated with a favorable prognosis. CEBPA mutations are seen in a subset of AML cases and are associated with a better prognosis. TP53 mutations are associated with an aggressive form of CLL and resistance to certain treatments. NOTCH1 mutations are found in a subset of CLL cases and are associated with an unfavorable prognosis [84]. The presence of the Philadelphia chromosome and the resulting BCR-ABL1 fusion gene is a defining genetic abnormality in CML. It drives the overproduction of abnormal white blood cells and is a target for specific therapies, such as tyrosine kinase inhibitors (TKIs). These are just a few examples of genetic mutations associated with leukemia. The specific mutations and their prognostic significance can vary among individual patients and different subtypes of leukemia. Comprehensive genetic testing, including next-generation sequencing techniques, is often used to detect these mutations and guide treatment decisions. The interpretation of genetic mutations requires the expertise of a hematopathologist or oncologist specialized in molecular genetics.

7.3.17 Lymphadenopathy

Lymphadenopathy, which refers to the enlargement of lymph nodes, can occur in leukemia but is not always present in all cases or all types of leukemia. The presence and extent of lymphadenopathy can vary depending on the specific type and stage of leukemia, as well as individual factors. Here are some general considerations regarding lymphadenopathy in leukemia. Lymphadenopathy can occur in ALL, particularly in cases with extensive involvement of lymphoid tissues. Enlarged lymph nodes may be detected during physical examination or observed on imaging scans. Lymphadenopathy is less common in AML compared to lymphoid malignancies. However, in some cases, AML can be associated with lymph node enlargement, usually as a result of extramedullary involvement or secondary effects. CLL is characterized by the accumulation of abnormal lymphocytes, which can infiltrate lymph nodes and cause lymphadenopathy. Enlarged lymph nodes, particularly in the neck, underarms, or groin, are a common feature of CLL. Lymphadenopathy is not a prominent feature of CML during the chronic phase. However, in advanced stages or blast crises, lymphadenopathy may occur due to extramedullary involvement [85]. It's important to note that the presence or absence of lymphadenopathy alone cannot definitively diagnose leukemia, as other factors and diagnostic criteria are considered. Other conditions, such as infections or other types of cancer, can also cause lymph node enlargement. The evaluation of lymphadenopathy in leukemia requires a comprehensive assessment, including physical examination, imaging studies, and potentially lymph node biopsy, to determine the cause and guide treatment decisions.

7.3.18 Underlying Medical Conditions

When a patient is diagnosed with leukemia, it is possible for them to have underlying medical conditions. Leukemia itself is a type of cancer that affects the blood and bone marrow, but it can coexist with other health conditions. Some individuals may have pre-existing medical conditions before the onset of leukemia, while others may develop additional health issues as a result of the disease or its treatment. Common comorbidities or underlying medical conditions seen in leukemia patients can include Immune system disorders which weakened immune systems, making them more susceptible to infections and other immune-related conditions. Certain blood disorders, such as myelodysplastic syndromes (MDS) or myeloproliferative neoplasms (MPN), can be present alongside or precede the development of leukemia. Individuals with chronic conditions like diabetes, hypertension, or heart disease can have these conditions in addition to leukemia. Some patients with leukemia may also have autoimmune disorders like rheumatoid arthritis, lupus, or autoimmune thyroid diseases. In rare cases, individuals with genetic syndromes, such as Down syndrome (trisomy 21) or Li-Fraumeni syndrome, may be predisposed to developing leukemia [86]. It's important to note that the presence of underlying medical conditions can influence the treatment options and approach to managing leukemia. Each patient's situation is unique, and their healthcare team will consider all relevant factors when creating a treatment plan tailored to their specific needs.

7.3.19 Organ Dysfunctions

In leukemia, the presence of organ dysfunction can occur due to various factors, including the direct infiltration of leukemia cells into organs, the effects of abnormal blood cell production on organ function, or complications associated with the disease or its treatment. The signs of increasing organ dysfunction in leukemia patients can vary depending on the specific type and stage of leukemia, as well as individual factors. Leukemia can affect liver function, leading to hepatomegaly (enlarged liver), jaundice (yellowing of the skin and eyes), elevated liver enzymes (as seen in blood tests), and impaired liver function tests. Leukemia-related complications or treatment can cause kidney damage, resulting in symptoms such as decreased urine output, swelling (edema), elevated creatinine levels (as seen in blood tests), and electrolyte imbalances. In some cases, leukemia cells can infiltrate the central nervous system, leading to symptoms such as headaches, visual disturbances, seizures, confusion, or neurological deficits. Leukemia-associated complications, such as infections or leukostasis (abnormal clustering of leukemia cells), can impact lung function, resulting in respiratory symptoms such as shortness of breath, cough, or pneumonia. Leukemia patients may experience cardiovascular complications, including arrhythmias (abnormal heart rhythms), congestive heart failure, or coagulation abnormalities that can lead to excessive bleeding or clotting. Leukemia-related complications or treatment can cause gastrointestinal symptoms such as abdominal pain, nausea, vomiting, diarrhea, or gastrointestinal bleeding. It's important to note that the signs of organ dysfunction can vary among individuals and depend on the specific circumstances of each case. Prompt medical evaluation and management are crucial in addressing organ dysfunction in leukemia patients. Healthcare professionals closely monitor organ function through clinical examination, laboratory tests, imaging studies, and specialized evaluations to detect and manage any signs of increasing organ dysfunction [87].

7.3.20 Extramedullary Involvement

Extramedullary involvement refers to the spread or infiltration of leukemia cells outside the bone marrow, affecting other organs or tissues. While the bone marrow is the primary site of leukemia cell production, some types of leukemia can exhibit extramedullary involvement. The signs of extramedullary involvement in leukemia patients can vary depending on the specific type and stage of leukemia, as well as individual factors. Enlarged lymph nodes (lymphadenopathy) may be observed in cases where leukemia cells infiltrate the lymphatic system, particularly in lymphoid leukemias such as ALL or CLL. Leukemia cells can infiltrate the Central Nervous System (CNS), leading to symptoms such as headaches, visual disturbances, seizures, cranial nerve palsies, or neurological deficits. CNS involvement is more commonly seen in acute leukemias, particularly ALL. Leukemia cutis is a condition characterized by the infiltration of leukemia cells into the skin. It can manifest as skin nodules, papules, plaques, or petechiae. Leukemia cutis is more commonly associated with AML but can occur in other types as well. Leukemia cells can infiltrate the gastrointestinal tract, leading to symptoms such as abdominal pain, gastrointestinal bleeding, diarrhea, or bowel obstruction. This extramedullary involvement can occur in various types of leukemia. In some cases of leukemia, the spleen and liver may become enlarged (hepatosplenomegaly) due to infiltration of leukemia cells. Leukemia cells can rarely infiltrate other organs, such as the lungs, kidneys, heart, or testes, leading to organ-specific symptoms or dysfunction. It's important to note that extramedullary involvement can occur in specific subtypes or stages of leukemia, and it may not be present in all cases [88]. The evaluation of extramedullary involvement requires a comprehensive assessment by healthcare professionals, including physical examination, imaging studies, and potentially biopsies or specialized tests to confirm the presence of leukemia cells in affected organs.

8 Uncovering Significant Contributions from Added Articles in the Field

The study conducted by Sharma et al. [89] focused on the application of blood cancer diagnosis and addressed the need for an automated technique in blood cancer detection. The objective was to develop a hybrid CNN (Convolutional Neural Network) model for ALL (Acute Lymphoblastic Leukemia) detection. The methods employed in the study included K-means clustering, resulting an accuracy of 97.01%. However, a limitation of the study was the need to increase the accuracy of the model. The Histogram of Oriented Gradients (HOG) descriptor with the firefly optimization algorithm was used as a feature extraction and selection mechanism from the Region of Blood Cell (ROBC). According to the authors, the proposed method demonstrated good results in terms of accuracy. However, the study did not mention commonly used validation techniques for classifiers such as Cross fold.

Hegde et al. [90] focused on detection of leukemia and the necessity for an automated technique in blood cancer detection. The proposed system utilized for the segmentation of white blood cells (WBCs) and classify them into normal and blasts cells. The methods employed in the study included thresholding, morphology, SVM (Support Vector Machine), and NN (Neural Network) classifier. The results showed a 98% accuracy for the segmentation of WBCs and a 92.8% accuracy for leukemia detection. However, a limitation of the study was the need for improvement in blasts detection. The key findings highlighted the use of thresholding, morphological operations, and filtering employed for segmentation, while classification was performed using a hybrid system consisting of SVM and NN. SVM classifier was validated using hold-out validation, but no validation technique was mentioned for the NN classifier. Furthermore, the combination of NN and SVM classifiers may require proper exploration of a suitable validation technique.

Sarmad et al. [91] aimed to detect acute lymphoblastic leukemia (ALL) and classify it into its subtypes, namely L1, L2, and L3. The need for accurate detection of leukemia subtypes was emphasized. The utilization of a deep convolutional neural network (CNN) architecture known as AlexNet was presented here. The results showed an accuracy of 98% for leukemia detection. This study has the limited number of database images used, and the absence of pre-processing and feature extraction techniques. The key finding of the study was the use of transfer learning was employed by using the AlexNet, which improved the classification performance. The system's performance could be enhanced through the application of pre-processing and other enhancement techniques. Exploration of other deep CNN architectures to identify the most suitable architecture for this particular application.

Shafique et al. [50] proposed a methodology for detecting and diagnosing Acute Lymphoblastic Leukemia (ALL), highlighting the necessity of a computer-aided framework for precise diagnosis. The primary objective was to segment white blood cells (WBCs), red blood cells (RBCs), and platelets, followed by feature extraction and classification into two classes: normal cells and blast cells. The study utilized the Zack algorithm for segmentation and an SVM classifier for classification. The results demonstrated an accuracy of 93.7% for the framework. However, a limitation of the study was the limited number of dataset images available, comprising only 108 images. Key findings indicated that combining color and shape features with the SVM classifier slightly improved performance compared to the K-Nearest Neighbors (KNN) approach. While implementing a deep classifier could enhance the framework, it would require a larger dataset. Furthermore, future studies could focus on detecting different types of ALL, such as L1, L2, and L3, to enhance the framework's capabilities.

The study done by Shahin et al. [92] was focused on the identification of white blood cells (WBC) with the aim of addressing the challenges of under-segmentation, over-segmentation, and the complexity of feature extraction methods. The main objective was to develop a comprehensive system that involved pre-processing, feature extraction and selection, classification, and the application of a Transfer Learning Approach (TLA) using the WBCsNet architecture. The methods utilized in the study included the use of Convolutional Neural Networks (CNN), Support Vector Machines (SVM), and the WBCsNet architecture. The results obtained showed an accuracy of up to 96% for WBC identification. This study has a limitation when dealing with larger datasets, as the accuracy decreased to 92.6%. Key findings in this experimentation is the combination of CNN and SVM classifiers, along with the implementation of the WBCsNet architecture, contributed to the improved accuracy. The performance of the system can be further enhanced, the dataset size should be increased, as deep learning approaches require a larger amount of data.

The another study involved Mishra et al. [93] was aimed to work on acute lymphoblastic leukemia (ALL) detection to address the need for an automated Computer-Aided Diagnosis (CAD) framework. The objective was to develop a comprehensive system that involved pre-processing, feature extraction, dimensionality reduction, and classification. Triangle thresholding, discrete orthogonal S-transform, and the Adaboost algorithm with the RF (ADBRF) classifier were utilized for the experimentation. The results obtained showed an accuracy of approximately 99% for ALL detection. However, a limitation of the study was identified as it only used one dataset, IDB1, with a limited number of cells (799), and it did not detect the subtypes of ALL (L1, L2, and L3). Key findings revealed that the ADABOOST RF classifier outperformed SVM and NN classifiers in this context. This framework could be extended for the detection of acute myeloid leukemia with its subtypes, and further improvements and optimizations could be explored to enable the detection of ALL subtypes.

Jha and Dutta [52] presented the system for the detection of ALL in blood smear images with single-cell, focused to develop an image processing framework that utilizes deep learning techniques to improve the accuracies of existing systems. The methods proposed in the study included a Mutual Information (MI) based hybrid model, a Deep Convolutional Neural Network (CNN) classifier with the chronological Sine Cosine Algorithm (SCA), and k-fold validation. The results obtained showed an accuracy of 98.7% for ALL detection. The limitation of the study was identified as it considered only single cells for analysis. The study concluded that the DCNN with SCA yielded good results compared to the current state-of-the-art approach using a Neural Network (NN) with hybridized work. Key findings highlighted the use of hybrid segmentation techniques involving fuzzy means and active contour. This work could be extended to include multiple cells and that the application of hybrid optimization algorithms could further enhance performance.

Moshavash et al. [53] studied and presented the diagnosis system of Acute Lymphoblastic Leukemia (ALL). The background highlighted the challenge of leukocyte segmentation under uneven imaging conditions. The objective was to perform leukocyte segmentation, feature extraction, and classification. This study included the Zack algorithm, Support Vector Machine (SVM), K-Nearest Neighbors (KNN), Naïve Bayes, and decision tree. The results obtained showed an accuracy of 97.6% for leukocyte segmentation. A limitation of the framework is in detecting overlapping cells and the inability to classify leukemia into its sub-classes. Key findings of the study included the combination of two ensemble classifiers for classification, with Classifier 1 using four different classifier combinations and Classifier 2 using SVM with different kernel functions. Additionally, the study compared two types of feature extraction methods, namely GLCM and LBP, and found that LBP achieved higher accuracy. This system was complex due to the utilization of two ensemble classifiers with different combinations and the use of two feature extraction methods.

Negm [94] focused on the detection and classification of Acute Lymphoblastic Leukemia (ALL). The background highlighted the critical and challenging nature of manual detection in leukemia cases. The methods utilized in the study included histogram equalization, the k-means algorithm, the watershed algorithm, decision tree, and neural network (NN) classifier. The results obtained showed an overall accuracy of 96.6% for the decision tree classifier and 96.76% for the NN classifier. A limitation of the study was the relatively small dataset size, consisting of only 115 images from a public dataset. The key finding of the study was that the NN classifier performed better than the decision tree, although the decision tree proved to be faster in terms of processing time. Further expansion of this algorithm is it should explore the different subtypes of both ALL and Acute Myeloid Leukemia (AML).

Mohamadreza [95] proposed the system of detection of ALL with the aim of improving the diagnosis system for this condition. The objective was to perform segmentation, feature extraction, and classification of leukemia cells. This work employed k-means algorithm, watershed transform, and SVM classifier. The results obtained showed an accuracy of up to 99% in the detection of ALL. A limitation of the study was the limited dataset, consisting of only 600 images, and popular datasets such as ALL-IDB1 and ALL-IDB2 were not used. The key findings of the study included the use of fractal features and feature reduction through the PCA algorithm. The inclusion of chaotic features at the feature extraction stage led to increased classification accuracy. The similar framework could be extended for the detection of Acute Myeloid Leukemia (AML) and its subtypes. However, the study did not explore the exact selection of features in depth, and further exploration regarding the complexity and processing time associated with the use of the PCA algorithm could be conducted.

Sukhia et al. [96] focused on the detection of ALL and the need for an automatic and innovative approach to improve the detection process. The methods utilized included the Diffused Expectation Maximization (DEM) algorithm, thresholding techniques, and a sparse classifier. The results obtained showed an accuracy of 94% in the detection of ALL. A limitation of the study was the limited dataset, consisting of 260 images, with 160 used for training and 100 for validation purposes. The key findings of the study indicated that local binary pattern (LBP), HD, and SFTA features achieved higher accuracy compared to color and shape features. The authors emphasized the significance of the feature extraction and selection stage and highlighted the effectiveness of the sparse classifier in achieving accurate results. A similar framework could be extended for the diagnosis of other types of leukemia.

Lavanya et al. [97] proposed the system of detection of anemia using image processing techniques, which are considered more reliable and cost-effective for analysis. The methods utilized in the study include edge detection, midpoint ellipse algorithm (MEA), histogram thresholding, diffused expectation maximization (DEM), Otsu's thresholding, and K-nearest neighbors (KNN) classifier. However, the results in terms of accuracy percentage were not evaluated, and the datasets used were not mentioned, which is a limitation of the study. The key findings highlight the utilization of ellipse detection and the diffused expectation (DE) algorithm for anemia detection, while the study did not consider different types of thalassemia. The mentioned framework could not be fully justified due to the lack of clear exploration of accuracy and datasets in the system evaluation.

Elsalamony [39] focused on the detection of healthy and unhealthy red blood cells (RBC) with the objective of detecting RBC, separating over-lapped cells, and performing classification. The methods employed morphological operations, circular Hough transform, watershed transform, and neural networks (NN). The results show an accuracy of greater than 97.8% in detecting different types of cells. However, the dataset used in the study is limited to 160 images, which is a limitation. The key findings of the study include the detection of various types of anemia, such as sickle cell, elliptocytosis, microcytes, and unknown shapes. The study also considers the green color as a parameter for detecting healthy cells. Different cell parameters, including area, convex area, perimeter, eccentricity, solidity, and ratio, are considered for distinguishing between healthy and unhealthy cells.

Rashid [46] explores the detection of thalassemia, aiming to develop an accurate diagnostic system for the disease. The study incorporates various techniques including global contrast enhancement, image color conversion, k-means clustering, median filtering, and the seed region growing area extraction (SRGAE) algorithm. Results indicate an average segmentation accuracy of 94.57%. However, a limitation arises from the dataset's small size, comprising only 60 images. The study primarily focuses on distinguishing between alpha, beta, and thalassemia trait images, achieving significant differentiation between these types. The SRGAE algorithm is shown to effectively extract the region of interest. The authors emphasize the simplicity and effectiveness of the contrast enhancement technique for image enhancement and underscore the importance of color conversion in the detection process. The inclusion of the SRGAE algorithm adds novelty to the study, with potential applications suggested for other microscopic analysis scenarios.

Sandanayake et al. [45] worked on the identification of thalassemia, aiming to develop an automated framework for detection. The methods utilized include gray scaling, thresholding, edge detection, artificial neural networks (ANN), and support vector machine (SVM) classifier. The results indicate an accuracy of 89% for female patients and 97% for male patients. However, it should be noted that not all types of thalassemia, which are determined by the shapes of red blood cells (RBC), are detected in this study. The authors consider six different parameters for thalassemia analysis, including RBC, hemoglobin (HB), mean corpuscular volume (MCV), mean corpuscular hemoglobin (MCH), and RBC distribution width (RDW). They suggest extending the research to detect and diagnose all types of thalassemia and exploring the severity of the disease through further investigations.

Tyas et al. [98] proposed a system for detecting minor thalassemia, with the aim of developing an automated framework for its detection. The methods employed include histogram equalization, morphological operations, and backpropagation neural network (NN). The results indicate an accuracy of 92.55%. However, it should be noted that the database used in the study is limited to only 256 images. The authors highlight the use of texture features, color features, and shape features for precise feature extraction. The utilization of 43 feature values for each cell demonstrates the emphasis on accurate feature extraction, although it may increase processing time.

Poostchi et al. [99] worked on malaria detection and cell counting, aiming to develop a faster and reliable method for malaria diagnosis. The objective involved RBC detection, feature computation, and cell classification. Methods included microscopic imaging with thin blood smear, multiscale Laplacian of Gaussian (LOG) combined with coupled edge profile active contours (C-EPAC), and the use of support vector machines (SVM) and artificial neural networks (ANN) for classification. The results indicate an accuracy of 98% for SVM and 99% for ANN. However, it should be noted that the dataset used in the study contains only 70 images. The authors highlight the combination of multiscale LOG and C-EPAC, and the superior accuracy achieved by ANN compared to SVM. Further exploration of the framework to be done for detecting additional types of parasites and assessing the severity of the disease by counting parasites and identifying their life stages.

Yildirim and Çinar [100] focused on WBC classification, which plays a crucial role in detecting various diseases, employed median and Gaussian filtering, along with a CNN classifier. The results indicate that the accuracy of the deep learning architectures used varies between 62 and 83%. However, it should be noted that the accuracy achieved in this study is lower compared to other popular frameworks. The key findings highlight the application of different deep learning architectures such as AlexNet, ResNet, DenseNet201, and GoogleNet for WBC classification. Both original data and pre-processed data are analyzed, with pre-processed data using

Gaussian and median filtering showing improved accuracy. The authors emphasized that pre-processing techniques contribute to accuracy improvement and suggest that increasing the dataset size could enhance the performance of the deep learning models.

Tantikitti et al. [44] proposed on the detection of dengue using image processing techniques. The need for a framework for viral disease detection, specifically dengue, is addressed. The objective involves blood cell classification and dengue virus detection. The methods used include color transformation, multi-level thresholding, and decision tree classification. The results show an accuracy of 72% for dengue detection and 92% for WBC classification. However, it should be noted that the database images are limited, and sub-types of the dengue virus are not detected. The key findings highlight the consideration of different parameters, such as the number and percentage of lymphocytes, phagocytes, and WBCs, as well as the percentage of hematocrit. The authors express their opinion that the decision tree algorithm, although slower, is more reliable compared to neural networks.

Different diseases diagnosis is proposed with computer-aided systems including thalassemia, anemia, dengue, malaria, and other diseases related to microscopic blood analysis. Out of these, leukemia always leads to life-threatening in the preceding stages. Hence from these articles, it could be highlighted that leukemia detection and diagnosis is the potential research area in the current scenario. Researchers have employed standard publicly available databases in most cases. The real images are to be considered and tested for the proposed frameworks to make the system reliable and robust, proving it to be used in real-life-scenario. Different techniques are employed by researchers consisting of traditional image processing [101], machine learning [102], and deep learning [103] for classification. Although, there is scope for improvement in accuracy and other performance metrics in the experimentations. There are many pros and cons of different techniques, which are to be considered when proposing the methodology for leukemia diagnosis. The preceding sub-section explores the generalized methodology flow for the detection of diseases considering microscopic imaging.

9 Generalized Methodology Flow of Disease Detection

Disease detection and diagnosis system utilizing a computer-aided approach consists of different stages shown in Fig. 11. The methodology presented here outlines a generalized approach for microscopic blood cell analysis, encompassing several distinct stages. These stages include image acquisition, image segmentation, feature extraction, and disease detection. Each stage plays a crucial role in the overall process of analyzing blood cell images and detecting diseases accurately. A trained pathologist collects a blood sample from the patient. To initiate the process, a blood smear is created by preparing a slide. Subsequently, the prepared slide is carefully examined using a high-quality microscope, resulting in the generation of an image. The image can be captured either directly using a camera or by utilizing an adapter connected

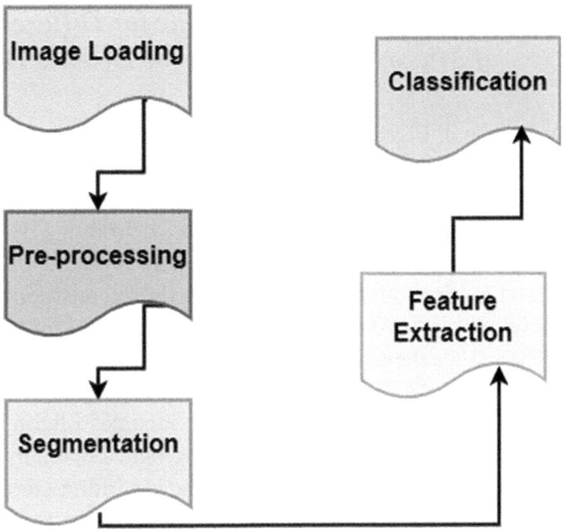

Fig. 11 Generalized methodology flow for disease detection

to the microscope. This captured image is then selected for further examination and analysis. In acquired images, it is common to encounter undesired regions and overlapping of various blood components. This image has been enhanced using a suitable image enhancement technique. Consequently, an image of high quality is obtained and ready for analysis. Subsequently, through pre-processing techniques, the different components of blood, such as red blood cells (RBCs), white blood cells (WBCs), plasma, and platelets, are effectively separated for further examination and analysis. Segmentation is performed based on the generalized characteristics of blood components.

This will distinguish the area of interest for further classification. RBC, WBC, and other components are further classified into sub-classes. This will help to specify a specific sub-class image for extracting features of blood cells and, depending on the analysis in subsequent stages such as classifier, disease detection. Following segmentation, various features are extracted by taking into account various blood components. The characteristics of various blood components, including WBC and RBC, encompass factors such as size, shape, color, and count. These attributes play a significant role in the analysis and evaluation of blood samples. The analysis of these characteristics will help to detect the disease or count the cells. A decision about the disease could be made based on the various features extracted. Different classifiers could be designed to make decisions.

9.1 State-Of-The-Art Methods for Different Stages of Microscopic Analysis for Disease Detection

As shown in Fig. 11, there are different stages of the computer-aided system for the classification of diseases. Initially, images of the patients are loaded into the system followed by pre-processing stage. After the preprocessing segmentation is done with the localization of the region of interest from the dataset images. Segmentation is followed by the feature extraction where prominent features are extracted and then final classification is done that outputs the infected person or normal person. Figure 11 shows a broad overview of the techniques used for these different stages. The next section takes the overview of the same.

Different AI Techniques for Disease Detection: A Broad Perspective In a broader perceptive preprocessing techniques includes filtering [104], edge detection [105], histogram equalization [106], morphological operations [107], and gray-scale transformation [108]. Depending upon the algorithm employed for the segmentation or classification, the pre-processing is done. The next stage is the segmentation of the dataset images in order to get the region of interest for the detection purpose. Different segmentation techniques are classified as, threshold methods [109], edge-based methods [110], watershed-based methods [111], region-based methods [112], PDE-based methods [113], clustering-based methods [114], and ANN-based methods [115]. Following Table 1 shows the pros and cons of the above-mentioned methods (Fig. 12).

In this case, the algorithm employed for classification or application determines the method to be used. Various features are extracted from segmented images and fed to the classifier for classification purposes. These features include color [116], texture [117], correlation [118], statistical [119], energy [120], entropy [121], directional moment [122], and gray level co-occurrences [123]. The importance of these features varies depending on the dataset used (Table 3).

Feature extraction is followed by the classification. Broadly the different classifiers are machine learning classifiers including SVM [124], Naïve Bayes (NB) [125], Regression [126], Decision Tree and Random Forest [127], KNN [128], and Neural Networks [129]. These classifiers could be compared with certain categories. The following Table 4 shows their comparison, with pros and cons.

As discussed earlier regarding the applications, pros, and cons of segmentation techniques, various features, and classification techniques, different researchers have utilized these methods with further exploration and modifications tailored to specific applications and datasets. The following subsection emphasizes the techniques employed by researchers, considering different stages of disease diagnosis.

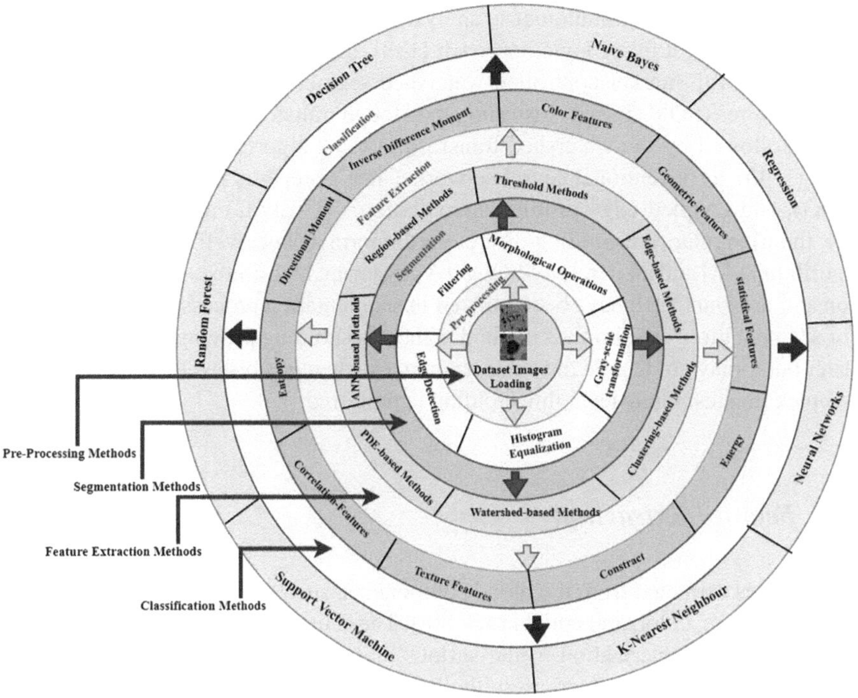

Fig. 12 Broad perceptive of disease detection methods

9.2 Image Pre-Processing

The pre-processing stage involves the application of various methods to enhance the quality and clarity of the acquired images. These methods include the utilization of the Self-dual multi-scale morphological toggle (SMTT) block [130], Wiener filtering, and median filtering [131]. Additionally, techniques such as Gaussian filtering, grayscale transformation [132] in linear, logarithmic, and power-law variations, histogram stretching [133], RGM image green color component [134], morphological operations, and edge detection [135] are employed to further refine the images and prepare them for subsequent analysis. Pre-processing depends upon the preceding stages of diagnosis, segmentation, and classification. As per the requirement of algorithm, it is performed.

9.3 Image Segmentation

Researchers have employed a variety of segmentation methods to delineate different components in the blood cell analysis process. These methods include the watershed transform [136], granulometric analysis, and mathematical morphology (MM) operations [137], a fuzzy logic approach [138], the zack algorithm [139], k-means clustering [140], marker-controlled watershed segmentation, stimulating discriminant measures (SDM) based clustering, the Hough transform [141], iterative thresholding followed by the watershed transform, Otsu's algorithm [142], edge thresholding [143], a conventional neural network, Chen's prepared Laplacian of Gaussian (LOG), and coupled edge profile active contours (C-EPAC) algorithm. Additionally, the triangular thresholding discrete orthonormal Stockwell transform (DOST) algorithm [144] and an iterative ISODATA clustering algorithm combined with rotation and invariant LBP have been utilized in segmentation processes by researchers. For segmentation, the simplest method is the thresholding proving to be simple and faster but it may not handle the complex images when it is bi-level. For handling the complex images, multi-level thresholding is preferred.

9.4 Feature Extraction

The features extracted from the blood components, such as WBC and RBC, encompass size, shape, color, and count [145–148]. These features play a crucial role in the analysis and characterization of the various blood components. Table 2 shows the features for different applications with their comparative analysis.

9.5 Classification

There are various classifiers for image classification that are used for microscopic imaging of blood cells. These include machine learning algorithms, which are listed below. Decision Tree Classifier [149], Random Forest [150], K-Nearest Neighbors (KNN) [112], Logistic Regression, Binary Logistic Regression model, Multinomial logistic regression, and Ordinal regression, Naive Bayes Algorithms including Gaussian, Multinomial, and Bernoulli Naive Bayes [128], Support Vector Machine (SVM) [151], Convolutional Neural Networks [152]. Various researchers utilize different classifiers for disease detection in microscopic blood analysis. The selection of a specific classifier depends on the particular disease under investigation, and there is no strict rule in the literature regarding the use of a particular classifier for a specific disease. In general, classifiers such as Support Vector Machines (SVM), decision trees, and Naive Bayes have well-documented performances. Moreover, these classifiers demonstrate effective performance even with smaller training databases. On the other hand, advanced machine learning classifiers like Neural Networks (NN) and its variants require larger databases for training, which may result in longer decision-making times. However, NN has shown promising results in disease classification and detection accuracy across various cases. Furthermore, researchers have proposed a diverse range of approaches in the field of medical

Table 2 Segmentation methods comparison

Sr. no	Method	Pros	Cons
1.	Threshold methods	Simple and fast to compute	Sensitive to noise and variations in illumination
		Effective for images with clear intensity separation between foreground and background	Difficult to handle images with complex backgrounds or foregrounds
2.	Edge-based methods	Effective in detecting object boundaries and contours	Sensitive to noise and weak edges
		Can handle images with complex backgrounds	May generate fragmented or incomplete segmentations
3.	Watershed-based methods	Can handle images with multiple objects and complex overlapping structures	Sensitive to noise and may generate oversegmentation
		Provide accurate object boundaries	Requires careful parameter tuning to obtain satisfactory results
4.	Region-based methods	Effective in segmenting objects with uniform or similar appearance	Sensitive to initial seed selection
		Can handle images with intensity variations and noise	May struggle with objects that have similar appearance but different structures
5.	PDE-based methods	Can handle images with intensity variations and noise	Computationally expensive and time-consuming
		Provide smooth and continuous segmentations	Sensitive to parameter selection and initialization
6.	Clustering-based methods	Effective in identifying groups of pixels with similar properties	Sensitivity to noise and outliers
		Can handle images with complex backgrounds and foregrounds	May require manual parameter tuning and initialization
7.	ANN-based methods	Can learn complex and non-linear relationships between pixels	Requires a large amount of training data
		Can handle images with diverse appearances and backgrounds	Computationally expensive during both training and inference

diagnosis, leveraging machine learning, deep learning, and artificial intelligence techniques. These methods include Convolutional Neural Networks (CNN) [153], Recurrent Neural Networks (RNN) [154], U-nets and their advanced modifications [155], as well as deep learning algorithms like feed-forward networks [156] and auto-encoders. Notably, for deep learning-based diagnosis, prominent models such as CNN, AlexNet, and ResNet have been employed. LeNet- 5, a pioneering 7-level

Table 3 Different features with applications

Sr. no	Features	Application	Comparison
1.	Color features	Object recognition, image classification, and color-based segmentation	Color features can capture information about hue, saturation, and brightness, enabling discrimination between objects based on their color characteristics. They are particularly useful when color is a distinctive attribute for distinguishing objects or when color-based analysis is required
2.	Texture features	Texture classification, material recognition, image retrieval	Texture features describe the spatial arrangement and variation of pixel intensities within an image. They are valuable for distinguishing between different surface textures or materials. Popular texture features include local binary patterns (LBP), gray-level co-occurrence matrices (GLCM), and Gabor filters
3.	Correlation features	Image registration, motion analysis, pattern recognition	Correlation features measure the similarity or dissimilarity between two signals or images. They are often used in tasks such as image alignment, tracking, and recognition, where matching or comparing patterns or templates is required
4.	Statistical features	Image segmentation, object detection, pattern recognition	Statistical features, such as mean, variance, skewness, and kurtosis, describe the statistical properties of image intensity distributions. They are widely used for tasks like image segmentation, where different regions can be characterized by their statistical properties
5.	Energy and entropy features	Image compression, image quality assessment, texture analysis	Energy and entropy features provide information about the distribution and randomness of pixel intensities within an image. They are commonly employed in tasks like image compression, where measuring the amount of information or the level of detail is important
6.	Directional moment features	Edge detection, shape analysis, texture analysis	Directional moment features capture the directional characteristics of image edges or texture patterns. They are often utilized for tasks that require analyzing the orientation or directionality of structures or features within an image
7.	Gray level co-occurrence features	Texture analysis, image classification, and pattern recognition	Gray level co-occurrence features capture the spatial relationships between pixel intensities in an image. They are useful for characterizing texture patterns by measuring the occurrence and distribution of pixel intensity pairs at different spatial offsets

Table 4 Applications of different classifier algorithms

Sr. no	Algorithm	Applications	Pros	Cons
	Support Vector Machine (SVM)	Classification tasks for both binary and multiclass problems	Effective in high-dimensional spaces	Computationally expensive for large datasets
		Image recognition, text categorization, and sentiment analysis	Robust against overfitting, especially with the kernel trick	Requires careful tuning of hyperparameters
			Versatile due to the use of different kernel functions	May not perform well with noisy data
	Naive Bayes (NB)	Text classification, spam filtering, sentiment analysis	Simple and computationally efficient	Assumes feature independence (naive assumption)
		Document categorization, recommendation systems	Performs well with high-dimensional data and a large number of features	May not perform well with highly correlated features
			Robust to irrelevant features	Sensitive to feature scaling
	Regression	Predictive modeling when the target variable is continuous	Easy to implement and interpret	Sensitive to outliers in the training data
		Stock price prediction, real-estate price estimation	Suitable for both simple and complex relationships between variables	Limited in handling categorical or discrete features
			Provides insights into feature importance. Cons:	Prone to overfitting with high-dimensional data
	Decision Tree	Classification and regression tasks	Easy to interpret and visualize	Prone to overfitting, especially with deep trees
		Medical diagnosis, credit risk assessment	Can handle both numerical and categorical data	Sensitive to small variations in the training data
			Does not require feature scaling. Cons:	Not suitable for capturing complex relationships between features
	Random Forest	Classification and regression tasks	Reduces overfitting by aggregating multiple decision trees	Computationally expensive during training and inference
		Anomaly detection, recommendation systems	Handles high-dimensional data and large datasets	Harder to interpret compared to single decision trees

(continued)

Table 4 (continued)

Sr. no	Algorithm	Applications	Pros	Cons
			Provides feature importance scores	May not perform well on imbalanced datasets
	k-Nearest Neighbors (KNN)	Classification and regression tasks	Simple and easy to implement	Computationally expensive during inference, especially with large datasets
		Recommender systems, anomaly detection	Non-parametric, thus suitable for complex data distributions	Sensitive to irrelevant features and feature scaling
			Performs well with small to medium-sized datasets. Cons:	Requires tuning of the k parameter
	Neural Networks	Image and speech recognition, natural language processing	Highly capable of capturing complex patterns and relationships	Requires a large amount of data for training
		Autonomous vehicles, game playing (e.g., AlphaGo)	Can handle high-dimensional and unstructured data	Computationally intensive, especially for deep architectures
			State-of-the-art performance in various domains	Prone to overfitting, necessitating regularization techniques

convolutional network, has also been utilized in this area. However, it is worth noting that the computational resources available can impose constraints, particularly when handling higher-resolution images that require larger and more convolutional layers. When compared to Letnet-5, Alexnet doubles the layer depth [157]. ZFNet was created as a result of the classifier's behaviour and the visualization of intermediary feature layers [158]. When compared to AlexNet, the convolutions' stride and filter diameters are both reduced. Deep CNN designs compromise the computational expenses with an increase in accuracy. The Inception/Googlenet design [159] takes care of this. VGGNet is quite appealing because it has 16 convolutional layers and a very homogeneous architecture. Remaining neural networks use skip connections or shortcuts to skip over some levels. The bulk of ResNet models employ batch normalization in addition to double- or triple-layer skips with nonlinearities (ReLU) [160]. Convolutional neural networks (CNNs), Graph neural networks (GNNS), Explainable deep learning neural networks (xDNNs), and Transfer learning (TL), among other more efficient deep learning net- works, perform better in medical diagnostics. A model built for one task is used as the foundation for a model on another task using the machine learning method known as transfer learning [161]. A type of deep learning algorithm called Graph Neural Networks (GNNs) is used to deduce data from graphs. Graph-level, edge-level, and node-level prediction tasks are all made easy by GNNs, which are neural networks that can be applied directly to graphs [162]. In deep learning uses, assembling learning may also prove effective. Ensembling is a method for combining different learning algorithms to obtain their combined results. Although most applications showed that individual deep learning models were competent, it is always possible to use a group of deep learning models to accomplish the same job as an assembly technique [163]. Image capture is essential for the detection and diagnosis of a specific illness using image processing techniques. Images may produce inaccurate predictions as a result of certain sounds and other infections. In order to improve the sample quality, some pre-processing techniques must be used after picture capture. Typically, samples are pre-processed to improve their quality in order to obtain accurate disease prediction findings. These processes include grey-scaling, pre-filtering, morphological operations, and edge detections [164]. Classification is done via machine learning and deep learning classifiers. As we go for deep learning, the requirement of dataset is always huge for proper training of the classifier. But in case of leukemia, the challenge is the limited sized dataset. Hence, the designing of classification methodology should be done considering the limited dataset availability [165]. The real-image-dataset is not utilized by most of the researcher which is necessary to make the system robust and to provide real-life-solutions [166].

9.6 Explainability

In numerous real-world issue domains, explainability is crucial [167]. Re-searchers have created a variety of software-based frameworks and systems to address societal problems in the actual world [168]. Computer vision and related disciplines address a wide range of issues, including social, technical, financial, commercial, medical, securities, and any additional multidisciplinary problems that may arise. In addition to various computer vision techniques, academics are looking into some complex and effective methods. Artificial intelligence, deep learning, and machine learning are some of these. The enhanced performance in terms of various measures is the best part about these newly developed AI algorithms [169]. Medical diagnosis is the most important real-world issue because it immediately impacts people's lives [4]. Any illness should have an accurate diagnosis so that a patient can receive the right treatment recommendations [170]. In the medical community, there are various methods to diagnose the disorder. Some diseases' signs are used to make a diagnosis [171]. However, the generalized symptoms are comparable in many diseases [172]. Therefore, it is essential to use various diagnostics methods for the same. Traditional image processing [173], computer vision [174], and sophisticated AI methods [175] are used for imaging diagnosis. Figure illustrates the strange trade-off between explainability and precision [176].

Explainability decreases as precision rises. In comparison to recently developed AI techniques, earlier created traditional signal and image processing techniques produce unsatisfactory results [177]. The tragic aspect of these methods, how- ever, is their inexplicability [178]. It was unable to describe how the algorithm works. Contrarily, traditional image and signal processing methods provide less accurate diagnoses while explaining the algorithms in a seemingly flawless manner [179]. As a result, researchers were compelled to employ AI techniques in their study on medical diagnosis and related topics [180]. For AI techniques to be trusted and widely used for diagnosis, it produced the need for explainable frameworks.

9.6.1 Three Stages of Explainability

The three phases of the XAI framework are depicted in Fig. 13 with the first phase being an explainable building process, the second phase is an explainable choice, and explainable decision process is the third phase. The framework's stage-by-stage description is provided below (Fig. 14).

Stage-1 (Facilitates Acceptance)

A multidisciplinary team of experts in their fields who are familiar with the AI they are creating can be formed with the help of the initial explainability stage. Techniques for offline explainability are essential for future AI adoption and offer opportunities to enhance the system.

Stage-2 (Fosters Trust with Users and Supervisors)

In business, daily living, and particularly in the diagnosis of medical conditions, trust is regarded as being of the utmost importance. The difference between an autonomous system and one that is "micro-controlled" by its users is what is thought of as the confidence. A system's value decreases in direct proportion to how much management it requires. When a system acts in accordance with our mental model,

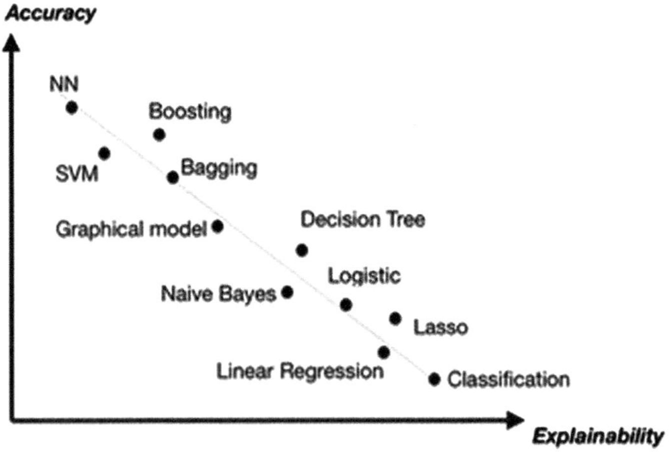

Fig. 13 Accuracy and explainability trade-off [181]

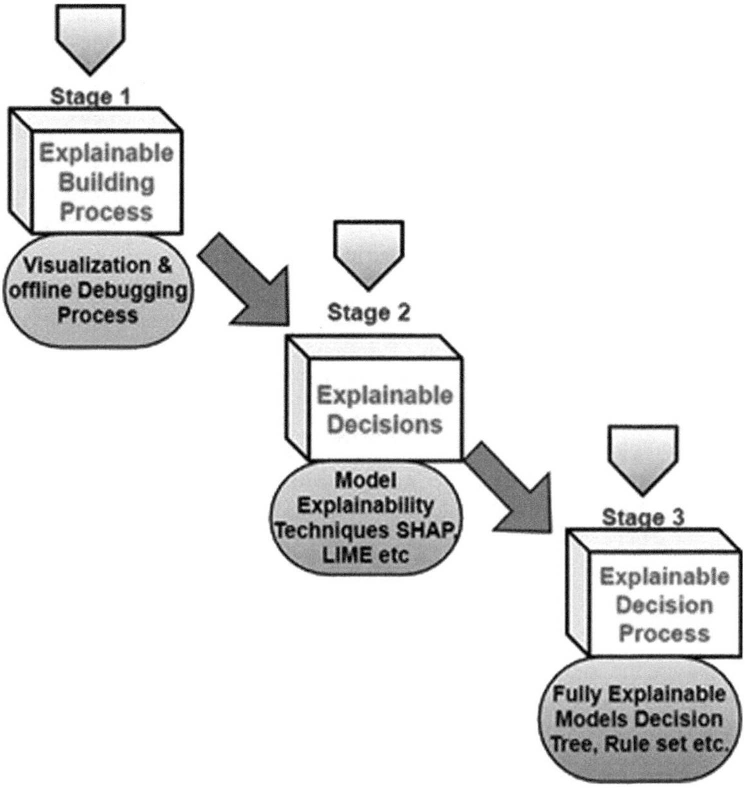

Fig. 14 Three stages of explainability

we grow to trust it because it won't surprise us. Users of a system who are aware of its limitations are more apt to benefit from it than those whose results are thought to be unreliable. Any model can be created more quickly with justifications. This is where the capacity to rationalize AI choices is useful. The second step is explained by this.

Stage-3 (Enables Interoperability with Business Logic)

The goals of Stages 1 and 2 are to help people comprehend how AIs work mentally. This enables people to critically evaluate how AIs operate and decide when to trust and accept their results, forecasts, or recommendations. You'll need to develop business logic that will gradually apply the same "reasoning" to a large number of AIs if you want to expand this up. Stage 3 focuses on achieving interoperability between AIs and other software, especially business logic software. Following section explores the popular explainable frameworks.

9.6.2 Explainable Frameworks

LIME Model

LIME is a technology created by researchers at the University of Washington to get excellent transparency into the happenings inside the algorithm. It provides localizable explanations that are model independent. It becomes more challenging to preserve the local realism for the models as the dimensions rise. The discovery of a model that replicates the original model in a localized way, on the other hand, is a much more manageable issue that LIME deals with. LIME takes interpretability into account during both the optimization process and the idea of an interpretable representation, enabling the addition of domain- and task-specific interpretability requirements [182]. A modular approach called LIME can describe any model's predictions clearly and precisely. The researchers suggested SP-LIME, which is used for choosing notable and unique predictions that present viewers with a comprehensive view of the model. The AI model's test findings are accepted. It consists of three stages that are local, model- neutral, and interpretable (Fig. 15).

What-If Tool

The TensorFlow team created the What-If Tool, an interactive visual tool for visualizing datasets and better comprehending the outputs of Tensor-Flow models. It is used to analyze the models that have been used. In addition to TensorFlow models, the What-If Tool can be applied to XG-Boost, Scikit Learn, and other models. When a model is deployed, this tool allows users to examine its performance on a dataset. It enables people to examine, compare, and research machine learning models, which helps them better understand a classification or regression model. Due to its simple user interface and absence of complex coding requirements, anyone can use it, including developers, product managers, researchers, and students [183]. Additionally, the dataset can be divided into segments according to features, and performance can be compared across different slices, revealing which subsets of data the model

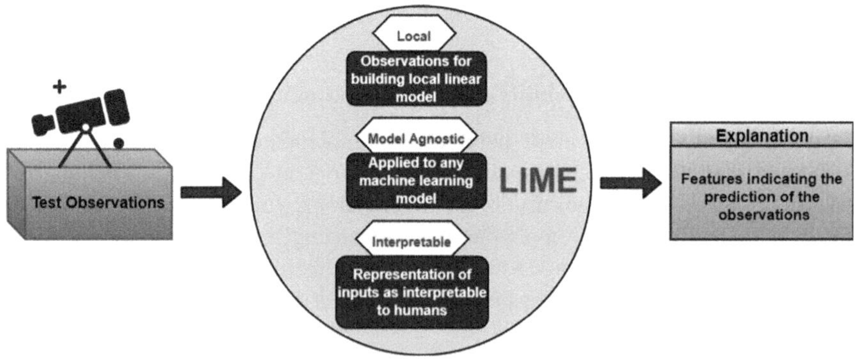

Fig. 15 LIME framework of explainability

performs best on and which it performs worst on. This feature is particularly helpful for ML fairness studies. With just a visual interface, the tool aims to give users a straightforward, perceptive, and effective way to play with a trained machine learning model on a set of data. The actions that the what-if utility is supposed to take are shown in Fig. 16.

Skater

In order to help with the development of interpretable machine learning systems, which are commonly needed for applications linked to the real-world, it is a single framework that enables Model Interpretation for all types of models. Skater is a free, open-source Python tool that breaks down the learned structures of a model that might otherwise be viewed as a "black box" on both a global and local scale [184].

SHAP

The average marginal input of a feature value across all likely coalitions is referred to as SHAP, also known as Shapley Additive explanations. SHAP aims to quantify the contribution of each attribute to the prediction of an instance x. Shapley values are computed using the coalitional game theory-based SHAP explanation method, where the feature values of a data instance act as coalition participants. The "payout" (i.e., forecast) of a prediction should be distributed among its different features according to Shapley values. Participants can be either single entities or collections of elements. Super-pixels, collections of pixels where the forecasts for an image's explanation are

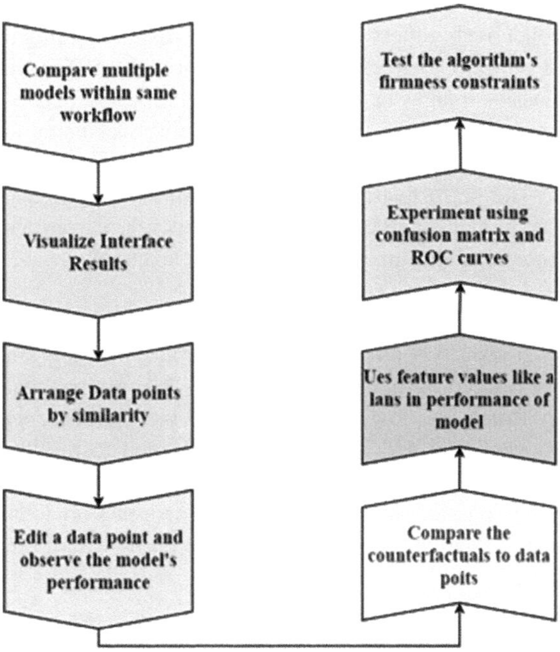

Fig. 16 Things could be performed by What-if-tool. (modified after: https://towardsdatascience.com/using-what-if-tool-to-investigate-machine-learning-models-913c7d4118f)

dispersed, play a role in this process. The Shapley value explanation, portrayed as an additive feature attribution method and a linear model, is one of SHAP's key contributions [185].

AIX360

It is possible to analyse datasets and machine learning models using this open-source, cost-free application. The Python-based AI Explainability 360 package provides a wide range of algorithms and proxies explainabil- ity measures for many different sections of explanations. A compendium of taxonomy terms and user instructions are available on the AIX360 web- site for users who are not experts in explainable AI. In order to satisfy the broad range of explainability methodologies required by different stake- holders, the AIX360 toolkit aims to provide a consistent, adaptable, and user-friendly programming interface and accompanying software architecture [186]. The purpose is to appeal to algorithm engineers and data scientists, who might not be specialists in explainability.

Activation Atlases

Activation Atlases, created by Google in collaboration with OpenAI, were a ground-breaking method for visualising the interactions between neural networks as well as how they evolve with experience and layer depth. This technique was developed to study the inner workings of convolutional vision networks and to derive a conceptual outline from the hidden layers of the network. The early feature visualisation study was primarily concerned with individual neurons. Activation atlases move beyond the observation of individual neurons to the visualisation of the space those neurons collectively reflect by collecting and displaying hundreds of thousands of examples of neurons interacting. When a neural network relies on incorrect correlations to classify images or when a feature is repeated between two groups, it can result in unexpected flaws that humans can discover using activation atlases [187]. Using this information, people can even "attack" the model by altering photos to deceive it. The performance of activation atlases was greater than anticipated, suggesting that human significance of neural network activations may exist. As a result, useful interpretability in vision models can be attained.

Rulex Explainable AI

First-order conditional logic principles that are simple to comprehend and use are offered by the company Rolex. The primary machine learning algorithm used by Rolex, the Logic Learning Machine (LLM), operates entirely differently from conventional AI. The system is designed to produce conditional logic rules that predict the best choice option in an understandable manner even for process experts. Rolex regulations have made every forecast obvious. Rolex rules are independent and overlapped, in contrast to decision trees and other rule-generating algorithms [188].

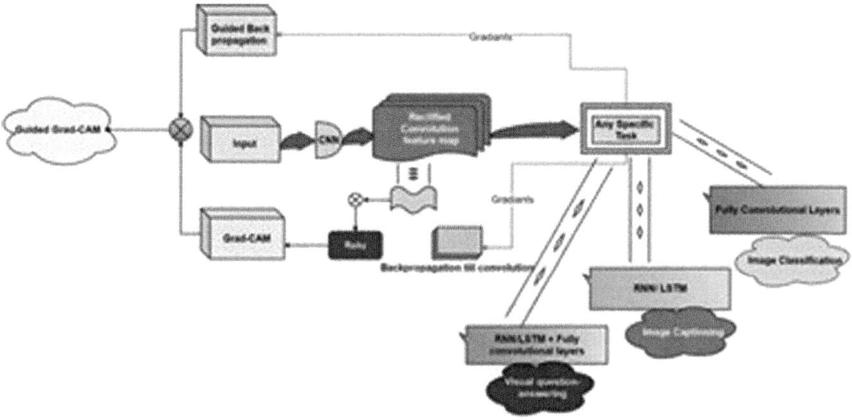

Fig. 17 Overview of Grad-CAM [190]

GradCAM

Using the gradients of any target concept flowing into the final convolutional layer, gradient-weighted class activation mapping (Grad-CAM) creates a coarse localization map highlighting the crucial areas in the picture for predicting the idea. It is a generalization of Class Activation Mapping (CAM), which can be applied to CNN models with fully connected layers but requires the use of a global average pooling layer for fully CNN models. The overview of the Grad-CAM explainability architecture is shown in Fig. 17. The activation maps for the relevant layers are acquired after the image is fed into the network with a target class. Back-propagating a one-hot signal with the desired class set to 1 to the corrected convolutional feature maps of interest yields the coarse Grad-CAM saliency map [189].

10 Discussions

Leukemia is the abnormality associated with the creation of blood by the bone marrow. The diagnosis of the same is crucial, as it is a life-threatening infection in its later stages. Diagnosis of leukemia is primarily done via microscopic imaging of blood slides, observing the attributes of the blood cells, which needs a very experienced pathologist. More significantly, there are different sub-types of leukemia that adds the complexities in the diagnosis of exact sub-type. The major sub-types included CML, CLL, ALL, and AML which differ in the diagnosis and treatment. Symptomatic diagnosis plays an additional role in the diagnosis of leukemia but this has the challenge of finding the exact dissimilarities in diagnosing and categorizing the sub-type of leukemia because the symptoms, blood tests and imaging scans are quite similar in these broader sub-types CML. CLL, AML, and ALL. So these different systematics and symptoms could be very challenging in the exact diagnosis of leukemia.

Moreover, the computer-aided-diagnosis could add significant contributions in the assured automatic diagnosis of diseases. It consisted of image loading, pre-processing, segmentation, and classification of the disease. The researchers proposed different methods for CAD system stages for the diagnosis of leukemia. Pre-processing stage will improve the image quality and will make the image suitable to pass to the segmentation stage of the diagnosis. It primarily include image enhancement, filtering and resizing of the image. Out of the different segmentation methods proposed by the researchers, the simplest method is the thresholding, which selects the particular threshold and gives the region of interest. Though, the thresholding methods are simpler for implementations, these methods required optimization to improve the time inexpensiveness and searching the perfect threshold. For the further improvement in the thresholding based segmentation, multi-level segmentation could play crucial role in the locating the region of interest in the microscopic images of blood for the detection of leukemia. Further for the classification purpose, the different machine learning and deep learning classifiers were employed by different researchers. The most challenging task during the classification is the limited dataset sample size, which leads to overfitting during the classification process. Also the sub-types of leukemia were not explored in many databases available publicly. Moreover, in addition to the publicly available databases, a good multi-class real-image-database is required to make the robust system of classification. A system with could be designed that work effectively for limited sized datasets with assured diagnosis utilizing the multi-class real-images of leukemia sub-types.

Explainability issues of deep learning frameworks also need to be viewed seriously during the implementation of deep learning frameworks for the diagnosis purpose. The trade-off between accuracy and explainability stated the fact that the explainability reduces with the increase in the classification accuracy. Researchers proposed different frameworks for the explainability of deep learning models. Out of the different XAI frameworks, the proper framework needed to be selected depending upon the application. In medical imaging applications, it is necessary to indicate the prominent features considered for the diagnosis decision. In medical diagnosis each and every local case is equally important, hence instead of global features, local features of each individual case are to be explained by the XAI framework. In this regard, local-interpretable-model-agnostic-explainations (LIME) framework is more appropriate for leukemia diagnosis explanability.

11 Conclusion

This chapter discussed the literature related to microscopic imaging, application areas where microscopic blood analysis is applicable, and details of the disease leukemia. A critical literature analysis is done considering the articles in the medical diagnosis field with consideration of microscopic analysis. Outcome of the survey indicated the disease leukemia has a good research potential and as it's a life-scaring disease, significant contribution needed in this area with respect to supportive diagnostic system. Leukemia is discussed with its different types, its systematics, and diagnosis procedure. Different AI techniques for the disease detection are surveyed and are critically reviewed to propose the research gaps, research questions, and accordingly the objectives are frame.

It's crucial to look at the traits of a black-box that might render an incorrect decision in support of the incorrect cause. It's a significant issue that could have a negative impact on the system's functionality once it is put into use in the actual world. The bulk of the techniques, especially the ones that rely on attribution, are open source implementations. Commercial interest is growing in explainability and, in particular, in the attribution methods that can be applied to a variety of business use cases. Deep learning models have greatly improved at describing their choices, especially those used for medical diagnosis. Model makers can overcome reliability issues by having a better understanding of the factors that affect decisions, which will help users gain confidence and make wiser choices. Nearly all of these methods seek to establish local explainability. The majority of deep learning interpretability algorithms create saliency maps, which highlight the impact of different image regions, using picture classification. The most complete and widely used approaches to explaining any black-box model are the LIME and SHAP techniques for visualizing feature interactions and feature importance. Grad-CAM, a visual explainability, has recently experienced some excellent growth in terms of interpretability and explainability.

References

1. B. Houwen, Blood film preparation and staining procedures. Clin. Lab. Med. **22**(1), 1–14 (2002)
2. A.S. Adewoyin, Peripheral blood film-a review. Ann. Ibadan Postgraduate Med. **12**(2), 71–79 (2014)
3. D.H. Kuan, C.C. Wu, W.Y. Su, N.T. Huang, A microfluidic device for simultaneous extraction of plasma, red blood cells, and on-chip white blood cell trapping. Sci. Rep. **8**(1), 15345 (2018)
4. N.M. Deshpande, S. Gite, R. Aluvalu, A review of microscopic analysis of blood cells for disease detection with AI perspective. Peer J. Comput. Sci. **7**, e460 (2021)
5. J.L. Vives Corrons, S. Albarède, G. Flandrin, S. Heller, K. Horvath, B. Houwen, J.C. Libeer, Haematology working group of the European external committee for external quality assurance programmes in laboratory medicine guidelines for blood smear preparation and staining procedure for setting up an external quality assessment scheme for blood smear interpretation. Part I: control material. Clin. Chem. Lab. Med. **42**(8), 922–926 (2004)
6. S. Abbas, Microscopic images dataset for automation of RBCs counting. Data Brief **5**, 35–40 (2015)
7. L.A. Bhavnani, U.K. Jaliya, M.J. Joshi, Segmentation and counting of WBCs and RBCs from microscopic blood sample images. Int. J. Image, Graph. Signal Process. **8**(11), 32 (2016)
8. M.F. Greaves, G. Brown, N.T. Rapson, T.A. Lister, Antisera to acute lymphoblastic leukemia cells. Clin. Immunol. Immunopathol. **4**(1), 67–84 (1975)
9. R.D. Meyer, P. Rosen, P., D. Armstrong, Phycomycosis complicating leukemia and lymphoma. Ann. Internal Med. **77**(6), 871–879 (1972)

10. B. Deschler, M. Lübbert, Acute myeloid leukemia: epidemiology and etiology. Cancer: Interdiscip. Int. J. Am. Cancer Soc. **107**(9), 2099–2107 (2006)
11. A. Ivey, R.K. Hills, M.A. Simpson, J.V. Jovanovic, A. Gilkes, A. Grech, D. Grimwade, Assessment of minimal residual disease in standard-risk AML. New Engl. J. Med. **374**(5), 422–433 (2016)
12. M. Gruber, I. Bozic, I. Leshchiner, D. Livitz, K. Stevenson, L. Rassenti, C.J. Wu, Growth dynamics in naturally progressing chronic lymphocytic leukaemia. Nature **570**(7762), 474–479 (2019)
13. H. Yao, E. Ashihara, T. Maekawa, Supportive therapies for myeloid leukemia including blood transfusion and growth factors. Nihon rinsho. Japan. J. Clin. Med. **67**(10), 1951–1957 (2009)
14. K.K. Sabarwal, R.P. Singh, Hazardous effects of chemical pesticides on human health–cancer and other associated disorders. Environmental Toxicology and Pharmacology Volume 63, October 2018, pp. 103–114. https://doi.org/10.1016/j.etap.2018.08.018
15. A. Sabarwal, K. Kumar, R.P. Singh, Hazardous effects of chemical pesticides on human health–cancer and other associated disorders. Environ. Toxicol. Pharmacol. **63**, 103–114 (2018)
16. V.J. Ramya, S. Lakshmi, A methodical review of diagnosis and classification of leukemia using blood microscopic images, in *2019 International Conference on Communication and Signal Processing (ICCSP)* (IEEE, 2019), pp. 0258–0262
17. M. Jones, D. Arnfield, P. Barber, How to get the most from blood samples: a guide to producing diagnostic blood smears. Vet. Nurse **11**(4), 184–189 (2020)
18. E. Kazemi-Rad, M. Mohebali, H. Hajjaran, S.A.S.A.N. Rezaei, S. Mamishi, Diagnosis and characterization of Leishmania species in Giemsa-stained slides by PCR-RFLP. Iran. J. Public Health **37**(1), 54–60 (2008)
19. D. Cruz, C. Jennifer, L.C. Castor, C.M.T. Mendoza, B.A. Jay, L.S.C. Jane, P.T.B. Brian, Determination of blood components (WBCs, RBCs, and Platelets) count in microscopic images using image processing and analysis, in *2017IEEE 9th International Conference on Humanoid, Nanotechnology, Information Technology, Communication and Control, Environment and Management (HNICEM)* (IEEE, 2017), pp. 1–7
20. R.T. Raphael, K.R. Joy, Segmentation and classification techniques of leukemia using image processing: an overview, in *2019 International Conference on Intelligent Sustainable Systems (ICISS)* (IEEE, 2019), pp. 378–384
21. P. Jagadev, H.G. Virani, Detection of leukemia and its types using image processing and machine learning, in *2017 International Conference on Trends in Electronics and Informatics (ICEI)* (IEEE, 2017), pp. 522–526
22. S. Agaian, M. Madhukar, A.T. Chronopoulos, Automated screening system for acute myelogenous leukemia detection in blood microscopic images. IEEE Syst. J. **8**(3), 995–1004 (2014)
23. V.J. Anselmo, B.E. Zawacki, Infrared photography as a diagnostic tool for the burn wound, in *Quantitative Imagery in the Biomedical Sciences II*, Vol. 40, pp. 181–188). SPIE (1974)
24. N. Toepfner, C. Herold, O. Otto, P. Rosendahl, A. Jacobi, M. Kräter, J. Guck, Detection of human disease conditions by single-cell morpho-rheological phenotyping of blood. elife **7**, e29213 (2018)
25. K.K. Anilkumar, V.J. Manoj, T.M. Sagi, A survey on image segmentation of blood and bone marrow smear images with emphasis to automated detection of Leukemia. Biocybern. Biomed. Eng. **40**(4), 1406–1420 (2020)
26. Q. Bani Baker, M.A. Alsmirat, K. Balhaf, M.A. Shehab, Accelerating white blood cells image segmentation using GPUs. Concurr. Comput. Pract. Exp. **33**(2), e5133 (2021)
27. F. Xing, L. Yang, Robust nucleus/cell detection and segmentation in digital pathology and microscopy images: a comprehensive review. IEEE Rev. Biomed. Eng. **9**, 234–263 (2016)
28. N. Salem, N.M. Sobhy, M. El Dosoky, A comparative study of white blood cells segmentation using otsu threshold and watershed transformation. J. Biomed. Eng. Med. Imaging **3**(3), 15 (2016)

29. M. Imran Razzak, S. Naz, Microscopic blood smear segmentation and classification using deep contour aware CNN and extreme machine learning, in *Proceedings of the IEEE Conference on Computer Vision and Pattern Recognition Workshops* (2017), pp. 49–55
30. Y. Liu, F. Cao, J. Zhao, J. Chu, Segmentation of white blood cells image using adaptive location and iteration. IEEE J. Biomed. Health Inform. **21**(6), 1644–1655 (2016)
31. F. Al-Hafiz, S. Al-Megren, H. Kurdi, Red blood cell segmentation by thresholding and Canny detector. Procedia Comput. Sci. **141**, 327–334 (2018)
32. J. Prinyakupt, C. Pluempitiwiriyawej, Segmentation of white blood cells and comparison of cell morphology by linear and naïve Bayes classifiers. Biomed. Eng. Online **14**, 1–19 (2015)
33. Z. Zhong, T. Wang, K. Zeng, X. Zhou, Z. Li, White blood cell segmentation via sparsity and geometry constraints. IEEE Access **7**, 167593–167604 (2019)
34. A.H. Chaudhary, J. Ikhlaq, M.A. Iftikhar, M. Alvi, Blood cell counting and segmentation using image processing techniques. Appl. Intell. Technol. Healthc. 87–98 (2019)
35. M. Sajjad, S. Khan, Z. Jan, K. Muhammad, H. Moon, J.T. Kwak, I. Mehmood, Leukocytes classification and segmentation in microscopic blood smear: a resource-aware healthcare service in smart cities. IEEE Access **5**, 3475–3489 (2016)
36. S.T. Mohamed, H.M. Ebeid, A.E. Hassanien, M.F. Tolba, Optimized feed forward neural network for microscopic white blood cell images classification, in *The International Conference on Advanced Machine Learning Technologies and Applications (AMLTA2019)*, Vol. 4 (Springer International Publishing, 2020), pp. 758–767
37. H. Miao, C. Xiao, Simultaneous segmentation of leukocyte and erythrocyte in microscopic images using a marker-controlled watershed algorithm. Comput. Math. Methods Med. **2018** (2018)
38. M.V. Bills, B.T. Nguyen, J.Y. Yoon, Simplified white blood cell differential: an inexpensive, smartphone-and paper-based blood cell count. IEEE Sens. J. **19**(18), 7822–7828 (2019)
39. H.A. Elsalamony, Healthy and unhealthy red blood cell detection in human blood smears using neural networks. Micron **83**, 32–41 (2016)
40. K. Alotaibi, *Sickle Blood Cell Detection Based on Image Segmentation* (South Dakota State University, 2016)
41. B. Javidi, A. Markman, S. Rawat, T. O'Connor, A. Anand, B. Andemariam, Sickle cell disease diagnosis based on spatiotemporal cell dynamics analysis using 3D printed shearing digital holographic microscopy. Opt. Express **26**(10), 13614–13627 (2018)
42. M. Poostchi, K. Silamut, R.J. Maude, S. Jaeger, G. Thoma, Image analysis and machine learning for detecting malaria. Transl. Res. **194**, 36–55 (2018)
43. Y. Duan, J. Wang, M. Hu, M. Zhou, Q. Li, L. Sun, Y. Wang, Leukocyte classification based on spatial and spectral features of microscopic hyperspectral images. Opt. Laser Techno. **112**, 530–538 (2019)
44. S. Tantikitti, S. Tumswadi, W. Premchaiswadi, Image processing for detection of dengue virus based on WBC classification and decision tree, in *2015 13th International Conference on ICT and Knowledge Engineering (ICT Knowledge Engineering 2015)* (IEEE, 2015), pp. 84–89
45. T. Sandanayake, A.T.P.M.N. Thalewela, H. Thilakesooriya, R.M.A.U. Rathnayake, S.A.Y.A. Wimalasooriya, Automated thalassemia identifier using image processing. Tech. Rep. 53696127 (2016)
46. N.Z.N. Rashid, M.Y. Mashor, R. Hassan, Unsupervised color image segmentation of red blood cell for thalassemia disease, in *2015 2nd International conference on biomedical engineering (ICoBE)* (IEEE, 2015), pp. 1–6
47. A.S. Negm, O.A. Hassana, A.H. Kandil, A decision support system for acute leukemia classification from digital microscopic images. J. Adv. Res. (2015)
48. S. Chin Neoh, W. Srisukkham, L. Zhang, S. Todryk, B. Greystoke, C. Peng Lim, N. Aslam, An intelligent decision support system for leukaemia diagnosis using microscopic blood images. Sci. Rep. **5**(1), 1–14 (2015)
49. H. Singh, G. Kaur, Automatic detection of blood cancer in microscopic images: a review. Dr. Balkrishan Int. J. Innov. Adv. Comput. Sci. **6**(4), 40–43 (2017)

50. S. Shafique, S. Tehsin, S. Anas, F. Masud, Computer-assisted acute lymphoblastic leukemia detection and diagnosis, in *2019 2nd International Conference on Communication, Computing and Digital systems (C-CODE)* (IEEE, 2019), pp. 184–189
51. L. Putzu, C. Di Ruberto, White blood cells identification and counting from microscopic blood images. World Acad. Sci. Eng. Technol. **73**, 363–370 (2013)
52. K.K. Jha, H.S. Dutta, Mutual information based hybrid model and deep learning for acute lymphocytic leukemia detection in single cell blood smear images. Comput. Methods Programs Biomed. **179**, 104987 (2019)
53. Z. Moshavash, H. Danyali, M.S. Helfroush, An automatic and robust decision support system for accurate acute leukemia diagnosis from blood microscopic images. J. Digit. Imaging **31**, 702–717 (2018)
54. S. Mishra, B. Majhi, P.K. Sa, Texture feature based classification on microscopic blood smear for acute lymphoblastic leukemia detection. Biomed. Signal Process. Control **47**, 303–311 (2019)
55. V. Acharya, P. Kumar, Detection of acute lymphoblastic leukemia using image segmentation and data mining algorithms. Med. Biol. Eng. Comput. **57**, 1783–1811 (2019)
56. N. Ahmed, A. Yigit, Z. Isik, A. Alpkocak, Identification of leukemia subtypes from microscopic images using convolutional neural network. Diagnostics **9**(3), 104 (2019)
57. A.T. Sahlol, P. Kollmannsberger, A.A. Ewees, Efficient classification of white blood cell leukemia with improved swarm optimization of deep features. Sci. Rep. **10**(1), 2536 (2020)
58. N.M. Deshpande, S.S. Gite, R. Aluvalu, A brief bibliometric survey of leukemia detection by machine learning and deep learning approaches. Lib. Philo. Pract. 4569 (2020)
59. M. Mohamed, B. Far, A. Guaily, An efficient technique for white blood cells nuclei automatic segmentation, in *2012 IEEE International Conference on Systems, Man, and Cybernetics (SMC)* (IEEE, 2012), pp. 220–225
60. R.D. Labati, V. Piuri, F. Scotti, All-IDB: The acute lymphoblastic leukemia image database for image processing, in *2011 18th IEEE international conference on image processing* (IEEE, 2011), pp. 2045–2048
61. J. Fischer, Der Ausstrichtutor: Konzeption eines internetbasierten Hämatologie-Lernprogramms und Entwicklung eines Prototypen (Doctoral dissertation, Universität Ulm) (2007)
62. S.H. Rezatofighi, H. Soltanian-Zadeh, Automatic recognition of five types of white blood cells in peripheral blood. Comput. Med. Imaging Graph. **35**(4), 333–343 (2011)
63. R. Gupta, S. Gehlot, A. Gupta, C-NMC: B-lineage acute lymphoblastic leukaemia: a blood cancer dataset. Med. Eng. Phys. **103**, 103793 (2022)
64. S. Bose, M. Deininger, J. Gora-Tybor, J.M. Goldman, J.V. Melo, The presence of typical and atypical BCR-ABL fusion genes in leukocytes of normal individuals: biologic significance and implications for the assessment of minimal residual disease. Blood, J. Am. Soc. Hematol. **92**(9), 3362–3367 (1998)
65. L.R. Goldin, O. Landgren, M.L. McMaster, G. Gridley, K. Hemminki, X.L. Li, L. Mellemkjaer, J.H. Olsen, S. Martha, M.S. Linet, Familial aggregation and heterogeneity of non-Hodgkin lymphoma in population-based samples. Cancer Epidemiol. Biomarkers Prevent. **14**(10), 2402–2406 (2005)
66. B.F. Eichhorst, K. Fischer, A.M. Fink, T. Elter, C.M. Wendtner, Goede, M. Bergmann, S. Stilgenbauer, G. Hopfinger, M. Ritgen, J. Bahlo, German CLL Study Group (GCLLSG), Limited clinical relevance of imaging techniques in the follow-up of patients with advanced chronic lymphocytic leukemia: results of a meta-analysis. Blood J. Am. Soc. Hematol. **117**(6), 1817–1821 (2011)
67. W. Cao, C. Liang, Y. Gen, C. Wang, C. Zhao, L. Sun, Role of diffusion-weighted imaging for detecting bone marrow infiltration in skull in children with acute lymphoblastic leukemia. Diagn. Interv. Radiol. **22**(6), 580 (2016)
68. J.F. Bruzzi, H. Macapinlac, A.M. Tsimberidou, M.T. Truong, M.J. Keating, E.M. Marom, R.F. Munden, Detection of Richter's transformation of chronic lymphocytic leukemia by PET/CT. J. Nucl. Med. **47**(8), 1267–1273 (2006)

69. M.C. Kincaid, W.R. Green, Ocular and orbital involvement in leukemia. Surv. Ophthalmol. **27**(4), 211–232 (1983)
70. A. Ogasawara, H. Matsushita, Y. Tanaka, Y. Shirasugi, K. Ando, S. Asai, H. Miyachi, A simple screening method for the diagnosis of chronic myeloid leukemia using the parameters of a complete blood count and differentials. Clin. Chim. Acta **489**, 249–253 (2019)
71. A.S. Davis, A.J. Viera, M.D. Mead, Leukemia: an overview for primary care. Am. Fam. Phys. **89**(9), 731–738 (2014)
72. J.M. Peters, M.Q. Ansari, Multiparameter flow cytometry in the diagnosis and management of acute leukemia. Arch. Pathol. Lab. Med. **135**(1), 44–54 (2011)
73. S. Yohe, Molecular genetic markers in acute myeloid leukemia. J. Clin. Med. **4**(3), 460–478 (2015)
74. J.H. Yang, Y. Kim, J. Lim, M. Kim, E.J. Oh, H.K. Lee, K. Han, Determination of acute leukemia lineage with new morphologic parameters available in the complete blood cell count. Ann. Clin. Lab. Sci. **44**(1), 19–26 (2014)
75. R. Rathee, M. Vashist, A. Kumar, S. Singh, Incidence of acute and chronic forms of leukemia in Haryana. Int. J. Pharm. Pharm. Sci. **6**(02), 323–325 (2014)
76. B. Psaila, J.B. Bussel, A.L. Frelinger, B. Babula, M.D. Linden, Y. Li, A.D. Michelson, Differences in platelet function in patients with acute myeloid leukemia and myelodysplasia compared to equally thrombocytopenic patients with immune thrombocytopenia. J. Thromb. Haemost. **9**(11), 2302–2310 (2011)
77. S. Joshi-Warpe, J. Wader, B.M. Warpe, Bone marrow aspiration and bone marrow biopsy in hematological disorders. Biomed. Pharmacol. J. **13**(2), 799–804 (2020)
78. T. Dharani, S. Hariprasath, Diagnosis of leukemia and its types using digital image processing techniques, in *2018 3rd International Conference on Communication and Electronics Systems (ICCES)* (IEEE, 2018), pp. 275–279
79. A. Baraka, M. Zakaria, Presepsin as a diagnostic marker of bacterial infections in febrile neutropenic pediatric patients with hematological malignancies. Int. J. Hematol. **108**, 184–191 (2018)
80. Bruserud, H.H. Aarstad, T.H.A. Tvedt, Combined C-reactive protein and novel inflammatory parameters as a predictor in cancer—what can we learn from the hematological experience? Cancers **12**(7), 1966 (2020)
81. R.H. Bottomley, C.A. Lovig, R. Holt, M.J. Griffin, Comparison of alkaline phosphatase from human normal and leukemic leukocytes. Can. Res. **29**(10), 1866–1874 (1969)
82. H. Döhner, S. Stilgenbauer, A. Benner, E. Leupolt, A. Kröber, K.D. Bullinger, M. Bentz, P. Lichter, Genomic aberrations and survival in chronic lymphocytic leukemia. N. Engl. J. Med. **343**(26), 1910–1916 (2000)
83. P.S. Patel, S.G. Adhvaryu, B.R. Baxi, Tumor markers in leukemia: evaluation of serum levels of different forms of sialic acid, Regan isoenzyme and lactate dehydrogenase. Int. J. Biol. Markers **6**(3), 177–182 (1991)
84. J.D. Rowley, Molecular genetics in acute leukemia. Leukemia **14**(3), 513–517 (2000)
85. M.S. Aljabry, S. Asiri, T. Elsafi, G. Elyamany, Generalized lymphadenopathy secondary to isolated extramedullary hematopoiesis as an initial manifestation of primary myelofibrosis. Hematol. Rep. **10**(2), 7588 (2018)
86. K. Hemminki, W. Huang, J. Sundquist, K. Sundquist, J. Ji, Autoimmune diseases and hematological malignancies: exploring the underlying mechanisms from epidemiological evidence, in *Seminars in Cancer Biology*, Vol. 64. (Academic Press, 2020), pp. 114–121
87. J.C. Fitzgerald, S.L. Weiss, S.L. Maude, D.M. Barrett, S.F. Lacey, J.J. Melenhorst, D.T. Teachey, Cytokine release syndrome after chimeric antigen receptor T cell therapy for acute lymphoblastic leukemia. Crit. Care Med. **45**(2), e124–e131 (2017)
88. V. Liso, G. Specchia, E.M. Pogliani, G. Palumbo, D. Mininni, V. Rossi, A. Biondi, Extramedullary involvement in patients with acute promyelocytic leukemia: a report of seven cases. Cancer: Interdiscip. Int. J. Am. Cancer Soc. **83**(8), 1522–1528 (1998)
89. A. Sharma, B. Buksh, Fuzzy logic and neural network for diagnosis of chronic disease

90. R.B. Hegde, K. Prasad, H. Hebbar, B.M.K. Singh, I. Sandhya, Automated decision support system for detection of leukemia from peripheral blood smear images. J. Digit. Imaging **33**, 361–374 (2020)
91. S. Shafique, S. Tehsin, Acute lymphoblastic leukemia detection and classification of its subtypes using pretrained deep convolutional neural networks. Technol. Cancer Res. Treatment **17**, 1533033818802789 (2018)
92. A.I. Shahin, Y. Guo, K.M. Amin, A.A. Sharawi, White blood cells identification system based on convolutional deep neural learning networks. Comput. Methods Programs Biomed. **168**, 69–80 (2019)
93. S. Mishra, B. Majhi, P.K. Sa, L. Sharma, Gray level co-occurrence matrix and random forest based acute lymphoblastic leukemia detection. Biomed. Signal Process. Control **33**, 272–280 (2017)
94. A.S. Negm, O.A. Hassan, A.H. Kandil, A decision support system for acute leukaemia classification based on digital microscopic images. Alex. Eng. J. **57**(4), 2319–2332 (2018)
95. M. Abbasi, S. Kermani, A. Tajebib, M. Moradi Amin, M. Abbasi, Automatic detection of acute lymphoblastic leukaemia based on extending the multifractal features. IET Image Proc. **14**(1), 132–137 (2020)
96. K.N. Sukhia, A. Ghafoor, M.M. Riaz, N. Iltaf, Automated acute lymphoblastic leukaemia detection system using microscopic images. IET Image Proc. **13**(13), 2548–2553 (2019)
97. T. Lavanya, S. Sushritha, Detection of sickle cell anemia and thalassemia using image processing techniques. Int. J. Innov. Sci. Res. Technol. **2**(12), 399–405 (2017)
98. D.A. Tyas, T. Ratnaningsih, A. Harjoko, S. Hartati, The classification of abnormal red blood cell on the minor thalassemia case using artificial neural network and convolutional neural network, in *Proceedings of the International Conference on Video and Image Processing* (2017), pp. 228–233
99. M. Poostchi, I. Ersoy, K. McMenamin, E. Gordon, N. Palaniappan, S. Pierce, S. Jaeger, Malaria parasite detection and cell counting for human and mouse using thin blood smear microscopy. J. Med. Imaging **5**(4), 044506–044506 (2018)
100. M. Yildirim, A. Çinar, Classification of white blood cells by deep learning methods for diagnosing disease. Rev. d'Intelligence Artif. **33**(5), 335–340 (2019)
101. H.P. Vaghela, H. Modi, M. Pandya, M.B. Potdar, Leukemia detection using digital image processing techniques. Int. J. Appl. Inform. Syst. **10**(1), 43–51 (2015)
102. P.K. Das, A. Pradhan, S. Meher, Detection of acute lymphoblastic leukemia using machine learning techniques, in *Machine Learning, Deep Learning and Computational Intelligence for Wireless Communication: Proceedings of MDCWC 2020* (Springer Singapore, 2021), pp. 425–437
103. R. Raina, N.K. Gondhi, Chaahat, D. Singh, M. Kaur, H.N., A systematic review on acute leukemia detection using deep learning techniques. Arch. Comput. Methods Eng. **30**(1), 251–270 (2023)
104. N.N. Ganvir, D.M. Yadav, Filtering method for pre-processing mammogram images for breast cancer detection. Int. J. Eng. Adv. Technol. **9**(1), 4222–4229 (2019)
105. M. Chhabra, R. Kumar, Comparison of different edge detection techniques to improve quality of medical images. J. Comput. Theor. Nanosci. **17**(6), 2496–2507 (2020)
106. S. Sood, H. Singh, M. Malarvel, Image quality enhancement for Wheat rust diseased images using Histogram equalization technique, in *2021 5th International Conference on Computing Methodologies and Communication (ICCMC)* (IEEE, 2021), pp. 1035–1042
107. M. Omar, A. Hossain, L. Zhang, H. Shum, An intelligent mobile-based automatic diagnostic system to identify retinal diseases using mathematical morphological operations, in *The 8th International Conference on Software, Knowledge, Information Management and Applications (SKIMA 2014)* (IEEE, 2014), pp. 1–5
108. J. Mukherjee, I.K. Maitra, K.N. Dey, S.K. Bandyopadhyay, D. Bhattacharyya, T.H. Kim, Grayscale conversion of histopathological slide images as a preprocessing step for image segmentation. Int. J. Softw. Eng. its Appl **10**(1), 15–26 (2016)

109. N.M. Deshpande, S. Gite, B. Pradhan, K. Kotecha, A. Alamri, Improved Otsu and Kapur approach for white blood cells segmentation based on LebTLBO optimization for the detection of Leukemia. Math. Biosci. Eng. (2022)
110. G. Caseneuve, I. Valova, N. LeBlanc, M. Thibodeau, Chest X-ray image preprocessing for disease classification. Procedia Comput. Sci. **192**, 658–665 (2021)
111. H. Algailani, M.E.S. Hamad, Detection of sickle cell disease based on an improved Watershed segmentation, in *2018 International Conference on Computer, Control, Electrical, and Electronics Engineering (ICCCEEE)* (IEEE, 2018), pp. 1–4
112. J. Singh, H. Kaur, Plant disease detection based on region-based segmentation and KNN classifier, in *Proceedings of the International Conference on ISMAC in Computational Vision and Bio-Engineering 2018 (ISMAC-CVB)* (Springer International Publishing, 2019), pp. 1667–1675
113. S. Sanjayprabu, R.S. Kumar, N. Bhuvaneswari, R. Karthikamani, Detection of Covid-19 using PDE based filter, GLSZM features and classifiers. Math. Stat. Eng. Appl. **72**(1), 1141–1152 (2023)
114. K. Parimanam, L. Lakshmanan, T. Palaniswamy, Hybrid optimization based learning technique for multi-disease analytics from healthcare big data using optimal pre-processing, clustering and classifier. Concurr. Comput. Pract. Exp. **34**(17), e6986 (2022)
115. H. Hamidi, A. Daraee, Analysis of pre-processing and post-processing methods and using data mining to diagnose heart diseases. Int. J. Eng. **29**(7), 921–930 (2016)
116. M. El Sghair, R. Jovanovic, M. Tuba, An algorithm for plant diseases detection based on color features. Int. J. Agric. Sci. **2**, 1–6 (2017)
117. S. Kolkur, D.R. Kalbande, Survey of texture based feature extraction for skin disease detection, in *2016 International Conference on ICT in Business Industry & Government (ICTBIG)* (IEEE, 2016), pp. 1–6
118. Y. Ding, C. Luo, C. Li, T. Lan, Z. Qin, High-order correlation detecting in features for diagnosis of Alzheimer's disease and mild cognitive impairment. Biomed. Signal Process. Control **53**, 101564 (2019)
119. S.A. Kumar, S. Sasikala, Disease detection in tomato leaves using machine learning and statistical feature fusion, in *2021 International Conference on Advancements in Electrical, Electronics, Communication, Computing and Automation (ICAECA)* (IEEE, 2021), pp. 1–6
120. H. Li, X. Wang, C. Liu, Q. Zeng, Y. Zheng, X. Chu, C. Karmakar, A fusion framework based on multi-domain features and deep learning features of phonocardiogram for coronary artery disease detection. Comput. Biol. Med. **120**, 103733 (2020)
121. M. Shi, C. Zhan, H. He, Y. Jin, R. Wu, Y. Sun, B. Shen, Renyi distribution entropy analysis of short-term heart rate variability signals and its application in coronary artery disease detection. Front. Physiol. **10**, 809 (2019)
122. B.F. Stanley, S.W. Franklin, Effective feature extraction for cerebral microbleed detection using edge emphasized weber maximum directional co-occurance matrix. J. Ambient. Intell. Humaniz. Comput. **14**(10), 13683–13696 (2023)
123. A.M. Khuzi, R. Besar, W.W. Zaki, N.N. Ahmad, Identification of masses in digital mammogram using gray level co-occurrence matrices. Biomed. Imaging Intervent. J. **5**(3) (2009)
124. F.F. Chamasemani, Y.P. Singh, Multi-class support vector machine (SVM) classifiers—an application in hypothyroid detection and classification, in *2011 Sixth International Conference on Bio-Inspired Computing: Theories and Applications* (IEEE, 2011), pp. 351–356
125. V.R. Balaji, S.T. Suganthi, R. Rajadevi, V.K. Kumar, B.S. Balaji, S. Pandiyan, Skin disease detection and segmentation using dynamic graph cut algorithm and classification through Naive Bayes classifier. Measurement **163**, 107922 (2020)
126. A.S. Oliver, K. Ganesan, S.A. Yuvaraj, T. Jayasankar, M.Y. Sikkandar, N.B. Prakash, Accurate prediction of heart disease based on bio system using regressive learning based neural network classifier. J. Ambient. Intell. Humaniz. Comput. **63**, 1–9 (2021)
127. A.T. Azar, H.I. Elshazly, A.E. Hassanien, A.M. Elkorany, A random forest classifier for lymph diseases. Comput. Methods Prog. Biomed. **113**(2), 465–473 (2014)

128. K. Chandel, V. Kunwar, S. Sabitha, T. Choudhury, S. Mukherjee, A comparative study on thyroid disease detection using K-nearest neighbor and Naive Bayes classification techniques. CSI Trans. ICT **4**, 313–319 (2016)
129. C.H. Weng, T.C.K. Huang, R.P. Han, Disease prediction with different types of neural network classifiers. Telematics Inform. **33**(2), 277–292 (2016)
130. S.J. Belekar, S.R. Chougule, WBC segmentation using morphological operation and SMMT operator: a review. Int. J. Innov. Res. Comput. Commun. Eng. **3**(1), 434–440 (2015)
131. S. Perumal, T. Velmurugan, Preprocessing by contrast enhancement techniques for medical images. Int. J. Pure Appl. Math. **118**(18), 3681–3688 (2018)
132. M. Yogeshwari, G. Thailambal, Automatic feature extraction and detection of plant leaf disease using GLCM features and convolutional neural networks. Mater. Today: Proc. (2021)
133. S.S. Ittannavar, R.H. Havaldar, Comparative study of mam logarithmic scale mogram enhancement techniques for early detection of breast cancer. Int. J. Technol. Sci. 5–8 (2016)
134. L.S. Wei, Q. Gan, T. Ji, Skin disease recognition method based on image color and texture features. Comput. Math. Methods Med. 2018 (2018)
135. Y. Zhao, W. Gui, Z. Chen, Edge detection based on multistructure elements morphology, in *2006 6th World Congress on Intelligent Control and Automation*, Vol. 2 (IEEE, 2006), pp. 9795–9798
136. J.B. Kim, H.J. Kim, Multiresolution-based watersheds for efficient image segmentation. Pattern Recogn. Lett. **24**(1–3), 473–488 (2003)
137. F.Y. Shih, *Image Processing and Mathematical Morphology: Fundamentals and Applications* (CRC Press, 2009)
138. A. Rastogi, R. Arora, S. Sharma, Leaf disease detection and grading using computer vision technology fuzzy logic, in *2015 2nd International Conference on Signal Processing and Integrated Networks (SPIN)* (IEEE, 2015), pp. 500–505
139. V. Shankar, M.M. Deshpande, N. Chaitra, S. Aditi, Automatic detection of acute lymphoblasitc leukemia using image processing, in *2016 IEEE International Conference on Advances in Computer Applications (ICACA)* (IEEE, 2016), pp. 186–189
140. C.U. Kumari, S.J. Prasad, G. Mounika, Leaf disease detection: feature extraction with K-means clustering and classification with ANN, in *2019 3rd international conference on computing methodologies and communication (ICCMC)* (IEEE, 2019), pp. 1095–1098
141. N. Guil, J. Villalba, E.L. Zapata, A fast Hough transform for segment detection. IEEE Trans. Image Process. **4**(11), 1541–1548 (1995)
142. S.K. Fazilov, O.R. Yusupov, K.S. Abdiyeva, Mammography image segmentation in breast cancer identification using the otsu method. Web Sci. Int. Sci. Res. J. **3**(8), 196–205 (2022)
143. D.M. Sharath, M.G. Rohan, S.A. Kumar, C. Prathap, Disease detection in pomegranate using image processing, in *2020 4th International Conference on Trends in Electronics and Informatics (ICOEI)*(48184) (IEEE, 2020), pp. 994–999
144. M.E. Pothen, M.L. Pai, Detection of rice leaf diseases using image processing, in *2020 Fourth International Conference on Computing Methodologies and Communication (ICCMC)* (IEEE, 2020), pp. 424–430
145. S. Arslan, E. Ozyurek, C. Gunduz-Demir, A color and shape based algorithm for segmentation of white blood cells in peripheral blood and bone marrow images. Cytometry A **85**(6), 480–490 (2014)
146. A. Şengür, Y. Akbulut, Ü. Budak, Z. Cömert, White blood cell classification based on shape and deep features, in *2019 International Artificial Intelligence and Data Processing Symposium (IDAP)* (IEEE, 2019), pp. 1–4
147. A. Sengür, Y. Akbulut, Ü. Budak, Z. Cömert, White blood cell classification based on shape and deep features, in *2019 International Artificial Intelligence and Data Processing Symposium (IDAP)* (IEEE, 2019), pp. 1–4
148. K.A.K. Al-Dulaimi, J. Banks, V. Chandran, I. Tomeo-Reyes, K. Nguyen Thanh, *Classification of White Blood Cell Types from Microscope Images: Techniques and Challenges. Microscopy Science: Last Approaches on Educational Programs and Applied Research*, vol. 8 (Microscopy Book Series, 2018), pp. 17–25

149. A.T. Azar, S.M. El-Metwally, Decision tree classifiers for automated medical diagnosis. Neural Comput. Appl. **23**, 2387–2403 (2013)
150. S. Ramesh, R. Hebbar, M. Niveditha, R. Pooja, N. Shashank, P.V. Vinod, Plant disease detection using machine learning, in *2018 International Conference on Design Innovations for 3Cs Compute Communicate Control (ICDI3C)* (IEEE, 2018), pp. 41–45
151. N.R. Bhimte, V.R. Thool, Diseases detection of cotton leaf spot using image processing and SVM classifier, in *2018 Second International Conference on Intelligent Computing and Control Systems (ICICCS)* (IEEE, 2018), pp. 340–344
152. M. Zolfaghari, H. Sajedi, A survey on automated detection and classification of acute leukemia and WBCs in microscopic blood cells. Multimedia Tools Appl. **81**(5), 6723–6753 (2022)
153. P. Arena, A. Basile, M. Bucolo, L. Fortuna, Image processing for medical diagnosis using CNN. Nucl. Instrum. Methods Phys. Res., Sect. A **497**(1), 174–178 (2003)
154. A.N. Jagannatha, H. Yu, Bidirectional RNN for medical event detection in electronic health records, in *Proceedings of the 2016 Conference of the North American Chapter of the Association for Computational Linguistics: Human Language Technologies* (San Diego, California, 2016), pp. 473–482
155. G. Du, X. Cao, J. Liang, X. Chen, Y. Zhan, Medical image segmentation based on u-net: a review. J. Imaging Sci. Technol. **64**(2) (2020)
156. B. Gaonkar, D. Hovda, N. Martin, L. Macyszyn, Deep learning in the small sample size setting: cascaded feed forward neural networks for medical image segmentation, in medical imaging: computer-aided diagnosis. Med. Imaging Comput.-Aided Diagnosis **9785**, 97852I (2016)
157. P. Hao, J.H. Zhai, S.F. Zhang, A simple and effective method for image classification, in *2017 International Conference on Machine Learning and Cybernetics (ICMLC)*, vol. 1 (IEEE, Ningbo, China, 2017), pp. 230–235
158. M.Z. Alom, T.M. Taha, C. Yakopcic, S. Westberg, P. Sidike, A state-of-the-art survey on deep learning theory and architectures. Electronics **8**(3), 292 (2019)
159. A.K. Singh, B. Ganapathysubramanian, S. Sarkar, A. Singh, Deep learning for plant stress phenotyping: trends and future perspectives. Trends Plant Sci. **23**(10), 883–898 (2018)
160. J. Yu, S.M. Sharpe, A.W. Schumann, N.S. Boyd, Deep learning for image-based weed detection in turfgrass. Eur. J. Agron. **104**, 78–84 (2019)
161. K. Weiss, T.M. Khoshgoftaar, D. Wang, A survey of transfer learning. J. Big Data **3**(1), 1–40 (2016)
162. J. Zhou, G. Cui, S. Hu, Z. Zhang, C. Yang, Graph neural networks: a review of methods and applications. AI Open **1**, 57–81 (2020)
163. Y. Cao, T.A. Geddes, J.Y.H. Yang, P. Yang, Ensemble deep learning in bioinformatics. Nat. Mach. Intell. **2**(9), 500–508 (2020)
164. R.B. Jeyavathana, R. Balasubramanian, A.A. Pandian, A survey: analysis on pre-processing and segmentation techniques for medical images. Int. J. Res. Sci. Innov. **3**(6), 113–120 (2016)
165. T. Mustaqim, C. Fatichah, N. Suciati, Deep learning for the detection of acute lymphoblastic leukemia subtypes on microscopic images: a systematic literature review. IEEE Access (2023)
166. N.M. Deshpande, S. Gite, B. Pradhan, A. Alamri, C.W. Lee, A new method for diagnosis of leukemia utilizing a hybrid DL-ML approach for binary and multi-class classification on a limited-sized database. Comput. Model. Eng. Sci. (2023)
167. P. Linardatos, V. Papastefanopoulos, S. Kotsiantis, Explainable AI: a review of machine exp machine learning interpretability methods. Entropy **23**(1), 18 (2021)
168. O.H. Barros, *Business Information System Design Based on Process Patterns and Frameworks* (University of Chile, Santiago, Industrial Engineering Department, 2004)
169. S.F. Weng, J. Reps, J. Kai, J.M. Garibaldi, N. Qureshi, Can machine-learning improve cardiovascular risk prediction using routine clinical data? PLoS One **12**(4) (2017)
170. E.P. Balogh, B.T. Miller, J.R. Ball, *Improving Diagnosis in Health Care* (National Academies Press, Washington DC, 2015). https://doi.org/10.17226/21794
171. J. Corner, J. Hopkinson, D. Fitzsimmons, S. Barclay, M. Muers, Is late diagnosis of lung cancer inevitable? Interview study of patients' recollections of symptoms before diagnosis. Thorax **60**(4), 314–319 (2005)

172. T. Shiba, T. Watanabe, H. Kachi, T. Koyanagi, N. Maruyama, Distinct interacting core taxa in co-occurrence networks enable discrimination of polymicrobial oral diseases with similar symptoms. Sci. Rep. **6**(1), 1–13 (2016)
173. C.C. Tucker, S. Chakraborty, Quantitative assessment of lesion characteristics and disease severity using digital image processing. J. Phytopathol. **145**(7), 273–278 (1997)
174. N.A. Khan, H. Pervaz, A.K. Latif, A. Musharraf, Unsupervised identification of malaria parasites using computer vision, in *2014 11th International Joint Conference on Computer Science and Software Engineering (JCSSE)* (IEEE, Chonburi, Thailand, 2014), pp. 263–267
175. S. Kaur, J. Singla, L. Nkenyereye, S. Jha, D. Prashar, Medical diagnostic systems using artificial intelligence (AI) algorithms: principles and perspectives. IEEE Access **8**, 228049–228069 (2020)
176. N. Frost, M. Moshkovitz, C. Rashtchian, *ExKMC: Expanding Explainable k-Means Clustering* (2020). arXiv preprint arXiv:2006.02399
177. S. Robertson, H. Azizpour, K. Smith, J. Hartman, Digital image analysis in breast pathology—from image processing techniques to artificial intelligence. Transl. Res. **194**, 19–35 (2018)
178. M. Kuzlu, U. Cali, V. Sharma, Ö. Güler, Gaining insight into solar photovoltaic power generation forecasting utilizing explainable artificial intelligence tools. IEEE Access **8**, 187814–187823 (2020)
179. E. Pintelas, M. Liaskos, I.E. Livieris, S. Kotsiantis, P. Pintelas, Explainable machine learning framework for image classification problems: case study on glioma cancer prediction. J. Imaging **6**(6), 37 (2020)
180. F. Filippetti, G. Franceschini, C. Tassoni, P. Vas, AI techniques in induction machines diagnosis including the speed ripple effect. IEEE Trans. Ind. Appl. **34**(1), 98–108 (1998)
181. A. Duval, Explainable artificial intelligence (XAI). MA4K9 Scholarly Report, Mathematics Institute, The University of Warwick (2019), pp. 1–53
182. R. Hughes, C. Edmond, L. Wells, M. Glencross, L. Zhu, et al., *Explainable AI (XAI) an Introduction to the XAI Landscape with Practical Examples*. SIGGRAPH Asia 2020 Courses, pp. 1–62 (2020)
183. E. Tjoa, C. Guan, Quantifying explainability of saliency methods in deep neural networks (2020). arXiv preprint arXiv:2009.02899
184. Aamananandrai (2022). https://dev.to/amananandrai/6-explainable-ai-xai-frameworks-for-transparency-in-ai-3koj
185. N. Agarwal, S. Das, Interpretable machine learning tools: a survey, in *IEEE Symposium Series on Computational Intelligence (SSCI)* (Singapore, 2020), pp. 1528–1534
186. A. Das, P. Rad, Opportunities and challenges in explainable artificial intelligence (XAI): a survey (2020). arXiv preprint arXiv:2006.11371
187. V. Arya, R.K. Bellamy, P.Y. Chen, A. Dhurandhar, M. Hind, AI explainability 360: an extensible toolkit for understanding data and machine learning models. J. Mach. Learn. Res. **21**(130), 1–6 (2020)
188. L.M. Fenoy, A. Ciontos, *Performance Evaluation of Explainable AI Methods Against Adversarial Noise*. https://projekter.aau.dk/projekter/files/334478864/MasterThesis.pdf
189. R.R. Selvaraju, M. Cogswell, A. Das, R. Vedantam, D. Parikh, et al., Grad-CAM: Visual explanations from deep networks via gradient-based localization, in *Proceedings of the IEEE International Conference on Computer Vision* (Venice, Italy, 2017), pp. 618–626
190. X.H. Li, Y. Shi, H. Li, W. Bai, Y. Song, et al., *Quantitative Evaluations on Saliency Methods: An Experimental Study* (2020). arXiv preprint arXiv:2012.15616

Non-invasive Stress Recognition Framework Using Consumer Internet of Things in Smart Healthcare Applications

Ritu Tanwar, Ghanapriya Singh, and Pankaj Kumar Pal

Abstract Current healthcare models predominantly focus on symptom-based assessments during intermittent physician visits, lacking continuous monitoring capabilities. Acknowledging the uniqueness of individuals and the necessity for ongoing stress monitoring, this chapter introduces a novel wearables-based stress recognition framework tailored for smart healthcare applications leveraging Consumer Internet of Things (CIoT). The framework's design aims to cater to individual needs and facilitate treatment planning accordingly. The key to the framework's efficiency is the integration of lightweight machine learning (ML) models at the edge layer, enabling real-time stress prediction while optimizing edge device resources and processing speed. Embracing a consumer-centric approach, this study seamlessly combines edge and cloud layers, with the cloud layer employing a multimodal explainable deep learning model. The performance evaluation against seven state-of-the-art stress recognition approaches demonstrates a 3% accuracy enhancement, highlighting the effectiveness of the proposed methodology. At the edge layer, a Support Vector Machine (SVM) model achieves a remarkable 97% accuracy, showcasing its superiority over other ML models. On the cloud layer, an explainable deep learning model attains 98% accuracy, emphasizing the significance of electrodermal activity and motion signals in stress recognition. By embracing a hybrid edge/cloud approach, the proposed framework delivers non-invasive physiological signal analysis tailored to consumer preferences, surpassing single-layered frameworks in performance and adaptability. This chapter establishes a foundational decision-making criterion for result interpretation and advanced consumer-centric analysis,

R. Tanwar (✉) · P. K. Pal
Department of Electronics Engineering, National Institute of Technology, Srinagar, Pauri, Uttarakhand 246174, India
e-mail: ritu.tanwarphd2021@nituk.ac.in

P. K. Pal
e-mail: pankajpal86@nituk.ac.in

G. Singh
Department of Electronics and Communication Engineering, National Institute of Technology, Kurukshetra, Haryana 136119, India
e-mail: ghanapriya@nitkkr.ac.in

B. Pradhan and S. Mukhopadhyay (eds.), *IoT Sensors, ML, AI and XAI: Empowering A Smarter World*, Smart Sensors, Measurement and Instrumentation 50,
https://doi.org/10.1007/978-3-031-68602-3_13

paving the way for personalized and efficient stress management in smart healthcare ecosystems.

Keywords CIoT · Wearable technology · Physiological signals · Smart healthcare · Stress recognition

1 Introduction

The healthcare sector has recently seen the emergence of telemedicine, or wireless communication between a doctor and a patient via wearable technology. This makes it possible to monitor chronic conditions, including heart diseases, hypertension, and long-term stress, in real-time and everyday scenarios, which are difficult in physician- centric healthcare services [1]. Numerous diseases, such as depression, eating disorders, behaviour changes, schizophrenia and personality disorders, are strongly linked to stress [2]. Additionally, it can impact human brain physiology, increasing the risk of diabetes and heart-related issues [3].

Medically, given the stress' impacts, stress management is crucial because of its reactive nature. However, physician-centric healthcare approaches are inadequate in managing stress-related issues. For instance, the physician centric paradigm necessitates that a patient travels to a clinic several times each week, incurring high costs for the patient and leading to high withdrawal rates. Moreover, this model faces the significant problem of limited observations, which can result in ineffective diagnosis and treatment. Recent developments in smart healthcare based on integrating smartphones, wearables, and artificial intelligence (AI) technologies provide new opportunities for addressing stress-related problems [4–6]. These approaches can help patients achieve better results by providing a comprehensive, collaborative and non-invasive based treatment that incorporates patient observations, monitoring real-time stress remotely and providing treatment personally.

In this era of technology and connectivity, individuals can use non-invasive multiple technologies to support their day-to-day requirements. For this, the internet of things (IoT), machine learning (ML), deep learning (DL) and AI are technological advancements as solutions to various real-world problems, such as the healthcare system. Healthcare systems will benefit from increased efficiency and effectiveness owing to ML, DL, AI and IoT [7]. IoT is any device that can connect to the internet, communicate with other devices, and exchange data. The mass production of such embedded devices has surpassed numerous milestones, each underestimating the effect and quick growth of actively communicating IoT devices through the internet. These devices are mostly used in consumer markets, government agencies, and critical infrastructure [8]. As enterprises and smart homes adopt these cutting-edge technologies, the IoT paradigm is expanding, especially in the consumer sector.

Essentially, IoT in the context of consumer applications, use cases, and devices is referred to as consumer IoT or CIoT as shown in Fig. 1. The CIoT is a vast network of interconnected smart gadgets, including wearables, smartphones, televisions, and health monitoring equipment. The CIoTs' features include connectivity, scalability, miniaturization, cloud computing and AI and their applications are diverse such as remote accessibility, assets tracking, home security, automated tasks, and wearable technology [9]. Increased CIoT usage generates enormous amounts of data, which ML and DL efficiently process to generate a plethora of insightful information that has the potential to transform the healthcare industry, most notably healthcare services. Stress is triggered when a person is exposed to hazardous physical or mental surroundings. This causes a fight or flight reaction, leading to the release of hormones that alter the human being's central nervous system. Under stress, heart rate is often faster, respiration becomes quicker, and muscles are often tighter. Stress can be indicated by physiological parameters such as the electrocardiogram (ECG), electrodermal activities (EDA), accelerometer (Acc.), respiratory (Resp.), electromyogram (EMG), and body temperature (Temp.) [10]. Overall, stress affects the human brain, heart, skin response, blood, muscle, and respiration-related activities, which can be measured by electroencephalograph (EEG), ECG, EDA, photoplethysmography (PPG), EMG and respiratory (Resp.) physiological signals, respectively [10].

The ML models for assessing and categorizing stress have been developed using physiological signals in recent stress recognition studies. Tanwar et al. and Tanwar et al. proposed effective methods of physiological parameters feature extraction using

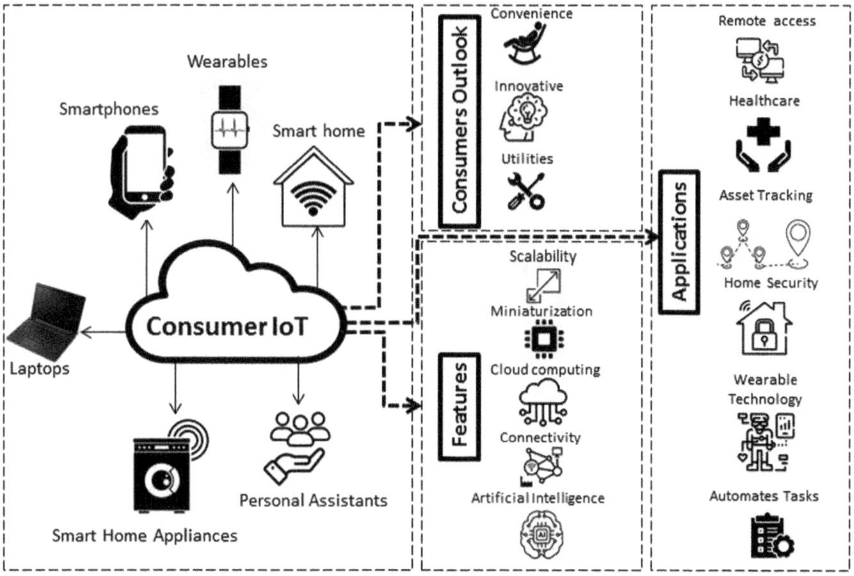

Fig. 1 Consumer Internet of Things (IoT) devices, their features and applications and consumer outlook towards CIoT. Here, smartphones, wearables, smart home, laptops, personal assistants and smart home appliances are Consumer IoTs. Consumer outlooks towards consumers IoTs are convenience, innovation, and utilities and Consumer IoTs features are scalability, miniaturization, cloud computing, connectivity and artificial intelligence. Applications of consumer IoTs include remote access, healthcare, asset tracking, home security, wearable technology and automation tasks

ML and DL techniques such as convolutional neural networks (CNN) [11, 12]. Recent research has focused on DL methods such as CNN and LSTM in addition to implementing ML techniques such as decision trees and random forest [13]. Despite this substantial corpus of recent studies, today's solutions still suffer from various limitations and shortcomings. As existing stress recognition models are episodic, a person's mental state may not be accurate due to insufficient information before and after stress recognition clinical examination. Therefore, CIoT-based real-time stress monitoring is crucial. Moreover, CIoT-assisted monitoring and AI-powered evaluations enable just-in-time solutions that can transform the present approach based on reactive and symptoms into a proactive solution.

A CIoT-based model furthermore offers personalization that incorporates person-to-person variations. Existing approaches, however, are not suitable for use with endpoint CIoTs. These methods are unsuited for endpoint IoT devices because of their complexity and high computing effort requirements, which are severely constrained in terms of processing resources and memory. Additionally, an acceptable trade-off between edge and cloud computing is required due to high latency and consumer privacy; this trade-off ML model accuracy and energy consumption [14]. Another challenge is non-transparent ML models, which consumers find difficult to know the reason behind the model prediction. Lastly, the accuracy of current methods is decreased by the lack of readily available labelled training data. CIoT-assisted monitoring and AI-powered evaluations enable just-in-time solutions that can transform the present approach based on reactive and symptoms into a proactive solution. This study overcomes the shortcomings mentioned above by describing a hybrid edge/cloud computing-based CIoT-based solution for smart healthcare that allows real-time and noninvasive monitoring and early stress recognition in everyday scenarios. The main objectives of this chapter are:

1. To present a general edge or cloud based CIoT approach using wearables physiological modalities for stress recognition in consumers,
2. To conduct a thorough performance investigation of the effective ML and DL models using wearables and compare them for consumers' stress recognition. The ML models are selected from the literature and compare support vector machine (SVM), ensemble modeling, adaBoost, gradBoost and logistic regression (LR) for the edge layer device and DL model (CNN) at cloud layer. This may serve as foundation for evaluating stress recognition model performance based on ML and DL models using different wearables in consumers.
3. To explain the stress recognition results to the consumers, we empirically demonstrated a DL with explainable artificial intelligence (XAI) method at the cloud layer and addressed the non-transparency of ML and DL models for stress recognition.
4. To perform stress recognition using CIoT (e.g., smart watch) in enhancing smart healthcare and compare with the approaches that use either only cloud or only edge computing solutions.
5. To validate the approach, WESAD dataset is used and evaluated stress recognition frameworks' performance based on different wearables and compare with

existing seven state of the art approaches in the literature for stress recognition models.

The rest of the study is organized into five sections. Section 2 details the general overview of the non-invasive CIoT framework for stress recognition. Section 3 presents the details about the ML, DL and explainable AI methods implemented for stress recognition in different layers of the CIoT framework. Section 4 discusses the experimental findings obtained with the implemented models, and Sect. 5 concludes the study.

2 Related Work

2.1 Stress Recognition

Stress recognition has been a hot research domain for many years, especially after wearable technology advancement. Stress recognition has evolved to be a key domain in affective computing in recent years because it strongly correlates with a person's mental and physical well-being. Human–computer interaction has extensively used human stress or emotion among various other domains. Basically, human emotion or stress recognition can be implemented using either facial and speech or physiological signals. Physiological signals are chosen more frequently than facial expressions or speech signals in healthcare services because they show a high correlation with a person's physical and mental health. Additionally, it is almost impossible to manipulate these signals. The most often used physiological modalities in stress recognition are ECG, EMG, GSR, Temp, BVP, PPG and Respiration [14].

Researchers have previously recognized and tracked human emotion or stress for various purposes using these modalities. Movie clips, images, videos, and some mathematical tasks were used to stimulate human stress. For example, Ayata et al. proposed a human emotion recognition system utilizing wearables that records PPG and GSR signals [15]. Here, emotions are triggered using music. Likewise, video clips were used to trigger emotion and emotions were recognized using facial expressions and ECG signals.

Although much research has been done on using EEG to analyze emotions, it is not as wearable or practical as other physiological signals. EEG devices are typically difficult with portability and use. The modern wearables on the marketplace, which can be used to quickly acquire physiological data (EDA, EMG, Resp., and Temp.) and important biological information, do not have this issue [10]. Monitoring biosignals other than EEG using multimodal signals yields exciting results, and the presented research does this remarkably well.

2.2 Wearable Technology

Wearable technology offers more affordable ways for people and healthcare providers to address a range of health conditions. These non-invasive technologies can be designed by integrating multiple sensors with human wearables, including watches, wristbands, necklaces, shirts, handbags, shoes, and hats. Information about the patient's health and the environment is acquired using the attached sensor [16]. After that, this data is transferred to the servers or databases. Utilizing healthcare applications, certain wearables can link to smartphones as well. Numerous research studies have been published in the literature demonstrating the use of a smartphone, computers, and these wearable devices in real-time supervision [15]. For instance, Shimmer wearable device monitors ECG and GSR, and Samsung wearables track PPG, temp, and acc. signals [17].

Wan et al. designed an IoT-enabled health monitoring system with embedded sensors for blood pressure, skin temperature, and heartbeat to provide healthcare services remotely [18]. In similar research, wearable device integration in a wireless sensor network has been proposed by Castillejo et al. as a solution for activity recognition in a mobile e-health application for remote patient monitoring [19]. IoT-enabled wearables were used to monitor physiological signals, including ECG and EMG in order to collect important patient data. These wearables' interconnection with a smartphone application boosts the device's computation power. Additionally, the application can be utilized to quickly analyze and visualize the collected data.

2.3 Consumer Internet of Things

As the field of consumer IoT develops, it is redefining the idea of coupling medical equipments as well as offering a wide range of applications to enhance healthcare. Additionally, CIoT has a huge potential for offering smart services, including remote patient monitoring, tracing, diagnosis, and alarm generating, operating and controlling medical equipment, and dispensing smart medications. CIoT provides us with the capability to respond quickly and better to medical situations anywhere [20]. Wearables around a patient's body are connected to CIoT, which then monitors and analyses multimodal physiological health and psychological information.

To incorporate smart healthcare applications in CIoT frameworks, a number of consumer-based wearable technologies include smart platform integrations, such as smart clothing. Smart clothing relies on physiological signals to comprehend human activities such as cardiac, respiratory and movement [21]. However, given that domain is still in the innovation and emerging phase, advances in design approach and augmentation of the end services, intelligent behaviour, and smart integration of multimodal health information are necessary to spur more research and attention in this area.

All consumer IoT things, including wearables and devices, are coupled and integrated to comprehend their biological and psychological surroundings, collect and process the data they have acquired and the knowledge they have extracted, and develop the ability to adapt. With a minimal amount of human involvement, the framework is capable of making intelligent decisions [22]. Numerous consumer"things" based IoT systems are proposed in the literature for a variety of applications. A study provides an IoT framework to aid in the growth of smart cities and significantly increase their sustainability [23]. In [24], the authors propose a three-layer IoT framework that integrates physiological parameters for home security to attain good performance and high intelligence. Other research presented elderly monitoring smart system that could process physiological information and could provide assistance to the elders [25].

2.4 Smart Healthcare

Smart healthcare's social and economic benefits are piquing the interest of many government agencies, private businesses, and scholars from various sectors. IoT-cloud technology integration has thus led to the emergence of several studies, frameworks, and applications pertaining to smart healthcare [26, 27]. Medical professionals, as well as healthcare personnel, can examine healthcare data under such a smart healthcare framework in real-time. A study [28] presented a framework for smart healthcare in which citizens of smart cities can utilize smart devices to navigate their way to healthcare facilities. Ullah et al. [29] proposed a connected smart healthcare system that analyzed electronic health records. For diabetes patients, Obeidat and Ammar [30] presented a smart glucose tracking model that accounted for routine activities and places. Samani and Zhu [31] suggested the idea of a robot-controlled autonomous ambulance to treat patients with life-threatening cardiac arrhythmias. Firouzi et al. [32] proposed a three-layer IoT framework that integrates physiological parameters for home security to attain good performance and high intelligence. The adequate use of the internet, communication, wearable technology, data and analytics and cloud systems is all-important for a viable smart healthcare environment [32].

Our proposed approach for smart healthcare applications addresses the various requirements and difficulties mentioned in the literature and offers a method for deploying an automatic stress recognition model. Specifically, we leverage physiological signals, cloud, CIoT, and wearables. To highlight the comparison between the related stress recognition research studies and our proposed study, we present the overall stress recognition-based work in terms of the dataset used and wearables used, including the method used, explainability of the model, multimodality and IoT/CIoT architecture. Table 1 shows that prior research has a number of drawbacks. As far as we are aware, the majority of earlier solutions were not used in the IoT environment. Hossain et al. [33] analysis of the IoT included cloud-based learning algorithm and prediction. These solutions, however, have issues with high latency and energy utilization while transmitting data from end users' IoT devices to the cloud. We

Table 1 Some studies based on wearable technology and stress recognition. MLP, CNN, LSTM, RF, SVM, LR, XAI represent multilayer perceptron, convolution neural network, long short-term memory, random forest, support vector machine, logistic regression, and explainable artificial intelligence respectively

References	Dataset	Wearables	Method	Explainability	Multimodality	IoT/ CIoT
[34]	IEMOCAP	Motion Capture	MLP, CNN, LSTM	N	Y	N
[35]	CASE	Body sensors	LSTM	N	Y	Y
[36]	SWELL, WESAD	Kinect 3D sensor, other body sensors	RF	N	Y	N
[37]	WESAD	Empatica E4, RespiBan	CNN, RF, Autoencoder, Transfer learning	N	Y	N
[38]	Manually constructed	Fossil Gen4 Explorist	SVM, RF, AdaBoost, GradBoost, LR	N	N	N
[39]	WESAD	Empatica E4, RespiBan	CNN	N	Y	N
Proposed model	WESAD	Empatica E4, RespiBan	SVM, AdaBoost, GradBoost, Ensemble modeling, XAI + CNN	Y	Y	Y

provide a robust, user convenient and hybrid CIoT solution for stress recognition to address these issues that excels seven chosen state-of-the-art approaches.

3 CIoT Framework for Stress Recognition for Smart Healthcare Systems

Figure 2 shows a general CIoT framework for stress monitoring and providing necessary assistance and service. CIoT framework consists of four layers: the physical layer (data collection), edge layer, cloud layer and application layer.

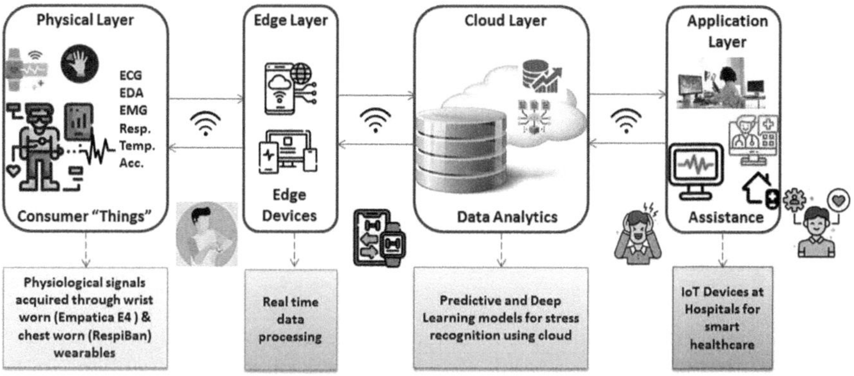

Fig. 2 General CIoT-based stress recognition framework. Consumer"Things" comes under physical layer where physiological signals are acquired using wearables, edge devices (smart watches) at the edge layer for data processing, predictive and deep learning models at the cloud layer, and healthcare assistance using IoT devices in hospitals for smart healthcare at the application layer

3.1 Data Collection Layer

Different consumer things, such as wearables, are shown in the physical layer (Fig. 2), wherein raw data is collected by the consumers' wearables and transferred to the next edge layer.

3.2 Edge Layer

Different edge servers/devices make up the edge computing layer, which links the wearables and cloud layer via the internet, takes raw input data, processes, and performs the given task. The massive number of IoT and wearable devices and their variety of uses shows the value of multi-service edge servers in the network. This layer carries out tasks like job migration across various edge servers, data storage, and processing. For instance, here in this work, collected data is processed, and stress is predicted using consumers' wearables edge devices such as a smartphone and chest worn device.

3.3 Cloud Layer

The cloud computing layer establishes the non-invasive stress recognition and chooses optimum physiological signals responsible for causing stress. The cloud computing layer can be used to address relevant parameters such as leading causes and model features. Additionally, it serves storage access, making it possible for users to access the analysis report whenever needed. Thus, decisions regarding stress assessment will be made with instant access and more information. Here, for advanced analysis of data, the cloud layer is used with the help of internet connectivity, where the data is stored and analyzed. Details are provided as per the monitored stress, which can be understandable by consumers as well.

3.4 Application Layer

The application layer includes the assessment of the severity and diagnosis of stress. The cloud computing layer's results for the severity of stress recognition can be transmitted to the physician, who can track and observe the patient's stress levels and provide assistance accordingly. For example, suppose a consumer is found to be dealing with chronic stress, then with the help of doctors and consultants. In that case, the consumer is given treatment accordingly in the application layer.

4 Proposed Techniques

4.1 Physiological Data Acquired from Wearable Technology

In this study, we use physiological signals for stress recognition such as ECG, EDA, EMG, 3-axis Acc., Temp. and respiration.

4.2 Framework

We design the CIoT flowchart for stress recognition, shown in Fig. 3, from a user viewpoint. The physiological data is collected using wearables. The user first receives the stress recognition results from the consumers devices (e.g., smartwatch) using ECG, EMG, EDA, Acc., Temp., Resp., BVP. At the edge device, the results are predicted using ML (SVM, AdaBoost, GradBoost, LR) and ensemble algorithms. If a user wants more advanced analysis, which a user can understand, then data from edge device is transmitted to the cloud through internet for advanced data processing. At the cloud layer, the prediction is based on the DL and XAI model. More accurate and explainable results are then generated for the user.

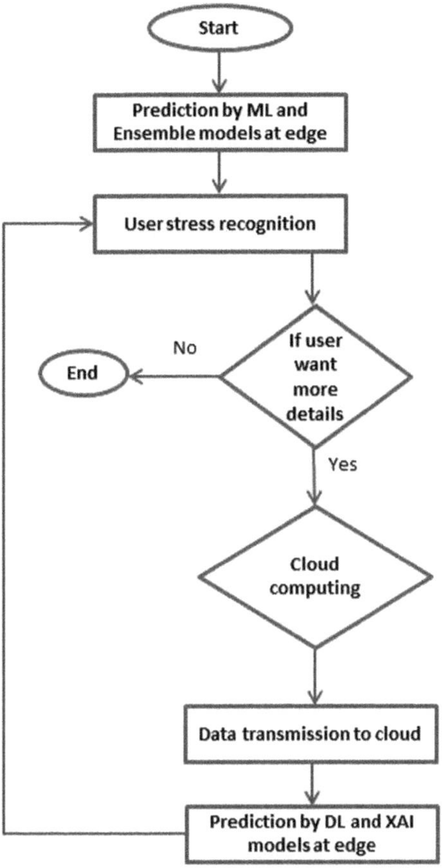

Fig. 3 CIoT flowchart for stress recognition, where predictions are made using ML at edge layer and explainable predictions at the cloud layer

4.3 *Stress Recognition on the Edge Computing Layer*

SVM, AdaBoost, GradBoost, LR and ensemble modeling were used on the edge computing layer of the proposed framework as shown in Fig. 2.

4.4 *Stress Recognition on the Cloud Computing Layer*

The data is uploaded into the cloud to ensure user satisfaction about the prediction, and for advanced analysis. CNN and SHAP models are employed to provide more accurate and explainable results that a user understands.

4.4.1 Convolution Neural Network

CNN's are a type of neural network that combines convolutional, pooling, and fully connected layers; they are frequently employed for image categorization. The data is subjected to convolution between a number of filters in the convolutional layer, producing feature maps. By either picking the maximum value or the neighbourhood's average, the pooling layer subsamples data in non-overlapped neighbours on the maps. The input data is subjected to linear transforms in the fully connected layer. The outcome of the completely connected layer becomes nonlinear due to the activation function. Backpropagation is used to adjust all parameters in order to minimize a pre-determined loss function according to the output and the ground truth. We developed a CNN (Fig. 4) using the architecture described in Table 2, which was inspired by the CNN model presented in [14].

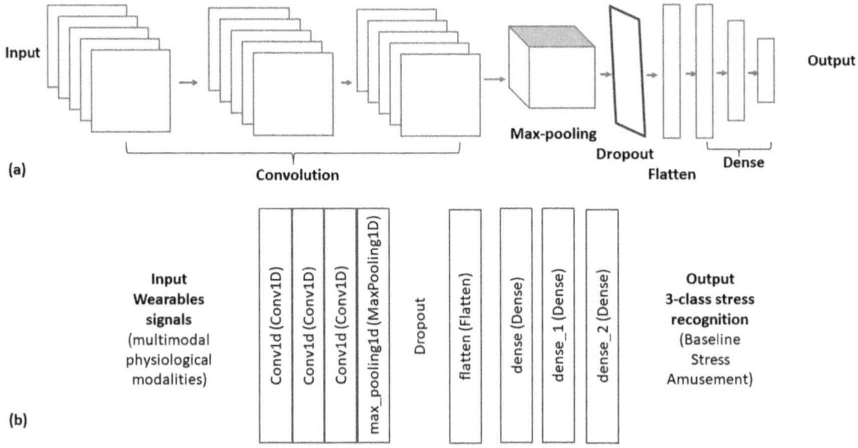

Fig. 4 **a** Architecture of CNN model used and **b** Layers distribution in CNN model

Table 2 Summary of proposed CNN Model

Layer (type)	Output Shape	Param #
conv1d (Conv1D)	(None, 8, 32)	128
conv1d 1 (Conv1D)	(None, 8, 64)	6208
conv1d 2 (Conv1D)	(None, 8, 128)	24,704
max pooling1d (MaxPooling1D)	(None, 4, 128)	0
dropout (Dropout)	(None, 4, 128)	0
flatten (Flatten)	(None, 512)	0
dense (Dense)	(None, 256)	131,328
dense 1 (Dense)	(None, 512)	131,584
dense 2 (Dense)	(None, 3)	1539
Total params: 295,491		
Trainable params: 295,491		
Non-trainable params: 0		

4.4.2 Explainable Artificial Intelligence

XAI is deployed over several healthcare decision models in the healthcare system to address reliable analytics. It handles health records and clinical assessment, reduces sensor bias in healthcare, classifies disorders, and segments patient populations [12].

Adding an XAI module that supports the output decision of the system facilitates simple debugging and enhances the effectiveness of the proposed model. XAI enhances AI algorithms' transparency in order to support the predictions made by the model. XAI guarantees that DL algorithms comply with the established criteria and offer perfect justifications to end users for their conclusions and predictions [40]. Challenging the system's assessment, it also enables the system to improve the method to lessen bias. As a result, it makes healthcare models more secure and equitable and gives people more confidence in their ability to make decisions. In the context of this study, multimodal physiological signals are considered to recognize an individual suffering from stress. It is designed to quantify the contribution of each physiological signal, i.e., Shapley value for each modality, in order to classify the stressed class of individuals fairly.

5 Experimental Results

WESAD, a benchmark stress recognition dataset [10], is chosen to compare the proposed model with seven state-of-the-art approaches. These approaches have used linear discriminant analysis (LDA), random forest (RF), CNN, recurrent neural network (RNN), and deep neural networks. Additionally, we made a comparison between the studies deploying either cloud or edge-based strategies and our proposed study. As given in Table 1 in Sect. 3, the proposed work achieves better results utilizing cloud or edge computing layers and providing explainability to the users.

Typical measures for evaluating classifier performance include accuracy (A), recall (R), precision (P) and F1 score (F) as stated in Eqs. 1–4. Recall estimates the proportion of correct predictions in data that are all estimated to be positive, while precision evaluates the proportion of accurate predictions in the positive class. The precision and recall scores are combined to form the F1 score. The purpose is to ensure the overall good performance of a classifier.

$$A = \frac{TP + TN}{TP + TN + FP + FN} \tag{1}$$

$$P = \frac{TP}{TP + FP} \tag{2}$$

$$R = \frac{TP}{TP + FN} \tag{3}$$

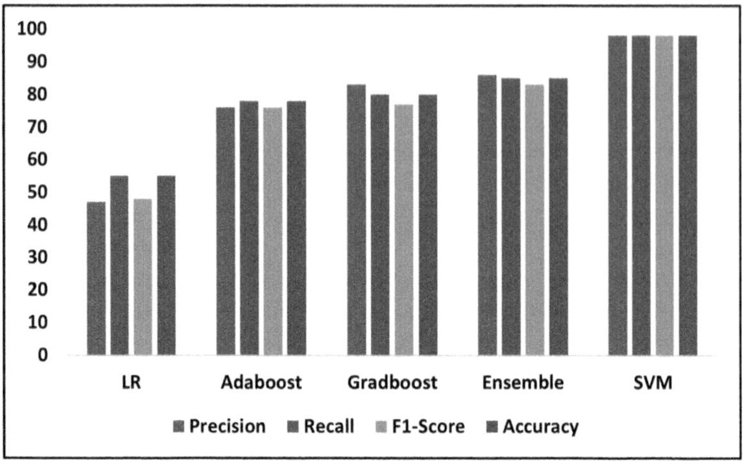

Fig. 5 Performance measures for LR, AdaBoost, GradBoost, Ensemble and SVM method. Here LR stands for Logistic regression and SVM for support vector machine

$$F = 2 * \frac{P * R}{P + R} \tag{4}$$

5.1 Performance Evaluation

5.1.1 At the Edge Layer

Figure 5 represents the evaluation metrics value corresponding to each deployed method on data collected from consumers' wearables. Here, the LR method shows the least values of performance measures (A = 55%, P = 47%, R = 55%, F = 48%) followed by the AdaBoost (A = 78%, P = 76%, R = 78%, F = 76%), GradBoost (A = 80%, P = 83%, R = 80%, F = 77%) and ensemble methods (A = 85%, P = 86%, R = 85%, F = 83%). SVM is the best-performing ML model here. Given the model performances in the Fig. 5, we selected SVM as the edge layer model. Moreover, SVM is a quick, memory-efficient model and as at the edge layer, we have limited resources, so it is suitable to use SVM at the edge.

5.1.2 At the Cloud Layer

As described in the previous section, we employed CNN and XAI models at the cloud layer. Table 3 presents the CNN model performance measures with an A = 98%, R = 99%, F = 99% and P = 99%. Overall, the CNN model at the cloud outperforms the

Table 3 CNN model performance measures

Model	A (%)	P (%)	R (%)	F (%)
CNN	98	99	99	99

Table 4 Comparison of the proposed model with previous WESAD-based studies

References	Modalities	Method	Classes	A (%)	F (%)	R (%)	P (%)
[10]	All	LDA	2	93.12	91.47	-	-
[42]	All	RF	3	85	86	-	-
[43]	EDA	CNN	-	92.85	89	-	-
[44]	All	DNN	3	87.7	83	-	-
[45]	All	DNN	2	96.6	-	-	-
[46]	All	CNN	4	94.77	95	95	95
[47]	All	RNN	2	96.5	94	93	96
This work	All	XAI + CNN	3	98	99	99	99

models deployed at the edge layer. We then compare the effectiveness of our models in the cloud with seven other approaches that use WESAD physiological signals and corresponding labels (Table 4). Our models in the cloud perform better, with at least a 2% increase in accuracy and the F1 score when compared to results from previous research studies that utilized the same performance metrics. Due to the high computational and memory requirements for neural network training, it is appropriate to carry out the demanding computation in the cloud because it supports scalability. Additionally, utilizing the cloud brings up several opportunities to enhance our approach, like working with medical institutions to analyze data or gathering edge data for additional analysis.

5.1.3 Modalities Importance and Explainability of Stress Recognition

Using CNN as predictive model, the relative importance of predictors to the likelihood of stress was evaluated, and the kernel SHAP model results that affected the model were integrated. To examine and compare the impact of the attributes, we calculate CNN's mean SHAP values. SHAP values are beneficial for illustrating how each attribute affects a particular prediction. Figure 6a demonstrate the effects of the attributes on all subjects, with each point denoting the effect of the attribute on the samples. The SHAP value, which is displayed on the x-axis, shows the average marginal contribution of the selected attribute to outcome through all potential coalitions. A SHAP value of 0 denotes no contribution, a value below 0 denotes a negative contribution, and a value above 0 denotes a positive contribution. Positive contribution denotes attributes with a significant importance to the outcome prediction, whereas negative contribution denotes attributes with a low importance to the

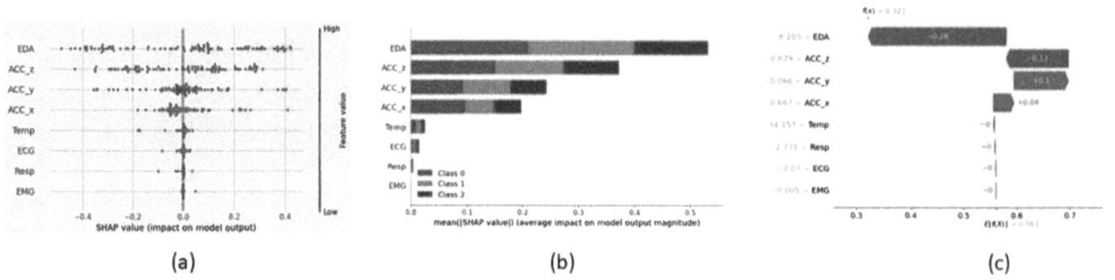

Fig. 6 **a** Average attribute importance (local explanations) for the CNN model, **b** SHAP based summary plot of entire model and **c** SHAP based waterfall plot for model explanations locally

outcome prediction. We observe that each attribute is ranked based on its significance as represented in the Fig. 6a, with the top attributes contributing more towards the model's prediction than that of the bottom ones. The most crucial attribute here is the"EDA," whereas the"EMG" attribute is the least crucial.

There are two coordinates, left and right, for the y-axis. While the right y-axis coordinate shows the value of the attributes from low to high, the left y-axis coordinate reflects the attributes prioritised by relevance in descending order [41]. The colour indicates how well (in red) or poorly (in blue) the attribute predicts. As observed in Fig. 6b, EDA modality has a significant and positive influence on the prediction of the probability of stress. The EDA, 3-axis accelerometer, temperature and ECG are the three most significant factors that are strongly connected with stress in CNN. Figure 6c also describes the local explanations plot, known as waterfall plot, of stress recognition model using SHAP. Here, the SHAP values are indicated along the x-axis and attributes, sorted in a descending order of their absolute SHAP values, along the y-axis.

We illustrated and visualised an individual interpretation of the model prediction using a summary plot [48], as shown in Fig. 6b, to further describe and explain the prediction of stress. The summary plot, shown in Fig. 6b, uses SHAP values to display the distribution of the effects that each attribute has on the prediction model. It ranks attributes according to the aggregate values of SHAP value magnitudes across all subjects data. The mean effect on output magnitude is shown along the x axis in Fig. 6b. The attributes are displayed on the y axis. Each of the three colours corresponds to one of the three mental states, where class 0, 1 and 2 corresponds to baseline, stress and amusement states respectively.

Figure 6c also describes the local explanations plot, known as waterfall plot [49]. Here, the SHAP values are indicated along the x-axis and attributes, sorted in a descending order of their absolute SHAP values, along the y-axis. Further, we compare the proposed research with earlier studies that utilized other stress-based datasets. As mentioned in Sect. 2, study [10] and study [45] classified stress or no stress with an accuracy of 93.12% and 96.6%, respectively. Our approach performs better than the earlier models by about 2% in accuracy and 5% in F1 score.

6 Conclusion

In this chapter, we have proposed wearables-based stress recognition system for consumers using CIoT and their different physiological signals. The system provides promising results using ML models and explainable DL model using different wearables. ML models have been implemented at edge layer as they are less computational and lightweight. The experimental findings demonstrated the proposed model has overall good performance in terms of accuracy, sensitivity, recall and f1-score with explainability. Accuracy, and f1-score are 98% and 99% respectively for wearables based physiological modalities (ECG, EDA, EMG, Acc., Resp., and Temp.) using CNN with XAI method. Explainable DL based technology has high accuracy and explainability to the consumers as compared to the traditional ML models. This framework can be implemented in the healthcare domain by using clinical specialists' feedback to update the model and creating a recommendation system for the users.

Data Availability Not Applicable.

Conflict of Interest No conflicts of interest.

References

1. N. Sultan, Reflective thoughts on the potential and challenges of wearable technology for healthcare provision and medical education. Int. J. Inf. Manage. **35**(5), 521–526 (2015)
2. T. Esch, G.B. Stefano, G.L. Fricchione, H. Benson, The role of stress in neurodegenerative diseases and mental disorders. Neuroendocrinol. Lett. **23**(3), 199–208 (2002)
3. A. Steptoe, M. Kivimäki, Stress and cardiovascular disease. Nature Rev. Cardiol. **9**(6), 360–370 (2012)
4. S.K. Islam, A. Fathy, Y. Wang, M. Kuhn, M. Mahfouz, Hassle-free vitals: biowireless for a patient-centric health-care paradigm. IEEE Microwave Mag. **15**(7), 25–33 (2014)
5. H. Demirkan, A smart healthcare systems framework. It Professional **15**(5), 38–45 (2013)
6. S.C. Sethuraman, P. Kompally, S.P. Mohanty, U. Choppali, Mywear: a novel smart garment for automatic continuous vital monitoring. IEEE Trans. Consum. Electron. **67**(3), 214–222 (2021)
7. H.K. Bharadwaj, A. Agarwal, V. Chamola, N.R. Lakkaniga, V. Hassija, M. Guizani, B. Sikdar, A review on the role of machine learning in enabling iot based healthcare applications. IEEE Access **9**, 38859–38890 (2021)
8. D. Harkin, M. Mann, I. Warren, Consumer iot and its under-regulation: findings from an Australian study. Policy Internet **14**(1), 96–113 (2022)
9. M.Q. Aldossari, A. Sidorova, Consumer acceptance of internet of things (iot): Smart home context. J. Comput. Inform. Syst. **60**(6), 507–517 (2020)
10. P. Schmidt, A. Reiss, R. Duerichen, C. Marberger, K. Van Laerhoven, Introducing wesad, a multimodal dataset for wearable stress and affect detection, in *Proceedings of the 20th ACM International Conference on Multimodal Interaction*, pp. 400–408 (2018)
11. R. Tanwar, O.C. Phukan, G. Singh, S. Tiwari, Cnn-lstm based stress recognition using wearables (2022)

12. R. Tanwar, G. Singh, P.K. Pal, Fuser: fusion of wearables data for stress recognition using explainable artificial intelligence models, in *2023 14th International Conference on Computing Communication and Networking Technologies (ICCCNT)*, pp. 1–6. (IEEE, 2023)
13. R. Tanwar, O.C. Phukan, G. Singh, P.K. Pal, S. Tiwari, Attention based hybrid deep learning model for wearable based stress recognition. Eng. Appl. Artif. Intell. **127**, 107391 (2024)
14. S. Jiang, F. Firouzi, K. Chakrabarty, E.B. Elbogen, A resilient and hierarchical iot-based solution for stress monitoring in everyday settings. IEEE Internet Things J. **9**(12), 10224–10243 (2021)
15. D. Ayata, Y. Yaslan, M.E. Kamasak, Emotion recognition from multimodal physiological signals for emotion aware healthcare systems. J. Med. Biol. Eng. **40**, 149–157 (2020)
16. V. Montesinos, F. Dell'Agnola, A. Arza, A. Aminifar, D. Atienza, Multi-modal acute stress recognition using off-the-shelf wearable devices, in *2019 41st Annual International Conference of the IEEE Engineering in Medicine and Biology Society (EMBC)*, pp. 2196–2201. (IEEE, 2019)
17. M. Wu, J. Luo, Wearable technology applications in healthcare: a literature review. Online J. Nurs. Inform **23**(3) (2019)
18. J. Wan, M. AAH Al-awlaqi, M. Li, M. O'Grady, X. Gu, J. Wang, N. Cao, Wearable iot enabled real-time health monitoring system. EURASIP J. Wirel. Commun. Netw. **2018**(1), 1–10 (2018)
19. P. Castillejo, J.-F. Mart´ınez, L. L´opez, G. Rubio, An internet of things approach for managing smart services provided by wearable devices. Int. J. Distrib. Sensor Networks **9**(2), 190813 (2013)
20. J. Manga, V.K. Sonti, Internet of things-empowered next-generation health-care systems, in *Integrating AI in IoT Analytics on the Cloud for Healthcare Applications*, pp. 201–218. (IGI Global, 2022)
21. H.K. Thakkar, S.R. Chowdhury, A.K. Bhoi, P. Barsocchi, Applications of wearable technologies in healthcare: an analytical study, in *5G IoT and Edge Computing for Smart Healthcare*, pp. 279–299. (Elsevier, 2022)
22. Y. Athavale, S. Krishnan, Biosignal monitoring using wearables: observations and opportunities. Biomed. Signal Process. Control **38**, 22–33 (2017)
23. J. Jin, J. Gubbi, S. Marusic, M. Palaniswami, An information framework for creating a smart city through internet of things. IEEE Internet Things J. **1**(2), 112–121 (2014)
24. H. Yu, Z. Zhou, Optimization of iot-based artificial intelligence assisted telemedicine health analysis system. IEEE access **9**, 85034–85048 (2021)
25. M. Hosseinzadeh, J. Koohpayehzadeh, M.Y. Ghafour, A.M. Ahmed, P. Asghari, A. Souri, H. Pourasghari, A. Rezapour, An elderly health monitoring system based on biological and behavioral indicators in internet of things. J. Ambient Intell. Humanized Comput. 1–11 (2023)
26. H. Cai, B. Xu, L. Jiang, A.V. Vasilakos, Iot-based big data storage systems in cloud computing: perspectives and challenges. IEEE Internet Things J. **4**(1), 75–87 (2016)
27. A. Botta, W. De Donato, V. Persico, A. Pescap´e, Integration of cloud computing and internet of things: a survey. Future Generation Comput. Syst. **56**, 684–700 (2016)
28. S.P. Mohanty, U. Choppali, E. Kougianos, Everything you wanted to know about smart cities: the internet of things is the backbone. IEEE Consum. Electron. Mag. **5**(3), 60–70 (2016)
29. M.R. Ullah, M.A.R. Bhuiyan, A.K. Das, Ihemha: Interactive healthcare system design with emotion computing and medical history analysis, in *2017 6th International Conference on Informatics, Electronics and Vision & 2017 7th International Symposium in Computational Medical and Health Technology (ICIEV-ISCMHT)*, pp. 1–8. (IEEE, 2017)
30. Y. Obeidat, A. Ammar, A system for blood glucose monitoring and smart insulin prediction. IEEE Sens. J. **21**(12), 13895–13909 (2021)
31. H. Samani, R. Zhu, Robotic automated external defibrillator ambulance for emergency medical service in smart cities. IEEE Access **4**, 268–283 (2016)
32. F. Firouzi, A.M. Rahmani, K. Mankodiya, M. Badaroglu, G.V. Merrett, P. Wong, B. Farahani, Internet-of-Things and big data for smarter healthcare: from device to architecture, applications and analytics. (Elsevier, 2018)
33. M.S. Hossain, G. Muhammad, M.F. Alhamid, B. Song, K. Al-Mutib, Audio-visual emotion recognition using big data towards 5g. Mobile Netw. Appl. **21**, 753–763 (2016)

34. S. Tripathi, S. Tripathi, H. Beigi, Multi-modal emotion recognition on iemocap with neural networks. arxiv 2018. arXiv preprint arXiv:1804.05788
35. M. Awais, M. Raza, N. Singh, K. Bashir, U. Manzoor, S.U. Islam, J.J. Rodrigues, Lstm-based emotion detection using physiological signals: Iot framework for healthcare and distance learning in covid-19. IEEE Internet Things J. **8**(23), 16863–16871 (2020)
36. K. Nkurikiyeyezu, A. Yokokubo, G. Lopez, The effect of person-specific biometrics in improving generic stress predictive models. arXiv preprint arXiv:1910.01770 (2019)
37. S. Ishaque, N. Khan, S. Krishnan, Comprehending the impact of deep learning algorithms on optimizing for recurring impediments associated with stress pre-diction using ecg data through statistical analysis. Biomed. Signal Process. Control **74**, 103484 (2022)
38. R. Dai, C. Lu, L. Yun, E. Lenze, M. Avidan, T. Kannampallil, Comparing stress prediction models using smartwatch physiological signals and participant self-reports. Comput. Methods Programs Biomed. **208**, 106207 (2021)
39. D. Gupta, M. Bhatia, A. Kumar, Resolving data overload and latency issues in multivariate time-series iomt data for mental health monitoring. IEEE Sens. J. **21**(22), 25421–25428 (2021)
40. D. Saraswat, P. Bhattacharya, A. Verma, V.K. Prasad, S. Tanwar, G. Sharma, P.N. Bokoro, R. Sharma, Explainable ai for healthcare 5.0: opportunities and challenges. (IEEE Access, 2022)
41. N. Nordin, Z. Zainol, M.H.M. Noor, L.F. Chan, An explainable predictive model for suicide attempt risk using an ensemble learning and shapley additive explanations (shap) approach. Asian J. Psychiatr. **79**, 103316 (2023)
42. J. Lin, S. Pan, C.S. Lee, S. Oviatt, An explainable deep fusion network for affect recognition using physiological signals, in *Proceedings of the 28th ACM International Conference on Information and Knowledge Management*, pp. 2069–2072 (2019)
43. R.K. Sah, H. Ghasemzadeh, Stress classification and personalization: getting the most out of the least. arXiv preprint arXiv:2107.05666 (2021)
44. A. Kumar, K. Sharma, A. Sharma, Hierarchical deep neural network for mental stress state detection using iot based biomarkers. Pattern Recogn. Lett. **145**, 81–87 (2021)
45. M. Gil-Martin, R. San-Segundo, A. Mateos, J. Ferreiros-Lopez, Human stress detection with wearable sensors using convolutional neural networks. IEEE Aerosp. Electron. Syst. Mag. **37**(1), 60–70 (2022)
46. S. Ghosh, S. Kim, M.F. Ijaz, P.K. Singh, M. Mahmud, Classification of mental stress from wearable physiological sensors using image-encoding-based deep neural network. Biosensors **12**(12), 1153 (2022)
47. A. Souza, M.B. Melchiades, S.J. Rigo, G.D.O. Ramos, Mostress: A sequence model for stress classification, in *2022 International Joint Conference on Neural Networks (IJCNN)*, pp. 1–8. (IEEE, 2022)
48. S.M. Lundberg, G. Erion, H. Chen, A. DeGrave, J.M. Prutkin, B. Nair, R. Katz, J. Himmelfarb, N. Bansal, S.-I. Lee, From local explanations to global understanding with explainable ai for trees. Nature Mach. Intell. **2**(1), 56–67 (2020)
49. M. Staniak, P. Biecek, Explanations of model predictions with live and breakdown packages. arXiv preprint arXiv:1804.01955 (2018)

Explainable Multi-task Learning Approach for Skin Lesion Classification

Keny Patel, Nitya Mehta, S. Easwaran, Rahee Walambe, Shilpa Gite, Biswajeet Pradhan, and Bunny Saini

Abstract The early diagnosis of skin cancer has significantly improved with the use of computer-aided techniques and deep learning (DL) models. However, existing methods often struggle with issues of interpretability and adaptability, which are crucial for clinical application. To address these limitations, we employed a Multi-Task Learning (MTL) approach that simultaneously performs classification and segmentation of skin lesions. This approach not only improves the accuracy and robustness of the models but also enhances their interpretability by incorporating Explainable AI (XAI) techniques into the MTL framework. Our convolutional-deconvolutional based MTL model, tested on the HAM-10000 dataset, demonstrated enhanced classification accuracy and interpretability with an Accuracy of 91.56% and IoU Score of 87.98%. The model outperformed state-of-the-art models, showing a marked enhancement in classification accuracy. Importantly, the use of MTL facilitates a reduction in model complexity while achieving superior performance, making this approach both powerful and efficient. From a practical standpoint, the replicability of our MTL framework is a key advantage, providing a scalable model for researchers and clinicians. The methodology's adaptability to different imaging datasets underscores its potential utility across various dermatological conditions. Future research could leverage this framework to further refine diagnostic accuracy in other complex imaging tasks, enhancing the scope of AI in medical diagnostics.

K. Patel · N. Mehta · S. Easwaran
Department of Computer Engineering, Symbiosis Institute of Technology, Symbiosis International (Deemed University), Pune 412115, India

R. Walambe · S. Gite (✉) · B. Saini
Symbiosis Centre for Applied Artificial Intelligence (SCAAI), Symbiosis Institute of Technology, Symbiosis International (Deemed University), Pune 412115, India
e-mail: shilpa.gite@sitpune.edu.in

B. Pradhan (✉)
Centre for Advanced Modelling and Geospatial Information Systems (CAMGIS), School of Civil and Environmental Engineering, Faculty of Engineering & IT, University of Technology Sydney, Sydney, NSW 2007, Australia
e-mail: Biswajeet.Pradhan@uts.edu.au

B. Pradhan and S. Mukhopadhyay (eds.), *IoT Sensors, ML, AI and XAI: Empowering A Smarter World*, Smart Sensors, Measurement and Instrumentation 50,
https://doi.org/10.1007/978-3-031-68602-3_14

Keywords Multi-task learning · Explainable AI · Skin lesion classification · Class activation heatmaps · Deep learning · Healthcare

1 Introduction

Early detection of skin cancer is crucial due to the high risk factor associated with developing the disease the increasing number of cases [1]. Skin lesions have been examined and detected using a wide range of strategies since they are crucial for determining the type of skin cancer [2]. The uncontrolled growth of cells results in the formation of a mass known as a tumor, which may lead to cancer. The tumors can be classified majorly as benign (when they can develop but not spread elsewhere) and malignant (when they are able to develop and spread to other physiological regions). The differentiation of lesions based on this classification is a challenging task [3]. The development of computer-aided diagnosis using imaging techniques has garnered increasing research interest [4]. With advancements in healthcare technology, various deep learning methodologies have significantly contributed to the diagnosis, detection, and prevention of skin cancer [5]. The biomedical domain of healthcare is a sensitive field of research, necessitating that outcome be supported by scientific observations to be valid and trustworthy. Therefore, evaluating and understanding the robustness of these AI-based methodologies is crucial factor for validating results and gaining interpretable insights [6].

2 Related Work

Several promising approaches have been reported in the literature, focusing on the development of AI based techniques for skin lesion classification and diagnosis. In this section, the important work in this domain is discussed.

2.1 Diagnosis of Skin Lesions from Images

Dermoscopic algorithms like Triage Amalgamated Dermoscopic Algorithm (TADA) [7], BRAFF checklist [8], etc. employ criteria based on distribution, color, and structure, such as starburst patterns, white structures, etc., to examine the lesions and detect skin cancer with high diagnostic accuracy and sensitivity. These dermoscopic algorithms largely depend on the extraction of meaningful features utilizing domain knowledge expertise. Apart from the dermoscopic algorithms, certain image processing methods are also available. Derbel et al. [9] have defined an image processing technique based on the ABCD rules for the same purpose. Anitha et al. [10] performed the task using Otsu thresholding and operations based on morphology to extract meaningful and descriptive features. Advancements such as quantitative multispectral imaging [11] have also been used for performing skin lesion analysis.

2.2 *Deep Learning for Diagnosis of Skin Lesions*

Given the recent advancements in deep learning, which have demonstrated notable performance in the medical and healthcare fields, various machine learning models and deep learning architectures [12] such as Support Vector Machines (SVM), Artificial Neural Network (ANN), Deep Convolutional Neural (DCNN) architecture, U-Net, Fully Convolutional network (FCN), etc. have been used to analyze the skin lesions and determine their cancerous nature. Different supervised learning approaches have been proposed to improve the task of diagnosing skin lesions. Elattar et al. [13] proposed an ensemble learning framework to leverage the strengths of various networks and models. In [14], a discussion on saliency detection from a sparse model representation is presented, which distinguishes the lesion region from the rest and utilizes Bayesian methodology to refine the boundary and shape of the lesion. Nasir et al. [15] employed a serial-based method for feature extraction following image segmentation using a hybrid technique. After feature extraction, a Boltzmann entropy method selects features and passes them to SVM for classification.

2.3 *MTL in Healthcare Domain*

Recent studies have shown the use of shared learning among various tasks to improve the efficiency of different models. Chu et al. [16] applied the methodology of Multi-Task Learning (MTL) to weak supervised learning in order to improve the quality of segmentation tasks, particularly focusing on liver and skin lesions. Liu et al. [18] employed a task similarity measure to facilitate knowledge sharing for rapid adaptation using the MIMIC-III dataset [17]. Their work indicates that the relevance of features enhances interpretability to some extent. Coppola et al. [19] demonstrated the use of gated mechanisms to enable the sharing of intelligent features that are pertinent to specific tasks rather than sharing features across tasks in MTL. They utilized the Derm7pt dataset [20] from the Interactive Atlas for their research.

Ngo et al. [21] used U-Net module along with a multi fibre network is for segmenting brain tumors using inputs from MRI modalities. Khan et al. [22] used a saliency segmentation heat map and a Moth Flame Optimization algorithm to effectively select discriminant features, thereby reducing dimensionality across multiple datasets fed into the classifier model. Research by Zhang et al. [23] on breast tumour showed that benign or malignant classification is aided by a likelihood score of malignancy obtained through regression and classification branches, yielding BI-RAD descriptors as output. They curated their dataset comprising 1,192 images by combining two existing datasets, BUSIS and BUSI, and observed that regression helped in handling inherent noise while learning was improved by BI-RAD descriptor branches. Xie et al. [24] used coarse segmentation maps to improve classification and segmentation performance when used as input. They used images from three datasets: ISIC-2017 [25], ISIC additional dataset, and PH2 [26]. Bootstrapping weak localization through transfer learning aided in accurately identifying lesion locations and types.

Thus, it can be observed that performance of various classifier frameworks has been improved by integrating tasks such as regression, segmentation, etc. [16, 18, 19, 21, 23, 25]. The relevancy of features examined using segmentation or localisation methods has contributed to a greater interpretability of the results. While CNN-based architectures have been widely used due to their applicability for computer vision, the extraction of specific elements from a lesion's image—critical for achieving improved skin cancer diagnosis—requires powerful hardware resources with high graphical processing unit (GPU) capability [27]. A notable challenge in training deep learning-based skin cancer detection models is the scarcity of high processing power.

2.4 *Explainable AI for Skin Lesion Diagnosis*

Machine centric strategies that quantify explainability of performance using decision-making impact analysis have been introduced. Lin et al. [28] proposed an evaluation of quantitative performance using Impact Score and Impact coverage, revealing that methods like Expected Gradient and GSInquire performed better in visual perception tasks. Thepplication of XAI methods has been adapted and tailored to increase the relevancy of the approach to the problem being solved. Metta et al. [29] customized an existing XAI method to differentiate between different types of skin lesions. They performed latent space analysis which shows repetitions of image over space. A real user expert survey was conducted by them showed that without consistent validation, explainable methods are not useful. A domain-specific explainable AI was proposed by Stieler et al. [30] where they synthesized Local Interpretable Model-Agnostic Explanation (LIME) with the dermatoscopy ABCD rule to differentiate between melanocytic and non-melanocytic skin lesions. They established both medically relevant and non-relevant LIME algorithm perturbations to underscore the importance of explanation.

Zia et al. [31] modified pre-trained MobileNetV2 and DenseNet201 models by stacking three convolutional layers at the end, modifying the classification head using batch normalization, and modifying the final layer based on the class problem, showing increased performance due to these modifications. The study undertaken by Nunnari et al. [32] measured the extent to which skin lesion visual features can be identified using saliency maps. They explored the behavior of deep learning architecture to find reference measurement values and thresholds, determining the limit of this approach.

Based on the literature review presented above, the following research questions have been formulated: (a) what features utilized for the localization of skin lesions can aid in determining their malignancy or benignity? (b) Is the shared feature learning representation computationally feasible and relevant to the outcome? To address these research gaps, this paper endeavors to investigate the explainability of a multi-task deep learning approach integrating image segmentation as an auxiliary task to classify skin lesions into benign and malignant categories. The elucidation of this approach serves to validate its efficacy and sheds light on the informational aspects of the shared features, thereby contributing to a deeper understanding of the classification process.

The proposed work encompasses several key components. Firstly, it involves the development of a Multi-Task Learning (MTL) based classifier framework that incorporates segmentation as an auxiliary task. This framework is aimed at improving classification accuracy while maintaining computational efficiency. Secondly, the implementation of this framework is executed to demonstrate its superiority over single-task learning methods in terms of training time and trainable parameters. Additionally, the use of Grad-CAM as an XAI approach validates the relevance of shared knowledge representation between classification and segmentation tasks, thereby enhancing the trustworthiness of the implementation. In terms of novelty and contribution, this work offers visual explainability of shared features within the MTL approach for skin lesion classification, leveraging segmentation as a secondary task to differentiate between benign and malignant lesions. Furthermore, the validation of the proposed methodology is substantiated through a comparative analysis with single-task learning approaches, and meaningful insights derived from visual explanations are aligned with a dermoscopic checklist, enriching the understanding of the classification process. The paper is organized into nine sections. Section 2 describes the existing works in the domain of MTL in healthcare and research related to its applicable explainable AI techniques. It also presents the research questions and gap analysis. Section 3 gives detailed information about our proposed methodology and its implementation. Section 4 presents the results obtained using our methodology and its comparison with the corresponding single-task learning approaches. Section 5 discusses the relevancy of the approach and talks about the insights derived by use of Explainable AI. Section 6 provides a summary and conclusion of the work presented in this paper and explores the scope of this work that can be extended in the future.

3 Data and Methodology

The primary techniques implemented in this research are discussed in this section. The fundamental concepts used in the proposed methodology, namely, MTL, convolutional neural network, deconvolutional neural network, ResNet-34, U-Net and Grad-CAM for Explainable AI are discussed in detail. The training process of the baseline approach along with the proposed architecture of the custom convolutional-deconvolutional model which is used to achieve the final results are also presented.

MTL is a machine learning process where a model is trained utilizing relevant information from several tasks simultaneously. Task here represents a desired objective to be attained on the given input [33]. The principle behind this methodology is that because human brains synthesize information from many activities, it develops and fosters new skills in a faster way by considering the fundamentals and abstracts of prior knowledge. As shown in Fig. 1, a MTL methodology is employed on an encoder-decoder based neural network to perform the task of classification with segmentation as an auxiliary task. It consists of four major components: a CNN-based backbone shared between both the tasks, a branch for performing segmentation, another branch for performing binary classification and a Grad-CAM method applied to the component of the shared parameters that generates Class Activation Heatmaps.

- Based on this baseline architecture, two models were developed: a U-Net based Multi-Task Learning (MTL) model with ResNet-34 as its backbone, and a custom model composed of convolutional layers and corresponding de-convolutional layers performing Up Sampling. The subsequent subsections provide detailed explanations of all components and important concepts utilized in the proposed architecture of these models.

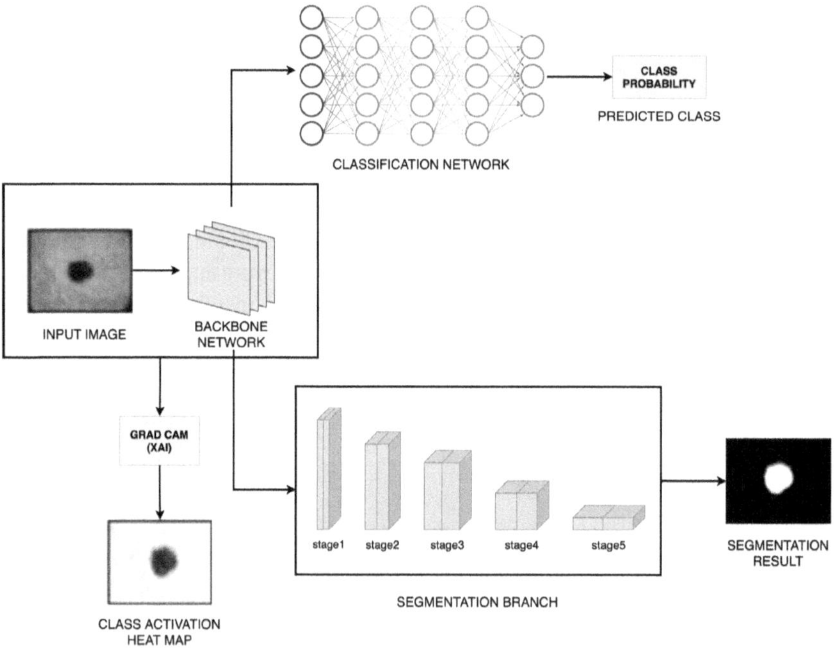

Fig. 1 Proposed model architecture

Table 1 Revised class and number of samples

Original Class Name	No of samples	Revised Class Name
Actinic keratoses	327	Benign
Basal cell carcinoma	514	Malignant
Benign keratosis-like lesions	1099	Benign
Dermatofibroma	115	Benign
Melanocytic Nevi	6705	Benign
Vascular lesions	142	Benign/Malignant
Melanoma	1113	Malignant

3.1 Dataset and Preprocessing

The HAM-10000 dataset [34] was used, comprising images of pigmented skin lesions along with metadata stored in a CSV file containing information about each image and corresponding disease labels. The dataset originally contains 10,015 skin lesion image samples belonging to one of the seven classes: Actinic keratosis, Benign keratosis-like lesions, vascular lesions, dermatofibroma, Basal cell carcinoma, melanocytic nevi and melanoma. For the purpose of binary classification, we segregated the instances into benign and malignant classes as shown in Table 1. Vascular lesion images were dropped because it contained both benign and malignant images and under-sampling was performed on Melanocytic Nevi. In total, 3368 data instances were used, out of which 1741 samples belonged to the benign class and 1627 belonged to the malignant class. The original samples were of size 600×450. The samples were resized to 256×256 and in order to perform normalization, pixel values of each image were multiplied by a factor of 1/255.0 to rescale it in a range of 0–1.

3.2 MTL

The core objective of MTL is to leverage the information shared between multiple tasks in order to help each task learn more accurately and hence improve performance with reduced no. of parameters in a more robust and a better generalized manner [35]. This approach is grounded on the assumption that most tasks are interconnected, and thus, jointly learning multiple tasks by exploiting shared features yields superior performance compared to independent learning. MTL achieves this by employing a shared representation of features to train tasks simultaneously. Moreover, MTL is often integrated with other learning methods to further enhance the performance of learning tasks [36].

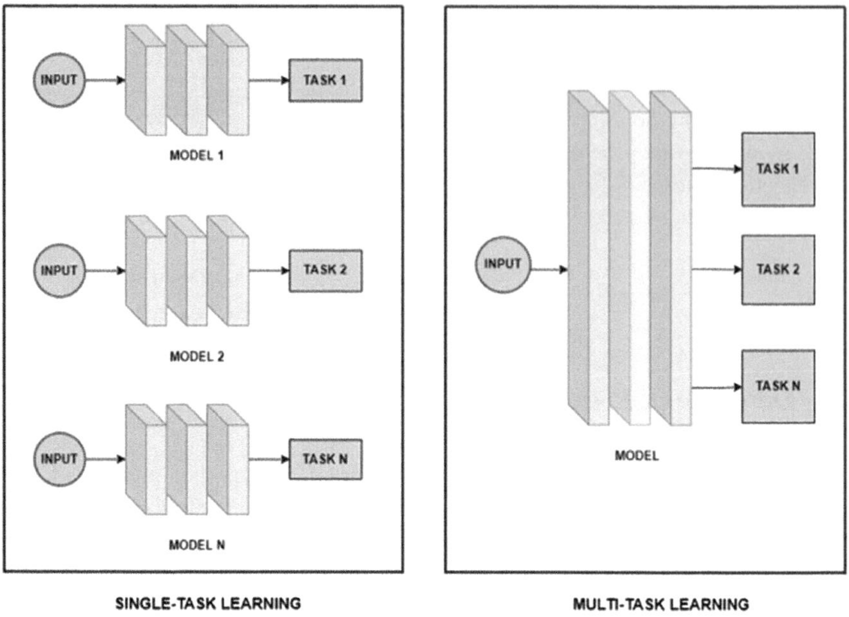

Fig. 2 Overview of MTL framework

Formally, 'n' number of non-identical tasks that are related to each other are trained together simultaneously in MTL to get the combined knowledge representation during learning as opposed to single task learning where each model performs only one of the n tasks. (Refer Fig. 2).

3.3 Convolutional Neural Network

Convolutional Neural Network (CNN) is a deep neural network and is an adaptation of Artificial Neural Network. It has three layers: convolutional layer, pooling layer and a fully connected layer (FC), which are stacked together [37]. The convolution is the first layer in which features are extracted from the input image; it learns image features by using input data of small squares. Down-sampling is done by a pooling layer which further reduces the activated parameters. Fully-connected layers from activations produce class scores that are used for classification.

The objective of convolution is to highlight some property of input, which is referred as a feature map, a feature implying specific character of input extracted by the kernel. Therefore, convolution can be defined as a feature extraction process that happens by moving a kernel mask across a signal array and at each location kernel response is calculated.

3.4 De-Convolutional Neural Network

While convolution with no padding produces a smaller output, deconvolution produces a larger output. It maps the data values to a larger set of data value feature resolution. (Known as up sampling/ transposed convolution). Hence feature maps of input images can be recovered while learning. Simply put, deconvolution reconstructs spatial resolution and performs convolution [38].

3.5 ResNet-34

ResNet-34 is a Residual network with 34 parameters, consisting of one convolution and pooling followed by four layers of the same behavior. Each layer performs a 3 × 3 convolution with a fixed feature map of [64, 128, 256, 512] dimension, and it bypasses the input every two convolutions. During the entire layer height and width dimensions remain constant.

3.6 U-Net

The U-Net architecture (Fig. 3) is a convolutional neural network designed for biomedical image segmentation shaped like 'U' and symmetric with two major parts [39]. The first half is encoder, which typically is a pre-trained network (like ResNet-34) where convolution blocks are applied along with Max Pooling that performs Down-sampling so that input image is encoded into feature representation at different levels. Second half is a decoder whose goal is to semantically project the encoder's learned discriminative features (lower resolution) onto the pixel space (higher resolution) for a dense classification. Up-sampling and concatenation are done which are followed by regular convolution operations.

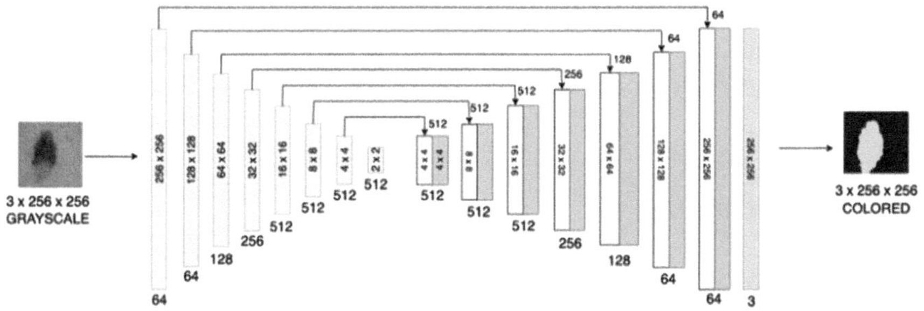

Fig. 3 U–Net model architecture

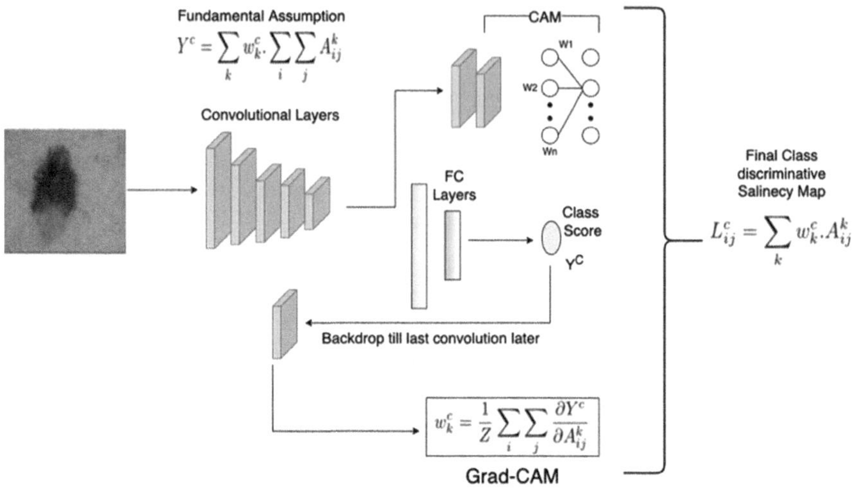

Fig. 4 Architecture of Grad-CAM

3.7 *Grad-CAM for Explainable AI*

Gradient-weighted Class Activation Mapping (Grad-CAM) [40] uses the gradients of the target from the final convolutional layer to create a localization map that highlights the critical areas in the image that are used for prediction or decision. This localization map is obtained by computing the differentiable activation with convolutional layers' feature map activations. To obtain the importance weight scores of neurons, the flowed back gradients are globally averaged pooled over the dimensions of height and width. The basic math behind calculation of importance weight scores (also referred to as class discriminative scores) is shown in Fig. 4. The activations in any layer of deep networks can be explained using the above approach thus implying which features of the image are significant to determine the outcome of the task. Grad-CAM works well on pretrained networks and can be used feasibly with a custom target layer. As there is a single target region in an instance, it is suitable for gaining insights via visual explanations. It is a feature-oriented approach that reveals where a determination of target is being made with regard to the input features by producing activation heatmaps [41].

3.8 *Model Training*

The pretrained ResNet-34 model on ImageNet dataset has been used as backbone for U-net and has been branched into a classification layer where a pooling layer, flatten layer and fully connected layers are used to give the output for binary classification. This baseline model was trained for 150 epochs with a batch size of 32 but it was overfitting as there was a 20% gap between validation metric and training metric. The solution to this was to perform data augmentation, use regularization (l2 norm) and do hyperparameter tuning.

Thus, the loss function of the MTL model can be defined as:

$$L_j = cL_c + sL_s + o \tag{1}$$

where L_c is the categorical cross-entropy loss function and L_s is the Dice loss.

Neural network parameters and c, s, o are the hyperparameters.

Data augmentation consisted of minor transformations such as horizontal/vertical flips, rotation and scaling to the original samples that made the model more robust. Adam optimizer was used with an initial learning rate of 0.00001 and a dropout layer with 0.15 value was added.

After performing these necessary changes, the model was retrained and the gap reduced to 8% but still it was still overfit in nature. Another issue was that the convergence of the classification branch was faster when relatively compared to the segmentation branch. To overcome this, we implemented a basic functional model that has a series of convolutional locks acting as an encoder followed by its corresponding decoder part (Figs. 5, 6, 7 and 8).

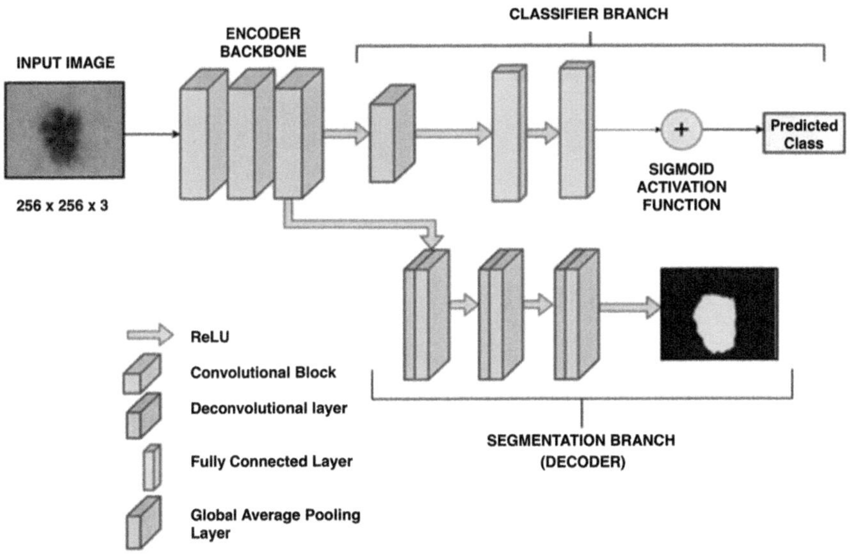

Fig. 5 Convolution-deconvolution based MTL model

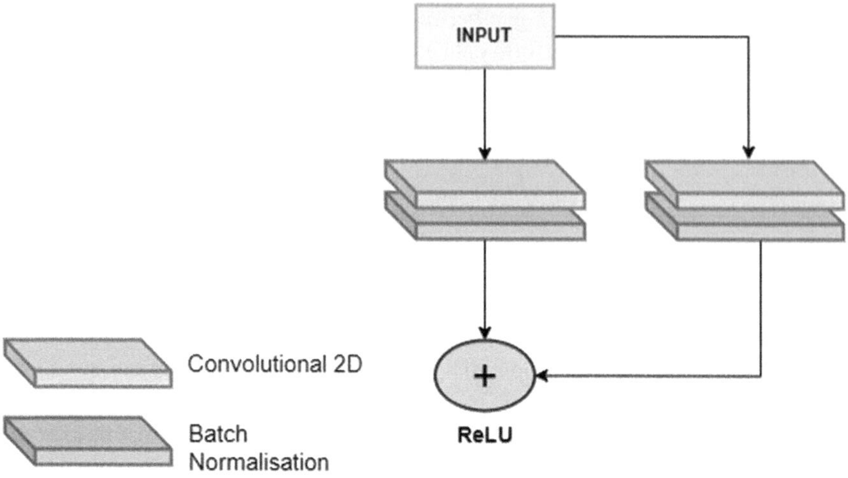

Fig. 6 Representation of convolution block

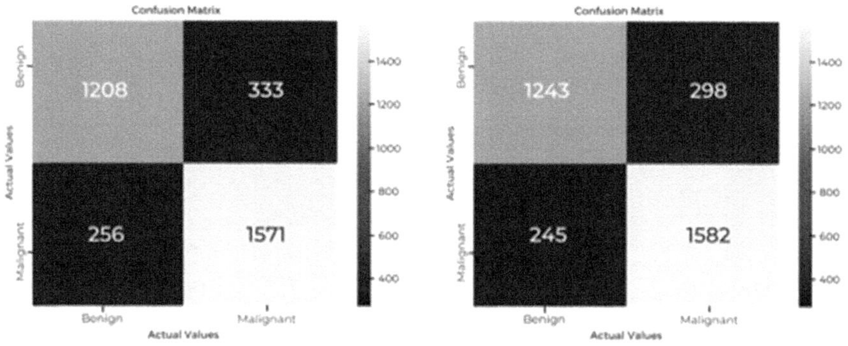

(a) STL Confusion Matrix (b) MTL Confusion Matrix

Fig. 7 Confusion matrix

4 Experiments and Results

In this study, a MTL based model was used to segment and classify the skin lesions into two categories. Furthermore, to establish how the different tasks when trained together affects the results, an XAI method, namely, Grad-CAM algorithm was used. To evaluate the performance of the model, it was tested on 654 samples of testing data belonging to the HAM-10000 dataset. The metrics used for evaluation of STL models are accuracy, precision and sensitivity for classification and IOU Score for segmentation model. Categorical cross entropy and dice loss were used as loss functions for STL classification and segmentation respectively [42]. The above listed

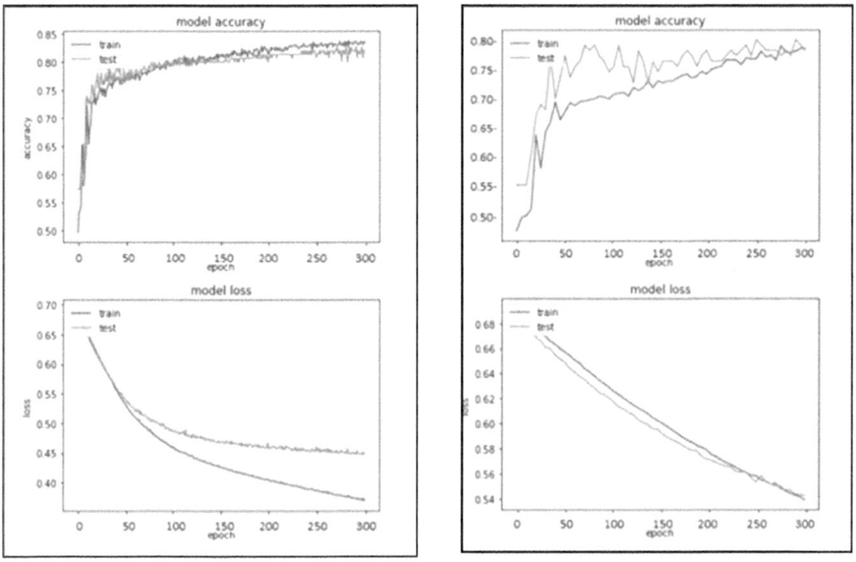

(a) STL Classification

(b) STL Segmentation

Fig. 8 Metric curves for STL

metrics were used for the MTL model as well. The loss function being used for MTL is described in previous section. Table 2 shows the detailed results of the implementation. The baseline MTL model i.e., U-Net with ResNet-34 as backbone gave an overall training accuracy of 92.01% but was overfitting in nature as the testing accuracy was 85.81%. Hence, our custom MTL model composed of convolution and deconvolution layers was used that yielded an overall accuracy of 91.56%.

Table 2 Results for STL and MTL for specified tasks

Task	Metric	STL	MTL
Classification	Accuracy	86.39%	91.56%
	Precision	78.39%	81.99%
	Sensitivity	82.51%	79.15%
	Categorical cross entropy (loss)	0.526	0.398
Segmentation	IOU Score	80.93%	87.98%
	Dice loss	0.55	0.43
Classification	Trainable parameters	29,515,809	25,134,291
Segmentation		36,490,261	
Classification	Training time	164 min (300 epochs)	147 min (180 epochs)
Segmentation		243 min (300 epochs)	

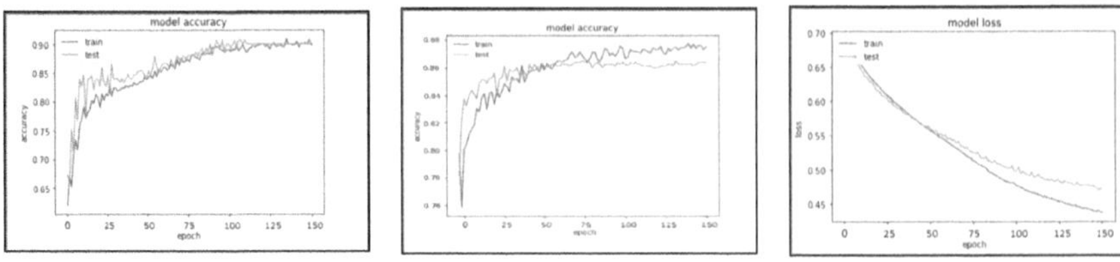

(a) MTL Classification (b) MTL Segmentation (c) MTL Loss

Fig. 9 Metric curves for MTL

4.1 Comparison with Single Task Learning

The performance of both MTL models was compared to that of the single task learning model in terms of parameters, time taken for training and evaluation metrics. A point to be noted is that MTL models showed convergence under 100 epochs as opposed to single task learning models that showed convergence under 180 epochs. The metric curves (plot between evaluation metric and number of epochs) illustrated in Figs. 9 and 10 makes this quite clear.

As seen in Table 2, the number of parameters and training time are reduced by almost 55–60% and the performance of the model improves by 5% on classification. This increase in the evaluation metric values is with a relatively low bias and low variance as evident in Figs. 9 and 10. The reason for the reduction in no. of parameters is that as MTL is governed by a combined loss function, the complexity of the model needs to be reduced which also helps to regulate the convergence rate of both the tasks. As seen in the confusion matrix (Fig. 8), (considering benign as positive and malignant as negative) the number of true positives and true negatives show an increase in the value and therefore, the accuracy increases. The number of false positives reduces more as compared to false negatives. Figure 9 shows the metric curves for the STL for both classification and segmentation. In Fig. 10 the curves for MTL are shown.

4.2 Segmentation Results

As we can see in Table 3 there are distinct differences in size, boundary and symmetry between segmented masks of class benign and malignant. These differences derived from segmentation task are used by classification task in the model.

Table 3 Segmentation masks

Predicted Masks	Predicted Class
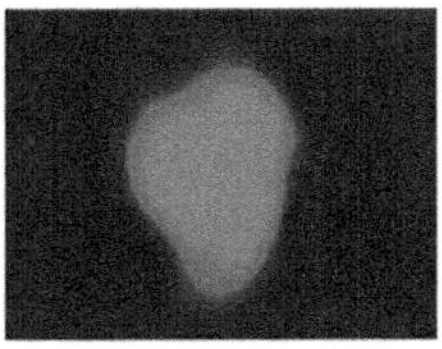	Benign
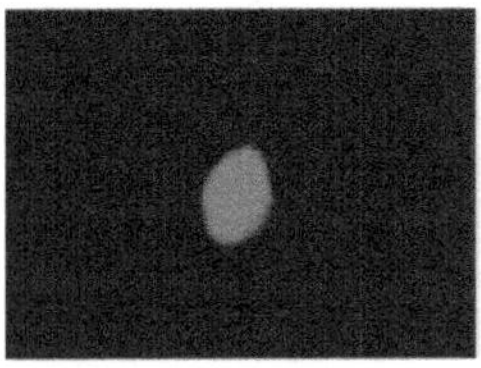	Benign
	Malignant
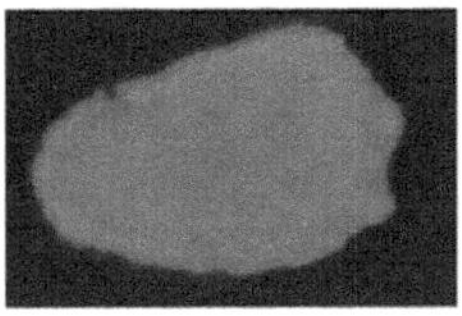	Malignant

4.3 *XAI Observations*

Grad-CAM method was applied on the model with the target layer being the final layer of the layers that are shared in the network of the model. It is done so to interpret what features were shared among both the tasks by viewing their activations and deduce if the shared feature learning is generalizing well while showing improved performance. Heatmaps were produced by this method and the inferences derived from this method are discussed further. The feature representation learning shared between both the tasks involves localizing the boundaries of the lesion and determining its basic shape.

As seen in Table 4, a clear difference is observed in the heatmaps of both the classes.

- The edges visible in boundaries of the benign skin lesions are smooth as compared to those of malignant, which are highly uneven.

Table 4 Comparison of activation heatmaps

Characteristic	Class Benign	Class Malignant
Edges	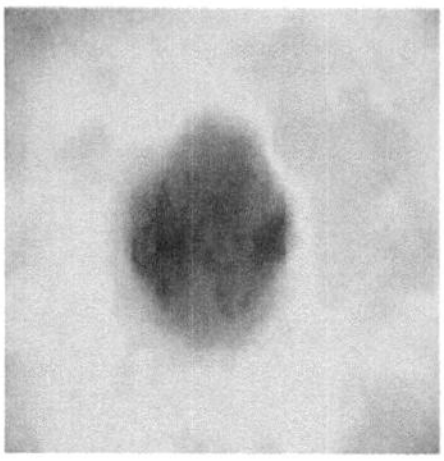	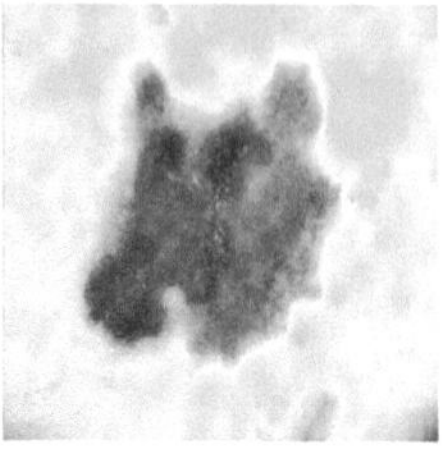
Size	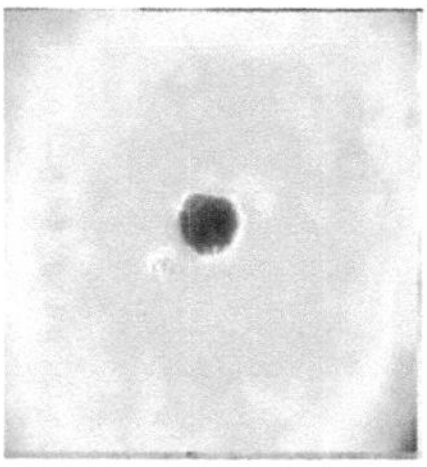	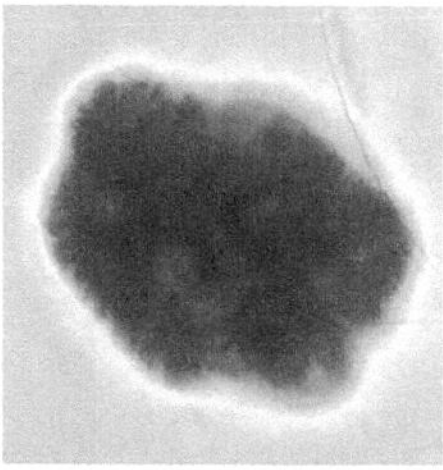
Symmetry	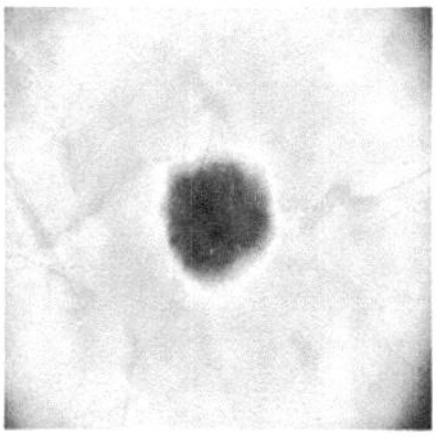	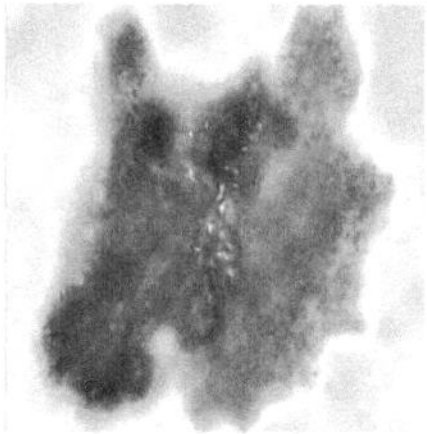

- The size of the fundamental shape of the malignant lesion is larger than that of benign.
- Regions in benign heatmaps are almost symmetrical in nature as opposed to that in malignant heatmaps which are highly asymmetrical.

An important point to be noted is that from a segmentation perspective, the shared features help to determine the boundary of the region containing skin lesions. This boundary further helps in establishing the properties of regularity in edges, size and symmetry of the lesion that forms the baseline for distinguishing between both the classes, hence helping to improve the performance of classification. The results of the Grad-CAM method confirms that the task of localization of the lesion spot is has contributed intricacies to the feature-learning process that has shown a positive impact on the task of classification. This shared learning is based on the mechanism of attention focusing, eavesdropping and regularization. Since subsequent tasks will offer further evidence for the relevance or irrelevance of those features, MTL can aid the model in concentrating on the feature representation that really matter. This is referred to as attention focusing. Some features, say F are easy to grasp for a task A yet complex to attain for another task B. MTL, in this case, is allowing the model to eavesdrop, i.e. learning the feature F through task B and using it for task via hints, i.e. predicting the features as an auxiliary task. An inductive bias is introduced while following MTL methodology that helps in regularization. All these mechanisms collectively contributed to improved generalization of the features and hence, also to the improved results.

4.4 Comparison with Existing Results

Garg et al. [43] proposed a CNN model for to classify it in different categories. Transfer learning method was applied on HAM-10000 dataset using ResNet network baseline model to increase the classification accuracy of images which resulted in 88% weighted average precision, 74% recall average and accuracy of 90.51%. Five pre-trained models were used by Kazi et al. [44] along with their proposed model for comparative analysis on HAM-10000. Inception V3 model gave the highest accuracy of 85.25% among pre-trained model, while the proposed model of three convolutional layer, with ReLU as an activation for each layer and SoftMax activation function in the final layer gave maximum accuracy of 90.55%. Similarly, Ioannis et al. [45] trained and tested 11 CNN architecture, from which DesNet169 gave the best results of 92.25% accuracy and 93.27% F1-score. A CNN model was built on top of AlexNet pre-trained model by Ameri et al. [46] that was modified to give 84% classification accuracy. SGDM was used to train the model with initial learning rate of 0.0001.

Polat et al. [47] used a combination of CNN and one versus all on HAM-10000 dataset to perform classification. End-to-end approach was used to train the model from raw images. As compared to CNN alone with accuracy of 77%, this model showed 92.90% accuracy. Yang et al. [48] proposed a multi-task DCNN model based on GoogleNet and U-Net which incorporated segmentation as a task. It gave IOU Score for lesion segmentation as 0.724 and the AUC on individual lesion classifications were 0.880 and 0.972.

Six models with each model having two variants one with all layers trained and another with fully trained additional layer and one frozen convolutional layer was applied on ISIC 2017 Challenge dataset by Ralf Raumanns et al. [49]. It showed that ROC curve of baseline model was 0.794 while that of multi-task was 0.808 for combined ensembles and 0.811 for individual crowdsourced features. Irek Saitova et al. [50] proposed a Y-Net architecture model to improve performance by using segmentation as an auxiliary task and labelling different modality with segmentation masks. The model allowed sharing of weights between classification and segmentation branches. This model showed accuracy of 0.84 and 0.527 F1 score while the baseline ResNet had 0.83 accuracy and 0.502 F1 score. A 2D MT-CNN model consisting of two branches, zrain tumor's nodule classification as main task and image reconstruction branch as an auxiliary task was proposed by Penghua Zhai et al. [51] to preserve information obtained in CNN hierarchical structure that is useful for identification of malignant module. In the LUNA-16 dataset, the method outperformed state-of-the-art models with a false positive rate of 3.2% and an AUC of 97.3%.

Our MTL method performs binary classification with reduced no. of parameters and lesser training time gives a comparatively better and a comparable accuracy value of 91.56% and IoU Score of 87.98%.

5 Discussion

The primary aim of using a MTL approach was to improve the performance with reduced no. of trainable parameters in lesser amount of time as compared to that of single task learning approach. U-Net based MTL with ResNet-34 as its backbone was chosen as the baseline model. The MTL framework showcased in this study achieved an accuracy of 91.56%, with a sensitivity of 79.15% and a precision of 81.99%. These metrics not only demonstrate the efficacy of the MTL model in classifying skin lesions but also highlight its superiority over traditional STL models, which only reached an accuracy of 86.39%. Moreover, the comparison between STL and MTL models revealed that the latter not only achieved superior performance metrics but also exhibited a more favorable convergence behavior. This observation suggests that MTL frameworks may be inherently more capable of handling the complex, intertwined features of medical imaging data, due to their ability to leverage shared representations. It is essential to consider that these shared features, which span across tasks such as segmentation and classification, likely offer a richer contextual understanding of the images, leading to more robust model performance. Though it showed an increase in desired performance evaluation metrics, it suffered from the problem of overfitting and a significant difference in convergence rates of both the tasks. For instance, while the model demonstrated rapid convergence, there was an imbalance in the rate at which the segmentation and classification tasks learned. This discrepancy suggests that further tuning of task-specific parameters might be required to ensure that both tasks benefit equally from the shared learning setup, thereby preventing one task from dominating the learning process. Hence, another model having less complexity based on convolutional-deconvolutional architecture was introduced that solved the issue of overfitting and both the tasks had comparable convergence rates.

For inferencing which features of the input are shared between the two tasks contributing to the collective generalized learning of the model and how the collective learning impacts the overall outcome, visual explanations in the form of class activation heatmaps were generated using the Grad-CAM technique. Interestingly, the differences observed between the class activation heatmaps of benign lesions and malignant lesions (defined by boundary detection for segmentation task) are presented in the ABCD checklist used by dermatologists in diagnosis of skin cancer [52]. The heatmaps generated by Grad-CAM clearly delineated the areas of focus, showing more pronounced and irregular boundaries in malignant lesions as compared to smoother, more uniform features in benign cases. This visual confirmation supports the numerical results, providing a tangible validation of the model's decision-making process. Additionally, the IoU score for segmentation tasks achieved by the MTL model stood at 87.98%, which is a substantial improvement over the STL models, emphasizing the MTL's enhanced capability in handling complex imaging tasks. This visualization not only reinforced the model's credibility but also offered a pathway to understanding the model's behavior in clinical settings, where explainability is as crucial as accuracy. Thus, the learning parameters shared between the tasks are relevant and are majorly contributing to the performance improvement.

The tasks used for simultaneous learning may not always improve the performance of the model [53]. When an XAI method to produce visualization heat maps was used as the second auxiliary task, the model performance had no significant improvement or deterioration which suggested that the task is not contributing to the joint learning. Non-comparable convergence rates of the shared tasks provide a high possibility of the model being an overfit in nature.

6 Conclusion

In conclusion, this work implemented a deep learning model based on MTL methodology to enhance the segmentation performance and binary classification of skin lesions. Utilizing Explainable AI alongside, the shared features between the two tasks were interpreted, confirming their relevance. The HAM-10000 Dataset was utilized for training and testing, with samples segregated into benign and malignant classes. The MTL approach involved skin lesion segmentation and binary classification, effectively reducing parameters compared to single-task learning. Two models were employed, with the convolutional-deconvolutional model outperforming the U-Net based model due to its better generalization. Explainable AI results revealed shared features, particularly pixels along lesion boundaries, corresponding well with dermatological ABCD checklist criteria. However, not all tasks learned simultaneously improved performance, emphasizing the importance of careful task selection for effective optimization. While existing datasets for MTL in the biomedical domain are limited, future work should focus on curating datasets that support co-existing tasks. Additionally, the development of intrinsic techniques for visual explanations within MTL frameworks holds promise for model interpretation in healthcare applications.

Acknowledgements The work is supported by the Centre for Advanced Modelling and Geospatial Information Systems (CAMGIS), University of Technology Sydney.

Declarations of Conflict of interest The authors declared that they have no conflicts of interest to this work.

References

1. Skin Cancer Facts and Statistics. The Skin Cancer Foundation, www.skincancer.org/skin-cancer-information/skin-cancer-facts
2. Jones, Owain T., et al., Recognising Skin Cancer in Primary Care. Adv. Ther. **37**(1) Springer Science and Business Media LLC, pp. 603–16 (2019)
3. A.A. Marghoob, A. Scope, The Complexity of Diagnosing Melanoma. J. Investig. Dermatol. **129**(1), 11–13 (2009)
4. J.A. Jaleel, et al., Computer Aided Detection of Skin Cancer. in *2013 International Conference on Circuits, Power and Computing Technologies (ICCPCT)* (IEEE, 2013)
5. M.A. Kassem, et al., Machine Learning and Deep Learning Methods for Skin Lesion Classification and Diagnosis: A Systematic Review. Diagnostics **11**(8), 1390 (2021)
6. F. Doshi-Velez, K. Been, Towards a rigorous science of interpretable machine learning. arXiv preprint arXiv:1702.08608 (2017)
7. T. Rogers, et al., Triage Amalgamated Dermoscopic Algorithm (TADA) for Skin Cancer Screening. Derm.Logy Pract. & Concept. **7**(2), Mattioli1885 (2017)
8. A. Lallas, et al., The BRAAFF Checklist: A New Dermoscopic Algorithm for Diagnosing Acral Melanoma. Br. J. Derm.Logy **173**(4), Oxford UP (OUP), 1041–49 (2015)
9. N.S. Zghal, D. Nabil, Melanoma skin cancer detection based on image processing. Curr. Med. Imaging Former. Curr. Med. Imaging Rev. **16**(1) Bentham Science Publishers Ltd., 50–58 (2020)
10. H. Priya, G. Asha, et al. (2018) Identification of Melanoma in Dermoscopy Images Using Image Processing Algorithms. in *2018 International Conference on Control, Power, Communication and Computing Technologies (ICCPCCT)* (IEEE, 2018)
11. S. Bozsányi, et al., Quantitative multispectral imaging differentiates melanoma from seborrheic keratosis. Diagnostics **11**(8), MDPI AG, 1315 (2021)
12. M.A. Kassem, et al., Machine learning and deep learning methods for skin lesion classification and diagnosis: A systematic review. Diagnostics **11**(8) MDPI AG, 1390 (2021)
13. A.H. Shahin, et al., Deep ensemble learning for skin lesion classification from dermoscopic images. in *2018 9th Cairo International Biomedical Engineering Conference (CIBEC)* (IEEE, 2018)
14. E. Ahn, et al., Saliency-based lesion segmentation via background detection in dermoscopic images. IEEE J. Biomed. Health Inform. **21**(6), Institute of Electrical and Electronics Engineers (IEEE), 1685–93 (2017)
15. M. Nasir, et al., An Improved Strategy for Skin Lesion Detection and Classification Using Uniform Segmentation and Feature Selection Based Approach. in *Microscopy Research and Technique*, ed. by P. Bianchini, vol. 81, no. 6, (Wiley, 2018), pp. 528–43
16. T. Chu, et al., Improving weakly supervised lesion segmentation using multi-task learning. in *Medical Imaging with Deep Learning* (PMLR, 2021)
17. A.E.W. Johnson, et al., MIMIC-III, a Freely Accessible Critical Care Database. in *Scientific Data*, vol. 3, no. 1 (Springer Science and Business Media LLC, 2016)
18. L. Liu, et al., Multi-task learning via adaptation to similar tasks for mortality prediction of diverse rare diseases. in *AMIA Annual Symposium Proceedings*, vol. 2020 (American Medical Informatics Association, 2020)
19. D. Coppola, et al., Interpreting Mechanisms of Prediction for Skin Cancer Diagnosis Using Multi-task Learning.in *2020 IEEE/CVF Conference on Computer Vision and Pattern Recognition Workshops (CVPRW)* (IEEE, 2020)
20. J. Kawahara, et al., Seven-Point Checklist and Skin Lesion Classification Using Multitask Multimodal Neural Nets. in *IEEE Journal of Biomedical and Health Informatics*, vol. 23, no. 2 (Institute of Electrical and Electronics Engineers (IEEE), 2019), pp. 538–46
21. D.-K. Ngo, et al., Multi-Task Learning for Small Brain Tumor Segmentation From MRI. Appl. Sci. **10**(21), MDPI AG, 7790 (2020)

22. M.A. Khan, et al., Skin Lesion Segmentation and Multiclass Classification Using Deep Learning Features and Improved Moth Flame Optimization. Diagnostics **11**(5), MDPI AG, 811 (2021)
23. B. Zhang, et al., Bi-Rads-Net: An explainable multitask learning approach for cancer diagnosis in breast ultrasound images. in *2021 IEEE 31st International Workshop on Machine Learning for Signal Processing (MLSP)* (IEEE, 2021)
24. Y. Xie, et al., A mutual bootstrapping model for automated skin lesion segmentation and classification. in *IEEE Transactions on Medical Imaging*, vol. 39, no. 7 (Institute of Electrical and Electronics Engineers (IEEE), 2020), pp. 2482–93
25. N.C.F. Codella, et al., Skin lesion analysis toward melanoma detection: A challenge at the 2017 international symposium on biomedical imaging (ISBI), hosted by the international skin imaging collaboration (ISIC). in *2018 IEEE 15th international symposium on biomedical imaging (ISBI 2018)* (IEEE, 2018)
26. T. Mendonça, et al., PH 2-A dermoscopic image database for research and benchmarking. In *2013 35th annual international conference of the IEEE engineering in medicine and biology society (EMBC)* (IEEE, 2013)
27. M. Dildar, et al., Skin cancer detection: A review using deep learning techniques. Int. J. Environ. Res. Public Health **18**(10), MDPI AG, 5479 (2021)
28. Z.Q. Lin, et al., Do explanations reflect decisions? A machine-centric strategy to quantify the performance of explainability algorithms. arXiv preprint arXiv:1910.07387 (2019)
29. C. Metta, et al., Explainable deep image classifiers for skin lesion diagnosis. arXiv preprint arXiv:2111.11863 (2021)
30. F. Stieler, et al., Towards domain-specific explainable AI: Model interpretation of a skin image classifier using a human approach. in *2021 IEEE/CVF Conference on Computer Vision and Pattern Recognition Workshops (CVPRW)* (IEEE, 2021)
31. M. Zia U. Rehman, et al., Classification of skin cancer lesions using explainable deep learning. Sensors **22**(18), 6915 (2022)
32. F. Nunnari, K. Md Abdul, S. Daniel, On the overlap between grad-cam saliency maps and explainable visual features in skin cancer images. in *Machine Learning and Knowledge Extraction: 5th IFIP TC 5, TC 12, WG 8.4, WG 8.9, WG 12.9 International Cross-Domain Conference, CD-MAKE 2021, Virtual Event, August 17–20, 2021, Proceedings 5. Springer International Publishing* (2021)
33. M. Crawshaw, Multi-task learning with deep neural networks: A survey. arXiv preprint arXiv: 2009.09796 (2020)
34. P. Tschandl, et al., The HAM10000 Dataset, a large collection of multi-source dermatoscopic images of common pigmented skin lesions. Sci. Data **5**(1), Springer Science and Business Media LLC, (2018)
35. R. Caruana, Multitask learning. Mach. Learn. **28**, 41–75 (1997)
36. Y. Zhang, Y. Qiang, An overview of multi-task learning. Natl. Sci. Rev. **5**(1), Oxford UP (OUP), 30–43 (2017)
37. D. Stutz, Understanding convolutional neural networks. in *Seminar on Current Topics in Computer Vision and Machine Learning* (2014)
38. H. Noh, H. Seunghoon, H. Bohyung, Learning deconvolution network for semantic segmentation. in *Proceedings of the IEEE international conference on computer vision* (2015)
39. N. Navab, et al., Medical image computing and computer-assisted intervention–MICCAI 2015: 18th International Conference, Munich, Germany, October 5–9, 2015. in *Proceedings, Part III*, vol. 9351 (Springer, 2015)
40. R.R. Selvaraju, et al., Grad-CAM: Visual explanations from deep networks via gradient-based localization. in *2017 IEEE International Conference on Computer Vision (ICCV)* (IEEE, 2017)
41. P.P. Angelov, et al., Explainable artificial intelligence: an analytical review. Wiley Interdiscip. Rev.: Data Min. Knowl. Discov. **11**(5), e1424 (2021)
42. M. Hossin, Md.N. Sulaiman, A review on evaluation metrics for data classification evaluations. Int. J. Data Min. & Knowl. Manag. Process. **5**(2), 1 (2015)

43. R. Garg, et al., Decision support system for detection and classification of skin cancer using CNN. in *Advances in Intelligent Systems and Computing* (Springer Singapore, 2020), pp. 578–86
44. K.Md.M. Haider, et al., An enhanced CNN model for classifying skin cancer. in *Proceedings of the 2nd International Conference on Computing Advancements, USA* (ACM, 2022)
45. I. Kousis, et al., Deep learning methods for accurate skin cancer recognition and mobile application. Electronics **11**(9), MDPI AG, 1294 (2022)
46. A. Ameri, A deep learning approach to skin cancer detection in dermoscopy images. J. Biomed. Phys. Eng. **10**(6), Salvia Medical Sciences Ltd., (2020)
47. K. Polat, O.K. Kaan, Detection of skin diseases from dermoscopy image using the combination of convolutional neural network and one-versus-all. J. Artif. Intell. Syst. **2**(1) Institute of Electronics and Computer, 80–97 (2020)
48. X. Yang, et al., A novel multi-task deep learning model for skin lesion segmentation and classification. arXiv preprint arXiv:1703.01025 (2017)
49. R. Raumanns, et al., Multi-task learning with crowdsourced features improves skin lesion diagnosis. arXiv preprint arXiv:2004.14745 (2020)
50. I. Saitov, et al., Dermoscopic attributes classification using deep learning and multi-task learning. Procedia Comput. Sci. **178**, Elsevier BV, 328–36 (2020)
51. P. Zhai, et al., Multi-Task Learning for Lung Nodule Classification on Chest CT. in *IEEE Access*, vol. 8 (Institute of Electrical and Electronics Engineers (IEEE), 2020), pp. 180317–27
52. S. Krishna, A.S. Remya Ajai, Analysis of Three Point Checklist and ABCD Methods for the Feature Extraction of Dermoscopic Images to Detect Melanoma. in *2019 9th International Symposium on Embedded Computing and System Design (ISED)* (IEEE, 2019)
53. T. Standley, et al., Which tasks should be learned together in multi-task learning? in *International Conference on Machine Learning* (PMLR, 2020)

IoT Operational Intelligence

Ollencio R. J. D'Souza, Subhas C. Mukhopadhyay, and Quan Z. Sheng

Abstract IoT, sensors, networks, data communications, microcontrollers, and microprocessors are frequently used terms in successfully functioning processes. However, terms like "process optimisation", "intelligent devices", and "Inference Models" using Machine Learning (ML) are new and related to the functional integrity of systems and may need context, so they add value to our understanding of the domains they perform in. This chapter explains sensors or sensor clusters, ML & the relevance of inference. We explore the intelligent edge using microcontroller devices enhanced by "inference models", using machine learning placed next to where data is generated/collected. We explore enhanced microcontroller technology with inference models that analyse data at the edge, generating "edge intelligence" and improving outcomes by sending analysed, verified results data over trusted (IoT) wired or wireless networks to meet the needs of real-time domains. We establish that systems work efficiently and effectively with credible feedback from sensors. By customising the training of intelligent devices, we show how "edge" devices generate "actionable intelligence" at the edge to optimise workloads and reduce the repetitive work humans need to do manually. With the "invention" of Large Language Models (LLMs), knowledge is placed at our fingertips. Both humans and machines now have access to knowledge using published APIs. We explore the benefits of emerging technologies in solving operational problems in real-time domains such as health and emergencies (fire).

O. R. J. D'Souza
School of Engineering, Macquarie University, North Ryde, NSW 2109, Australia
e-mail: ollencio.dsouza@hdr.mq.edu.au

S. C. Mukhopadhyay (✉)
School of Engineering, Faculty of Science and Engineering, Macquarie University, North Ryde, NSW, Australia
e-mail: subhas.mukhopadhyay@mq.edu.au

Q. Z. Sheng
School of Computing, Faculty of Computing, Macquarie University, North Ryde, NSW, Australia
e-mail: Michael.Sheng@mq.edu.au

B. Pradhan and S. Mukhopadhyay (eds.), *IoT Sensors, ML, AI and XAI: Empowering A Smarter World*, Smart Sensors, Measurement and Instrumentation 50,
https://doi.org/10.1007/978-3-031-68602-3_15

Keywords Machine learning (ML) · Artificial intelligence (AI) · Edge intelligence (EI) · Large language models (LLMs) · Situational awareness · Process automation · Actionable intelligence · Intelligent sensors · Embedded devices

1 Introduction

The Internet of Things or IoT affects our lives, from measuring our water usage to letting us know how often someone came to our front door when we were not there! Every enterprise follows a process to achieve its objective. Each process should fit into another process so that the process chain delivers planned objectives with no disruption. All critical parameters in all processes must be monitored using sensors to ensure they operate between their threshold limits and schedules. IoT plays a significant role in parametric sensing and data transfer from the source (one point) to (another), such as the operations control centre [1].

To enhance operational Intelligence, sensors of different capabilities are placed at various points of a multilayered process chain to collect data from the operational process for analysis and reporting. When analysed, sensor data provides insights into the process's performance, typically called "situational awareness [2]" Fig. 1.

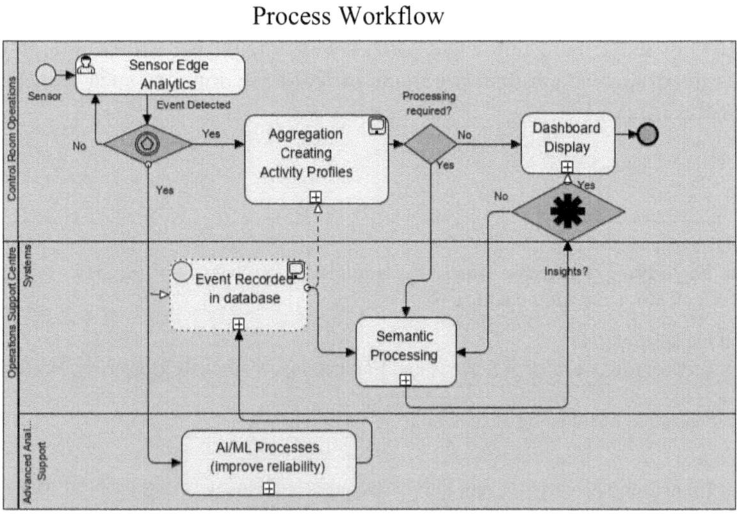

Fig. 1 A gnenric operational process workflow showing "departments"

The critical role of data analytics is to detect operational issues known as "anomalies". Corrective action can be initiated based on the live status reports of the real-time data analysis. We, therefore, create intelligent automation tools in software and embedded hardware. The term "embedded" signifies that some "code" has to be added to the hardware. This code enhances the device's capability to meet design and performance expectations. They are generally called "smart" devices because what is put into them has given them features/functions to deliver unique functionality. Growth in "smart" edge devices also delivers programmability and automation by providing significant autonomous performance, especially in manufacturing domains [3].

The emergence of sophisticated robotic capabilities that mimic and even surpass human-level functionality has made it imperative that we learn how to harness these innovative "embedded" creations to benefit ourselves and humanity. At the core of the technology revolution, we have trained our inventions to levels with self-correcting capability. In the following sections, we will use examples as much as possible to reinforce facts so they are helpful to the reader.

The section flow will be as follows;

(a) Introduction to embedded (configurable) sensors.
(b) Microcontroller-sensor clusters.
(c) Using ML to create "smart" sensors.
(d) How sensors "corroborate, collaborate and confirm" an event (ML).
(e) A healthcare project example/principle that can be applied to others.

2 Introduction to Embedded (Configurable/Intelligent) Sensors

This section will explore the role of "intelligent sensors", "sensor clusters", and process workflows. A guide to how customised functions are developed, the power of machine learning (ML) and Artificial Intelligence (AI) and their role in intelligent sensor creation. There is a need for semantically driven responses based on unique domain ontologies. Other notable developments that deliver autonomous functionality at the "intelligent edge" (explained later with examples) help reduce repetitive workloads [1, 3].

Today, sensors have progressed from a single condition on/off sensor to a multi-sensor microcontroller-based device. Machine learning has given us the tools to mathematically modify the internal logic of a microcontroller to maximise a particular behaviour in a device known as an "inference" model. The enhanced "sensory" functionality allows us to "detect" unique input characteristics and to stop "anomalies" unique to the application from affecting reliability. For example, a fire safety system at a site can be improved by upgrading detection priorities and process performance (Fig. 2), such as detecting "smokeless" fires (hot, thermal waft, below ignition threshold) [1, 4].

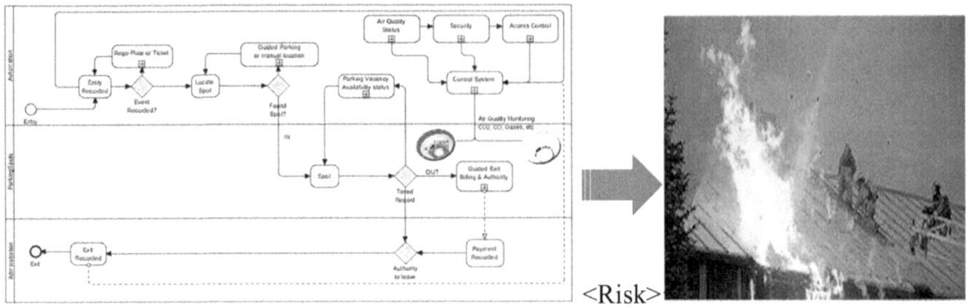

Fig. 2 Intelligent sensor systems and process workflow to manage fire risk

3 Examples of Embedded IoT Logic-Based Devices and Configurations

Example-1: Algorithms are chosen to drive unique sensing functionality in several areas of process activity, especially doing something consistently, repeatedly and accurately. "Models" differ from algorithms because the device behaviours are not algorithmic but learned from training. The "detection behaviour" or functionality is derived from repeatedly enabling the device to perform a detection function by example, correctly and consistently, using its ability to enhance its own "inference" capability from the training provided [5].

(Note: new terms used in this paragraph; "algorithms", "models", "inference", and "process"—are explained below) [6].

In Fig. 3, we have an algorithmically developed sensor in a logic linkup, which must pass through a decision block governed by human-assembled logic to function in a particular way. Sensors of different types are used, and trigger times are "coded in" to mimic a process requirement. When these conditions have been satisfied, the decision block looks at the output of the sensor block. It filters it with "decision logic", "analysis thresholds", "sum of parameters over time", and "trigger timer logic". Eventually, it sends the data over an IoT network such as LoRaWAN. If we look at this diagram, we will understand some critical characteristics of a "process" block or chip that defines the functionality of a node in an IoT network [7].

Explanation of Terms:

a) "Algorithms" are a developed **logic instruction set** that controls the functionality of a sensor block
b) "Models" are **self-contained trained instruction sets** that are created by training a model in a host device and then uploading the reduced model into a microcontroller firmware to change the "behaviour" (functionality) of the microcontroller to meet the detection outcomes desired.
c) "**analytics at the edge**" generates results after analysing data at the collection point. (Not a load of individual data streams per sensor)

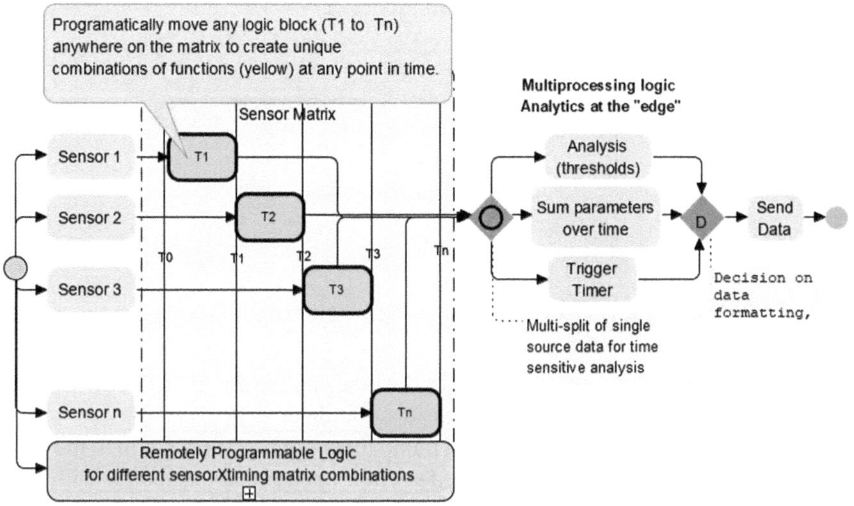

Fig. 3 Inside a multi-sensor algorithmically driven smart sensor

Example 2:

Smart cities have many essential field tasks that all mesh together to meet the smart city objectives. Garbage collection is just one such council task in the overall process. Although this may seem like a simple task, that of a truck being driven around collecting garbage from bins filled by the community to be disposed of at a "dump site" (also called Recycling Centers)... The complexity of garbage collection can only be addressed by a well-organised process. The process is designed to meet the many regulations the local Government sets [8].

In Fig. 4, we have a process encompassing critical steps in the Government's sanitation operation, such as garbage collection and monitoring sewage overflows synchronised via an IoT network [9]. The sensor technology and IoT network reduce the effort of council workers who rely on bin status to manage their collection run.

Fig. 4 Bins equipped with sensors that report back status over IoT

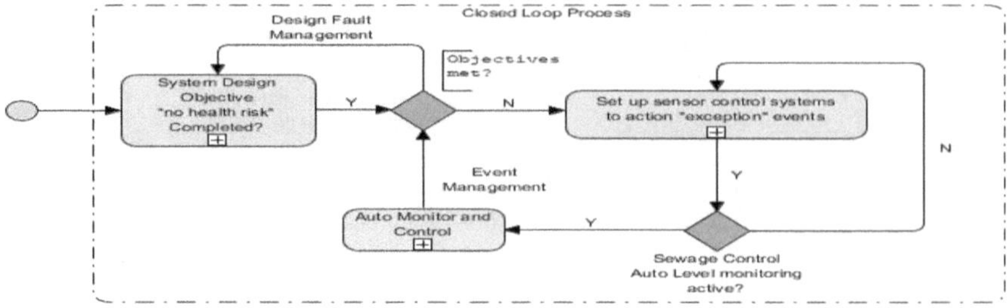

Fig. 5 Garbage and Sewer services

Another critical sanitation concern is to sense sewage so that if an overflow is due to blockage, an alert and a suitable response are initiated (Fig. 5) [10].

The configuration could also trigger a pump that pumps sewage sump to the main sewer. An intelligent city IoT anomaly detection operational dashboard will track key parameters essential to their operational focus.

4 Traditional IoT System Architecture and Operation

The traditional technique uses individual sensors to generate data, which is then sent to a data repository over an IoT network, where it is first stored. The data is then cleaned because of the poor quality received over a stressed network. The data is then analysed. The results are based on standard analytic procedures that provide "event" data such as date, time, pressure, humidity, light levels, etc. Subsequently, action instructions are generated after a review of the analysis results. We must warn that this configuration is not intended for "real-time domains such as Fire, Health, Traffic, etc."[8, 11].

In Fig. 6, A typical SmartCity system configuration and dashboard are shown [12, 13]. The individual configuration and scope are very networking intensive (open system) and prone to reliability failures due to network congestion and other issues that might not receive the data in temporal sync. Such data, when analysed, provides the wrong sequence and priority, which generally causes management and response issues.

In Fig. 7, the configuration is smaller and has a clearly defined operating environment (a closed system) [14].

Two LoRa gateways collect information from multiple nodes spread all over the city. The data is held in the gateways briefly to enable other resourced hardware to "collect" the buffered data. It allows the data to be collected over slow networks, giving adequate time for the data to be collected [15].

Such configurations are easy to roll out, but data integrity is difficult to maintain. In Fig. 7, the IoT configuration drawing is presented, showing sensor stream collection via two public IoT gateways and then stored on a data storage server, which

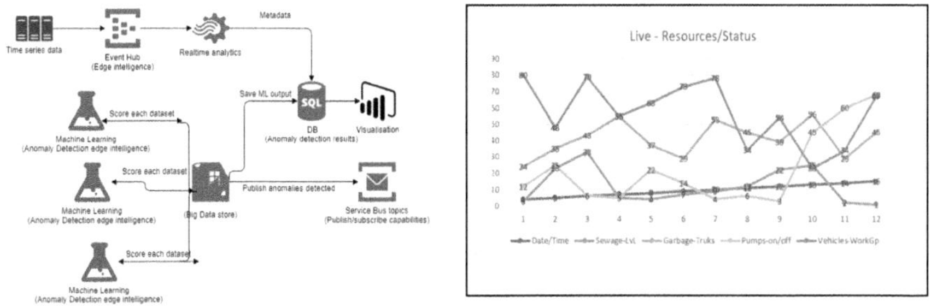

Fig. 6 A typical smart city system configuration stores all information in a database for record-keeping and analysis

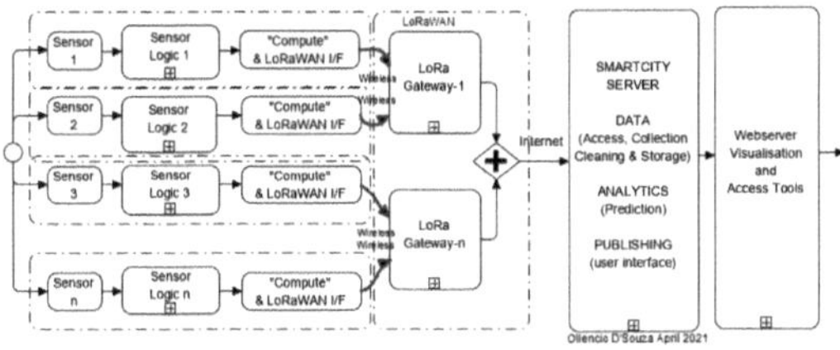

Sensors generate data
Sent to central repository
Data collection and cleaning
Data analysis and results
Develop response
Send instructions

Fig. 7 Traditional IoT configuration all data collected at one location, analysed, and static results placed on a dashboard

has to be checked regularly to ensure data has been received. If not, the "retrieve" process is initiated [16]. This process eventually collects all the data, but the break in communications, the stop-start process of collecting data, will not likely satisfy real-time application demands presented via visualisation on a dashboard Fig. 8.

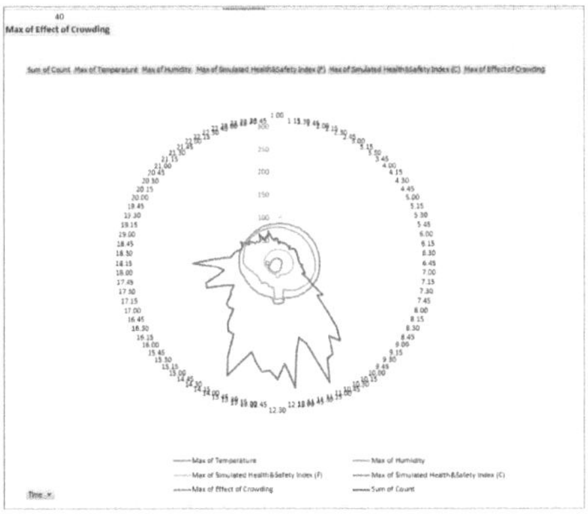

The 24hr live axial graph presents several threshold markers as rings. The data rendered on this graph demonstrates **live** status of parametric measures (temperature, people count, humidity.etc), harmless as a single manifestation but as a Nexus they would break through legislated health and safety thresholds.

Fig. 8 Live dahboard for actionable intelligence

5 Examples and Project Highlights

Data is collected from events or systems managing sites to give us an idea of the situational risk faced (after data analysis), and a response for each risk scenario is developed [14].

The new system architecture must improve the process and be "fit for purpose" or suitable for real-time domains like health, fire, security, and safety. The data is analysed to provide real-time credible "situational awareness". The "information" is not just an alarm trigger but the situational analysis outcome, semantically processed as "actionable intelligence" delivered directly to dashboards and first responder situational awareness notifications, enabling responders to mitigate the risk Fig. 6.

We have explored applications used in other jurisdictions and domains to understand the expanse of the application base we are dealing with and the volume of data carried over various networks linking these sites. The operator must assess if they have the resources to continue expanding individual sensor data generation, transmission, storage, and analysis [17].

Questions to be asked are:

1. How long does it take for critical "data" to travel across IoT networks for "fit-for-purpose" (i.e. real-time) response to be delivered to the end user?
2. Are current overstretched "IoT networks able to deliver (real-time) "actionable intelligence" to the end user in time to avert a crisis?
3. Is the transported data of a quality standard to be trusted?
4. Is this entire process of delivering results in time for critical events like a fire or terrorist attack?

The answer to most of these questions is no because networks have been oversubscribed and overwhelmed [10, 18]. Storing data, cleaning it, and then sending it for analysis is very time-consuming and untrustworthy, and it will only worsen as more sensors come online.

6 Why "Intelligence", and What is an Intelligent Sensor?

In the previous real-world examples, weaknesses of two different IoT configurations were explored, especially concerning data transfer to a repository that could not guarantee the integrity of the received data [2].

Edge intelligence is the outcome of data analysis at the edge where the data is collected, and the results are sent directly to the first responders [3, 19, 20].

Edge intelligence platforms use microcontroller hardware with the trained inference model uploaded to its firmware. It provides the inputs of the multiple parametric sensors on board to the inference engine. The inference engine is created using machine learning. The 32-bit model is created on a cloud-based platform using your data. This model is then reduced in resolution to 8-bit or 4-bit replicas of the original inference model built in the cloud service to meet the requirements of the microcontroller hardware you have chosen from the approved list. It functions as an analytic device when the inference model is uploaded to its firmware [21].

Table 1 describes the functionality the sensor clusters built onboard the selected microcontroller enables.

The sensors on board the microcontroller and the inference model deliver helpful functionality based on the training provided [22, 23] (Fig. 10).

In the first instance, as in Fig. 9, all required sensors are on board the microcon troller. When the inference model is uploaded, it receives inputs from the sensors, analyses them, and creates detection outputs as it was trained to do [24].

The Arduino 33 BLE sense is just one of the Edge Impulse (a cloud inference model creation platform service provider) compatible microcontroller devices on the market today, and it has the essential sensors on board. There are many such microcontroller devices, each with their unique capabilities. The common factor in

Table 1 Rich tapestry of sensing functions

SNo	Sensor title	Features extractable from each sensor
1.0	Temperature	Temperature is a critical parameter for human health, fire safety, etc
2.0	Humidity	Humidity is critical to measure for human health and comfort
3.0	Pressure	Pressure is a critical support parameter for health, safety, security issues
4.0	Light levels	Essential parameters especially in sleep health
5.0	Proximity	Detecting motion 24 × 7 is essential to trigger visual classification
6.0	Colour	Differentiating colour spectrum—essential colour temperature app,
7.0	Acceleration	Essential to understand trajectory and speed for "roughness" measure
8.0	Gyroscope	Essential to measure twists, turns and other specific movements
9.0	Magnetometer	Essential for geo positioning in an environment
10.0	Fusion	Complex detections possible at different associations and combinations
11.0	Audio	Most activity results in "noise" the kind of noise can expose a lot

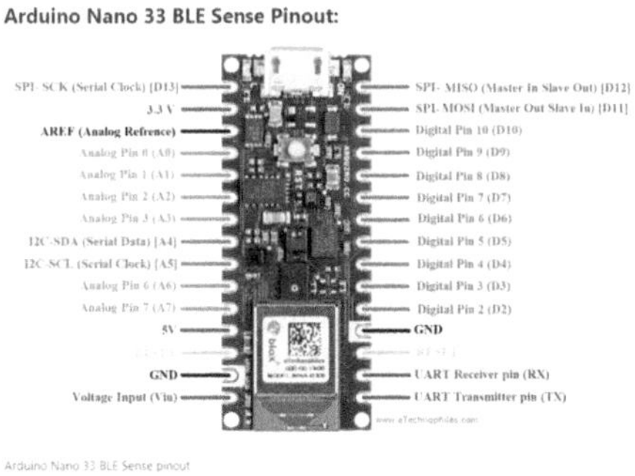

Microcontroller

*Wide range of sensors on board the microcontroller.
*Adequate resources on board (cluster), memory(RAM)
*Arduino IDE programmable
*ML "inference" model upload
*Bluetooth radio for secure comms

Fig. 9 A microcontroller with 12 sensing capabilities using on board sensors

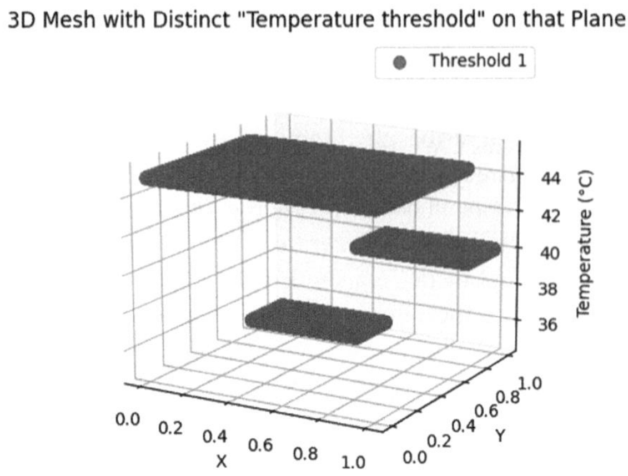

Fig. 10 A 3D perspective of fire status data

all of them is that they can accept a "trained" inference model and apply the altered analytic functionality to the sensors on board as per the trained model. This mode is called "Transfer" learning.

In Fig. 6, a 3D plot created from the data collected shows smoke and fire detection sensor data that graphically captures and presents live situational awareness. The analysis produces 3D area maps showing the temperature, strike area and thresholds reached. The graph (Z axis) shows areas and sizes under threat to enable response teams to prepare adequately. The 3D data analysis produces accurate representations in real-time of the presence of a fire, the location and area it occupies, and the temperature thresholds reached.

The workflow and system architectures have changed significantly to incorporate "intelligence" throughout the workflow.

7 How This 3D Visualisation Capability Helps with Fires

Australia's summer season is dramatic because dense vegetation, high winds, crowds, and intense heat lead to colossal fires in minutes.

So, having a LIVE dashboard to visualise real-time trends is essential.

Results and the proper Visualisation of Key facts on the data:

1. Any parameter can break through a set threshold without causing a severe issue.
2. But when the temperature rises, added heat is dangerous. It is not fatal, but when does high wind or humidity join? Together, they could be lethal!
3. But when a crowd is all together with high temperature, little space and no wind—the prediction of a calamity would be accurate.

In Fig. 11, 2D Linear visualisation is shown in multiple parametric line graphs.

In Fig. 12 3D temperature.

In Fig. 13 Aggression & Energy Level.

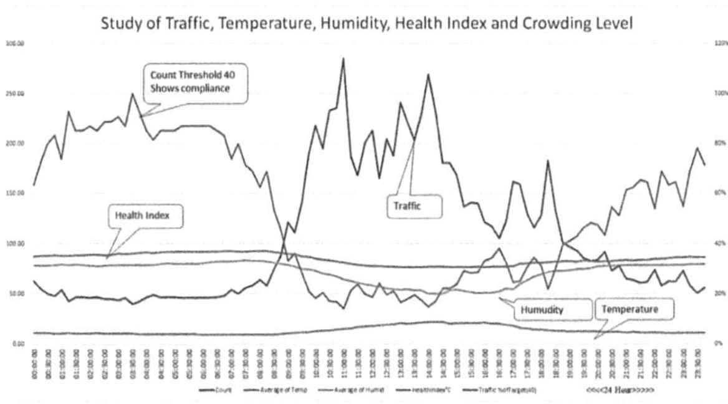

*Each parametric measure is mapped live in temporal sync.

*The legislated health index threshold is also mapped

*The count of people gathered in that area is regulated to avoid a dangerous parametric NEXUS (IoT system generating data)

Fig. 11 The NEXUS of parametric measures is more critical

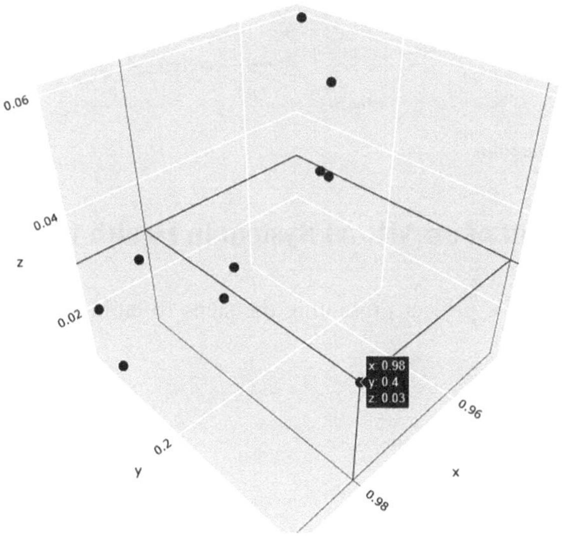

Fig. 12 3D Temperature

In Fig. 12 How do we measure aggression in an aged care facility?.
In Fig. 13 3D and "point cloud"(bubble size) represents intensity.
Live 3D mapping is always a concentrated view of actual status.

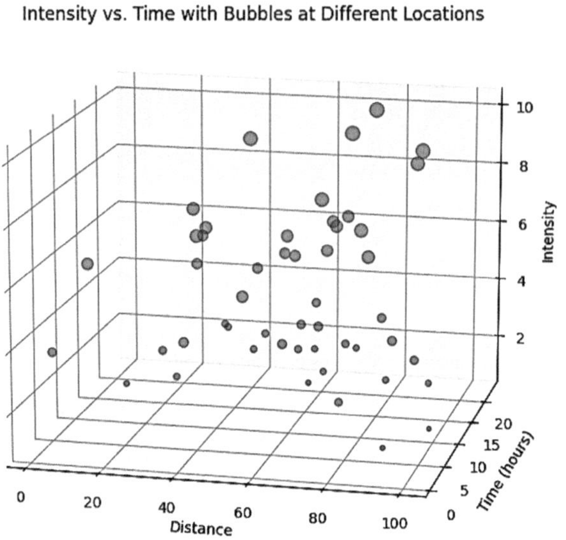

Fig. 13 Gyro/Acceleration

8 Development of an ML/AI System in Health Care

The example is a live project presenting the steps to build your inference model (Fig. 14).

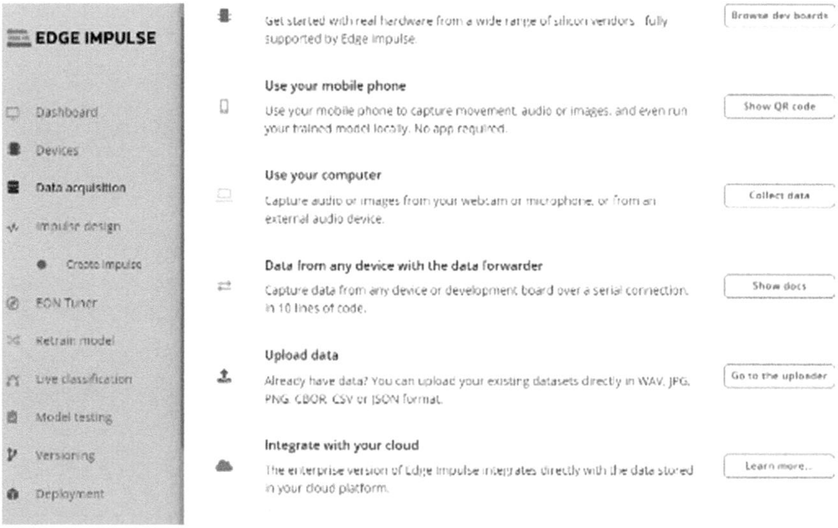

Fig. 14 Cloud service data ingest interface

Step 1: Choose your provider—mostly FREE—unless you are commercial.
Step 2: What are you creating—(our example—automated health care status).
Step 3: Collect a lot of data straight into the cloud data ingest interface.
Step 4: The system will automatically split your collected data into 3sets (Train, Test and Validate). Use the train set to train.

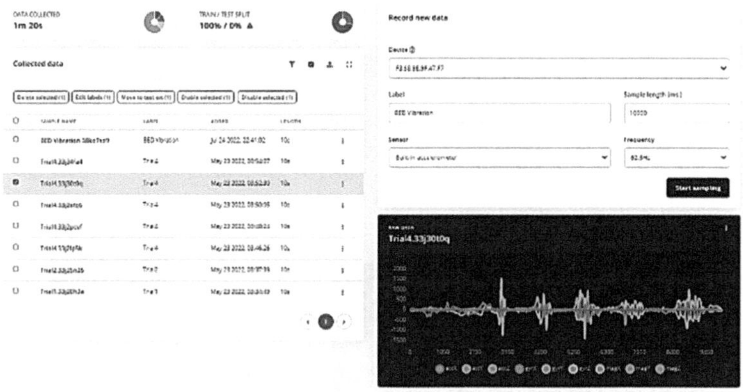

The system will guide you in completing all training data collection.

Step 5: Build your model (again, help from the user interface).
Step 6: You have chosen your microcontroller, so connect it via two kinds of interfaces provided—serial-USB-direct or download and then upload to the firmware.
Step 7: Reset the board, and you are good to go to field testing (use test segment data).

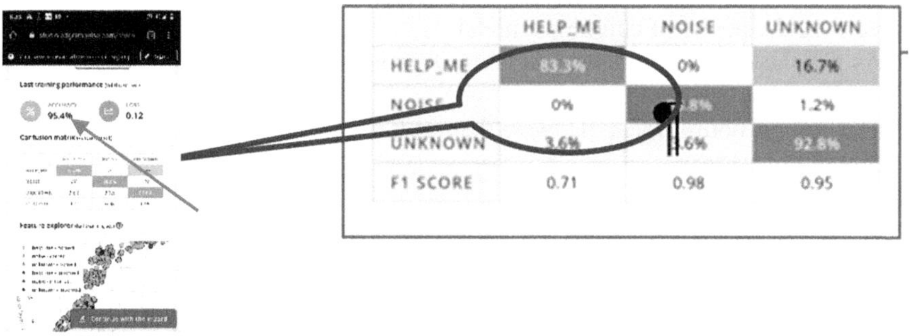

Step 8: Verifying your model before going to live trials. Use confusion matrix score.

At 80%, you might have to trust the model less or retrain it to be better at lower scores.

Other tools provided, in most cases, enhance the model.

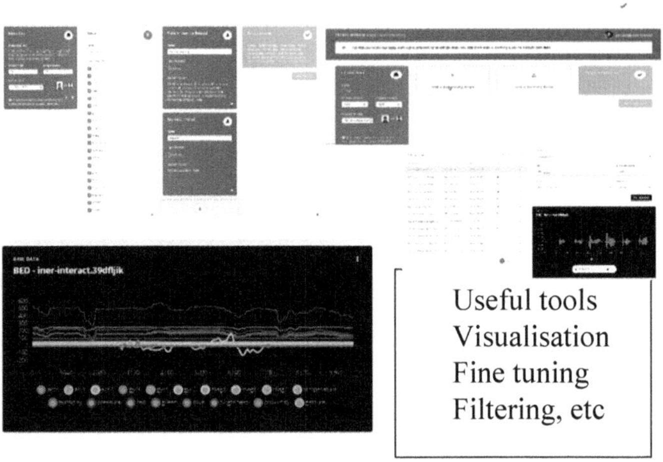

9 Results

The ability to create a model anytime you need one is an excellent opportunity to experience the power of building a model.

As per the confusion matrix, close to 90% is almost always achievable if you follow the steps.

The graphs, both 2D and 3D, were the result of the models built.

Project successes:

1. In Healthcare

1. **Temperature:** Expressed in degrees Celsius (°C).
2. **Relative Humidity:** Expressed in percentage (%).
3. **CO2 (Carbon Dioxide) Levels:** Expressed in parts per million (ppm).
4. **TVOCs (Total Volatile Organic Compounds):** Expressed in (µg/m³) or parts per billion (ppb).
5. **Acceleration:** Expressed as change of speed of movement in 3D space (X, Y & Z directions).
6. **Gyroscopic Movement:** Expressed as values in movement in 3D space (X, Y & Z directions).
7. **Magnetometer:** Expressed as values in movement in 3D space (X, Y & Z directions)
8. **Noise Levels:** Expressed in decibels (dB), a unit that measures the intensity of sound.

The collection of data was targeted using trained inference models.

This standard is currently operational for tests generating data that might become de facto.

10 Conclusion and Future

Inference model creation using machine learning is becoming more accessible, and approved microcontroller devices are getting cheaper and have better functionality. Large language models are the repository of all knowledge and accept a "prompt" or, in other words, a properly framed question that enables the AI system to respond in plain English (or anything else it is trained on). This will revolutionise health care from medication to health procedure compliance.

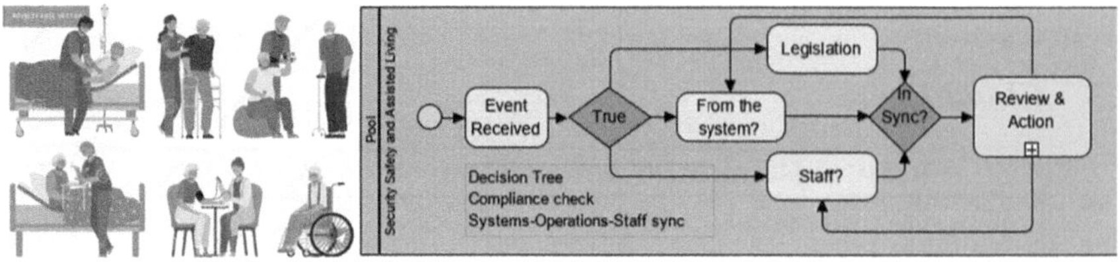

Currently, LLMs (short for Large Language Models) intelligent machines, when asked, using an adequately framed "prompt, understand the request and extract the correct answer from their knowledgebase. However, if the prompt is not constructed well enough, we could end up with an unsuitable answer called a "hallucination".

Prompt Engineering is a new tool/skill that maximises the benefits of having such a powerful tool at our disposal. New APIs enable humans and machines to access the LLMs to find relevant information to complete a task. In a modern construct of the term to describe this activity, it is also known as knowledge engineering to some. Humanity can utilise this unique global knowledge repository managed by these intelligent machines through self-learning from the people who created it.

References

1. O. D'Souza, S. C. Mukhopadhyay, M. Sheng, Health, security and fire safety process optimisation using intelligence at the edge. Sensors **22**(21), 21 (2022) https://doi.org/10.3390/s22218143
2. N. Ahmed, Intelligence at the Extreme Edge: A Survey on Reformable TinyML. ArXiv Cornell Univ., Apr. 2022, Accessed: Dec. 09, 2023. [Online]. Available: https://www.academia.edu/109509075/Intelligence_at_the_Extreme_Edge_A_Survey_on_Reformable_TinyML
3. O. D'Souza, S. Mukhopadhyay, M. Sheng, Semantically Processed Sensor Data in Health Care, Legislation Compliant, Ontologies, in *Sensing Technology*, eds. by N.K. Suryadevara, B. George, K. P. Jayasundera, S. C. Mukhopadhyay, in Lecture Notes in Electrical Engineering (Cham: Springer Nature Switzerland, 2023), pp. 135–148. https://doi.org/10.1007/978-3-031-29871-4_16
4. A. Adnan, A. Salam, A. Arifin, M. Rizal, *Forest Fire Detection using LoRa Wireless Mesh Topology*. 2018, p. 187. https://doi.org/10.1109/EIConCIT.2018.8878488
5. Tracking sleep quality with embedded ML and the Arduino Nano 33 BLE Sense, Arduino Blog. Accessed: Jan. 12, 2023. [Online]. Available: https://blog.arduino.cc/2023/01/10/tracking-sleep-quality-with-embedded-ml-and-the-arduino-nano-33-ble-sense/
6. A. R. Ahad, A. D. Antar, M. Ahmed, *IoT Sensor-Based Activity Recognition: Human Activity Recognition*, in Intelligent Systems Reference Library (Cham: Springer International Publishing AG, 2020)
7. S. Pereira Detro, E.A.P. Santos, H. Panetto, E.D. Loures, M. Lezoche, C. Cabral Moro Barra, Applying process mining and semantic reasoning for process model customisation in healthcare. Enterp. Inf. Syst., 14(7), 983–1009 (2020). https://doi.org/10.1080/17517575.2019.1632382
8. M. Ellena, M. Breil, S. Soriani, The heat-health nexus in the urban context: A systematic literature review exploring the socio-economic vulnerabilities and built environment characteristics. Urban Clim. **34**, 100676 (Dec.2020). https://doi.org/10.1016/j.uclim.2020.100676

9. I. Garcia, C. Pacheco, J. Calvo-Manzano, Using a web-based tool to define and implement software process improvement initiatives in a small industrial setting. IET Softw. **4**(4), 237–251 (Aug.2010). https://doi.org/10.1049/iet-sen.2009.0045
10. B. Cieślik, P. Konieczka, A review of phosphorus recovery methods at various steps of wastewater treatment and sewage sludge management. The concept of 'no solid waste generation' and analytical methods. J. Clean. Prod. **142**, 1728–1740 (2017). https://doi.org/10.1016/j.jclepro.2016.11.116
11. S. Aheleroff, X. Xu, R.Y. Zhong, Y. Lu, Digital Twin as a Service (DTaaS) in Industry 4.0: An Architecture Reference Model. Adv. Eng. Inform. **47**, 101225 (Jan.2021). https://doi.org/10.1016/j.aei.2020.101225
12. F. Akhter, S. Khadivizand, H. R. Siddiquei, M.E.E. Alahi, S. Mukhopadhyay, IoT enabled intelligent sensor node for smart city: Pedestrian counting and ambient monitoring. Sensors **19**(15), 15 (2019). https://doi.org/10.3390/s19153374
13. D. Ivanov, A. Chezhegov, M. Kiselev, A. Grunin, D. Larionov, Neuromorphic artificial intelligence systems. Front. Neurosci. **16** (2022). Accessed: Aug. 24, 2023. [Online]. Available: https://www.frontiersin.org/articles/https://doi.org/10.3389/fnins.2022.959626
14. O. D'Souza, S. Mukhopadhyay, F. Akhter, S. Khadivizand, and E. Memar, Extracting operational insights from everyday IoT data, generated by IoT sensors over LoRaWAN, pp. 241–249 (2021). https://doi.org/10.1007/978-981-16-3368-3_23.
15. R. Alake, How You Should Read Research Papers According To Andrew Ng (Stanford Deep Learning Lectures), Medium. Accessed: Aug. 12, 2020. [Online]. Available: https://towardsdatascience.com/how-you-should-read-research-papers-according-to-andrew-ng-stanford-deep-learning-lectures-98ecbd3ccfb3
16. F. Xue, W. Lu, Z. Chen, C.J. Webster, From LiDAR point cloud towards digital twin city: Clustering city objects based on Gestalt principles. ISPRS J. Photogramm. Remote Sens. **167**, 418–431 (Sep.2020). https://doi.org/10.1016/j.isprsjprs.2020.07.020
17. S.S. Khairat, A. Dukkipati, H.A. Lauria, T. Bice, D. Travers, S.S. Carson, The Impact of Visualization Dashboards on Quality of Care and Clinician Satisfaction: Integrative Literature Review. JMIR Hum. Factors **5**(2), e22–e22 (2018). https://doi.org/10.2196/humanfactors.9328
18. T. T. Network, "The Things Network," The Things Network. Accessed: Jun. 11, 2021. [Online]. Available: https://www.thethingsnetwork.org/community
19. M. Akter, N. Moustafa, T. Lynar, and I. Razzak, Edge intelligence: Federated learning-based privacy protection framework for smart healthcare systems. IEEE J. Biomed. Health Inform. 1–11 (2022). https://doi.org/10.1109/JBHI.2022.3192648.
20. I. Bosch, A. Serrano, L. Vergara, Multi-sensor Network System for Wildfire Detection Using Infrared Image Processing. Sci. World J. **2013**, e402196 (Jun.2013). https://doi.org/10.1155/2013/402196
21. D. Ardagna, M. Mecella, and J. Yang, Eds., *Business Process Management Workshops: BPM 2008 International Workshops, Milano, Italy, September 1–4, 2008. Revised Papers*, vol. 17, in Lecture Notes in Business Information Processing, vol. 17 (Berlin, Heidelberg: Springer Berlin Heidelberg, 2009). https://doi.org/10.1007/978-3-642-00328-8
22. M. Abdulkarem, K. Samsudin, F.Z. Rokhani, M.F.A. Rasid, Wireless sensor network for structural health monitoring: A contemporary review of technologies, challenges, and future direction. Struct. Health Monit. **19**(3), 693–735 (May2020). https://doi.org/10.1177/1475921719854528
23. R. David et al., TensorFlow lite micro: Embedded machine learning for tinyml systems. Proc. Mach. Learn. Syst. **3**, 800–811 (2021)
24. C. Li, V.M. Lubecke, O. Boric-Lubecke, J. Lin, A review on recent advances in Doppler radar sensors for noncontact healthcare monitoring. IEEE Trans. Microw. Theory Tech. **61**(5), 2046–2060 (2013)

Techniques to Implement Blockchain in Internet of Medical Things (IoMT)

Rashmi Singh, Damodar Prasad Tiwari, Ankur Taneja, Rajnish Choubey, Tanuj Kumar Yadav, and Mohd. Suleman Siddiqui

Abstract The large amount of data that is generated through the sensors and used in the health care system with the help of Internet of Medical Things (IoMT) technology is highly sensitive and is facing a lot of security issues to maintain its integrity and privacy. Looking into the risk and security issues associated with IoMT researchers are concentrating in developing techniques to implement blockchain concepts at different layers of IoMT technology. IoMT till now has been working on centralized data whereas blockchain works on decentralized data. With the difference in technology implementation of blockchain in IoMT is quite challenging and has limitations. We present in this chapter the various reasons for implementation of blockchain in IOMT, the architecture of IoMT with the blockchain concepts, various limitation of the implementation and the future work. In order to implement blockchain in IoMT various methods are used. In this paper a complete classification of various methods based on the utilization to handle various issues like data security, privacy, interoperability, identity management, smart contract execution, consensus, integration etc. is done. Amongst all the methods four major methods are identified namely Ethereum-Based Contributions, Modified Consensus Protocol, Modified Cryptographic Technique and Hyperledger-Based Contributions. A review of all the methods of the above-mentioned techniques is done along with a complete description on how these methods are implemented. In this chapter a complete comparative analysis has been done method wise to show their contribution of security enhancement by blockchain technology in IoMT.

Keywords IoMT · Blockchain Ethereum-based contributions · Modified consensus protocol · Modified cryptographic technique and hyperledger-based contributions

---

R. Singh (✉) · D. Prasad Tiwari · A. Taneja · R. Choubey · T. K. Yadav · Mohd. S. Siddiqui
Bansal Institute of Science and Technology, Bhopal, India
e-mail: rashmi@bistbpl.in

B. Pradhan and S. Mukhopadhyay (eds.), *IoT Sensors, ML, AI and XAI: Empowering A Smarter World*, Smart Sensors, Measurement and Instrumentation 50,
https://doi.org/10.1007/978-3-031-68602-3_16

1 Introduction

The term "Internet of Medical Things" (IoMT) defines a network of internet-connected medical equipment, wearables, sensors, and other healthcare-related machineries. These networked devices gather, exchange, and examine data to offer insightful information, allow for remote monitoring, and improve healthcare delivery. IoMT has made it possible to transform healthcare by improving patient care, enhancing production, cutting costs, and giving people more control over their own health. IoMT lets medical specialists to keep an eye on patients' vital signs and health conditions from a distance, allowing for quick interventions and cutting down on the number of trips to the hospital. IoMT enables telemedicine consultations, allowing patients to obtain medical supervision and care from the convenience of their homes. The Internet of Things (IoMT) encompasses a variety of wearable devices, including smartwatches, fitness trackers, and medical-grade sensors The enormous amount of data produced by IoMT devices can be analyzed with the help of machine learning algorithms [1] to spot patterns, trends, and potential health problems, resulting in enhanced decision-making and personalized healthcare recommendations. IoMT can improve asset management, streamline systems, and optimize hospital operations, which will result in more effective healthcare delivery. IoMT technology can help with the management of chronic disorders by offering ongoing monitoring and individualized treatment strategies. IoMT solutions can assist patients in maintaining their drug regimens by reminding them and monitoring compliance. IoMT data can be used for medical research and drug development, resulting to advancements in healthcare IoMT offers many advantages, but it also has issues with data security, privacy, intrusion [2, 3] interoperability, and regulatory compliance. In this quickly emerging sector, protecting patient data and ensuring that IoMT devices can interconnect with one another without interruption are essential problems. The IoMT environment is anticipated to expand and revolutionize numerous facets of healthcare as technology continues to progress, enhancing patient outcomes and reshaping the healthcare sector as a whole.

In paper we review the various challenges that are encountered to implement blockchain technology in IoMT to overcome the security challenges that IoMT is facing. In Sect. 2 we identify the various challenges associated with IoMT. Section 3 will explain the various advantages and challenges of implementing blockchain in IoMT. Section 4 describes the revised architecture of IoMT using Blockchain. In Sect. 5 a detailed review of various techniques to implement Blockchain in IoMT is discussed. Out of all the techniques 4 major techniques namely Ethereum-Based Contributions, Modified Consensus Protocol, Modified Cryptographic Technique and Hyperledger-Based Contributions are discussed in detail and comparative analysis is done. Finally, the chapter will conclude in Sect. 6.

2 Challenges of IoMT

The Internet of Medical Things (IoMT) has the power to completely alter the healthcare industry [4], but it also comes with a number of complications that must be solved before it can be successfully and safely united and implemented. The following are some of the major IoMT challenges:

Data security, ethical issues & permissions: IoMT devices gather and transmit private and sensitive patient health data. It is decisive to guarantee the security and privacy of delicate data in order to avoid illegal access, data breaks, and possible misappropriation. To protect patient information, healthcare organizations must use strong encryption, access controls, and cybersecurity safeguards. IoMT involves ethical issues with patient permission, data ownership, and the possibility for data abuse. Building trust in IoMT technology requires transparency and patient sovereignty in data gathering and usage [5]

Interoperability: The IoMT architecture includes several hardware and software products from various manufacturers that often use dissimilar communication protocols [6]. Data sharing and seamless incorporation of IoMT solutions with current healthcare systems may be vulnerable by a lack of interoperability.

Regulatory Compliance: Depending on the nation and location in which they are utilized, IoMT devices are subject to different regulatory regulations and standards [7]. For manufacturers and healthcare providers, it can be difficult and time-consuming to comply with healthcare regulation and ensure that quality and safety necessities are followed.

Accuracy and Reliability: Medical judgements based on IoMT data must be trustworthy and precise. Any errors or issues with IoMT devices could result in erroneous diagnoses or treatment plans, potentially jeopardizing the safety of the patient [8].

Internet Connectivity: IoMT devices rest on the internet connectivity to transmit data in real-time. Connectivity and network problems are very common and need to be handled timely for proper conduction. The timely delivery of crucial healthcare information could be hampered by connectivity problems, network outages, or bandwidth restrictions [9].

Heavy power consumption: The management of power usage is crucial because many IoMT devices are battery-operated [10].

Cost and Resource Allocation: Implementing IoMT solutions can be expensive, particularly in locations with limited resources or tiny healthcare institutions. Allocating funds for IoMT system purchases, maintenance, and staff training might be difficult [11].

User Acceptance and Training: Patients and healthcare professionals must feel at ease using IoMT devices and deciphering the data they provide. In order to fully enjoy the advantages of IoMT and prevent adoption resistance, proper training and education are needed [12].

Regulatory Approval and Time to Market: Creating and securing regulatory clearances for IoMT devices can take a while, and they may even take a while to market.

Collaboration between healthcare professionals, device manufacturers, policymakers, and cybersecurity specialists is necessary to address these issues [13]. Finding answers to these problems will be essential for realizing the full promise of IoMT in enhancing patient experiences and healthcare outcomes as the IoMT landscape changes.

3 Advantages and Challenges of Implementing Blockchain in IoMT

Data security: With the use of blockchain technology, data is transparently recorded and secured to prevent unauthorized access. Blockchain operates as a distributed ledger where data is replicated across multiple nodes in the network [14]. This decentralized nature eliminates a single point of failure, making it harder for hackers to compromise the entire system by attacking a central server [14]. Once data is added to a blockchain, it becomes nearly impossible to alter or delete it [15]. This immutability ensures that sensitive medical data remains tamper-proof and can be trusted by authorized parties. Blockchain employs cryptographic techniques to secure data. Data stored on the blockchain is encrypted, and a hash function generates unique fixed-size representations of the data. Even if the data is visible on the blockchain, its original content remains secure and private. To add data to the blockchain, a consensus mechanism (e.g., Proof of Work, Proof of Stake) is employed, where participants in the network must agree on the validity of the data. This process ensures that only verified and authorized data is added, reducing the chances of malicious or erroneous data infiltration [14]. Smart contracts are self-executing contracts with predefined rules. They facilitate automated processes based on predefined conditions. In IoMT, smart contracts can ensure that data is only accessed and shared with the appropriate parties, maintaining strict access controls. Blockchain-based identity management systems can be utilized in IoMT to establish a secure and unique identity for each device or user in the network. This prevents unauthorized access and protects against identity-related attacks.

Data Integrity and Trust: Once data is stored on the blockchain, it cannot be changed without the network's participants' agreement. This immutability ensures that sensitive medical data remains tamper-proof and can be trusted by authorized parties [16].

Interoperability and Data Sharing: Because smart contracts were developed Various IoMT devices, systems, and healthcare providers can share data in a secure and easy manner thanks to blockchain [17]. Blockchain can enhance interoperability among different IoMT devices and systems. Standardization of data formats and communication protocols within the blockchain network can improve data security by reducing vulnerabilities arising from disparate systems.

Patient Empowerment and Control: Blockchain enables patients to have more control over their health data because the information is private to the patient [18]. With blockchain, patients can have better control over their medical data. They can grant access to their data to specific healthcare providers and revoke it when necessary, ensuring data privacy and complying with regulations like GDPR.

Simplified Procedures and Lower Costs: As blockchain eliminates reliance on centralized third parties, many administrative procedures become more streamlined, which will lower costs. Blockchain's transparent and auditable nature allows for easy tracing of data access and modifications. This feature helps in identifying any unauthorized activities and maintaining a comprehensive audit trail [14].

Challenges to Implement Blockchain

Scalability: Due to the computational and storage demands of maintaining a decentralized ledger, blockchain scalability continues to be an issue. IoMT generates enormous amounts of data from various devices, therefore ensuring quick transaction processing and scalability is essential for handling the network load and making decisions in real time [19].

Standards and Interoperability: It is difficult to achieve interoperability across various blockchain networks, current healthcare systems, and devices [20]. The smooth sharing of data and collaborative work are hampered by the absence of standard protocols, data formats, and consensus processes.

Privacy and Regulatory Compliance: Using blockchain technology while upholding patient privacy and data protection laws (such as GDPR and HIPAA) is a difficult task. It takes careful design and implementation to strike a balance between transparency and immutability and the requirements for data anonymization, access control, and consent management.4. Architecture of IoMT using Blockchain [21].

Infrastructure and Technical Knowledge: Implementing blockchain in IoMT involves specialized technical knowledge, including an understanding of cryptography, consensus protocols, and the creation of smart contracts. Furthermore, the infrastructure needed to operate blockchain networks, which includes strong processing capabilities, network connectivity, and security measures, can be taxing.

Legacy System Integration: It can be difficult to integrate blockchain with current legacy systems in healthcare organizations. It takes careful planning, system integration, and change management to move and integrate various data sources, platforms, and business processes to a blockchain-based architecture.

Energy Consumption: Some Proof of Work consensus algorithms used in blockchain technology use a lot of energy. Energy-efficient consensus procedures or alternative blockchain designs must be investigated for IoMT devices with limited energy resources.

4 Architecture of IoMT Using Blockchain

The architecture of an Internet of Medical Things (IoMT) system using blockchain classically involves multiple layers and different components working collectively to ensure secure and efficient data management. Here is a high-level overview of the architecture which explains how blockchain is involved in IoMT to increase its security.

IoT Devices: A network of IoT devices, such as wearables, implanted devices, medical sensors, and other linked medical equipment, forms the basis of the system. These devices gather patient health information and might also include built-in cryptographic tools for digital signatures and data encryption.

Data Gathering and Transmission: The Internet of Things (IoT) devices collect data on patients' medical histories, vital signs, and other pertinent information. To be processed further, this data is sent securely to a gateway or a centralized data hub.

Data Processing and Smart Contracts: In the third layer, edge computing or cloud-based platforms are used to process and analyze the gathered data. You can use smart contracts, which are self-executing contracts that are coded on the blockchain, to automate some processes, enforce data access policies, and administer the communication between many parties.

Blockchain Network: The blockchain network, which acts as a decentralized and immutable ledger for storing and managing the medical data, is the foundational element of the system. There are various blockchain systems available, each with its own characteristics and applicability for particular use cases, such as Ethereum, Hyperledger Fabric, or private permissioned blockchains.

Consensus Mechanism: To verify and decide on the order of transactions, the blockchain network depends on a consensus mechanism. By doing so, malicious behaviors are stopped and it is ensured that the state of the blockchain is agreed upon by all nodes in the network. Depending on the blockchain platform selected, Proof of Work (PoW), Proof of Stake (PoS), or other consensus methods may be utilized.

Identity and Access Management: Blockchain-based identity management systems can offer safe and verified patient identities, enabling patients to manage consent for data exchange and preserve control over their health data. This layer makes ensuring that only authorized people or organizations can access sensitive information.

Interoperability and Integration: Standards like HL7 (Health Level Seven) or FHIR (Fast Healthcare Interoperability Resources) can be used to promote interoperability across various healthcare systems. Blockchain can act as a standardized and secure layer for data interchange, enabling smooth communication and information sharing between various systems.

User Interfaces and Applications: To give healthcare professionals, patients, and other stakeholders access to the IoMT system, user interfaces and applications have also been developed. The interfaces in question could include dashboards, These interfaces could be dashboards, mobile applications, or web management and analytics sites (Fig. 1).

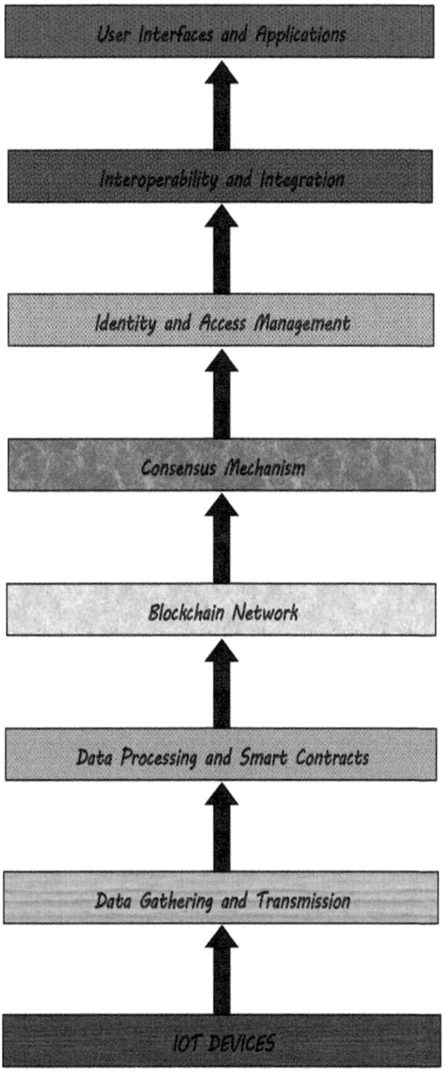

Fig. 1 Architecture of IoMT using Blockchain

It's important to note that the specific architecture may vary based on the requirements of the IoMT system, the chosen blockchain platform, and the regulatory considerations. Implementation details, including the choice of consensus mechanism, data encryption methods, and integration with existing healthcare infrastructure, will also depend on the specific use case and organizational needs.

5 Techniques to Implement Blockchain in IoMT

An overview of the methods and their categories utilized to incorporate blockchain into IoMT is provided by this classification. It draws attention to the variety of strategies used in the IoMT sector to handle issues with data security, privacy, interoperability, identity management, smart contract execution, consensus, and integration with existing system. The particular methods and formats picked for integration rely on the needs, use cases, and the blockchain platform or technology chosen (Table 1).

The most current studies that have used blockchain on IoMT are presented in this section. We categorize these studies in accordance with the most cutting-edge method for incorporating blockchain into IoMT.

5.1 Ethereum-Based Contributions

In order to apply blockchain in the Internet of Medical Things (IoMT) ecosystem, Ethereum, a well-known blockchain platform, has seen a number of contributions and innovations. Here are some significant Ethereum-based IoMT contributions:

5.1.1 Smart Contracts

On the Ethereum blockchain, smart contracts are self-executing contracts with established rules. The use of smart contracts will eliminate the involvement of third party intervention [22] Within the IoMT ecosystem, smart contracts allow for automation and programmability. They are useful for automating procedures, upholding contracts, and enabling the performance of particular actions brought about by predetermined circumstances. The smart contract functionality of Ethereum enables automated and safe data sharing in IoMT. Access control policies, permission management procedures, and data sharing agreements between various healthcare organizations can all be defined using smart contracts. The privacy and control of the patient are maintained while enabling secure and transparent data transmission.

Table 1 Classification of methods and their categories utilized to incorporate blockchain into IoMT

Data security and privacy techniques	Interoperability techniques	Identity and access management techniques	Smart contracts and business logic techniques	Consensus mechanisms	Integration with existing systems
a. Encryption:	a. Standardization:	a. Decentralized Identity:	a. Smart Contracts:	a. Modified Consensus Protocols:	a. APIs and Integration Frameworks:
Symmetric Encryption	HL7 FHIR (Fast Healthcare Interoperability Resources)	Self-Sovereign Identity (SSI)	Ethereum-based Smart Contracts	Proof-of-Authority (PoA)	RESTful APIs
Asymmetric Encryption	DICOM (Digital Imaging and Communications in Medicine)	Verifiable Credentials	Solidity Programming Language	Practical Byzantine Fault Tolerance (PBFT)	Messaging Protocols (e.g., MQTT)
Homomorphic Encryption	b. Middleware Solutions:	b. Authentication and Authorization:	b. Oracles:	Delegated Proof-of-Stake (DPoS)	b. Legacy System Integration:
b. Access Control:	Integration Middleware	Digital Signatures	Blockchain Oracles	b. Consensus Algorithms:	Adapter Patterns
Role-Based Access Control (RBAC)	Interoperability Frameworks	Multi-Factor Authentication (MFA)	External Data Integration	Proof-of-Work (PoW)	Enterprise Service Bus (ESB
Attribute-Based Access Control (ABAC)				Proof-of-Stake (PoS)	
c. Consent Management:				Practical Byzantine Fault Tolerance (PBFT)	
Smart Contracts					
Decentralized Identity Management					

5.1.2 Ethereum Virtual Machine (EVM)

On the Ethereum network, smart contracts are executed using the Ethereum Virtual Machine, a runtime environment and its a very important component of Ethereum blockchain platform It offers a sandboxed environment for smart contract code execution that is safe and deterministic. In order to implement Smart contracts on the EVM services like Oracle will be needed to provide data to smart contracts from the IoT devices. The EVM increases flexibility in developing IoMT applications by enabling developers to create and deploy smart contracts in a variety of programming languages along with that smart contract auditing will be needed to identify mistakes before deployment. [23]. As the data processing will be too resource intensive for EVM use of off chain solutions can be done to perform task outside the blockchain. Standardization, Interoperability and regulatory compliance need to be considered while implementing smart contracts.

5.1.3 ERC Standards

The monetization of health data in IoMT has been made possible by utilizing Ethereum's tokenization capabilities. On Ethereum-based marketplaces, tokenized assets representing health data can be bought, sold, and monetized. By doing this, people can still own and control their health data and gain from its value. Common interfaces and protocols for various token types and capabilities on the Ethereum blockchain are defined by Ethereum Request for Comment (ERC) standards. They are a set of technical specifications and guidelines for various aspects of the Ethereum blockchain ecosystem. For instance, the commonly used ERC-20 standard is used for fungible tokens and can be used to represent tokenized assets within the IoMT ecosystem such as for tokenize of patient health data or medical device access rights with ERC-777 being the improvised version of ERC -20. ERC-721 and ERC-1155 is the standard for non-fungible tokens (NFTs) and can be used to represent individual medical records, test results, or ownership of medical devices. In order to prevent phishing attacks as data security is of utmost important in IoMT ERC-948 is used. ERC-223 defines standards to include a mechanism to prevent tokens from being accidentally lost due to incorrect transfers and can be used for secure token transfer in reference to IoMT. ERC-1400 and ERC-1404 standards offer useful features on security of tokens and compliance with regulatory requirements. ERC-1410 is used to implement more complex token ecosystems in IoMT applications like partitioning and managing tokens which are based on different rules. In order to manage and transfer medical documents securely ERC-1643 and ERC-1644: provide guidelines for document management and transfer. [24]. In the IoMT, these standards can be used to manage ownership, represent assets, or track particular medical equipment or information.

5.1.4 Integration with External Systems

The IoMT ecosystem's external systems can be integrated with Ethereum's blockchain. Ethereum-based smart contracts can interface with off-chain data sources to collect data from medical devices and onboarded onto blockchain with the help of legacy systems, or APIs to access real-world data, authenticate users, or start actions through the use of oracles or middleware [25]. The seamless integration of blockchain technology with current healthcare infrastructure is made possible by this interoperability. Ethereum, Hyperledger Fabric, and Corda are different platforms that offer varying levels of customization and scalability in order to align blockchain platforms with IoMT.

5.1.5 Permissioned and Private Networks

Ethereum is well-known for its open network, but it can also support blockchain networks that are permissioned or private. When it comes to IoMT, data privacy and Permissioned networks limit access to authorized participants because security and regulatory compliance are essential. They control the access to the blockchain and its data. This maintains data privacy and secrecy while ensuring that only vetted organizations—such as healthcare providers, researchers, and regulators—can join in the blockchain network. They enable to design the network architecture to comply with the regulations of HIPAA, making it easier to manage patient data The private networks help to implement selective data disclosure mechanisms. So that only authorized parties can access specific data. For applications that require quick responses private networks give good performance in this situation as the competition is reduced. Permissioned networks have flexibility to choose and customize consensus mechanisms based on the use case which will lead to faster transaction times and lower energy consumption. Although private networks have limited access to external data, but with the help of oracles trusted external data can be fetched securely for use within the network also the cost of executing transaction on a private network is less as compared to public Ethereum networks. Private networks allow for faster development and testing of IoMT applications without the constraints and uncertainties of public networks. Hybrid solutions are the best as they t combine the benefits of public and private networks. Consortium networks are also applicable in IoMT as they enable multiple organizations to participate, validate transactions, and maintain the network collectively.

5.1.6 Decentralized Identification Solutions

Ethereum-based initiatives like uPort or Sovrin offer decentralized identification solutions that make it possible to handle identities securely and autonomously. These technologies enable safe authentication, consent management, and data exchange inside the IoMT ecosystem while giving people control over their identity and personal data. uPort uses Ethereum's blockchain to create tamper-proof identities that users can control directly. It also allows users to share selected data in order to increase privacy and thus provides full control over the data. uPort leverages the Ethereum network's security and immutability to ensure the integrity of identity records. Sovrin is a global public utility which employs a permissioned blockchain approach to create a decentralized identity network in order to provide trust and security. Sovrin offers a DID method known as "sov" that adheres to the W3C DID specification. Sovrin employs a consensus mechanism that allows multiple organizations to participate in maintaining the identity network while adhering to certain standards [26].

5.1.7 Scalability and Layer-2 Solutions

Both the cost and throughput of transactions on Ethereum's main net are constrained by scalability issues. Different layer-2 solutions, including sidechains (like Polygon) or state channels (like Raiden Network), can be utilized to handle a huge volume of traffic to address this using the security guarantees provided by Ethereum to conduct off-chain transactions. For IoMT applications with significant transaction volumes, layer-2 solutions allow scalability and lower transaction costs. Rollups are Layer-2 solutions that bundle multiple transactions off-chain and then submit a summary of these transactions to the Ethereum main net. Optimistic Rollups assume that transactions are valid by default but provide a mechanism for challenging invalid transactions. They offer high throughput by processing transactions off-chain and then submitting summarized data to Ethereum where as ZK-Rollups use Zero-Knowledge Proofs (ZKPs) to ensure the validity of transactions without revealing the transaction data. They offer strong privacy and scalability benefits. Other layer 2 channels like state channel, plasma sidechains, aim to alleviate the scalability issues of Ethereum while maintaining its security and decentralization principles [27].

In [16], A private Ethereum-based totally structure is proposed to put into effect clever contracts in order to control the users/devices requests and control get entry to based totally on a fixed of attributes along with the credentials, position and the area. It uses IPFS for facts garage. An interPlanetary file (IPFS) is used to save patient health information and gadgets technical information.

In [28], A personal blockchain-based totally system for scientific records management has been proposed in. It really works on Ethereum smart contracts to control facts get entry to permission among entities including patients, hospitals, doctors, studies companies and different stakeholders. The smart agreement incorporates clever representations of scientific records consisting of permissions, file ownership metadata and records integrity. The clinical report records are saved in external server (off-chain) and a cryptographic hash of the document is saved at the blockchain ensuring information integrity. The proposed system removes mining for simplification.

In [29], authors evolved a cloud-based totally framework to reveal the development of a neurological sickness using IoMT gadgets. They used cloud computing to shop and technique IoMT statistics and deploy Ethereum-based totally Blockchain Community to securely exchange and percentage records between healthcare customers.

In [30], authors proposed a permissioned blockchain-based totally structure for relaxed far flung patient tracking. They used Ethereum to put into effect smart contracts so as to research statistics and send alerts to affected person and healthcare providers. They proposed the use of sensible Byzantine Fault Tolerance (PBFT) as an opportunity to PoW consensus version. The proposed architecture lacks techniques to satisfy demanding situations associated with IoMT-Blockchain integration. And in SMEAD [31], an Ethereum-based architecture for remotely tracking diabetes.

Sufferers, smart contracts are used to control get entry to data.

5.2 Modified Consensus Protocol

Implementing a modified consensus protocol is crucial when utilizing blockchain in the Internet of Medical Things (IoMT) to address specific requirements and challenges. While specific research studies may vary, here are some examples of modified consensus protocols explored for implementing blockchain in IoMT:

5.2.1 Proof-Of-Authority (PoA)

PoA is a modified consensus protocol that focuses on identity and reputation rather than computational power. In IoMT, PoA can be used to ensure the validation and verification of transactions by trusted healthcare providers or organizations. It reduces the energy consumption and computational requirements associated with traditional Proof-of-Work (PoW) consensus protocols. In order to implement PoA the following steps are performed:

Firstly identification of trusted participants like hospitals, medical institutions, or regulatory bodies who will act as authorities in the PoA network id done, Then suitable blockchain platform that supports PoA consensus is selected. On this chosen platform the blockchain nodes for each authority are deployed. After this a genesis block configuration that specifies the initial parameters of the blockchain is created. Then a cryptographic key pairs which are are used to validate and create new blocks is generated for each authority. Now the blockchain nodes are configured to use PoA consensus algorithm. After this Bootstrap the network by connecting the nodes to each other. Ensure that each node is aware of the other authorities' addresses and public keys. Once a block is created, it must be finalized by a majority of the other authorities. This ensures consensus and the immutability of the blockchain. Regularly monitor the network to ensure that authorities are operating as intended. Maintain the network by rotating authority roles, updating software, and addressing any issues that may arise [32].

5.2.2 Practical Byzantine Fault Tolerance (PBFT)

PBFT is a consensus protocol that focuses on reaching agreement among a set of nodes, even in the presence of Byzantine faults. In IoMT, PBFT can be applied to achieve consensus among healthcare entities to ensure the integrity and consistency of the blockchain [33]. It is suitable for use cases where a limited number of trusted nodes participate in the consensus process. PBFT is applied as follows in IoMT blockchain:

The first three steps are same as PoA which include network Participants and Node Setup, Node Configuration and Genesis Block Creation. After this configure the blockchain network to use the PBFT consensus algorithm. In PBFT, nodes are categorized into roles ie primary (leader) and backup nodes. The primary node proposes new blocks, and backup nodes verify and agree on the proposed blocks. The nodes in PBFT use messages to communicate. These nodes also detect faults, such as node failures or malicious behavior. If the primary node becomes unresponsive, a backup node can trigger a view change to select a new primary node. Nodes collaborate to detect and recover from faulty behavior. Once the required number of commit messages is received, the block is considered committed. The transactions in the block are applied to the blockchain's state. Regularly monitoring of the network is done to detect and address any issues, such as node failures or malicious behavior.

5.2.3 Delegated Proof-of-Stake (DPoS)

DPoS is a consensus protocol that leverages a voting-based system to select a group of delegates responsible for validating transactions and producing blocks. In IoMT, DPoS can be employed to reduce the computational requirements and increase transaction throughput by delegating the consensus process to trusted entities within the healthcare ecosystem [34]. In DPoS, participants can vote to elect a set of delegates or validators who will be responsible for confirming transactions and producing new blocks. Delegates should be entities that are trusted within the IoMT ecosystem. Participants in the IoMT network can submit transactions to any delegate. Delegates validate and include these transactions in the blocks they produce. Once a delegate produces a block, other delegates validate its contents and reach a consensus on its validity. This confirmation process helps maintain the integrity of the blockchain.

5.2.4 Federated Byzantine Agreement (FBA)

FBA is a consensus protocol that enables a network of participating nodes to reach agreement through a voting-based system. In IoMT, FBA can be used to achieve consensus among healthcare organizations, research institutions, and regulatory bodies. It provides fault tolerance and resilience against Byzantine faults, ensuring the security and reliability of the blockchain network.

5.2.5 Proof-of-Elapsed-Time (PoET)

PoET is a consensus protocol [5] that leverages a trusted execution environment, such as Intel's Software Guard Extensions (SGX), to determine the leader for block creation. In IoMT, PoET can be utilized to reduce energy consumption and increase scalability by enabling efficient and secure block creation based on trusted execution environments [35]. IoMT participants, such as medical devices, hospitals, and practitioners, run nodes that are equipped with trusted execution environments (TEEs). TEEs are isolated environments within a device's hardware that ensure secure and tamper-resistant execution. PoET requires each TEE to wait for a randomly determined amount of time before proposing a new block. The first TEE to finish this waiting period is selected as the leader for the current round. Each TEE calculates a random wait time based on its cryptographic identity and a network-wide timer. This ensures that every TEE has an equal chance of being selected as the leader, while avoiding the energy-intensive nature of PoW. The TEE that finishes waiting first becomes the leader and is responsible for proposing a new block of transactions. Other TEEs validate the proposed block, ensuring its correctness. Once a consensus is reached among the network's TEEs on the validity of the proposed block, it is added to the blockchain.

5.2.6 Proof-of-Activity (PoAcy)

PoAcy is a consensus protocol that combines PoW with Proof-of-Activity (PoA) to ensure both energy efficiency and security. It requires nodes to demonstrate their participation in a network by performing useful activities. In IoMT, PoAcy can be explored to incentivize healthcare devices to contribute computational resources to the consensus process while ensuring security and efficiency.

These are modified consensus protocols that have been investigated for implementing blockchain in the IoMT domain. Each protocol has its own characteristics, benefits, and limitations, and their suitability depends on the specific requirements and constraints of the IoMT system being considered. Ongoing research aims to develop and customize consensus protocols to address the unique challenges and requirements of IoMT applications.

In an effort to match the IoMT specificities, some works like [36] have proposed to alter the consensus protocol. In [37], authors proposed a consortium block chain-based architecture with the intention to file records generated from IoMT in a secure way at the same time as ensuring the patient's privacy. The proposed structure implements a patient agent software (PA) that defines the Blockchain functionalities.

5.3 Modified Cryptographic Technique

When implementing blockchain in the Internet of Medical Things (IoMT), various modified cryptographic techniques can be employed to enhance security, privacy, and efficiency [38]. Here are some examples:

5.3.1 Homomorphic Encryption

Cloud computing and cloud storage have contributed to a massive shift in records processing and its use. Availability and accessibility of assets with the reduction of enormous work is one of the fundamental motives for the cloud revolution. With this cloud computing revolution, outsourcing programs are in remarkable call for. The customer makes use of the provider by using importing their data to the cloud and eventually receives the result via processing it. Its blessings users substantially, however it additionally exposes touchy records to 1/3-celebration carrier vendors. In the healthcare enterprise, affected person fitness statistics are digital statistics of affected person's scientific history saved by means of hospitals or fitness care companies. Patient fitness facts are stored in statistics centers for storage and processing. Earlier than doing computations on records, conventional encryption strategies decrypt the statistics of their original form. As end result, sensitive scientific facts are lost. Homomorphic encryption can defend touchy records with the aid of permitting statistics to be processed in an encrypted shape such that handiest encrypted statistics is available to carrier carriers. [39]

5.3.2 Zero-Knowledge Proofs

Zero-expertise proofs are a totally thrilling and charming cryptographic concept and show to be useful in diverse applications, particularly in privateness and blockchain generation. 0-knowledge proofs contain two parties: a prover and a verifier. The prover makes a declaration that his or her proof is valid, which the verifier needs to approve, without the prover [40].

An extraordinary technique to cozy personal records is ZKP, that's a cryptographic approach that have been used to offer privateness by design inside the context of DLT and blockchain. Quickly, ZKP lets in an entity called prover to argue to some other celebration, referred to as verifier, that a decided assertion is actual without revealing extra records than strictly important to convince her [41].

5.3.3 Multi-Party Computation

Multi-celebration computation (MPC) allows a couple of events to at the same time compute a end result without revealing their person inputs. Inside the context of

IoMT, MPC may be used to analyze scientific statistics across a couple of assets even as retaining privateness. It permits collaborative evaluation and choice-making without directly exposing uncooked information to all events concerned [42].

5.3.4 Threshold Cryptography

Threshold cryptography involves distributing cryptographic keys amongst multiple parties in a manner that requires a threshold wide variety of members to collaborate with the intention to decrypt or perform cryptographic operations. This method can enhance the safety of blockchain-based totally IoMT systems with the aid of preventing a unmarried factor of failure and defensive sensitive keys from being compromised.

Threshold cryptography lets in one to share the energy of a cryptosystem. Threshold cryptosystems are distinct from threshold schemes in which the strength to regenerate a mystery secret is shared. A normal threshold scheme isn't without delay appropriate for threshold signatures. The use of a threshold scheme at once might require the shareholders to send their shares to a relied-on man or woman who could follow the cryptosystem for them [43].

5.3.5 Ring Signatures

Ring signatures enable a user to sign a message on behalf of a collection without revealing which precise member of the organization performed the signing. In IoMT, ring signatures can be used to provide nameless get admission to to data or authenticate transactions without exposing the identity of the taking part entity [44].

5.3.6 Secure Multi-party Computation (SMPC)

SMPC protocols enable a couple of events to mutually compute a function on their private statistics whilst retaining the facts personal. This approach may be carried out in IoMT for comfortable records evaluation and collaborative choice-making among multiple healthcare entities, making sure privateness and statistics confidentiality. The gadget permits cozy records garage in dispensed databases and touchy statistics is by no means found out in plaintext throughout the authentication system. The communique among exceptional components of the device is secured the use of each symmetric and asymmetric cryptographic primitives [45].

5.3.7 Verifiable Computation

Verifiable computation allows a party to outsource computations to a third party and obtain a proof of the correctness of the computation results. In IoMT, this technique can be used to securely delegate computationally intensive tasks, such as data analysis, to external parties while maintaining data privacy and integrity.

The authors in [46] use some functions of the usual version of blockchain to provide privacy and facts integrity when sharing IoMT facts. They use the hashing technique and advocate a newly encryption set of rules to encrypt the transactions containing private and touchy information about sufferers. The principal benefit of this algorithm is its capability to cover big quantity of uniquely identified clinical items and its very low time complexity which fits the actual time requirement of IoMT. All transactions are saved in a blockchain maintained by means of the healthcare companies.

In [30], Authors proposed a custom designed blockchain-based framework suitable for IoMT devices. First, the proposed blockchain is personal: nodes need to be certificated so as to be a part of the network and send transactions. Second, authors put off the POW consensus protocol. These modified cryptographic techniques provide additional security and privacy measures when implementing blockchain in IoMT systems. By employing these techniques, healthcare organizations can enhance data protection, ensure privacy-preserving operations, and foster trust in the ecosystem.

5.4 Hyperledger-Based Contributions

Hyperledger, an open-source blockchain platform, has seen various contributions and developments for implementing blockchain in the Internet of Medical Things (IoMT) ecosystem. Here are some notable Hyperledger-based contributions in the context of IoMT:

5.4.1 Hyperledger Fabric

Hyperledger fabric is a permissioned blockchain framework that provides a flexible and modular architecture appropriate for IoMT applications. It lets in for satisfactory-grained access manage, confidentiality, and privateness of scientific records. Cloth's aid for clever contracts and channels allows at ease data sharing and transaction processing among more than one healthcare entities [47].

5.4.2 Hyperledger Indy

Hyperledger Indy focuses on decentralized identity management, which is crucial in IoMT. It provides a robust framework for self-sovereign identity (SSI), enabling individuals to control their personal health information and granting selective access to healthcare providers based on verifiable credentials [5].

5.4.3 Hyperledger Sawtooth

Hyperledger Sawtooth is a modular platform for constructing, deploying, and jogging disbursed ledgers. It gives flexibility in consensus algorithms, permitting IoMT packages to pick out the most suitable consensus protocol for their unique necessities. Sawtooth's assist for Ethereum digital system (EVM) clever contracts permits interoperability with Ethereum-primarily based structures [48].

5.4.4 Hyperledger Aries

Hyperledger Aries is a toolkit that provides a foundation for building interoperable and verifiable identity solutions. It offers standardized protocols for exchanging verifiable credentials, enabling seamless integration of identity management systems within IoMT applications. Aries can enhance privacy, security, and trust in the exchange of health-related data.

5.4.5 Hyperledger Caliper

Hyperledger Caliper is a benchmarking tool that helps evaluate the performance of blockchain platforms. It enables organizations in the IoMT domain to assess the scalability and efficiency of Hyperledger-based solutions. Caliper assists in optimizing the performance of blockchain networks for real-world IoMT deployments.

5.4.6 Hyperledger Grid

Hyperledger Grid is a framework for growing deliver chain solutions. Within the healthcare quarter, it may be utilized to decorate the traceability and provenance of scientific devices, pills, and components, making sure their authenticity and high-quality. Grid facilitates transparency and efficiency in deliver chain management for IoMT programs [49].

In [50], the authors proposed an IoT-blockchain based totally architecture to permit healthcare remote tracking. The architecture carries forms of blockchain: (1) Clinical gadgets Blockchain to shop medical statistics generated through medical gadgets for the duration of remedy duration, (2) session Blockchain maintained by using hospitals to permanently save sufferers records. The transactions are tested and established the use of clever contracts (Chain codes in material) carried out via endorsing friends Following sensible Byzantine Fault Tolerance set of rules. The authors evolved a consumer interface to visualize the patient health records.

These contributions demonstrate the potential of Hyperledger as a blockchain platform in IoMT. They address critical aspects such as identity management, data privacy, access control, interoperability, and supply chain management. Ongoing research and development efforts continue to refine and expand the capabilities of Hyperledger-based solutions to meet the unique requirements of the IoMT domain.

The following table gives the complete method wise contribution of security enhancement by blockchain technology of all the four techniques that has been discussed so far (Table 2).

Table 2 Method wise contribution of security enhancement by blockchain technology

Ethereum-based contributions	Modified consensus protocol	Modified cryptographic technique	Hyperledger-based contributions
Smart Contracts for Data Sharing	Device Authentication-Proof of Authority (PoA)	Homomorphic Encryption for Secure Data Processing	Medical Data Sharing and Interoperability: Hyperledger Fabric
Tokenization for Health Data Monetization	Practical Byzantine Fault Tolerance (PBFT) for Real-Time Data Consensus	Zero-Knowledge Proofs for Data Authentication	Identity and Access Management: Hyperledger Indy
Decentralized Identity Management-uPort or Sovrin	Delegated Proof of Stake (DPoS) for Data Validation	Multi-Party Computation for Secure Data Sharing	Clinical Trials and Research Collaboration: Hyperledger Sawtooth,
Health Level 7 (HL7) for Interoperability Standards	Proof of Elapsed Time (PoET) for Energy-Efficient Device Consensus:	Threshold Cryptography for Key Management	Supply Chain Management: Hyperledger-based frameworks such as Hyperledger Fabric or Hyperledger Besu
Permissioned and Private Networks for Privacy	Hierarchical Consensus for Scalability and Privacy	Verifiable Data Auditing for Integrity Verification	Claims Processing and Insurance: Hyperledger Fabric or Hyperledger Indy
Layer-2 Solutions for scalability	Federated Consensus for Data Sharing and Collaboration	Attribute-Based Encryption for Fine-Grained Access Control:	Device Lifecycle Management: Hyperledger Caliper

6 Conclusion

Numerous studies have focused on implementing Blockchain in the Internet of Medical Things (IoMT) due to the stringent and severe security requirements in the healthcare industry. The majority of them were concentrating on offering data protection, integrity, secrecy, and authentication. They suggested other use cases, such as managing medical data and remotely monitoring patients (RMP). Our analysis of the research reveals the different methods for incorporating Blockchain in the IoMT and the different layers on which these methods are applicable. A detailed study of major techniques are discussed in this chapter and it was also figured out that the majority of them did not address the massive data streams produced by IoMT devices with limited resources or suggest technological changes to the Blockchain architecture to address these issues.

References

1. R.S.P.K. Mannepalli, Survey on feature reduction techniques of intrusion detection system. (IJERCT) 2(3) (2020) ISSN: 2582–5488
2. R.S.P.K. Mannepalli, Cloud malicious threat detection using convolution filter and EBPNN. *5th International Conference on Information Systems and Computer Networks (ISCON) IEEE GLA University*, Mathura, India (2021)
3. R.S.P.K. Mannepalli, Cloud malicious threat detection by features from intelligent water drop set and EBPN. IJARET 11(12), 868-877 (2020)
4. D. Mohan, L. Alwin, P. Neeraja, K. Deepak Lawrence, V. Pathari, A private Ethereum blockchain implementation for secure data handling in Internet of Medical Things. J. Reliable Intell. Environ. (2021)
5. R.S.P.K. Mannepalli, Invasive weed optimization algorithm based trained neural network for cloud malicious threat detection. *2021 IEEE International Conference on Technology, Research, and Innovation for Betterment of Society (TRIBES)* (2021)
6. The Impact of Digital Technologies on Public Health in Developed and Developing Countries. Springer Science and Business Media LLC (2020)
7. A. Mishra, T.S. Jabar, Y.I. Alzoubi, K.N. Mishra, Enhancing privacy-preserving mechanisms in cloud storage: a novel conceptual framework. Concurrency and Computation: Practice and Experience (2023)
8. A. Prabhakar, T. Anjali, TCON—a lightweight trust-dependent consensus framework for blockchain. *2019 11th International Conference on Communication Systems & Networks (COMSNETS)* (2019)
9. R. Singh, A. Ahamed Ansari, AI ethics and challenges in healthcare. Handbook on "Handbook on Augmenting Tele health Services" CRC Press, Taylor & Francis Group, U.K (2024)
10. B. Natarajan, P. Abilash Kumar, S. Aboorva, A Blockchain based approach for privacy preservation in healthcare IoT, in Gunjan, V., Garcia, Diaz V., Cardona, M., Solanki, V., Sunitha, K. (eds.) ICICCT 2019, pp. 465–473. Springer, Singapore (2020). https://doi.org/10.1007/978-981-13-8461-552
11. A. Dwivedi, G. Srivastava, S. Dhar, R. Singh, A decentralized privacy-preserving healthcare blockchain for IoT. Sensors **19**, 326 (2019)
12. O. Attia, I. Khoufi, A. Laouiti, C. Adjih, An IoT-blockchain architecture based on hyperledger framework for healthcare monitoring application, in *2019 10th IFIP International Conference on New Technologies, Mobility and Security (NTMS)*, pp. 1–5 (2019)

13. S.K. Lo, Y. Liu, Q. Lu, C. Wang, X. Xu, H.-Y. Paik, L. Zhu, Towards trustworthy AI: Blockchain-based architecture design for accountability and fairness of federated learning systems, IEEE Internet Things J. 1 (2022), https://doi.org/10.1109/JIOT.2022.3144450
14. R. Alajlan, N. Alhumam, M. Frikha, Cybersecurity for blockchain-based IoT systems: a review. Appl. Sci. (2023)
15. G. Habib, S. Sharma, S. Ibrahim, I. Ahmad, S. Qureshi, M. Ishfaq, Blockchain technology: benefits, challenges, applications, and integration of blockchain technology with cloud computing. Future Internet **14**, 341 (2022). https://doi.org/10.3390/fi14110341
16. P. Gope, A.K. Das, N. Kumar, Y. Cheng, Lightweight and physically secure anonymous mutual authentication protocol for real-time data access in industrial wireless sensor networks. IEEE Trans. Industr. Inf. **15**(9), 4957–4968 (2019)
17. R. Singh, A. Ahamed Ansari, AI enabled internet of medical things in healthcare. Handbook on " Heterogeneous Computational Intelligence in Internet of Things" CRC Press, Taylor & Francis Group, U.K (2023)
18. M.A. Rahman, M.S. Hossain, M.S. Islam, N.A. Alrajeh, G. Muhammad, Secure and provenance enhanced internet of health things framework: a blockchain managed federated learning approach. IEEE Access **8**, 205071–205087 (2020). https://doi.org/10.1109/ACCESS.2020.3037474
19. Y. Liu, W. Yu, Z. Ai, G. Xu, L. Zhao, Z. Tian, A blockchain-empowered federated learning in healthcare-based cyber physical systems. IEEE Trans. Netw. Sci. Eng. 1 (2022), https://doi.org/10.1109/TNSE.2022.3168025
20. Z. Wang, L. Cai, X. Zhang, C. Choi, X. Su, A COVID-19 auxiliary diagnosis based on federated learning and blockchain. Comput. Math. Methods Med. **2022**, 1–12 (2022). https://doi.org/10.1155/2022/7078764
21. M.Y. Jabarulla, H.-N. Lee, A blockchain and artificial intelligence-based, patient-centric healthcare system for combating the COVID-19 pandemic: opportunities and applications. Healthcare 9, 1019 (2021), https://doi.org/10.3390/healthcare9081019
22. M.N. Sadiku, K.G. Eze, S.M. Musa, Smart contracts: a primer. J Sci Engine Res. **5**(5), 538–541 (2018)
23. D. Metcalf, S. Milliard, S.D. Gomez, M. Schwartz, Wearables and the internet of things for health: wearable, interconnected devices promise more efficient and comprehensive health care. IEEE Pulse **7**, 35–39 (2016)
24. L.-H. Lee, T. Braud, P. Zhou, L. Wang, D. Xu, Z. Lin, A. Kumar, C. Bermejo, P. Hui, All one needs to know about metaverse: A complete survey on technological singularity, virtual ecosystem, and research agenda (2021) arXiv preprint arXiv:2110.05352
25. Md. Ashraf Uddin, A. Stranieri, I. Gondal, V. Balasubramanian, A survey on the adoption of blockchain in IoT: challenges and solutions. Blockchain Res. Appl. **2**(2), 100006 (2021), ISSN: 2096-7209, https://doi.org/10.1016/j.bcra.2021.100006
26. A.G. Nabi, Comparative Study on Identity Management Methods Using Blockchain. Ph.D. Thesis, University of Zurich, Zurich, Switzerland (2017)
27. N.P. Triantafyllidis, Developing an Ethereum Blockchain Application. Ph.D. Thesis, University of Amsterdam, Amsterdam, The Netherlands (2016)
28. A.K. Das, M. Wazid, N. Kumar, A.V. Vasilakos, J.J. Rodrigues, Biometrics-based privacy-preserving user authentication scheme for cloud-based industrial internet of things deployment. IEEE Internet Things J. **5**(6), 4900–4913 (2018)
29. X. Li, J. Peng, J. Niu, F. Wu, J. Liao, K.-K.R. Choo, A robust and energy efficient authentication protocol for industrial internet of things. IEEE Internet Things J. **5**(3), 1606–1615 (2018)
30. V. Malamas, T. Dasaklis, P. Kotzanikolaou, M. Burmester, S. Katsikas, A forensics-by-design management framework for medical devices based on blockchain, in *2019 IEEE World Congress on Services (SERVICES)*, vol. 2642– 939X, pp. 35–40 (2019)
31. A. Khatoon, A blockchain-based smart contract system for healthcare management. Electronics **9**, 94 (2020)
32. M. Debe, K. Salah, M.H.U. Rehman et al., Lot public fog nodes reputation system: a decentralized solution using ethereum blockchain IEEE. Access **7**, 178082–178093 (2019)

33. L. Ismail, H. Materwala ,A review of blockchain architecture and consensus protocols: use cases, challenges, and solutions. Symmetry **11**(10), 1198 (2019)
34. Q.H. Zhou, H.W. Huang, Z.B. Zheng et al., Solutions to scalability of blockchain: a survey IEEE. Access **8**, 16440–16455 (2020)
35. A.F. Zorzo, H.C. Nunes, R.C. Lunardi, et al., Dependable IoT using blockchain-based technology. *2018 Eighth Latin—American Symposium on Dependable Computing*; 8–10 Oct 2018; Foz do Iguacu, IEEE, Brazil. Piscataway, NJ, USA, pp. 1–9 (2018)
36. D.C. Nguyen, P. Pathirana, K. Nguyen, A mobile cloud based iomt framework for automated health assessment and management, vol. 2019 (2019)
37. K.N. Griggs, O. Ossipova, C.P. Kohlios, A.N. Baccarini, E.A. Howson, T. Hayajneh, Healthcare blockchain system using smart contracts for secure automated remote patient monitoring. J. Med. Syst. **42**, 1–7 (2018)
38. B.G. Gebraselase, B.E. Helvik, Y. Jiang, Suitability of blockchains to enable and support networking functions. *Proceedings of the 2019 4th International Conference on Cloud Computing and Internet of Things* - CCIOT 2019 (2019)
39. K. Munjal, R. Bhatia, A systematic review of homomorphic encryption and its contributions in healthcare industry. Complex Intell. Syst. **9**, 3759–3786 (2023). https://doi.org/10.1007/s40747-022-00756-z
40. J. Hasan, Overview and Applications of Zero Knowledge Proof (ZKP). 8. 5 (2019)
41. E. Morais, T. Koens, C. van Wijk et al., A survey on zero knowledge range proofs and applications. SN Appl. Sci. **1**, 946 (2019). https://doi.org/10.1007/s42452-019-0989-z
42. S. Goldwasser, Multi-party computations: past and present. *Proceedings of the Annual ACM Symposium on Principles of Distributed Computing*, 1–6 (1997).https://doi.org/10.1145/259380.259405
43. Y. Desmedt, Threshold cryptography. Eur. Trans. Telecommun. **5**, 449–458 (2010). https://doi.org/10.1002/ett.4460050407
44. M. Demircioglu, S. Akleylek, M. Cenk, Efficient GeMSS based ring signature scheme. Malaysian J. Comput. Appl. Mat. **3**, 38–42 (2020). https://doi.org/10.37231/myjcam.2020.3.1.41
45. W. Alghamdi, R. Salama, M. Sirija, A. Abbas, K. Dilnoza, Secure multi-party computation for collaborative data analysis. *E3S Web of Conferences*, 399 (2023). https://doi.org/10.1051/e3sconf/202339904034
46. S. Mohan, R. Shubha, A. Marks, V. Iyer, SMEAD: a secured mobile enabled assisting device for diabetics monitoring, pp. 1–6 (2017)
47. M. Krstić, L. Krstić, Hyperledger frameworks with a special focus on Hyperledger Fabric. Vojnotehnicki glasnik. **68**, 639–663 (2020). https://doi.org/10.5937/vojtehg68-26206
48. B. Ampel, M. Patton, Performance Modeling of Hyperledger Sawtooth Blockchain, 59–61 (2019).https://doi.org/10.1109/ISI.2019.8823238
49. A. Goranovic, M. Meisel, S. Wilker, T. Sauter, Hyperledger Fabric Smart Grid Communication Testbed on Raspberry PI ARM Architecture, 1–4 (2019).https://doi.org/10.1109/WFCS.2019.8758000
50. M.A. Uddin, A. Stranieri, I. Gondal, V. Balasubramanian, Blockchain leveraged decentralized iot ehealth framework. Internet Things **9**, 100159 (2020)
51. R. Kumar, A.A. Khan, S. Zhang, W. Wang, Y. Abuidris, W. Amin, J. Kumar, Blockchain-federated-learning and deep learning models for COVID-19 detection using CT imaging, 2020, arXiv:2007.06537 [cs, eess]. http://arxiv.org/abs/2007.06537. Accessed November 17, 2020

Offloading Strategies and Computing Paradigms in IoT: A Survey

Abhishek Hazra, Poonam Maurya, Alakesh Kalita, and Indranil Sarkar

Abstract The increasing number of Internet of Things (IoT) devices has led to a rise in the demand for data processing, which requires new and effective ways of offloading and computing. This chapter explores various offloading mechanisms and computing paradigms that are used in the context of IoT. At the beginning, we provide an overview of standard data offloading methods such as cellular offloading, wireless mesh, WiFi offloading, fem-to-cell offloading, WiMAX offloading, and IP flow mobility offloading. We analyze each of these methods in terms of their strengths and limitations, offering insights into the various tools that are available for optimizing data processing in IoT and improving connections and efficiency. We also examine several computing paradigms that are designed for IoT, including cloudlet computing, cloud computing, fog computing, serverless computing, virtual machines, and peer-to-peer computing, providing a significant understanding of their contributions to optimizing computational resources in IoT systems. At last, we conclude this chapter by summarizing the key findings and insights, emphasizing the significance of the discussed offloading strategies and computing paradigms.

Keywords Task offloading · Offloading strategies · Computing paradigm · Internet of things

A. Hazra
Indian Institute of Information Technology SriCity, Sri City, India
e-mail: a.hazra@ieee.org

P. Maurya
Aalborg University, Aalborg, Denmark
e-mail: poonamm@es.aau.dk

A. Kalita
Singapore University of Technology and Design, Singapore, Singapore
e-mail: alakesh_kalita@sutd.edu.sg

I. Sarkar (✉)
University of Oslo, Oslo, Norway
e-mail: indranis@uio.no

B. Pradhan and S. Mukhopadhyay (eds.), *IoT Sensors, ML, AI and XAI: Empowering A Smarter World*, Smart Sensors, Measurement and Instrumentation 50,
https://doi.org/10.1007/978-3-031-68602-3_17

1 Introduction

The widespread adoption of IoT devices has ushered in a new era of extensive connectivity and data generation, significantly transforming our interactions with the digital world [1]. The vast amount of data produced by these devices, coupled with the increasing demand for real-time processing and analysis, poses significant challenges to existing computing paradigms. Data offloading emerges as a critical solution in this scenario, offering a means to enhance the efficiency and scalability of IoT systems. In the context of IoT, data offloading involves relocating computationally demanding tasks and data processing from restricted edge devices to more powerful computing resources, either at the network edge or in the cloud [2]. This intentional redistribution of computing tasks aims to alleviate the resource constraints of IoT devices by optimizing performance and energy usage. The significance of data offloading extends beyond its ability to enhance the functionality of resource-constrained devices, as it also addresses the evolving requirements of IoT applications (Fig. 1).

The significance of rapid data processing cannot be overstated in the context of IoT, where devices range from sensors and actuators to embedded systems with diverse computing capabilities. Real-time processing, low-latency connectivity, and energy efficiency are key factors driving the demand for new solutions. Data offloading is a crucial enabler for meeting these requirements since it enables devices to transfer complex computational tasks to more capable computing resources, resulting in faster response times and energy savings [3].

Task offloading methods are crucial in enhancing the computing efficiency of IoT devices by mitigating their resource limitations [4, 5]. One of the methods employed is cellular offloading, which leverages cellular networks to handle computing duties, hence guaranteeing dependable and fast connectivity for the purpose of real-time

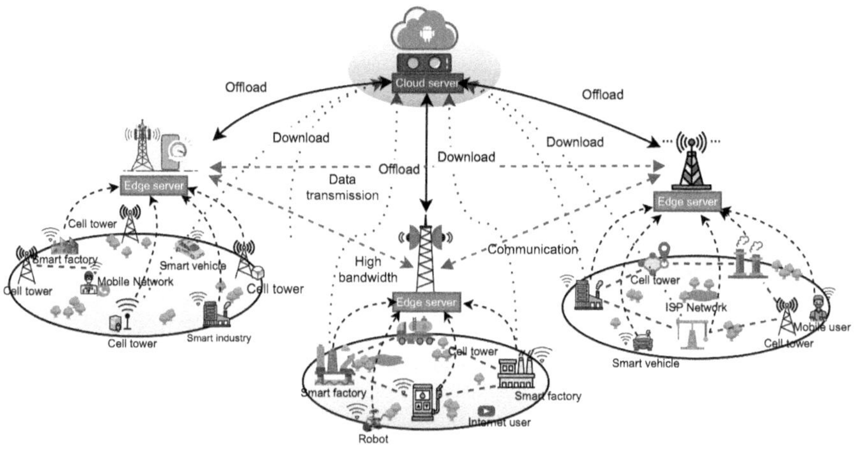

Fig. 1 Illustration standard IoT offloading

data processing [6]. The term "wireless" refers to the transmission of information or communication signals without the use Mesh offloading is a technique that enables the establishment of interconnected mesh networks, hence improving the efficient allocation of tasks in large-scale IoT deployments. The deployment of femtocells for offloading is a strategy aimed at improving cellular coverage in targeted regions, particularly for utilising IoT applications within buildings. The implementation of these solutions collectively aids in the resolution of various difficulties associated with the IoT, hence providing benefits such as enhanced coverage, scalability, and decreased congestion. Nevertheless, it is important to acknowledge that these technologies do have certain drawbacks. These drawbacks encompass factors such as the intricate nature of the network, the possibility of delays in data transmission, and worries regarding interference. Consequently, it becomes imperative to thoroughly evaluate these limits in accordance with the specific demands of the application at hand.

This chapter discusses the numerous offloading mechanisms and computing paradigms that have been developed in response to the challenges posed by the rapidly expanding IoT ecosystem. Our aim is to provide a comprehensive overview of the various offloading mechanisms, including their benefits, limitations, and applicability to a wide range of IoT scenarios. By analyzing the current state of the field, we aim to shed light on the evolving landscape of data offloading in IoT and its implications for the future of interconnected systems.

2 Standardized Data Offloading Technologies

Over the past few years, the increase in sensing devices, smartphones, and laptops drastically increased network traffic exponentially over the communication medium [7]. This growth of internet traffic across mobile and wireless networks has resulted in an urgent need for traffic offloading for high-speed data service and optimal network efficiency [8]. As a consequence, numerous technological solutions for handling internet traffic have arisen. WiFi, WiMAX, and IP flow are among the emerging key technologies. The increase in internet traffic also poses several problems for wireless networks. Technologies, including offloading core functional networks and content synchronization are getting more popular. The main purpose of this section is to introduce state-of-the-art IoT offloading techniques addressing both technical and industrial perspectives.

2.1 *Cellular Offloading*

Cellular offloading [9] refers to an approach employed in the field of task offloading, wherein cellular networks are utilized to augment the computing capacities of IoT devices. This strategy involves transferring computationally heavy activities from IoT devices to faraway servers using cellular connections, enabling efficient processing and resource optimization. This approach offers notable benefits in situations when IoT devices possess restricted computational capabilities or storage capacities. The advantages encompass greater real-time processing capabilities, decreased energy usage for IoT devices, and improved scalability [10]. Nevertheless, cellular offloading is not without its constraints, including the possibility of latency caused by the network, reliance on network accessibility, and apprehensions regarding the privacy and security of data. One example of practical applications can be seen in the context of smart cities, where devices equipped with sensors delegate data processing responsibilities to central servers in order to facilitate effective monitoring and management.

2.2 Wireless Mesh

Wireless Mesh task offloading is a strategic approach dependent on a network topology characterized by interconnected devices that work together to expand network coverage and enable the distribution of tasks. This strategy involves the collaborative sharing of computing resources among IoT devices within the mesh network, creating a distributed computing environment where tasks can be offloaded to one another. Wireless mesh task offloading exhibits several prominent attributes, namely self-healing capabilities, dynamic routing, and redundancy, which all contribute to the establishment of resilient and dependable network connectivity. The effective execution and integration of wireless mesh networks for the purpose of task offloading necessitate meticulous evaluation of node positioning, network density, and communication protocols in order to enhance performance. This methodology is particularly suitable for use in contexts such as smart homes or industrial environments, where a network of interconnected devices works together to effectively allocate computational duties and improve the overall efficiency of the system.

2.3 WiFi Offloading

With the tremendous gain in smart devices and massive data load, users are moving towards a new, more realistic technology, Wireless Fidelity (WiFi). WiFi is an IEEE 802.11-based wireless communication approach mainly used for broadcasting network traffic. WiFi technology supports heavy traffic congestion in the coverage areas compared to traditional internet technologies, such as UMTS, HSPA, and LTE. WiFi provides reliable wireless communication in indoor and outdoor citywide networks among the other communication technologies. WiFi comes with a natural solution for offloading smartphone data to overcome the degradation of cellular network performance and overcome the heavy internet workload in congested areas [11].

WiFi increases users' Quality of Experience (QoE) by providing high-speed internet access. Moreover, WiFi provides attractive solutions to users as it facilitates the transmission of internet traffic from a profitable radio-based licensed spectrum to an open unlicensed spectrum. Several studies [12] show that deploying multiple Base Station (BS) and increasing Network capacity is not a profitable solution from service providers' perspective as increasing network capacity increases CAPEX and OPEX. Instead, transferring data traffic and offloading through a WiFi network would benefit and provide better user QoE (Fig. 2).

Regardless of the degree of convergence between cellular and WiFi networks, there are three major approaches for providers to offload internet traffic using WiFi technology. The first strategy is unmanaged traffic forwarding, in which whenever a user's device comes into the proximity of a WiFi network, the user bypasses the core cellular network and forwards data through the WiFi infrastructure. On the other side, the user takes the support of a cellular network for transferring voice data. This technique doesn't allow any additional network infrastructure to be installed physically. However, this approach has some shortcomings. For example, whenever a user's device comes into the vicinity of a WiFi network, service providers lose control over the device and cannot offer the services that the user subscribed to. Despite several limitations, this technique can also be implemented in practical solutions for internet forwarding due to its engaging user experience and flexibility. In recent times network operators started deploying their own WiFi infrastructure in heavy network congestion areas to remove users' traffic data.

In the managed traffic forwarding approach, network operators have full control over the subscriber without integrating WiFi and cellular traffic infrastructure. With this technique, subscribers should follow a controlled data offloading strategy, and it

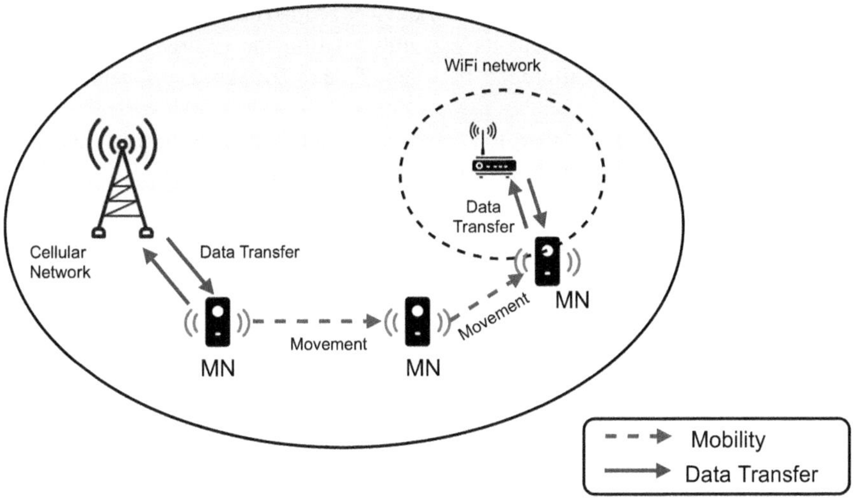

Fig. 2 Illustration WiFi offloading

is accomplished by installing a smart connection-aware gateway by which internet users transmit their data. However, automated traffic forwarding allows the provider to direct access over the user device and capacity to communicate any registered services whenever the user's device comes into the range of WiFi network, which is also achieved over the convergence of cellular networks and WiFi networks, making a network bridge between two heterogeneous architectures.

Basically, WiFi offloading strategies can usually be categorized into two groups: (1) opportunistic traffic data offloading, in which a user device offloads data only when an Mobile Node (MN) opportunistically comes into the proximity of a WiFi Access Point (AP), and (2) delayed traffic data offloading, in which a user device first wait and then offload data with some delay expecting the proximity of a new AP. To better understand these two technologies, we adopt an analytical model from [13] and examine the desired amount of traffic offloaded using both opportunistic offloading and delayed offloading. Consider a $\mathcal{T}$ time-slotted system in which an MN able to establish a data connection to both cellular network and WiFi network equally. Further, we consider that MN uses the cellular network all the time. However, MN shifts its data traffic whenever it comes into the proximity of a WiFi AP (Fig. 3).

2.3.1 Opportunistic Traffic Offloading

Denote $\mathcal{T}_c$ and $\mathcal{T}_w$ be the cellular residence time and WiFi residence time, respectively. Now according to author Shu et al. [13] we can analyze three different cases of opportunistic task offloading as follows.

- $\underline{Case\,1:}$ if $\mathcal{T}_d < \mathcal{T}_r$, then IoT devices complete their execution within the desired session.
- $\underline{Case\,2:}$ if $\mathcal{T}_r \leq \mathcal{T}_d < \mathcal{T}_r + \mathcal{T}_w$, then IoT devices offload their internet data through a WiFi network and complete its execution within the desired session. Here offloading time is the difference between $\mathcal{T}_d$ and $\mathcal{T}_r$. As $\mathcal{T}_d$, $\mathcal{T}_r$ and $\mathcal{T}_w$ are independent.
- $\underline{Case\,3:}$ if $\mathcal{T}_d \geq \mathcal{T}_r + \mathcal{T}_w$, then IoT devices offload their internet data through a WiFi network and complete its execution after the desired session.

Since all the cases are disjoint, the amount of total amount of traffic offloading can be calculated by adding *Case 1*, *Case 2* and *Case 3*, respectively.

2.3.2 Delayed Traffic Offloading

Similarly, we can analyze the offloading rate of the DTO strategy using three scenarios as shown in Fig. 4. Denote $\mathcal{D}$ be the maximum delay bound before the session starts, then the corresponding three cases of DTO are as follows.

1. $\underline{Case\,1:}$ If $\mathcal{D} + \mathcal{T}_d < \mathcal{T}_r$, then IoT devices complete their performance within the desired session.

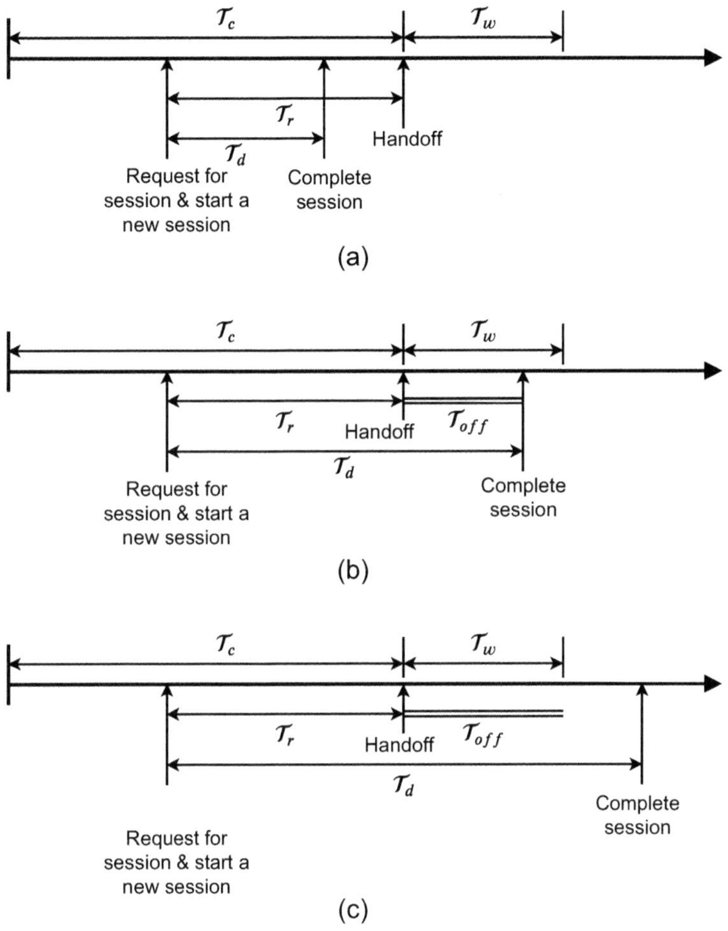

Fig. 3 OTO strategy

2. <u>*Case* 2 :</u> If $\mathcal{D} < \mathcal{T}_r$ and $\mathcal{T}_r \leq \mathcal{D} + \mathcal{T}_d < \mathcal{T}_r + \mathcal{T}_w$, then IoT devices offload their traffic data through a WiFi network and complete its execution within the desired session, whereas the session starts after the delay bound $\mathcal{D}$. Here the offloading time can be calculated as $\mathcal{D} + \mathcal{T}_d - \mathcal{T}_r$. As $\mathcal{T}_d$, $\mathcal{T}_r$ and $\mathcal{T}_w$ are independent.
3. <u>*Case* 3 :</u> When $\mathcal{D} < \mathcal{T}_r$ and $\mathcal{D} + \mathcal{T}_d \geq \mathcal{T}_r + \mathcal{T}_w$, then IoT devices offload their internet data through a WiFi network and complete its execution after the residence session. However, before starting the session, it has to wait for $\mathcal{D}$ delay bound.

Similar to the OTO strategy, the amount of total traffic offloading through DTO can be calculated by summing up all three cases *i.e., Case 1*, *Case 2* and *Case 2*. In this case, we considered three basic cases of opportunistic traffic offloading and delayed

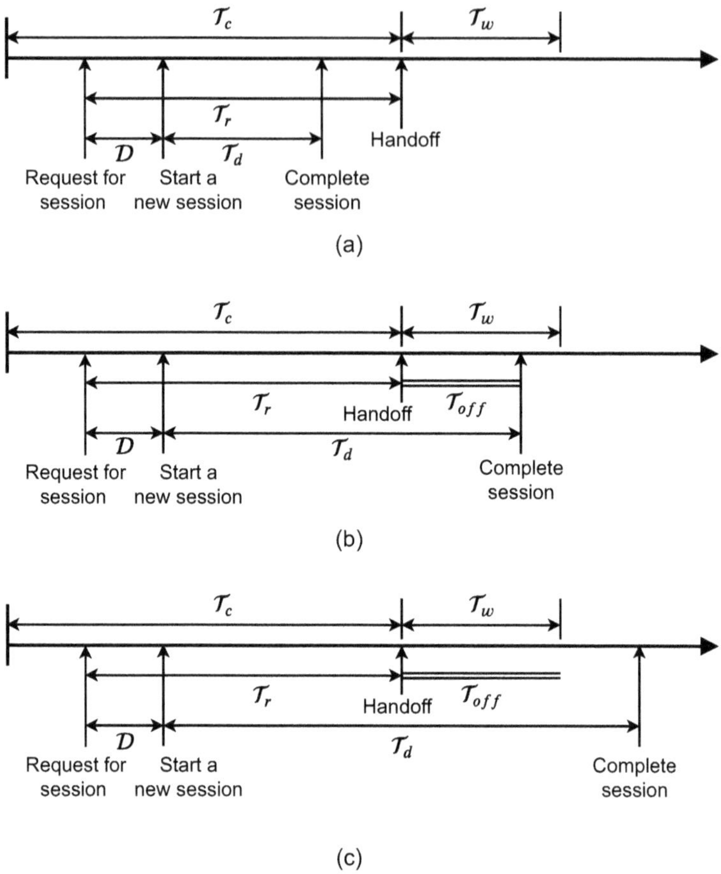

Fig. 4 DTO strategy

traffic offloading. A more detailed explanation of these models and performance analysis between them can be found in [13].

Opportunistic Traffic Offloading:
Denote $\mathbb{E}_{case1}^{OTO}\left[\mathcal{T}_{off}\right]$ be the expected WiFi offloading time using opportunistic offloading. To derive the expectation of opportunistic offloading, we need to explore the different possible cases presented in Fig. 3.

$\underline{Case\,1:}$ if $\mathcal{T}_d < \mathcal{T}_r$, then IoT devices complete its execution within the desired session. Thus the amount of traffic offloaded by OTO using *case 1* as.

$$\mathbb{E}_{case1}^{OTO}\left[\mathcal{T}_{off}\right] = 0 \tag{1}$$

$\underline{Case\,2:}$ if $\mathcal{T}_r \leq \mathcal{T}_d < \mathcal{T}_r + \mathcal{T}_w$, then IoT devices offload its internet data through WiFi network and complete its execution within the desired session. Here offloading time $\mathcal{T}_{off}$ is the difference between $\mathcal{T}_d$ and $\mathcal{T}_r$. As $\mathcal{T}_d$, $\mathcal{T}_r$ and $\mathcal{T}_w$ are independent, then the amount of traffic offloaded by OTO using *case 2* as.

$$\mathbb{E}_{case2}^{OTO}\left[\mathcal{T}_{off}\right] = \int_{\mathcal{T}_w=0}^{\infty}\int_{\mathcal{T}_r=0}^{\infty}\int_{\mathcal{T}_d=\mathcal{T}_r}^{\mathcal{T}_r+\mathcal{T}_w} \left(\mathcal{T}_d - \mathcal{T}_r\right) f_{d,r,w}\left(\mathcal{T}_d, \mathcal{T}_r, \mathcal{T}_w\right) d\mathcal{T}_d d\mathcal{T}_r d\mathcal{T}_w \quad (2)$$

$$= \int_{\mathcal{T}_w=0}^{\infty}\int_{\mathcal{T}_r=0}^{\infty}\int_{\mathcal{T}_d=\mathcal{T}_r}^{\mathcal{T}_r+\mathcal{T}_w} \left(\mathcal{T}_d - \mathcal{T}_r\right) f_d(\mathcal{T}_d) f_r(\mathcal{T}_r) f_w(\mathcal{T}_w) d\mathcal{T}_d d\mathcal{T}_r d\mathcal{T}_w \quad (3)$$

$\underline{Case\,3}:$ if $\mathcal{T}_d \geq \mathcal{T}_r + \mathcal{T}_w$, then IoT devices offload its internet data through WiFi network and complete its execution after the desired session. Thus the amount of traffic offloading by OTO using *case 3* as.

$$\mathbb{E}_{case3}^{OTO}\left[\mathcal{T}_{off}\right] = \int_{\mathcal{T}_w=0}^{\infty}\int_{\mathcal{T}_r=0}^{\infty}\int_{\mathcal{T}_d=\mathcal{T}_r+\mathcal{T}_w}^{\infty} \mathcal{T}_w f_d(\mathcal{T}_d) f_r(\mathcal{T}_r) f_w(\mathcal{T}_w) d\mathcal{T}_d d\mathcal{T}_r d\mathcal{T}_w \quad (4)$$

Since all the cases are disjoint, the amount of total traffic offloading by OTO using *case 1*, *case 2* and *case 2* as.

$$\mathbb{E}\left[\mathcal{T}_{off}\right] = \mathbb{E}_{case1}^{OTO}\left[\mathcal{T}_{off}\right] + \mathbb{E}_{case2}^{OTO}\left[\mathcal{T}_{off}\right] + \mathbb{E}_{case3}^{OTO}\left[\mathcal{T}_{off}\right] \quad (5)$$

Delayed Traffic Offloading

Similarly, we can derive DTO using three scenarios as shown in Fig. 4. Denote $\mathcal{D}$ the maximum delay bound before the session starts and the connection of nearby WiFi AP.

$\underline{Case\,1}:$ If $\mathcal{D} + \mathcal{T}_d < \mathcal{T}_r$, then IoT devices complete its performance within the desired session *i.e.*, $\mathcal{T}_{off} = 0$. Thus the amount of workload offloaded by DTO strategy using *case 1* as $\mathbb{E}_{case1}^{DTO}\left[\mathcal{T}_{off}\right] = 0$.

$\underline{Case\,2}:$ If $\mathcal{D} < \mathcal{T}_r$ and $\mathcal{T}_r \leq \mathcal{D} + \mathcal{T}_d < \mathcal{T}_r + \mathcal{T}_w$, then IoT devices offload its traffic data through WiFi network and complete its execution within the desired session whereas the session starts after the delay bound $\mathcal{D}$. Here offloading time is defined as $\mathcal{T}_{off} = \mathcal{D} + \mathcal{T}_d - \mathcal{T}_r$. As $\mathcal{T}_d$, $\mathcal{T}_r$ and $\mathcal{T}_w$ are independent, then the amount of traffic offloaded by DTO using *case 2* as.

$$\mathbb{E}_{case2}^{DTO}\left[\mathcal{T}_{off}\right] = \int_{\mathcal{T}_w=0}^{\infty}\int_{\mathcal{T}_r=\mathcal{D}-\mathcal{D}}^{\infty}\int_{\mathcal{T}_d=\mathcal{T}_r}^{\mathcal{T}_r+\mathcal{T}_w-\mathcal{D}} \left(\mathcal{D} + \mathcal{T}_d - \mathcal{T}_r\right) f_d(\mathcal{T}_d) f_r(\mathcal{T}_r) f_w(\mathcal{T}_w) d\mathcal{T}_d d\mathcal{T}_r d\mathcal{T}_w \quad (6)$$

$\underline{Case\,3}:$ When $\mathcal{D} < \mathcal{T}_r$ and $\mathcal{D} + \mathcal{T}_d \geq \mathcal{T}_r + \mathcal{T}_w$, then IoT devices offload its internet data through WiFi network and complete its execution after the residence session, however before starting the session it has to wait $\mathcal{D}$ delay bound. Thus the amount of traffic offloading by DTO using *case 3* as.

$$\mathbb{E}_{case3}^{DTO}\left[\mathcal{T}_{off}\right] = \int_{\mathcal{T}_w=0}^{\infty}\int_{\mathcal{T}_r=\mathcal{D}}^{\infty}\int_{\mathcal{T}_d=\mathcal{T}_r+\mathcal{T}_w-\mathcal{D}}^{\infty} \mathcal{T}_w f_d(\mathcal{T}_d) f_r(\mathcal{T}_r) f_w(\mathcal{T}_w) d\mathcal{T}_d d\mathcal{T}_r d\mathcal{T}_w \tag{7}$$

Similarly, the amount of total traffic offloading by DTO using *case 1*, *case 2* and *case 2* as.

$$\mathbb{E}\left[\mathcal{T}_{off}\right] = \mathbb{E}_{case1}^{DTO}\left[\mathcal{T}_{off}\right] + \mathbb{E}_{case2}^{DTO}\left[\mathcal{T}_{off}\right] + \mathbb{E}_{case3}^{DTO}\left[\mathcal{T}_{off}\right] \tag{8}$$

In this case, we considered three basic cases of opportunistic traffic offloading and delayed traffic offloading. A more detailed explanation of these models and performance analysis between them can be found in [13].

2.4 Core Network Offloading

Within the core network offloading paradigm, an innovative gateway strategically positioned in the data path between the Radio Network Controller (RNC) and the Serving GPRS Support Node (SGSN) takes center stage to efficiently redirect traffic bound for the Internet. This approach hinges on the gateway's ability to discern the nature of incoming session requests through various identifiers, with the access point name (APN) serving as a key discriminator. When the gateway identifies a session request earmarked for Internet-bound services, it dynamically alters the traffic path. In this redirection process, the packet-switched (PS) core network is circumvented, and the data is seamlessly forwarded directly to the nearest Internet peering point. This intelligent and dynamic rerouting mechanism optimizes network resources, streamlining the flow of Internet-bound traffic and enhancing the overall efficiency of the core network.

2.5 Fem-to-Cell Offloading

The concept of Fem-to-Cell Offloading [14] entails the utilization of femtocell networks to facilitate the process of job offloading inside the IoT framework. Femtocells refer to compact and low-power cellular base stations that have been specifically built to provide limited coverage within residential households or small commercial establishments. Femtocells are crucial components in the context of task offloading, since they provide a dedicated and efficient communication channel between IoT devices and the core network. This technique offers notable benefits in situations characterized by a high concentration of devices or in indoor settings where conventional cellular networks may encounter difficulties. The implementation of fem-to-cell offloading has been found to have a positive impact on various aspects of network performance. Specifically, it has been observed to enhance data rates, decrease latency, and improve overall network performance. Nevertheless, it is necessary to acknowledge and tackle certain obstacles, including interference, handover management, and network coordination, in order to ensure the successful implementation of the proposed deployment strategy. The successful deployment of fem-to-cell offloading in task distribution situations can be facilitated by employing intelligent handover algorithms, interference mitigation techniques, and optimal network planning. These solutions have the potential to enhance connection and performance for IoT applications.

2.6 WiMAX Offloading

WiMAX Offloading pertains to the application of WiMAX (Worldwide Interoperability for Microwave Access) technology for the purpose of task offloading within IoT situations. WiMAX, a wireless communication protocol, provides broadband wireless connectivity with a substantial coverage range. Task offloading involves the transfer of computational tasks from IoT devices to external resources, aiming to enhance performance and alleviate network congestion. WiMAX, as a wireless communication technology, offers an alternative connectivity solution for IoT devices. Effective data transmission may be achieved by leveraging WiMAX, hence addressing the congestion issue commonly experienced in conventional networks. In contrast to other offloading systems, WiMAX exhibits enhanced coverage and superior data rates, rendering it well-suited for situations involving the dispersion of devices across expansive regions. This technique improves the overall performance and efficiency of IoT networks by offering a resilient and high-capacity communication connection. However, it is crucial to consider factors such as the availability of spectrum, the costs associated with deployment, and the compatibility with pre-existing infrastructure when applying WiMAX offloading solutions. These considerations are essential in order to achieve optimal outcomes in a wide range of IoT applications (Fig. 5).

2.7 IP Flow Mobility Offloading

The concept of IP Flow Mobility Offloading entails the utilization of IP flow mobility as a means to assist the process of task offloading inside IoT environments. This methodology facilitates uninterrupted mobility management for IoT devices by granting them the capability to sustain their active connections during the process of shifting between diverse access networks. IP flow mobility is a mechanism that guarantees continuous communication during the process of transitioning between diverse network types. This mechanism serves to improve the overall dependability of task offloading schemes. The practical utilization of IP flow mobility offloading encompasses situations wherein IoT devices necessitate seamless transitioning between cellular networks, Wi-Fi, or alternative communication technologies, all

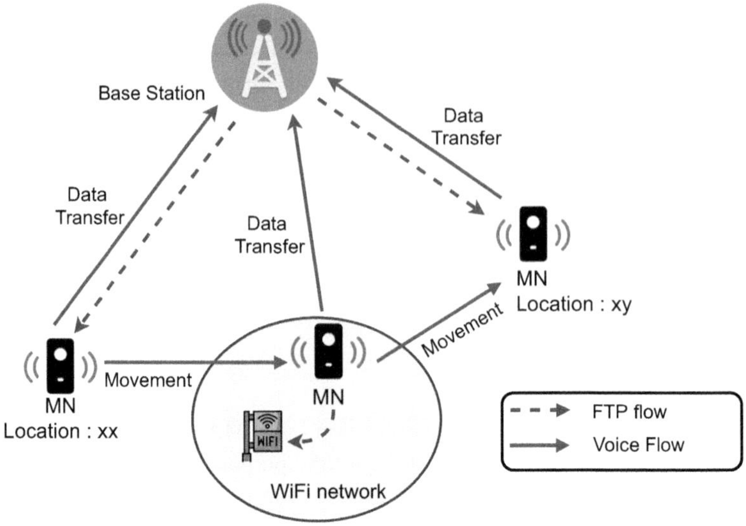

Fig. 5 IP flow for IoT data offloading

while ensuring uninterrupted data transmission. Although this particular strategy presents benefits in terms of preserving connectivity, it is imperative to solve the problems associated with security and performance optimization to guarantee the secure and efficient functioning of IoT devices within dynamic network environments. The successful deployment of IP flow mobility offloading in various IoT use cases relies heavily on the careful consideration of factors related to security and optimization.

3 Computing Paradigms for IoT

This section describes various computing paradigms and illustrates the necessity for such technologies in the IoT environment by classifying several critical factors. We then demonstrate how IoT platforms and edge computing paradigms are combined to manage real-time intelligence over edge networks. Note that computational intelligence is one of the essential features required to handle the ever-increasing volume of data returned by IoT nodes equipped with sensors in order to provide a comprehensive and valuable capability of detecting, analyzing, and classifying the required data.

3.1 Cloudlet Computing

Cloudlet computing, a concept introduced by CMU[1] (Carnegie Mellon University), shares numerous similarities with other computing paradigms such as Mobile Cloud Computing (MCC) and Mobile Edge Computing (MEC). A cloudlet refers to a computer or a group of computers with ample resources and a reliable Internet connection. These cloudlets are designed to be used by mobile devices in close proximity, essentially serving as compact data centers or miniature clouds located near mobile devices, ensuring a single network hop distance. The concept involves transferring computational tasks from mobile devices to cloudlets based on Virtual Machines (VMs) at the network edge. On the other hand, IoT technology enables the integration of billions of smart objects worldwide with cloudlet-based systems, enhancing their efficiency and analytical capabilities. Cloudlet computing proves to be a powerful and efficient computer-based paradigm in a distributed environment for processing and exploring IoT applications. As depicted in Fig. 6, cloudlet computing comprises diverse resources interconnected through high-speed networks. The primary advantage of this computing paradigm is that application software runs efficiently without concerns about computing resource limitations. This paradigm also enables applications to be deployed and executed in real time, as resources can be allocated and allocated dynamically. Furthermore, cloudlet computing is more secure and reliable than traditional cloud computing, as the application is distributed across various locations.

3.2 Cloud Computing

IoT technology allows billions of smart objects worldwide to be integrated with computer-based systems, which allows them to analyze and operate more efficiently [15]. One of the most powerful and efficient computer-based paradigms in a distributed environment for processing and exploring IoT applications is cloud computing. As illustrated in Fig. 6, cloud computing consists of diverse resources that are connected through high-speed networks. Users can choose from three essential cloud services: Software-as-a-Service (SaaS), Infrastructure-as-a-Service (IaaS), and Platform-as-a-Service (PaaS). Using the SaaS model, the software is deployed on computing servers as per the requirements of IoT applications. The main advantage of this model is that application software runs efficiently without having to install and manage it on IoT devices or smart devices. In the PaaS model, providers deliver a computing platform for IoT applications in the form of an operating system, an execution environment for a programming language, web servers, etc. IoT applications are deployed quickly on tons of servers without having to manage the underlying hardware and software layers. A few examples of PaaS clouds are Amazon Elastic Beanstalk, Cloud Foundry, Mendix, Engine Yard, Range Scape, and Google

[1] https://www.cmu.edu.

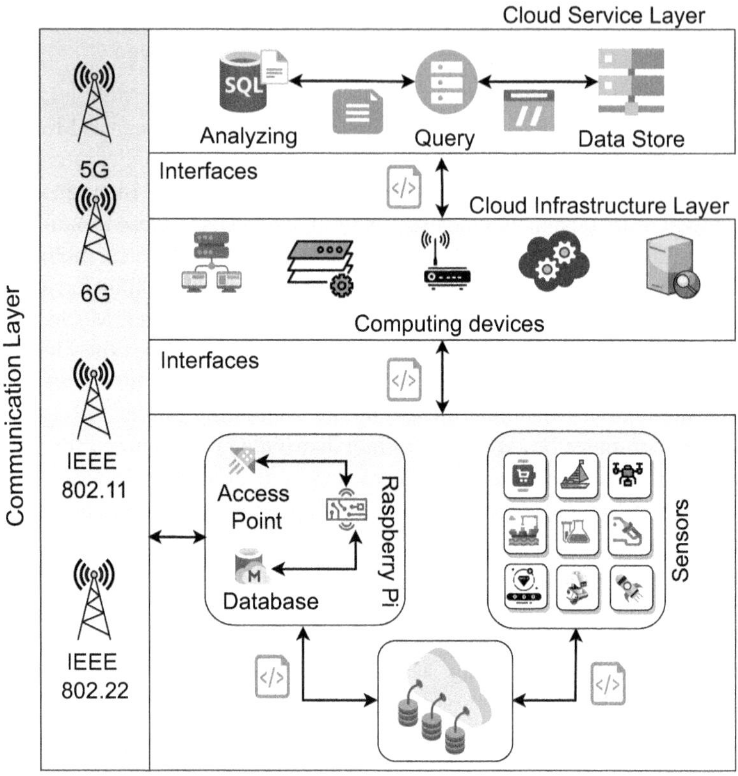

Fig. 6 Illustration of the cloudlet-IoT execution model

Engine. Finally, the IaaS cloud model offers thousands of computational servers with sufficient resource capacity for executing resource-intensive and request-based IoT applications. Each application should allow a suitable VM instance as per the requirements, such as Microsoft Azure and Amazon EC2.

3.3 Fog Computing

Fog computing is emerging as an alternative computing platform for decentralized environments to reduce CDC (Cloud Data Center) dependency and overcome cloud-related problems [16]. It supports a distributed architecture where computing devices are placed locally and distributed closer to IoT devices, minimizing latency and optimizing parameters. It is most appropriate for delay-dependent IoT applications.

Pioneered by Cisco in 2012 and standardized by the OpenFog Consortium, fog computing extends computation, communication, and storage facilities toward the network edge, featuring characteristics such as geographic distribution, heterogeneity, wireless access, low latency, and location awareness.

Fog devices create a highly virtualized platform similar to traditional CDCs, providing processing, networking, and storage services for IoT devices and users. According to Hazra et al. [17], fog computing encompasses ubiquitous, heterogeneous, and decentralized computing devices. It can process and store information locally without involving a third party and employs various wireless communication modes. Several research articles claim that fog computing supports a system-level horizontal architecture, distributing computing resources, including computation and storage capacity, and a reliable communication network to connect with IoT devices and CDCs for large applications [18, 19]. In addition to providing reliable wireless communication with end-device mobility, fog computing reinforces a heterogeneous set of resources, is geographically distributed, and supports real-time applications.

3.4 Serverless Computing

Specifically, IoT applications based on IoT protocols like LoRaWAN [20], Sigfox, NB-IoT [21], etc can be effectively deployed and performed on standard cloud infrastructure. Alternatively, cloud servers struggle to handle event-driven applications, appropriately limiting resource consumption [22]. However, a flexible serverless architecture has evolved in the cloud to process and analyze event-driven IoT applications rapidly. The serverless module frees providers from maintaining cloud servers and uses current cloud resources and services to process IoT applications. Existing cloud services include Amazon SNS, Google Cloud Pub/Sub, and DynamoDB, to name a few. Function-as-a-Service is a computing model that complements IaaS. A function is a software that meets IoT applications' requirements in a serverless environment. Each part can handle a single IoT request. They designate an application or job to be executed on an optimal cloud server for a specific IoT application. For many designers, serverless computing architectures present more flexibility, quicker time to release, and enhanced scalability, all at a decreased cost [23]. Several cloud activities, such as file uploading, message queueing, and direct HTTP requests have been allocated for processing background cloud applications in the serverless environment. The working of serverless computing is illustrated in Fig. 7.

3.5 Virtual Machines

Virtualization is the fundamental technology that makes the cloud environment feasible in a geographical context. Virtualization technology enables service providers to assign several virtual machine instances to tasks or applications without the

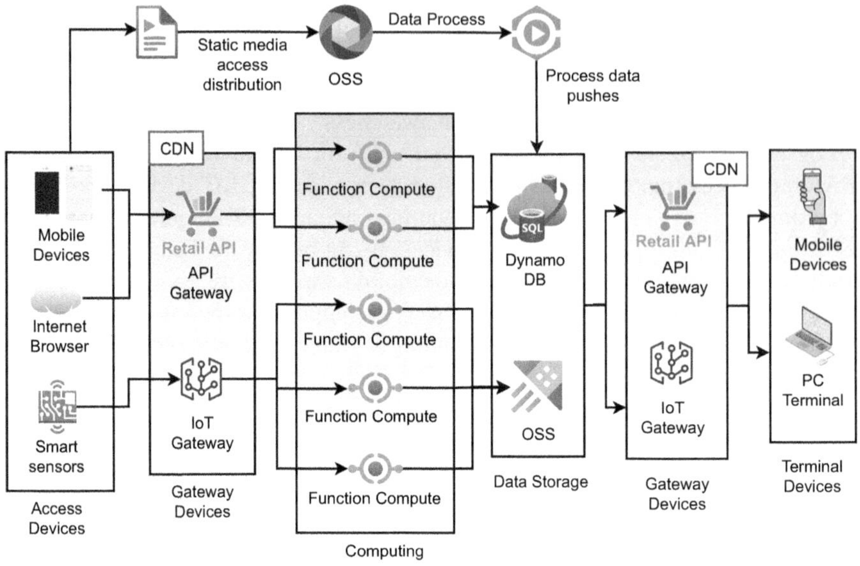

Fig. 7 Block diagram of serverless computing

users' knowledge. Virtualization technology, which improves resource efficiency and reduces task energy consumption, is largely responsible for the maturity of cloud computing. Cloud providers previously distributed resources to users as virtual machine instances. Virtual Machine instances virtualize computing resources at the hardware level, with each VM instance having its underlying OS and sharing the processing resources of a host server via '*Bear-Metal*' or '*Hypervisor*' technology. A hypervisor, also known as a virtual machine monitor, is software or firmware on a server that helps deploy and run virtual machines. Virtual machines have the advantage of virtualizing server resources, boosting job parallelism, and improving computing resource efficiency.

3.6 Peer-to-Peer Computing

Peer-to-peer (P2P) computer networks are another type of distributed system model. The central intuition of designing the P2P model is to link a group of computing devices with identical rights and responsibilities. In contrast to a traditional client-server system where a server is solely responsible for data processing, all the connected devices act like a client or server. A popular example of a P2P network is torrents. As P2P computing does not distinguish between client and server, it is difficult to back up the data and store it in a specific location. On the other hand, nodes participating in P2P computing must register their identification and available

Table 1 Comparison of computing paradigms for IoT

Feature	Cloudlet computing	Cloud computing	Fog computing	Serverless computing	Virtual machines	Peer-to-Peer computing
Location	Edge of the network, closer to IoT devices	Centralized data centers	Closer to the edge than traditional cloud, but not as close as edge devices	Centralized cloud infrastructure	Typically in data centers	Decentralized, spread across participating devices
Latency	Low latency due to proximity to IoT devices	Higher latency compared to edge computing	Low latency, closer to edge devices	Low latency as functions are executed closer to the data source	Higher latency compared to edge computing	Latency depends on the location of peers and network conditions
Scalability	Scalable, especially for edge-intensive applications	Highly scalable with elastic resources	Scalable, especially for distributed applications	Highly scalable, automatically scales with demand	Scalable, can run multiple VMs on a single physical server	Scalable, as the number of participating peers can increase
Resource Efficiency	Utilizes resources efficiently by offloading computation from IoT devices	Resource usage may not be as efficient due to centralized data centers	Efficient use of resources by distributing computation across the network	Efficient resource usage as resources are allocated dynamically	Resource usage can be less efficient due to virtualization overhead	Resource usage depends on the contribution of individual peers
Platforms	MAUI, CloneCloud, Mobile Fog, Dew Computing	Amazon Web Services (AWS), Microsoft Azure, Google Cloud Platform (GCP)	Cisco IOx, Microsoft Azure IoT Edge, OpenFog	AWS Lambda, Azure Functions, Google Cloud Functions	VMware, Hyper-V, KVM	BitTorrent, Bitcoin, Napster

service capacity with the network. The communication of services is done through the message discovery protocol. Notably, the P2P computing network sometimes breaks down due to the lack of internet connectivity. This computing environment is widely used in vehicular and industrial networks to process device-generated data in near-the-edge devices [24] (Table 1).

4 Conclusion

The landscape of Internet of Things (IoT) is marked by an unprecedented proliferation of devices, generating vast amounts of data and ushering in a new era of connectivity. As the demand for real-time processing and analysis escalates, addressing the challenges posed by resource-constrained IoT devices becomes imperative. This survey delves into the realm of task offloading mechanisms and computing paradigms, providing a comprehensive overview of strategies designed to enhance the efficiency and scalability of IoT systems.

The survey commences by exploring standardized data offloading technologies, including cellular offloading, wireless mesh, WiFi offloading, fem-to-cell offloading, WiMAX offloading, and IP flow mobility offloading. Each method is dissected, highlighting its strengths and limitations in the context of optimizing data processing, connectivity, and energy usage in IoT deployments. Despite their advantages, these technologies come with challenges such as network intricacies, latency concerns, and privacy considerations, emphasizing the need for a nuanced evaluation tailored to specific application requirements.

In the realm of computing paradigms, the survey investigates cloudlet computing, cloud computing, fog computing, serverless computing, virtual machines, and peer-to-peer computing. These paradigms offer diverse solutions to address the growing demands of IoT applications. Cloudlet computing, situated at the network edge, enables efficient task offloading, while cloud computing provides scalable and powerful solutions. Fog computing minimizes latency by distributing computing resources closer to IoT devices, and serverless computing offers flexibility and scalability without the burden of managing infrastructure. Virtual machines contribute to resource efficiency, and peer-to-peer computing fosters collaborative networks without a central server.

In conclusion, this survey underscores the pivotal role of task offloading and computing paradigms in optimizing IoT systems. The interplay of these strategies, considering their merits and challenges, paves the way for enhanced connectivity, real-time processing, and energy efficiency in the dynamic landscape of IoT. As IoT continues to evolve, a nuanced understanding of these mechanisms becomes paramount for researchers, developers, and stakeholders to navigate the complexities and unlock the full potential of interconnected systems.

References

1. A. Hazra, Promising role of visual IoT: challenges and future research directions. IEEE Eng. Manag. Rev., 1–7 (2023)
2. M. Adhikari, A. Hazra, S. Nandy, Deep transfer learning for communicable disease detection and recommendation in edge networks. IEEE/ACM Trans. Comput. Biol. Bioinformatics **20**(4), 2468–2479 (2023)
3. I. Sarkar, S. Kumar, Deep learning-based energy-efficient computational offloading strategy in heterogeneous fog computing networks. J. Supercomput. **78**(13), 15089–15106 (2022)
4. A. Hazra, P.K. Donta, T. Amgoth, S. Dustdar, Cooperative transmission scheduling and computation offloading with collaboration of fog and cloud for industrial iot applications. IEEE Internet Things J., 1 (2022)
5. I. Sarkar, M. Adhikari, N. Kumar, S. Kumar, Dynamic task placement for deadline-aware iot applications in federated fog networks. IEEE Internet Things J., 1 (2021)
6. A. Hazra, M. Adhikari, T. Amgoth, S. Srirama, Joint computation offloading and scheduling optimization of iot applications in fog networks. IEEE Trans. Netw. Sci. Eng., 1 (2020)

7. A. Hazra, M. Adhikari, D. Kumar, T. Amgoth, Fair scheduling and computation co-offloading strategy for industrial applications in fog networks. IEEE Trans. Netw. Serv. Manag., 1 (2023)
8. I. Sarkar, M. Adhikari, N. Kumar, S. Kumar, A collaborative computational offloading strategy for latency-sensitive applications in fog networks. IEEE Internet Things J. **9**(6), 4565–4572 (2022)
9. F.H. Panahi, F.H. Panahi, T. Ohtsuki, Intelligent cellular offloading with vlc-enabled unmanned aerial vehicles. IEEE Internet Things J. **10**(20), 17 718–17 733 (2023)
10. C. Sriharsha, C.S.R. Murthy, A novel cellular user offloading via uav-borne irs. IEEE Wirel. Commun. Lett. **12**(10), 1736–1740 (2023)
11. F. Zhou, L. Feng, M. Kadoch, P. Yu, W. Li, Z. Wang, Multiagent rl aided task offloading and resource management in wi-fi 6 and 5g coexisting industrial wireless environment. IEEE Trans. Indus. Inform. **18**(5), 2923–2933 (2022)
12. A. Aijaz, H. Aghvami, M. Amani, A survey on mobile data offloading: technical and business perspectives. IEEE Wirel. Commun. **20**(2), 104–112 (2013)
13. D. Suh, H. Ko, S. Pack, Efficiency analysis of wifi offloading techniques. IEEE Trans. Vehicular Technol. **65**(5), 3813–3817 (2015)
14. S. Mitra, S. Chattopadhyay, S.S. Das, Deployment considerations for mobile data offloading in lte-femtocell networks. Int. Conf. Signal Process. Commun. (SPCOM) **2014**, 1–6 (2014)
15. A. Hazra, P. Choudhary, O. Vivek, An advance mobility management scheme in wireless network, in *2018 9th International Conference on Computing, Communication and Networking Technologies (ICCCNT)*, pp. 1–5 (2018)
16. A. Hazra, P. Rana, M. Adhikari, T. Amgoth, Fog computing for next-generation internet of things: fundamental, state-of-the-art and research challenges. Comput. Sci. Rev. **48**, 100549 (2023)
17. A. Hazra, M. Adhikari, T. Amgoth, S.N. Srirama, Collaborative AI-enabled intelligent partial service provisioning in green industrial fog networks. IEEE Internet Things J., 1 (2021)
18. A. Hazra, M. Adhikari, T. Amgoth, Dynamic service deployment strategy using reinforcement learning in edge networks, in *2022 International Conference on Computing, Communication, Security and Intelligent Systems (IC3SIS)*, pp. 1–6 (2022)
19. I. Sarkar, M. Adhikari, S. Kumar, V. G. Menon, Deep reinforcement learning for intelligent service provisioning in software-defined industrial fog networks. IEEE Internet Things J., 1 (2022)
20. P. Maurya, A. Singh, A.A. Kherani, A review: spreading factor allocation schemes for lorawan. Telecommun. Syst. **80**(3), 449–468 (2022)
21. B.S. Chaudhari, M. Zennaro, S. Borkar, Lpwan technologies: emerging application characteristics, requirements, and design considerations. Future Internet **12**(3), 46 (2020)
22. S.R. Poojara, C.K. Dehury, P. Jakovits, S.N. Srirama, Serverless data pipeline approaches for iot data in fog and cloud computing. Future Gen. Comput. Syst. **130**, 91–105 (2022)
23. S. Sarkar, R. Wankar, S.N. Srirama, N.K. Suryadevara, Serverless management of sensing systems for fog computing framework. IEEE Sensors J. **20**(3), 1564–1572 (2020)
24. A. Talpur, M. Gurusamy, Machine learning for security in vehicular networks: a comprehensive survey. IEEE Commun. Surv. Tutorials **24**(1), 346–379 (2022)

Predicting Forest Canopy Height Using GEDI LiDAR Based Machine Learning Technique Over Similipal Biosphere, India

Chiranjit Singha and Satiprasad Sahoo

Abstract One of the elements of an ecosystem's functioning that is thought to be most important is its forest ecosystem. It offers a range of services, including climate regulation, watershed preservation, and the provision of food and materials. With the use of Sentinel 1A, Sentinel 2B, the SRTM digital elevation model, and GEDI LiDAR data, the final forest canopy height (FCH) map was created in Similipal Biosphere Reserve, Odisha, India. Fourteen parameters of varied relevance are predicted using three machine learning models (Random Forest (RF), eXtreme Gradient Boosting (XGB), and Classification and Regression Tree (CART) for FCH estimation. The XGB tree model (R2 = 0.524, RMSE = 6.553) and the CART method (R2 = 0.891, RMSE = 2.293) had the next highest goodness of fit, behind the RF model (R2 = 0.745, RMSE = 4.441). Another crucial element that may be utilized to track a forest's biomass and aboveground carbon storage is the height of the canopy. These findings imply that incorporating various variables and data sources enhances canopy cover and height prediction accuracy when compared to using a single data source. The findings of this study may be useful in developing forest management plans that promote the sustainable use of forest resources in near real time basis. Future research will include additional parameters, such as meteorological, vegetation biophysical, geo-environmental, and tree species data. Advanced methods, like hybrid machine learning and deep learning, as well as ground-based UAVs and ICESat platforms, will be helpful in estimating FCH quickly.

Keywords Forest canopy height · Sentinel-2B · Sentinel-1A · GEDI · Machine learning

C. Singha (✉)
Department of Agricultural Engineering, Institute of Agriculture, Visva-Bharati (A Central University), Sriniketan, Birbhum 731236, India
e-mail: singha.chiranjit@gmail.com

S. Sahoo
Prajukti Research Private Limited, Baruipur 743610, West Bengal, India

B. Pradhan and S. Mukhopadhyay (eds.), *IoT Sensors, ML, AI and XAI: Empowering A Smarter World*, Smart Sensors, Measurement and Instrumentation 50,
https://doi.org/10.1007/978-3-031-68602-3_18

1 Introduction

A forest ecosystem is considered to be one of the most critical components of an ecosystem's functioning [1]. It provides various services such as food and materials, protection of watersheds, and regulating climate. Unfortunately, human activities are threatening the survival of forests. Various international organizations have been working toward establishing programs and projects that will help manage and protect forest ecosystems. For instance, the UN has been supporting countries with the development of national forest monitoring systems. These systems rely on the data collected by the National Forest Inventories (NFI). Over the years, many national forestry agencies have been relying on Earth observation satellites to collect forest cover information for their programs. Unfortunately, due to the lack of technical and financial resources, these agencies are not able to consistently produce reliable forest cover information. One of the main factors that prevent them from producing reliable forest cover information is the lack of training materials. One of the most effective ways to improve the accuracy of the data collected by satellites is through the use of training data automation [2]. This method allows remote sensing analysts to collect and interpret data from satellites. They can then use the data to create a variety of models and classifications to improve the forest cover information [3]. Currently, the lack of timely forest structure information is a major issue that affects the quality of the data collected by satellites in tropical regions. This issue is important for analyzing deforestation and forest degradation, monitoring the changes in the ecosystem, and estimating the biomass of the forests. For decades, forest researchers have been collecting forest canopy features using field measurements. However, this method is time-consuming and costly. Instead, they can collect the data through ground measurements. Airborne laser scanning and terrestrial laser scanning like ICESat platform and NASA's Global Ecosystem Dynamics Investigation (GEDI) waveform-based LiDAR sensor are the main methods used by remote sensing researchers to map the forest canopy height [4, 5]. These methods provide more accurate data than traditional field measurements. However, they are expensive and have limited coverage in large areas. The data collected from the GEDI waveform database is used to create a responsive forest that can be used to predict future changes in the forest structure. Lang et al. [6] showed the first global, wall-to-wall forest top canopy height map at 10 m ground resolution with GEDI and Sentinel 2 data for the year 2020. The use of LiDAR data for forest canopy height (FCH) mapping is well documented in the various review works [7, 8]. The combination of ALOS PALSAR and Sentinel 2A is used for the forest structure using the different machine learning methods in Iran [9]. Healey et al. [10] used Landsat-based relative height estimates using a new GEDI asset on GEE for the forest canopy structure. Machine learning models coupled with Sentinel 2 and SRTM are effective for the prediction of tree canopy height as well as biomass estimation [11]. Pourshamsi et al. [12] estimated the tropical forest height using the SAR and LiDAR data through the four machine learning models namely Support Vector Machine (SVM), Random Forest (RF), Rotation Forest (RoF), and Canonical Correlation Forest (CCFs). The purpose of the present study is to estimate the forest canopy height by the RS technology using the RF regression, XGB tree and the CART algorithm through the Google Earth Engine (GEE) cloud.

2 Materials and Method

2.1 Study Area

One of the most important areas in India that have been identified as a biosphere reserve is the Similipal Biosphere Reserve. This region extends between latitude 20° 17′–22° 34′ N and longitude 85°40′–87°10′ E in the Eastern Ghat mountain range and the Himalayas (Fig. 1). The entire area of the Similipal biosphere reserve is located in a tilted plateau region that is located in the Mayurbhanj district of India. This region has a spatial extension of 2750 km. The difference in the topography of the region can range from 200 to 1168 m above sea level. The annual rainfall in the region ranges from 120 to 210 cm and a seasonal mean temperature of 9.8–33.5 °C. In 1979, the government of India designated the area as a wildlife sanctuary. Over the years, the government has been able to extend the area by about 2200 km. The dominant tree species where the dry deciduous sal (*Shorea robusta Gaertn. f.*) belongs in this region.

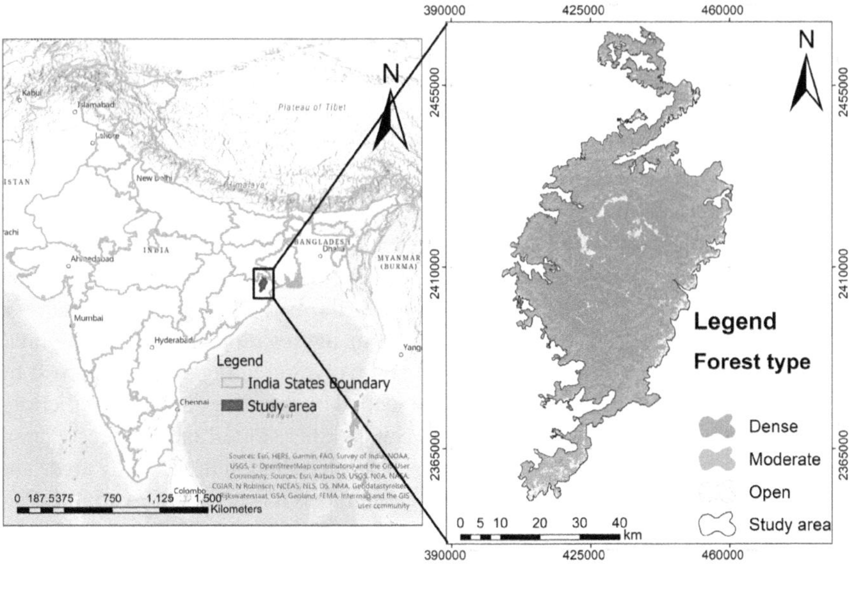

Fig. 1 Study area map

2.2 Remote Sensing Data Used

The current study implemented Sentinel-2B and Sentinel-1A ascending satellite imagery from April to June 2021 (Fig. 2). The 10-m ESA World Cover data acquired from the GEE earth data catalog (ESA/WorldCover/v100) for masking the forest areas in the Similipal region. Sentinel-2 provides optical satellite data in 13 spectral bands, from the visible and the near-infrared to the shortwave infrared. In the present study 10 spectral bands, excluding bands 1 (Coastal aerosol), 9 (Water vapor), and 10 (SWIR-Cirrus) were used (Table 1). Orthorectified surface reflectance satellite data were acquired from (https://scihub.copernicus.eu/). and a subset image was extracted from the acquired reflectance image using the area of interest. The cloud-free image is pre-processed with the QA band. Sentinel-1A C band GRD data are pre-processed using the Sentinel toolbox (SNAP) in the GEE cloud. The interquartile range of the VH signature in dB is resampling in a UTM45 North (Fig. 2). The median VV and VH backscatter signature (April to June) in the area ranges between −24.8 to 17.7 dB and −27.9 to −3.1 dB (Fig. 2). The remote sensing SRTM DEM data are used the created the elevation, slope, and aspect map of the entire study area. The elevation ranges were from 40 to 1200 m in the study area (Fig. 3). We demarcated five slopes in degree classes for forest canopy height estimation, namely 29–63° showed a high slope and $\leq$7.4° showed the low slope (Fig. 2). The GEDI L2A (LARSE/GEDI/GEDI02_A_002_MONTHLY) footprint ingested from the NASA Land Processes Distributed Active Archive Center (DAAC) with the help of GEE API interface. The variables of relative height (RH98) values value are used to for forest canopy height estimation in the study area.

2.3 Machine Learning Model

(I) *RF*: The Random Forest is a mostly used supervised learning technique that can be implemented for various problems such as classification and regression. It is based on the ensemble learning concept, which is a process that involves combining multiple classifiers to improve the performance of a model. RF objective function assigned by the bootstrap aggregation. The Random Forest is divided into two phases. The first one involves creating a random forest by combining the N decision tree and the second one involves making predictions for each tree. In the GEE code editor ee.Classifier.smileRandomForest classifier was used for the forest tree canopy height prediction in the study area. The details argument of random forest used numberOfTrees ~ 100, variablesPerSplit ~ 5, minLeafPopulation ~ 1, bagFraction ~ 0.5 respectively.

(II) *XGB tree*: XGBoost is mainly used for developing supervised learning programs. It can be used to predict a target variable based on the training data. Before we start working with trees, let's first review the various elements of supervised learning. XGB tree objective function depends on the training

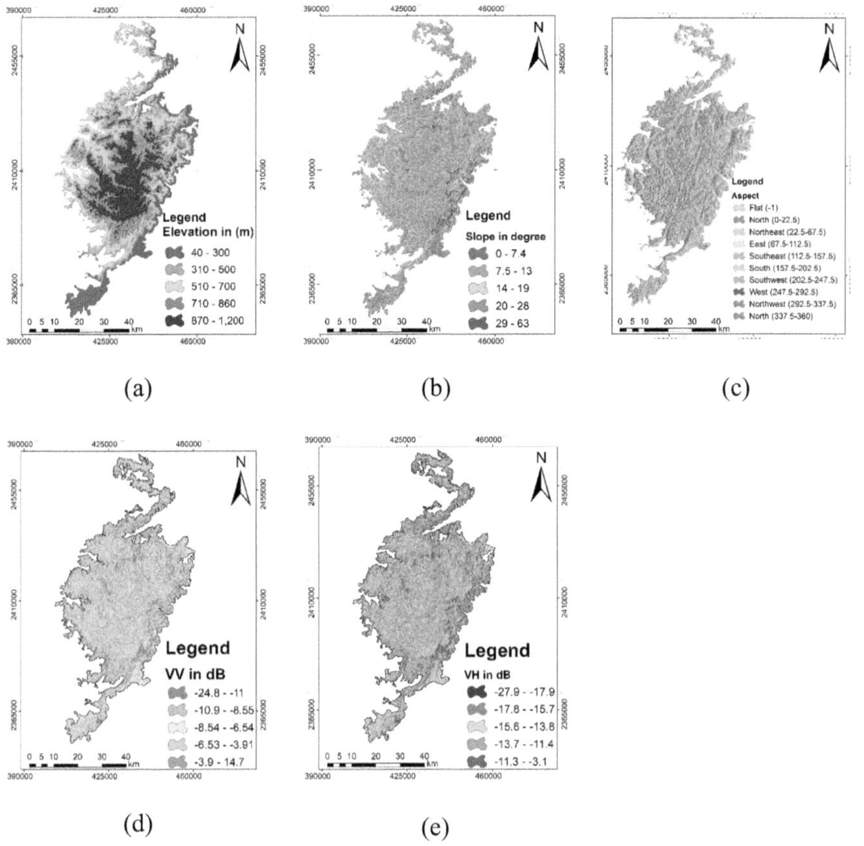

Fig. 2 Forest height influencing parameters **a** elevation, **b** slope, **c** aspect, **d** VV, **e** VH

loss and regularization parameters. The ee.Classifier.smileGradientTreeBoost classifier is implemented to run the XGB tree model in the GEE server. The tuning parameters of the XGB tree function used with numberOfTrees ~ 100, shrinkage ~ 0.005, samplingRate ~ 0.7, and loss ~ LeastAbsoluteDeviation (Table 1).

(III) *CART*: The classification algorithm known as the CART is a basic machine learning framework that can be used to build a decision tree based on Gini's impurity index. Step 1: Begin by creating a cluster of training instances that are trained to perform various tasks. Step 2: Select an attribute that is related to one of the splitting criteria. Step 3: Partition the instances according to the selected attribute namely maxNode, minLeafPopulation, etc. (Table 1). CART algorithm applied with the ee.Classifier.smileCart classifier.

The Machine learning regression model performance comparison with the statistical root means square error (RMSE), and determination coefficient (R2) with 70%

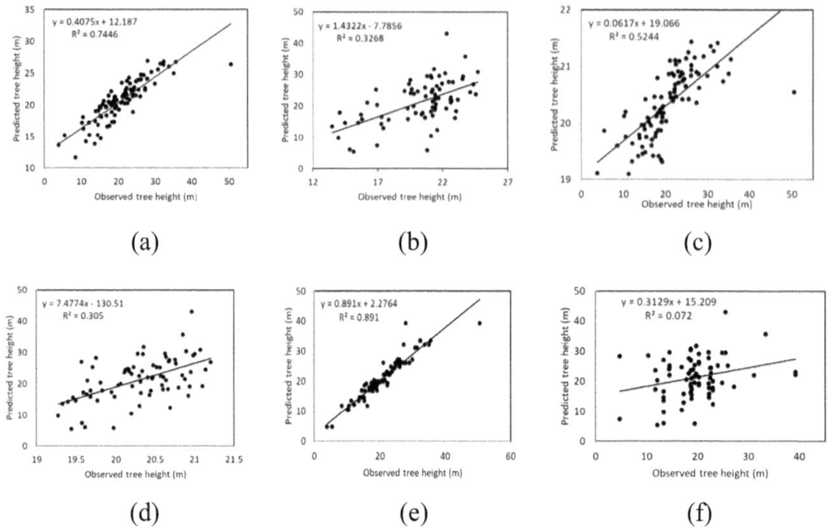

Fig. 3 Scatter plot of the observed and predicted tree canopy height in Similipal Biosphere **a** RF train **b** RF validation **c** XGB tree train **d** XGB validation **e** CART train **f** CART validation

for training and 30% for validation datasets (Eqs.1 and 2).

$$\text{RMSE} = \sqrt{\sum_{1}^{n} \frac{\left(X_i \sum \widehat{X}_i\right)^2}{n}} \tag{1}$$

$$R^2 = 1 - \frac{\sum_{i=1}^{n} \left(X_i - \widehat{X}_i\right)^2}{\sum_{i=1}^{n} \left(X_i - \overline{X}\right)^2} \tag{2}$$

where X_i is the observed training location of tree canopy height, $\widehat{X}_i$ is the predicted tree canopy height, $\overline{X}$ is the actual average value of tree canopy height, i is the equal index, n is the total number of validation sites.

3 Result

3.1 Model Training and Validation

The final forest canopy height (FCH) map was developed using the Sentinel 1A, Sentinel 2B, SRTM digital elevation model, and GEDI LiDAR data. The RF, XGB, and CART algorithms are used for the production of (FCH) maps. The details of the

Table 1 Details argument of the machine learning models parameters

Model name	Details argument tuning parameters
RF	numberOfTrees ~ 100, variablesPerSplit ~ 5, minLeafPopulation ~ 1, bagFraction ~ 0.5, maxNodes ~ default, seed ~ 0
XGB	numberOfTrees ~ 100, shrinkage ~ 0.005, samplingRate 0.7, maxNodes default:, loss ~ LeastAbsoluteDeviation, seed ~ 0)
CART	maxNodes ~ default, minLeafPopulation ~ 1

algorithms' performance comparison by the RMSE and R2 values are presented in Table 2. The results found that all three models had satisfactory performances using the training data. The highest goodness of fit was showed for the RF model (R2 = 0.745, RMSE = 4.441), followed by the XGB tree model (R2 = 0.524, RMSE = 6.553), and the CART algorithm (R2 = 0.891, RMSE = 2.293). Conversely the RF machine learning regression model well executed for FCH prediction validation datasets (R2 = 0.327, RMSE = 5.805), and was followed by the XGB tree model (R2 = 0.305, RMSE = 6.633). In contrast, the CART algorithm had the lowest performance (R2 = 0.072, RMSE = 7.917) (Fig. 4). It is perceived from Table 1 that, among the three machine learning models, the RF regression model had the highest performance with both on the training and the validation datasets for the FCH modelling.

All the prediction products were converted to a raster layer to export into the Google drive for further analysis. The FCH maps was categorized by five classes (Fig. 5) showing a range from 10 to 30 m (average = 20 m).

Three machine learning models are used to predict all the 14 parameters of variable importance for FCH estimation. The determination of the relative importance of the tree height prediction mapping B12, B2, and B4 parameters are the most important using the RF, XGB tree, and CART algorithm (Fig. 4). The VV and VH backscatter contribution to 3–6% for the FCH estimation while the elevation was 2–7% in the

Table 2 ML regression Model of training and validation result for forest canopy height prediction

Model	RF		XGB		CART	
	Train	Validation	Train	Validation	Train	Validation
RMSE	4.441	5.805	6.553	6.633	2.293	7.917
R2	0.745	0.327	0.524	0.305	0.891	0.072

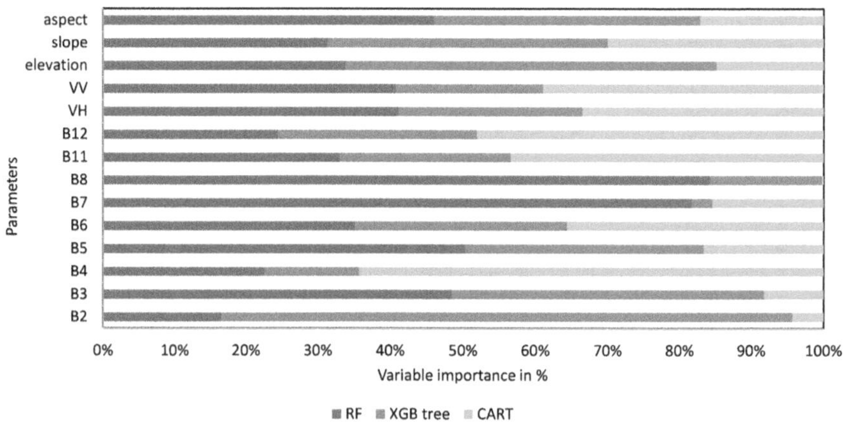

Fig. 4 Variable importance for the forest canopy predictors

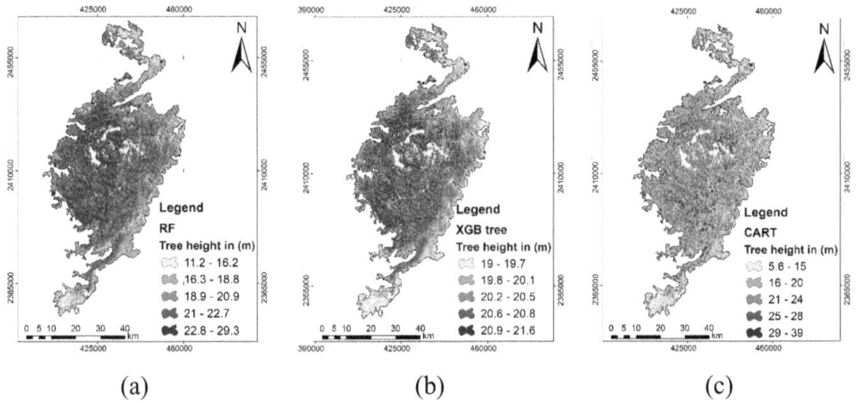

Fig. 5 Forest canopy height maps **a** RF, **b** XGB tree, and **c** CART

study area. According to the XGB tree, and CART algorithm the B7 parameters as less important in this case.

3.2 Forest Canopy Height Prediction

All the selected ML models generated the FCH prediction maps (spatial resolution ~ 10 m × 10 m) into five levels namely (i.e. very low, low, moderate, high, and very high). According to the RF model, 14.04% of the study region specified very high FCH from 22.8 to 29.3 m, whereas 25.81 and 19.31% of the region are medium and low FCH areas, respectively (Fig. 5). Based on the XGB tree model, the areal coverage of very low, low, medium, high, and very high forest canopy height areas was 251.09 (9.64%), 402.46 (15.45%), 722.51 (27.74%), 816.88 (31.37%), and 411.21 (15.79%) km^2 respectively (Fig. 5). The CART algorithm occurred that 812.33 km^2 (31.29%) of the study region has a very low FCH, whereas 288.62 (11.08%), 382.85 (14.70%), 414.33 (15.91%) and 706.03 (27.11%) km^2 has a very high, high, medium and low tree height area (Table 3). The clearcut Fig. 5 showed that the three ML models predicted FCH values are higher in the west, and central parts of the Similipal region, conforming to the high elevation; lower values were stretched in the boundaries of the northern, eastern, and southern regions.

4 Discussion

The height of the forest canopy is also an important factor that can be used to monitor the biomass and the aboveground carbon stock in a forest [13]. Through NASA's GEDI campaign, the agency has been collecting LiDAR data on the vertical

Table 3 Area coverage for forest canopy height

	RF		XGB tree		CART	
	Area (km^2)	Area (%)	Area (km^2)	Area (%)	Area (km^2)	Area (%)
Very low	392.77	15.08	251.09	9.64	812.33	31.19
Low	502.37	19.29	402.46	15.45	706.03	27.11
Mideum	672.07	25.81	722.51	27.74	414.33	15.91
High	671.22	25.78	816.88	31.37	382.85	14.70
Very heigh	365.71	14.04	411.21	15.79	288.62	11.08
Total	2604.15	100.00	2604.15	100.00	2604.15	100.00

structure of forests to measure their global importance [14]. Despite the availability of cloud platforms, such as Google Earth Engine, which provide a variety of tools for analyzing satellite data, traditional machine learning techniques still require a lot of flexibility to work effectively [15]. A total of 14 variables, including original spectral bands, SAR backscatter, and SRTM DEM data were used for FCH estimation in the Similipal Biosphere. The three ML models generated the FCH prediction maps. Furthermore, R2 and RMSE were employed to measure the quality of the FCH models. According to the RF model, 14.04% of the study region specified a very high FCH. The CART algorithm occurred that 812.33 km^2 (31.29%) of the study region has a very low FCH. The VV and VH backscatter contributed to 3–6% of the tree canopy height estimation while the elevation was 2–7% in the study area. The highest goodness of fit was showed for the RF model (R2 = 0.745, RMSE = 4.441), followed by the XGB tree model (R2 = 0.524, RMSE = 6.553), and the CART algorithm (R2 = 0.891, RMSE = 2.293). ML models derived from FCH maps represent the higher canopy height found in the western, and central regions of the study area, compatible with the high elevation. The lower FCH showed in the edges of the eastern, northern, and southern parts due to the more human interaction adopted in this region for the low density of the forest.

The future of the study includes more parameters namely meteorological, vegetation biophysical, geo-environmental, and tree species data with the help of advanced techniques such as hybrid machine learning deep learning and ground-based UAV and ICESat platforms are very useful for the rapid estimation of FCH.

5 Conclusion

The current study was able to develop a machine learning method through the GEE cloud for forest canopy height mapping that was able to retrieve the data collected by NASA's GEDI LiDAR, SRTM, and Copernicus ESA Sentinel-1 SAR and Sentinel-2 optical sensors. It was able to create a wall-to-wall map of the forest canopy height covering the entire study area. We conclude that supervised models (RF and XGB tree) performed better than the CART algorithm to estimate forest canopy height. In addition, the effectiveness of the cloud-based models applied in this study can update the collection of predictive mapping systems for forest canopy height modelling on a near real-time basis. The results of their work could help improve the efficiency of the monitoring and management of forests. It could also help provinces make informed decisions regarding the protection of their forests.

References

1. C. Singha, K.C. Swain, B.B. Sahoo, P. Ghosh, S.K. Swain, Assessment of bio diversity conservation using geospatial models. J. Pharmacognosy Phytochem. **8**(1), 1177–1186, ISSN: 2349–8234 (2019)
2. K.C. Swain, S.K. Swain, C. Singha, Status mapping of landuse/land cover change of forest resources. J. Energy Res. Environ. Technol. **4**(3), 234–239 (2017)
3. S.K. Swain, K.C. Swain, C. Singha, Forest cover change mapping using harmonic-land trendr models in google earth engine. Sylwan **164**(8), 472 (2020). http://www.sylwan.ibles.org/artykul.html
4. R. Dubayah, J.B. Blair, S. Goetz, L. Fatoyinbo, M. Hansen, S. Healey, M. Hofton, G. Hurtt, J. Kellner, S. Luthcke et al., The global ecosystem dynamics investigation: high-resolution laser ranging of the earth's forests and topography. Sci. Remote Sens. **1**, 100002 (2020)
5. P. Potapov, A. Tyukavina, S. Turubanova, Y. Talero, A. Hernandez-Serna, M.C. Hansen, D. Saah, K. Tenneson, A. Poortinga, A. Aekakkararungroj et al., Annual continuous fields of woody vegetation structure in the Lower Mekong region from 2000–2017 Landsat time-series. Remote Sens. Environ. **232**, 111278 (2019)
6. N. Lang, W. Jetz, K. Schindler, J.D. Wegner, A high-resolution canopy height model of the Earth. Comput. Vis. Pattern Recogn. (2022). https://doi.org/10.48550/arXiv.2204.08322
7. I. Fayad, N. Baghdadi, J.S. Bailly, N. Barbier, V. Gond, B. Hérault, M. El Hajj, F. Fabre, J. Perrin, Regional scale rain-forest height mapping using regression-kriging of spaceborne and airborne LiDAR data: application on French Guiana. Remote Sens. **8**(3), 240 (2016)
8. C.A. Silva, S. Saatchi, M. Garcia, N. Labriere, C. Klauberg, A. Ferraz, V. Meyer, K.J. Jeffery, K. Abernethy, L. White, K. Zhao, Comparison of small-and large-footprint lidar characterization of tropical forest aboveground structure and biomass: a case study from Central Gabon. IEEE J. Select. Top. Appl. Earth Observ. Remote Sens. **11**(10), 3512–3526 (2018)
9. S. Vafaei, J. Soosani, K. Adeli, H. Fadaei, H. Naghavi, T.D. Pham, D. Tien Bui, Improving accuracy estimation of forest aboveground biomass based on incorporation of ALOS-2 PALSAR-2 and Sentinel-2A imagery and machine learning: a case study of the Hyrcanian forest area (Iran). Remote Sens. **10**, 172 (2018). https://doi.org/10.3390/rs10020172
10. S.P. Healey, Z. Yang, N. Gorelick, S. Ilyushchenko, Highly local model calibration with a new GEDI LiDAR asset on google earth engine reduces landsat forest height signal saturation. Remote Sens. **12**, 2840 (2020). https://doi.org/10.3390/rs10020172
11. L. Chen, Y. Wang, C. Ren, B. Zhang, Z. Wang, Optimal combination of predictors and algorithms for forest above-ground biomass mapping from sentinel and SRTM data. Remote Sens. **11**(4), 414 (2019). https://doi.org/10.3390/rs11040414
12. M. Pourshamsi, J. Xia, N. Yokoya, M. Garcia, M. Lavalle, E. Pottier, H. Balzter, Tropical forest canopy height estimation from combined polarimetric SAR and LiDAR using machine-learning. ISPRS J. Photogramm. Remote. Sens. **172**, 79–94 (2021). https://doi.org/10.1016/j.isprsjprs.2020.11.008
13. L. Duncanson, J.R. Kellner, J. Armston, R. Dubayah, D.M. Minor, S. Hancock, S.P. Healey, P.L. Patterson, S. Saarela, S. Marselis et al., Aboveground biomass density models for NASA's global ecosystem dynamics investigation (GEDI) lidar mission. Remote Sens. Environ. **270**, 112845 (2022)

14. R. Dubayah, J.B. Blair, S. Goetz, L. Fatoyinbo, M. Hansen, S. Healey, M. Hofton, G. Hurtt, J. Kellner, S. Luthcke et al., The global ecosystem dynamics investigation: high-resolution laser ranging of the earth's forests and topography. Sci. Remote Sens. **1**, 100002 (2020)
15. N. Gorelick, M. Hancher, M. Dixon, S. Ilyushchenko, D. Thau, R. Moore, Google earth engine: planetary-scale geospatial analysis for everyone. Remote Sens. Environ. **202**, 18–27 (2017)

AI-Driven Environmental Monitoring Using Google Earth Engine

Anubhava Srivastava and Himanshu Sharma

Abstract Environmental monitoring is critical for assessing and managing ecosystem health, tracking climate change consequences, and protecting human well-being. Traditional monitoring approaches frequently encounter issues such as limited spatial coverage, high prices, and labor intensiveness. This article investigates the potential of AI-powered environmental monitoring as a transformative solution to these concerns. AI technologies, such as machine learning and computer vision, provide novel approaches to automate data collection, analysis, and interpretation from a variety of environmental sensors and sources, including satellite photography, drones, and ground-based sensors. We discuss recent advances in AI-powered environmental monitoring applications in a variety of areas, including air and water quality assessment, biodiversity monitoring, deforestation detection, and urban heat island mapping. Data preprocessing, feature extraction, model training, and validation are key components of AI-driven environmental monitoring systems that are thoroughly described. Furthermore, the abstract includes case examples that demonstrate the successful implementation of AI algorithms for real-time monitoring and early warning systems in various ecological scenarios. Data quality, model interpretability, scalability, and ethical concerns in using AI for environmental monitoring are all addressed. Strategies for increasing transparency, accountability, and stakeholder participation in AI-powered monitoring frameworks are offered.

Keywords Google Earth Engine · Classification · Machine learning

A. Srivastava · H. Sharma (✉)
Computer Science and Engineering Department, Sharda University, Greater Noida, India
e-mail: himanshugbpuat@gmail.com

A. Srivastava
e-mail: pgi18001@rgipt.ac.in

B. Pradhan and S. Mukhopadhyay (eds.), *IoT Sensors, ML, AI and XAI: Empowering A Smarter World*, Smart Sensors, Measurement and Instrumentation 50,
https://doi.org/10.1007/978-3-031-68602-3_19

1 Introduction

Emerging technologies have substantially expanded the field of environmental monitoring, with artificial intelligence (AI) playing a critical role in data collecting and processing. One significant platform that uses AI for environmental monitoring is Google Earth Engine (GEE) [1], a cloud-based geospatial analytic platform that allows for large-scale data processing of satellite imagery and other geospatial information. AI-powered environmental monitoring allows for real-time analysis and decision-making in a variety of areas, including air quality, water quality, biodiversity, urbanization, and climate-related phenomena. In air quality monitoring, AI algorithms analyze satellite and sensor data to detect pollutants such as PM2.5 and NO_2 [2], providing insights into health hazards and aiding pollution reduction efforts. AI-driven analysis of satellite data detects changes in water turbidity and chlorophyll [3] concentrations, which aids in aquatic ecosystem management. In biodiversity monitoring, AI models examine habitat attributes and species distributions from remote sensing data to detect changes in biodiversity hotspots and inform conservation strategies. In urbanization monitoring, AI tools detect changes in urban growth patterns and land cover over time, assisting with urban planning and infrastructure development. Furthermore, AI-driven climate monitoring evaluates climate factors and forecasts extreme weather events, hence improving disaster preparedness and adaption tactics. These applications highlight AI's transformative potential for proactive environmental management and sustainable development. This introduction gives an overview of how AI-powered environmental monitoring [4], particularly with Google Earth Engine, is transforming our ability to identify and respond to environmental concerns.

Google Earth Engine gives academics and practitioners access to a massive collection of satellite imagery and environmental datasets, allowing them to use AI algorithms for tasks like land cover classification, forest change detection, and water quality evaluation. The platform's scalability and processing capability make it perfect for developing machine learning models capable of automatically analyzing complex environmental phenomena at various spatial and temporal scales. Mapping forest decline, flood events, and urban expansion in real time with Google Earth Engine (GEE) requires a sophisticated and comprehensive strategy to evaluating large-scale geospatial data. GEE provides access to a wide range of satellite imagery, including Landsat and Sentinel-2, that can be processed using time-series analysis techniques such as vegetation indices (e.g., NDVI) [5] and machine learning algorithms (e.g., random forest) to detect changes in forest cover over time. The creation of unique change detection algorithms within GEE enables the automatic detection of deforestation and forest degradation. Flood mapping [6] in GEE uses radar (e.g., Sentinel-1) and optical satellite data to monitor changes in surface water extent, which are then integrated with flood detection methods such as supervised classification and thresholding to accurately identify flood zones. The integration of hydrological models into GEE allows for the simulation of flood dynamics as well as the prediction of possible flood events using precipitation and topographic data. For urban expansion mapping, GEE supports land use/land cover classification with machine learning models (e.g., support vector machines, convolutional neural networks) trained on labeled datasets to recognize urban areas and quantify changes in urban extent over time. GEE's time-series analysis of satellite imagery makes it easier to identify urban growth trends and assess the effects of urbanization on the surrounding terrain. Google Earth Engine's scale, automation, and accessibility enable researchers, governments, and organizations to monitor environmental changes in real time, allowing for more informed decision-making in sustainable land management, disaster preparedness, and urban planning.

2 Related Work

AI-driven environmental monitoring has received a lot of attention in recent years, with multiple studies demonstrating the effectiveness of artificial intelligence [7] (AI) algorithms in processing geospatial data for environmental assessment and management. In the context of air quality monitoring, researchers used AI algorithms to analyze satellite imagery and ground-based sensor data. For example, [8] used convolutional neural networks (CNNs) on the Google Earth Engine platform to estimate PM2.5 concentrations from satellite images, giving a low-cost way for monitoring air quality at regional scales. Similarly, [3] used GEE to create a deep learning model to estimate air pollution concentrations, demonstrating the power of AI-driven approaches to air quality prediction and management.

AI algorithms have been used to evaluate multispectral satellite data and detect pollutants in water. Zurqani et al. [9] used machine learning methods within GEE to map water turbidity and suspended sediment concentrations in river systems, allowing for continuous monitoring of water quality dynamics. Furthermore, [10] used GEE and Landsat data to analyze water quality metrics such as chlorophyll-a concentrations, indicating the efficacy of AI-driven methodologies for freshwater ecosystem monitoring. To monitor biodiversity, researchers used AI and remote sensing data to examine habitat attributes and species distribution patterns. Franklin et al. (2020) used GEE to create habitat suitability models for endangered species based on environmental variables obtained from satellite data, thereby facilitating conservation planning and management. Karagiannopoulou et al. [11] used deep learning algorithms in GEE to identify and categorize wildlife species in satellite data, demonstrating the potential of AI-driven methodologies for large-scale biodiversity surveys.

AI has helped map urban heat islands (UHIs) by analyzing thermal infrared photography and urban land cover data. Yağcı and İşcan [12] employed GEE and machine learning models to map UHI intensity and identify characteristics that contribute to urban heat islands, resulting in insights for urban planning and climate adaptation. Similarly, Abdullah et al. [13] used AI approaches within GEE to examine spatiotemporal variations in UHI patterns and their effects on public health, emphasizing the necessity of AI-powered environmental monitoring in solving urban climate concerns. In the field of forest loss mapping, researchers used GEE's processing power to detect deforestation and forest degradation. Yang et al. [14] pioneered the development of the worldwide Forest Change dataset using GEE, which provides annual updates on worldwide forest cover loss and gain based on Landsat imagery. Niraj et al. [15] used GEE's machine learning algorithms to detect and monitor illicit logging in tropical forests, demonstrating GEE's effectiveness for real-time forest monitoring and conservation. Flood mapping studies have used GEE's access to radar and optical satellite data to analyze flood extent and dynamics. Islam et al. [16] used GEE to map flood inundation in Bangladesh, combining remote sensing data with hydrological modeling to improve flood prediction and disaster management. Sinha and Sinha [17] used GEE for near-real-time flood mapping in metropolitan areas, demonstrating the platform's value in urban flood risk assessment and response planning Tassi and Vizzari [18]. In the context of urban expansion mapping, researchers have used GEE's capabilities to track land use changes and urban growth patterns. Yan et al. [19] created deep learning models within GEE to automatically map urban extent and expansion with Sentinel-2 images, proving the efficacy of AI-driven methodologies for large-scale urban mapping. Srivastava and Biswas [20], Srivastava et al. [21] used GEE to examine spatiotemporal patterns of urban expansion, revealing insights into urbanization dynamics and their effects on ecosystems and infrastructure.

3 Data and Method

Chamoli district, located in the northern regions of Uttarakhand, India, has a large geographical area ranging from latitudes 30.22° N–31.18° N and longitudes 78.22° E–80.06° E, displaying a diversified natural and environmental landscape typical of the Indian Himalaya.

The district's biological structure includes a gradient of vegetation zones that correlate to different altitudes. Chamoli's lower elevations are home to rich subtropical woods characterized by sal, teak, and bamboo. As elevation rises to temperate zones, forests begin to include oak, pine, and bright rhododendron species. As you ascend, you'll see subalpine and alpine meadows with juniper, birch, and dwarf rhododendron growing among rocky terrain and snow-capped peaks, including famed Himalayan summits like Nanda Devi and Trishul. Chamoli's environmental structure is influenced by fundamental geological processes, making it vulnerable to natural hazards including landslides and flash floods, which are exacerbated by climate change effects like as glacier retreat and changing precipitation patterns. The district's proximity to glacier-fed rivers such as the Alaknanda and Mandakini makes it vulnerable to recurrent floods caused by glacial lake outburst floods (GLOFs) and heavy monsoon rains. Furthermore, Chamoli has seen forest fires during the dry season, harming its unique ecology and creating challenges to biodiversity conservation efforts in the midst of current sustainable development initiatives. The satellite imagery datasets, specifically Landsat and Sentinel-2, within Google Earth Engine

(GEE), to quantify forest loss due to fire and flood events in the Chamoli district. These satellite datasets cover the study area and key time periods, assuring radiometric and geometric accuracy. To detect fire-affected areas, the Differenced Normalized Burn Ratio (dNBR) technique is used, which computes the difference between pre-fire and post-fire Normalized Difference Vegetation Index (NDVI) values. This calculation identifies changes in vegetation health that are suggestive of fire impacts. Similarly, flood-affected areas are identified using the Normalized Difference Water Index (NDWI), which uses near-infrared and green band reflectance to identify locations with higher water presence due to flood occurrences. A systematic time-series analysis of vegetation indices, including NDVI, is performed to quantitatively characterize forest cover dynamics before and after disturbances induced by fire and flood. Based on the observed disturbances, specific study zones within Chamoli district are delimited, and forest loss indicators such as total affected area and percentage of vegetation cover loss are calculated using GEE spatial analytic tools. Spatial masking techniques are used to narrow the research to areas of interest where forest loss has occurred due to fire or flood impacts.

4 Result

The use of AI-driven environmental monitoring using Google Earth Engine (GEE) in Chamoli district has yielded valuable insights into landscape dynamics and ecosystem changes over time. This study used satellite imagery analysis and machine learning techniques to provide a complete assessment of major environmental characteristics such as air quality, water quality, forest cover, urbanization, and climate-related events. The findings demonstrate the effectiveness of AI approaches in generating actionable insights for sustainable land management and conservation activities. A thorough study of the forest cover changes in Chamoli district from 2001 to 2023 was carried by utilizing satellite imaging datasets from Google Earth Engine. Chamoli lost a net of 383 hectares (ha) of tree cover during this time period, representing a 0.20% drop since 2000. However, it is worth noting that between 2000 to 2020, Chamoli added 3.03 thousand hectares (kha) of tree cover across the region, accounting for 7.7% of total tree cover gain reported in Uttarakhand over the same period.

Further investigation found that Chamoli lost 95 ha of tree cover from 2001 to 2022, primarily owing to fires, and an additional 260 ha due to other loss factors such as deforestation and land-use changes as shown in Fig.1. The year 2009 saw the most tree cover loss due to fires over this time period, with 25 ha lost to flames, accounting for 47% of total tree cover loss that year. Here Fig. 2 shows total tree cover loss over study area and Fig. 3 shows total tree cover loss and loss due to fire incident happen between year 2001–2023.

Further analysis of tree cover loss patterns reveals unique drivers of change. Between 2001 and 2023, Chamoli lost 95 ha of tree cover owing to fires, and an additional 260 ha due to other factors such as deforestation and land-use changes.

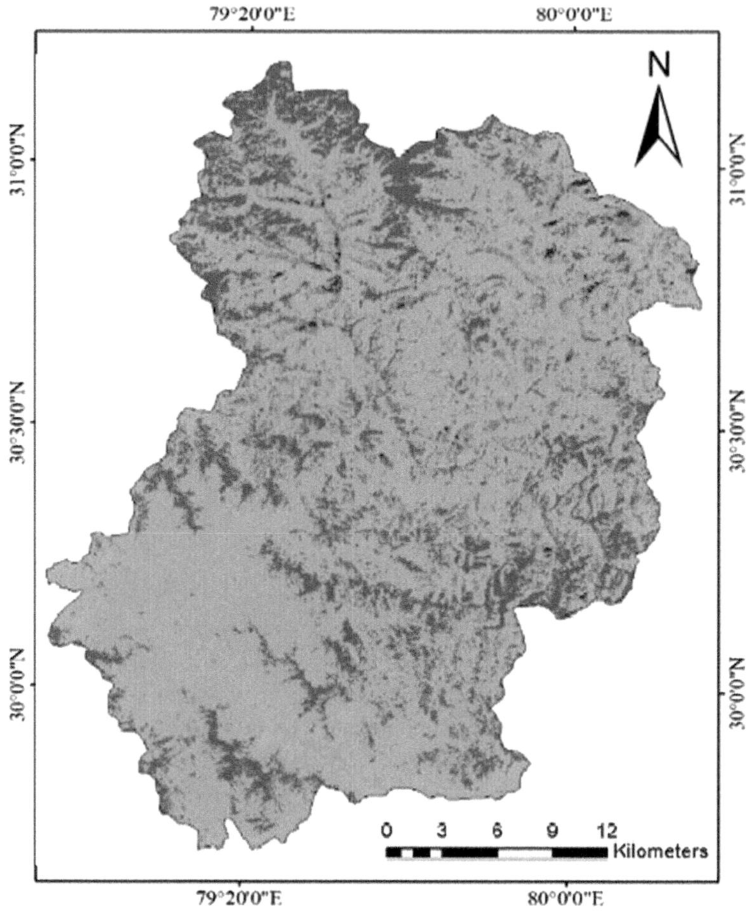

Fig. 1 Geographical location of study area

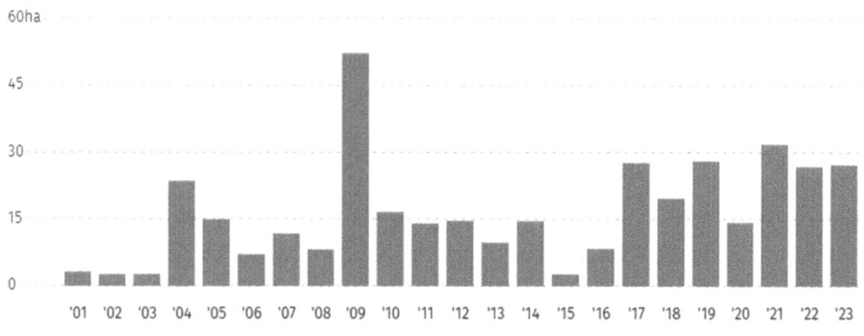

Fig. 2 Tree cover loss over the study area from year (2001–2023)

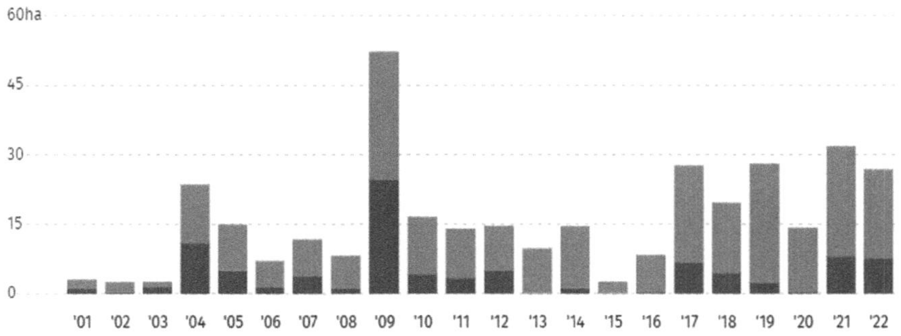

Fig. 3 Tree cover loss due to fire over the study area from year (2001–2023)

Notably, the year 2009 saw the most substantial tree cover loss due to fires, totaling 25 ha, or 47% of all tree cover loss documented that year.

From 2000 to 2023, Chamoli experienced a net change of 1.22 thousand hectares (kha) in tree cover, equivalent to a 0.47% change signifies the observed fluctuation in forested areas within Chamoli district over the specified timeframe, as assessed through satellite-based analysis and environmental monitoring. This net change represents the cumulative difference between the total tree cover extent at the beginning of 2000 and the tree cover extent at the end of 2023. The calculation accounts for both gains and losses in tree cover across the landscape.

The positive net change of 1.22 kha (1,220 ha) indicates an overall increase in tree cover within Chamoli district over the 23-year period. This increase, while relatively modest in percentage terms (0.47% change), underscores positive trends in forest recovery, afforestation efforts, and natural regeneration processes. Positive shifts in tree cover can result from ecological restoration initiatives, sustainable forestry practices, and land management strategies aimed at enhancing forest resilience and biodiversity as shown in Fig. 4.

The peak fire season in the study area normally begins in early January and lasts about four weeks. Notably, from January 1st to January 22nd, 2024, there were four VIIRS fire alerts registered, with only high confidence alerts included. This figure is abnormally high when compared to previous years, dating back to 2012. Chamoli has received 7 VIIRS fire alerts with high confidence so far in 2024, which is consistent with historical norms since 2012. In 2012, there were 81 VIIRS fire alerts, the most recorded number in a single year.

Recent monitoring efforts between April 14th and April 21st, 2024, recorded 39 VIIRS fire alerts in Chamoli, with 2.6% classed as high confidence alerts. Furthermore, between April 10th and April 17th, 2024, 41 deforestation alarms were registered, totaling less than one hectare. None of these deforestation notifications were classified as high confidence alerts by a single system, and none were highlighted by multiple detection systems.

Between April 19th, 2021 and April 15th, 2024, Chamoli district received a total of 408 VIIRS fire warnings shown in Fig. 5, highlighting the continuous importance

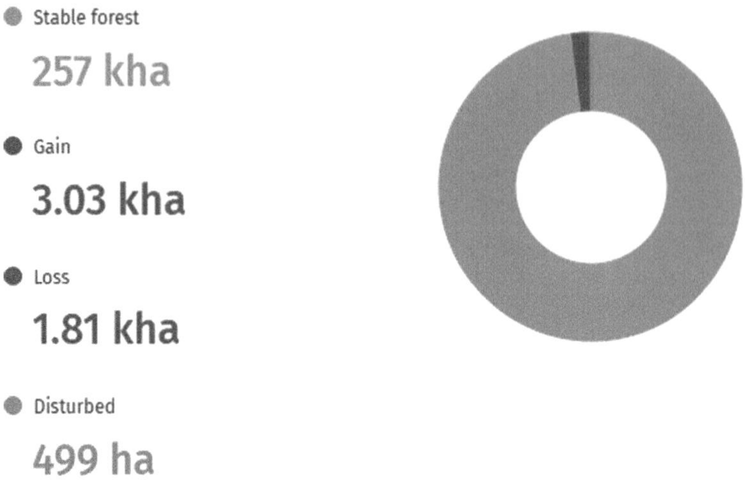

Fig. 4 Net change in forest computed over the study area (2001–2024)

of fire-related incidents in the region. These findings, based on satellite-based monitoring with the VIIRS system, provide critical data for assessing fire and deforestation trends in Chamoli, helping to inform environmental management and conservation strategies aimed at mitigating landscape disturbances and promoting ecosystem resilience.

From 2000 to 2023, Chamoli district saw a significant increase in tree cover of 3.03 thousand hectares (kha), accounting for approximately 7.7% of the total tree cover growth reported in Uttarakhand during this time period. This rise in tree cover indicates favorable trends in forest regeneration, afforestation activities, and natural vegetation rebound, which all contribute to increased carbon sequestration and ecological resilience. Between 2001 and 2023, Chamoli's woods experienced complicated carbon dynamics, including CO_2e emissions and removals. During this

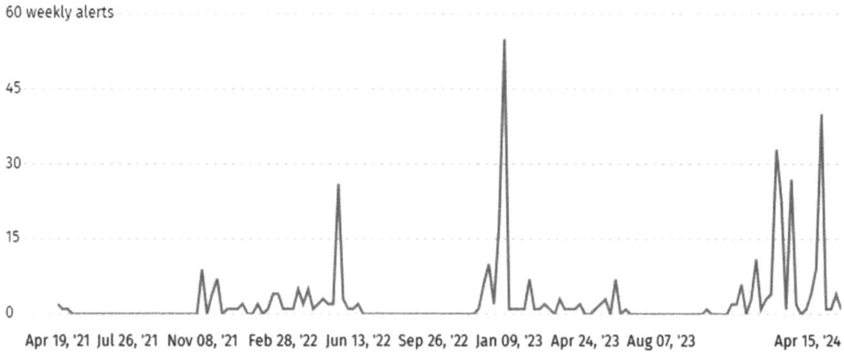

Fig. 5 Fire incident happens over the study area from April 2019 to April 2024

period, the woods emitted around 7.39 thousand metric tons of CO_2e each year owing to tree cover loss, while also absorbing a significant quantity of CO_2e, resulting in a net carbon sink of −624 thousand metric tons per year. Despite disturbances, Chamoli forests acted as a substantial carbon sink, absorbing more CO_2e than they released. From 2001 to 2023, tree cover reduction in Chamoli released an average of 7.39 thousand metric tons of CO_2e per year, totaling 163 thousand metric tons of CO_2e. Understanding these carbon dynamics is critical for determining the function of forests in climate regulation and carbon balance, emphasizing the need of sustainable forest management techniques in preserving and improving carbon sequestration capacity. In 2000, Chamoli had an aboveground live woody biomass density of 249 metric tons per hectare (t/ha), resulting in a total aboveground biomass of 48.0 million metric tons. This biomass serves as a vital carbon storage, highlighting the importance of healthy forest ecosystems in sequestering atmospheric carbon and buffering the effects of climate change.

5 Conclusion

Google Earth Engine makes it easier to integrate AI algorithms for satellite picture processing, allowing for a wide range of environmental monitoring applications. Land cover classification and change detection are examples of tasks that can be solved using machine learning models. Convolutional neural networks, for example, can accurately distinguish between different types of land cover, enabling for the monitoring of deforestation, urbanization, and natural disasters. Google Earth Engine, by using the power of AI, can effectively process massive amounts of data and deliver useful insights into environmental changes that would be impossible to examine manually.

The use of AI-driven environmental monitoring with Google Earth Engine provides numerous significant advantages. For starters, it improves the accuracy and efficiency of analyzing enormous amounts of environmental data, allowing researchers to spot trends and patterns that would otherwise go undetected. Real-time monitoring enables prompt reactions to environmental problems like illegal logging or urban encroachment. Furthermore, the cost-effectiveness of AI-driven solutions allows for continual monitoring of ecological systems, which promotes sustainable resource management methods.

Despite its benefits, AI-powered environmental monitoring using Google Earth Engine is not without hurdles and restrictions. Data quality concerns, such as cloud cover and insufficient resolution in satellite imagery, might have an impact on AI algorithms' accuracy and perception of environmental changes. Furthermore, complex environmental phenomena, such as the impact of climate change on ecosystems, provide hurdles for AI models in providing useful insights. To ensure the proper use of technology in environmental monitoring, ethical concerns such as data privacy and potential biases in AI systems must be addressed. To summarize, the integration of AI algorithms with Google Earth Engine has altered the landscape of environmental monitoring, providing unparalleled possibilities for evaluating and understanding our changing planet. While the advantages of AI-driven techniques are significant, it is critical to recognize and address the obstacles and limitations associated with this approach. Researchers and politicians can make educated decisions to conserve and maintain our ecosystem for future generations by combining the power of artificial intelligence with Google Earth Engine.

References

1. N. Gorelick, M. Hancher, M. Dixon, S. Ilyushchenko, D. Thau, R. Moore, Google earth engine: planetary-scale geospatial analysis for everyone. Remote Sens. Environ. **202**, 18–27 (2017). https://doi.org/10.1016/j.rse.2017.06.031
2. A. Novo, N. Fariñas-álvarez, J. Martínez-Sánchez, H. González-Jorge, J.M. Fernández-Alonso, H. Lorenzo, Mapping forest fire risk—a case study in Galicia (Spain). Remote Sens. **12**(22), 1–21 (2020). https://doi.org/10.3390/rs12223705
3. M.A. Brovelli, Y. Sun, V. Yordanov, Monitoring forest change in the amazon using multi-temporal remote sensing data and machine learning classification on Google Earth Engine, 1–21 (2020)
4. A. Srivastava, S. Umrao, S. Biswas, R. dubey, M.I. Zafar, FCCC: Forest Cover Change Calculator user ınterface for ıdentifying fire ıncidents in forest region using satellite data. Int. J. Adv. Comput. Sci. Appl. **14**(7), pp. 948–959 (2023). https://doi.org/10.14569/IJACSA.2023.01407103
5. A. Srivastava, S. Umrao, S. Biswas, Exploring forest transformation by analyzing spatial-temporal attributes of vegetation using vegetation ındices **14**(5) (2023)
6. Y. Wu et al., Landslide susceptibility assessment using frequency ratio, statistical index and certainty factor models for the Gangu County, China. Arab. J. Geosci. **9**(2), 1–16 (2016). https://doi.org/10.1007/s12517-015-2112-0
7. A. Srivastava, S. Bharadwaj, R. Dubey, V.B. Sharma, Mappıng vegetatıon and measurıng the performance of machıne learnıng algorıthm ın lulc classıfıcatıon ın the large area usıng sentınel-2 and landsat-8 datasets of dehradun as a test case. Mappıng vegetatıon and measurıng the performance of machıne sentınel-2 (2022). https://doi.org/10.5194/isprs-archives-XLIII-B3-2022-529-2022
8. R. Tombe, S. Viriri, Remote sensing image scene classification: advances and open challenges. Geomatics **3**(1), 137–155 (2023). https://doi.org/10.3390/geomatics3010007
9. H.A. Zurqani, J.S. Allen, C.J. Post, C.A. Pellett, T.C. Walker, Mapping and quantifying agricultural irrigation in heterogeneous landscapes using Google Earth Engine. Remote Sens. Appl. Soc. Environ. **23**, 100590 (2021). https://doi.org/10.1016/j.rsase.2021.100590
10. B. Kalantar, N. Ueda, M.O. Idrees, S. Janizadeh, K. Ahmadi, F. Shabani, Forest fire susceptibility prediction based on machine learning models with resampling algorithms on remote sensing data. Remote Sens. **12**(22), 1–24 (2020). https://doi.org/10.3390/rs12223682
11. A. Karagiannopoulou, A. Tsertou, G. Tsimiklis, A. Amditis, Data fusion in earth observation and the role of citizen as a sensor: a scoping review of applications, methods and future trends. Remote Sens. **14**(5) (2022). https://doi.org/10.3390/rs14051263
12. C. Yağcı, F. İşcan, Turkish journal of geographic ınformation systems GIS-based site suitability analysis of afforestation in Konya province, Turkey. Turkish J. Geogr. Inf. Syst. **3**(2), 89–95 (2021). https://dergipark.org.tr/en/pub/tucbis
13. A.Y. Abdullah, A. Masrur, M. Sarfaraz, G. Adnan, *Spatio-Temporal Patterns of Land Use/Land Cover Change in the Heterogeneous Coastal Region of Bangladesh between 1990 and 2017* (2017). https://doi.org/10.3390/rs11070790

14. X.X. Yang et al., A 30-m landsat-derived cropland extent product of Australia and China using random forest machine learning algorithm on Google Earth Engine cloud computing platform. GISci. Remote Sens. **57**(February), 85–91 (2020). https://doi.org/10.1016/j.isprsjprs.2018.07.017
15. K.C. Niraj, A. Singh, D.P. Shukla, Effect of the Normalized Difference Vegetation Index (NDVI) on GIS-enabled bivariate and multivariate statistical models for landslide susceptibility mapping. J. Ind. Soc. Remote Sens. **51**(8), 1739–1756 (2023). https://doi.org/10.1007/s12524-023-01738-5
16. K. Islam, M. Jashimuddin, B. Nath, T.K. Nath, Land use classification and change detection by using multi-temporal remotely sensed imagery: the case of Chunati wildlife sanctuary, Bangladesh. Egypt. J. Remote Sens. Sp. Sci. **21**(1), 37–47 (2018). https://doi.org/10.1016/j.ejrs.2016.12.005
17. S. Sinha, R. Sinha, Geomorphic evolution of Dehra Dun, NW Himalaya: tectonics and climatic coupling. Geomorphology **266**, 20–32 (2016). https://doi.org/10.1016/j.geomorph.2016.05.002
18. A. Tassi, M. Vizzari, Object-oriented lulc classification in google earth engine combining snic, glcm, and machine learning algorithms. Remote Sens. **12**(22), 1–17 (2020). https://doi.org/10.3390/rs12223776
19. Y. Yan, H. Lei, Y. Chen, B. Zhou, Article analyzing the dynamic spatiotemporal changes in urban extension across Zhejiang Province using NPP-VIIRS nighttime light data. Remote Sens. **14**(13) 2022. https://doi.org/10.3390/rs14133212
20. A. Srivastava, S. Biswas, Analyzing land cover changes over landsat-7 data using Google Earth Engine, in *Proceedings of 3rd International Conference on Artifical Intelligence and Smart Energy, ICAIS* (2023), pp. 1228–1233. https://doi.org/10.1109/ICAIS56108.2023.10073795
21. A. Srivastava, R. Dubey, S. Biswas, Comparison of sentinel and landsat data sets over Lucknow region using gradient tree boost supervised classifier. Lect. Notes Netw. Syst. **730**, 221–232 (2023). LNNS. https://doi.org/10.1007/978-981-99-3963-3_18

A Fusion Approach for Kannada Speech Recognition Using Audio and Visual Cue

R. Shashidhar, M. P. Shashank, G. Jagadamba, and Vinayakumar Ravi

Abstract The act of conveying thoughts and ideas to another person without using words or visible facial expressions is known as communication. It is a vital component of human existence since it enables interaction with the outside world. An audiovisual speech recognition system that uses image processing to aid speech recognition systems in the lip-reading process is proposed in this work. This helps the deaf and hearing-impaired by identifying the word using the Kannada audio database, a Dravidian language spoken by more than 60 million people in the state of Karnataka. Three components make up the proposed speech recognition system: the audio model, the visual model, and the fusion model. Mel-frequency Cepstrum coefficient is successfully used to extract audio information, and random forest employed a classification and achieved an accuracy of 86%. The feed-forward neural network was employed as a classifier and achieved a 57% accuracy rate. The Viola-Jones technique was utilised to extract the visual features. With the help of an artificial neural network and random forest algorithm, audio and visual fusion achieved an accuracy of 89%. It demonstrates that when the audio and the visual models are combined, the accuracy is higher than when each model is used alone.

Keywords Audiovisual speech recognition · Kannada · Random forest · Artificial neural network · Feed-forward neural network · Viola-Jones algorithm

R. Shashidhar · M. P. Shashank
Department of Electronics and Communication Engineering, JSS Science and Technology University, Mysuru 570006, India
e-mail: shashidhar.r@jssstuniv.in

M. P. Shashank
e-mail: shashankmp@jssstuniv.in

G. Jagadamba
Department of Information Science and Engineering, Siddaganga Institute of Technology, Tumakuru-03, India
e-mail: jagadamba@sit.ac.in

V. Ravi (✉)
Center for Artificial Intelligence, Prince Mohammad Bin Fahd University, Khobar, Saudi Arabia
e-mail: vravi@pmu.edu.sa; vinayakumarr77@gmail.com

B. Pradhan and S. Mukhopadhyay (eds.), *IoT Sensors, ML, AI and XAI: Empowering A Smarter World*, Smart Sensors, Measurement and Instrumentation 50,
https://doi.org/10.1007/978-3-031-68602-3_20

1 Introduction

The development of audiovisual voice recognition systems has been a crucial research area in recent years, enabling computers to approach artificial intelligence and mimic human perception of words and things. As compared to other recognition systems now in use, audiovisual speech recognition is improving in effectiveness, functionality, and robustness, making it an essential component of man–machine interfaces. Lip reading is a highly essential system approach implemented in a recognition system where numerous techniques can be used to increase the recognition model's performance in current circumstances. If we look back at lip-reading history, the first work on the subject was proposed [1] in 1954. Petajan later received the concept from the University of Illinois which became a popular method in the 1980s. Following that, other studies in the field of lip reading were conducted [2, 3].

In 1989, a pixel-based approach was introduced to detect lip reading using an Artificial Neural Network (ANN) [4] because audio signals are susceptible to noise. Hidden Markov Models (HMMs) were first utilized in the lip reading system for the sentence identification in 1993 [5], with a success rate of roughly 25%.

There are a few procedures to take here; the first is to recognize the speaker's face and the lip region which of higher interest. The second step to be followed is to reduce the image extracted data and then to extract the lip feature. The third step involves the classification of the visual data that is derived from the retrieved lip features using a high-efficiency classifier.

Face detection, which has a very broad variety of applications, is the first stage in many face-oriented technologies, such as face identification and recognition. The position of the entries and the sizes of each entity in a photograph are calculated using the object class identification technique known as face detection. The primary emphasis of face detection algorithms is the human face. Similar to image recognition, the face detection algorithm compares each component of the image. The comparison will be invalid if any features connected to the face change in the database. The extraction of the lip area is the most crucial area of interest in improving recognition rates. Many approaches for obtaining a facial picture from a face have been proposed. The object-oriented Viola-Jones algorithm converts the image to greyscale since there is less information to process. After a face is detected using a grayscale image, the Viola-Jones technique pinpoints the location of every coloured image. In fewer steps, several boxes would be able to recognise Haar-like characteristics, or features that are comparable to faces. The information from those indicated boxes aids in determining the exact location of the face inside an image.

The necessary features of the lips for visual recognition were retrieved using the Viola Jones algorithm. Some of these components extracted were: brightness, contrast, correlation, entropy, mean, variance, and standard deviation. The Feed Forward Neural Network (FFNN) was used to train and validate these retrieved features. Likewise, audio characteristics like Mel frequency Cepstral Coefficient (MFCC) and Zero Crossing Rate (ZCR) were effectively extracted, trained, and evaluated using Random Forest (RF). RF has been used for audio processing, whereas

FFNN has been used for visual processing. Models that combine two or more methodologies to understand and evaluate data are known as a kind of hybrid model. These models are utilized because they provide more detailed findings. A hybrid model that combines RF and ANN was used for audio and visual information integration.

Mohammad Mortazavi proposed the augmentation reality and audio visual speech recognition methods for 100 deaf people and got the accuracy of 80% [6]. It was necessary to operate the current Audiovisual Speech Recognition (AVSR) system in a set of very constrained and tightly controlled circumstances. The current AVSR technology operates in the acoustic channel in an effort to provide a higher quality output. By employing the retrieved visual cues from the mouth region, which largely concentrate on the speaker's lips independently for a higher recognition rate, alternative approaches are achieved here with some success. The suggested approach uses lip characteristics to perform Visual Speech Recognition (VSR) for Indian languages. ANN classifier with the Gray Level Co-occurrence Matrix (GLCM) algorithm for precise classification of visual characteristics from the input video. The proposed method's accuracy rate was 90% [5]. When the Maximum Likelihood Linear Regression (MLLR) based noise modification was performed to the audio HMM, the improvement was confirmed by employing visual elements. At 20 dB, the overall accuracy was 87% [7]. They recognized Myanmar consonants by locating the lip region. This technique can be used on lip motion sequences in a video input stream, as shown by the system. A total accuracy rate of 87.6% was attained using the CIELAB colour transformation technique, Moore neighbourhood tracing approach, and the SVM classifier [8].

Using Moore Neighborhood tracing and the linear SVM classifier, Thein proposed lip reading for Myanmar consonants [9]. The Visual Geometry Group Network produced the feature vectors that were used in the Fusion model's various Long Short Term Memory (LSTM) layers (VGGNet) as input to handle variable-length sequences. The level of precision attained was 76%. A unique visual feature extraction strategy for a VSR system employing the Convolutional Neural Network (CNN) was proposed. It illustrates the supervised learning methodology used to separate 40 phonemes spoken by six different speakers from a video stream using a mouth area picture sequence, with a recognition accuracy of 58%. The ability of a CNN to triumph in the Large Scale Visual Recognition Challenge (ILSVRC) competition serves as evidence of its generalisation potential. As a result, the system achieved a 60% accuracy [10].

Two HMM-based visual models were used to test and match the features: the traditional Gaussian Mixture Model (GMM)-HMM and the new Deep Neural Network (DNN)-HMM hybrid. The system's accuracy is 79.6% [11]. The word level is where the asynchrony between the audio and visual streams is provided. The Multi Stream-Asynchrony Deep Belief Net (MS-ADBN) model and the Multi Stream HMM model were contrasted by the author because the MS ADBN model significantly loses the audio visual stream asynchrony constraint. The overall accuracy for the Signal-to-Noise Ratio (SNR) ranges from 0 to 15 dB is 75% based on the MS ADBN model [12]. In many noisy environments and system architectures, dynamic weighing systems are used. The HMM-ANNs model, which has an accuracy of 87% and is primarily based on the change of probabilities shared by both the HMM and ANNs architectures, is used to create the system. For high and medium SNR, accuracy is roughly 72% [13]. This method can be applied to a variety of applications, including forensics, human–machine interface, security systems, and biometric authentication systems. This study develops both a speaker identification system that employs a Random Forest classifier algorithm to distinguish between different speakers and another speaker identification method that does the same [14]. The results of the experiment indicate that the CNN-RF model outperforms the conventional CNN model. Second, CNN RF models were integrated into the robot and Nao's record sound command box was upgraded. After a rigorous test, the proposed CNN-RF model successfully enables the Nao robot's basic speech-emotion detection capabilities [15].

They work in a way that is more like the human brain than traditional computer logic. This study shows different forms of Artificial Neural Networks that are briefly addressed, it can be stated that Recurrent Neural Network (RNN) obtains higher voice recognition rates than Multi-Layer Perceptron (MLP), but the training procedure is more sophisticated and dynamically sensitive, which can cause issues [16, 17]. Experiments were done utilizing spoken utterances from 30 speakers to validate the system's performance (20 males and 10 females). When using training speech utterances, the system achieved average recognition rates of up to 77% for 5-word utterances and 43% for 20-word utterances. For 20-word utterances, the unknown utterances led to an 18% recognition rate, and an 18% recognition rate was achieved with unknown utterances [18, 19]. The neural network has been recognized as the most effective classifier for emotion identification. CNN outperforms the decision tree by up to 72% in recognizing emotions, but the decision tree algorithm gives 38% and a 52% correctness for the smaller set of emotion [20, 21]. This work demonstrates how effective recurrent neural networks are in classifying voice signals. A small collection of characters could be recognized even with simplified representations. The performance of a recurrent neural network is superior to that of a multi-layer perceptron neural network, according to this article [22, 23].

The final aim of AVSR technology is to create a recognizer that can compete with humans in terms of performance. A neural network-based fusion method was presented to combine the two modalities' reliability metrics effectively and deliver noise-resistant identification performance under diverse noise situations [24, 25]. For the proposed work, accuracy of 76% was reported for the audio visual speech recognition system using a Feed Forward Neural network for a custom dataset [26, 27] and the recognition of the visual speech using VGG16 architecture for a specific custom dataset [28]. Various methods for lip reading and visual speech recognition are mentioned [29, 30]. The applications and, additionally major drawbacks of visual speech recognition are mentioned. Some researchers proposed deep learning models like HMM and the deep convolutional neural network to integrate audio and visual for English Languages [31–33]. Munaf proposed audio-visual emotion recognition for different subjects of different genders. The authors proposed the core principal component for analysis is feature extraction and Support Vector Machine. The results of visual speech recognition is 74.15%, audio speech recognition is 67.39% and audio-visual speech recognition got an accuracy of 80.27% [34]. Automatic speech recognition using the support vector machine. Authors proposed the appearance-based visual speech recognition [35]. Gangqiang discuss the latest lip reading works using machine learning and also discuss the datasets, feature extraction methods, and classification methods and compare the various methods [33].

The following is a list of the work's primary contributions:

(a) Establish a Kannada language database.
(b) Create an audio recognition algorithm.
(c) Create a visual recognition algorithm.
(d) Applying algorithms for integrated features while integrating both the audio and the visual.
(e) Integrated audio and visual speech using a hybrid algorithm.

The rest of the article goes in this sequence: The Sect. 1 explains the previous research works including the major drawbacks, Sect. 2 explains about the proposed methodology followed by Sect. 3 that explains the result of the method proposed and final Sect. 4 concludes with the proposed work with future scope.

2 Proposed Methodology

The two independent types of speech recognition—audio speech recognition and visual speech recognition—are combined in what is known as "audio-visual speech recognition." Fig. 1 depicts the method that was suggested for audio visual speech recognition.

The following procedures are included in the system.

Step 1: First, region of interest and face detection is done through the Viola-Jones algorithm.

Step 2: It includes audio feature extraction, which is done through MFCC, and training is done through an RF algorithm.

Step 3: Then, visual feature extraction is done where the lip extraction and object recognition are done through the Viola-Jones algorithm, and the Feed Forward Neural Network algorithm is used for training.

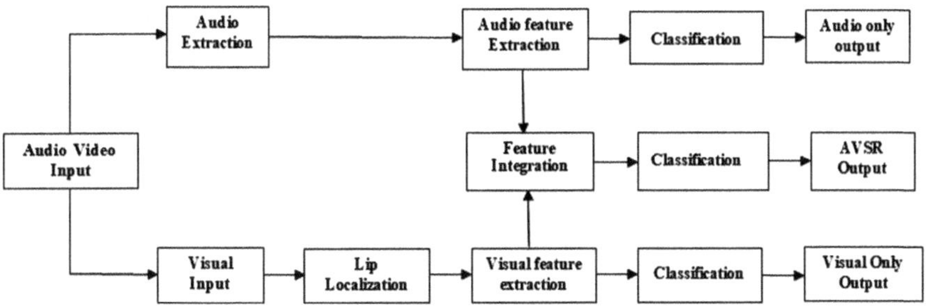

Fig. 1 Block diagram of the proposed model

Table 1 Features of the dataset collected

Parameters	Values
Resolution	1080 × 1920 p
Frames per second	60 FPS
Duration of video	1–1.20 PS
Storage size of video	10 MB

Step 4: Finally, the fusion model combines the audio and video features and provides the resultant model for speech recognition.

The dataset creation followed by the audio speech recognition and visual speech recognition, finally the integration of audio and visual speech recognition will be described in the following sub-sections.

2.1 Dataset Creation

Dataset is collected for the Kannada Words which has 1080 × 1920p resolution, 60 FPS frame rate, duration of the video is 1–1.20 s, and size of the video is around 10 MB. Table 1 shows the features of the dataset.

There are 400 samples altogether in this collection. 150 of the 250 samples in this dataset are used for testing, while the remaining 250 samples are used for training. Five male speakers and five female speakers between the ages of 18 and 30 were used to create this dataset. Five words are repeated five times by each speaker.

2.2 Audio Model

The implementation of the audio recognition part is done for the Kannada data set. Firstly, train the given data set by importing one.wav file, and the audio data is read. Preprocessing is the first phase in the process. It entails the elimination of undesirable elements, such as background noise. Audio feature extraction using the MFCC approach is the phase in which the preprocessed signals are given to the model and the feature vectors are generated.

The Fourier processed signal is passed through the Mel filter banks, a collection of band pass filters, to calculate the Mel spectrum. Mel is a unit of measurement based on how the human ear perceives frequency. It doesn't correspond linearly to the original frequency of the tone since the human auditory system doesn't seem to perceive pitch in a linear fashion. The Mel scale features linear frequency spacing below 1 kHz and logarithmic frequency spacing above 1 kHz. This is how Mel approximates physical frequency:

$$f_{Mel} = 2595 \log_{10}(1 + f/700) \tag{1}$$

Here, f gives physical frequency in Hz, fMel gives perceived frequency.

Classification of audio data is to be done by the random forest algorithm. The feature-extracted data is split into two sets which are trained sets and test sets. Then, this model is trained with 80% of the data and validated with the remaining 20%. Extracted MFCC features of trained sets are then classified using a generic random forest and saved the trained set as a.mat file as a trained model.

The RF algorithm is a type of supervised classification system that makes extensive use of decision trees that have been trained on a specific set of data. The random forest, which is based on computations that would average each tree's prediction, is essentially used to make final projections. By doing this, the over-fitting issue that might produce useless results when utilising other classification algorithms is resolved.

A random forest in Fig. 2 is composed of three independent decision trees that cooperate. The best example of ensemble learning is Random Forest, which draws conclusions from a variety of models for a more accurate outcome. To reach the conclusion, the algorithm mostly relies on each node's leaves outcome or final judgements. The accuracy of the model is often improved by averaging the outcomes of numerous separate decision trees.

The reduction in node impurity weighted by the likelihood of reaching that node is the standard formula for determining feature relevance. The node probability can be computed by dividing the total number of samples by the number of samples that reach the node. the more expensive the value. The feature is more significant the higher the value. To determine each decision tree's nodes' importance using the Gini Importance formula, let's assume that there are only two child nodes (binary tree):

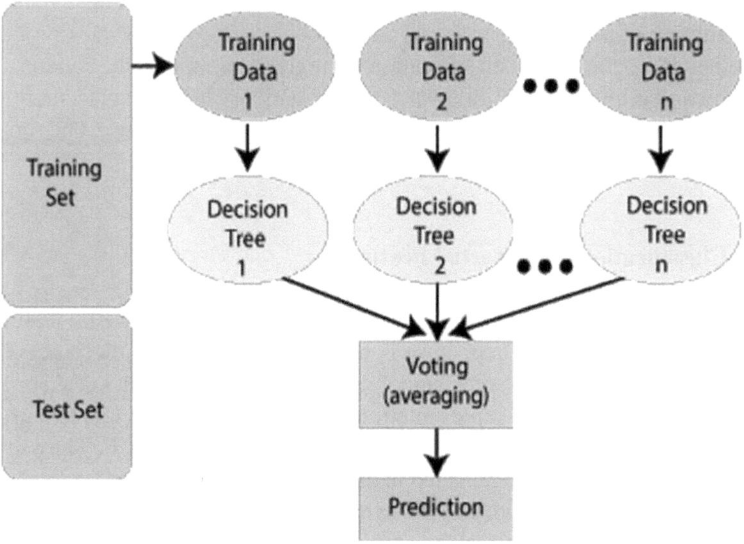

Fig. 2 Representation of the random forest

$$ni(j) = w_j C_j - w_{left(j)} C_{left(j)} - w_{right(j)} C_{right(j)} \tag{2}$$

where w(j) is the weighted number of samples that reach node j and ni(j) is the significance of node j. C(j) is the impurity value of node j, left(j) is the child node from node j's left split, and right(j) is the child node from node j's right split. The following formula is used to determine the significance of each feature on a decision tree:

$$i_i = \frac{\sum_{j:nodesplitsonfeaturesi}^{i} ni_j}{\sum_{k\ belongsallnodes}^{i} ni_k} \tag{3}$$

fi(i) = the importance of feature i. ni(j) = the importance of node j.

Then, by dividing by the total value of feature importance, these can be normalized to a value between 0 and 1

$$normfi_i = \frac{(fi_i)}{\sum_{j\ belongsallfeatures}^{i} fi_j} \tag{4}$$

The feature's average across all trees determines its final relevance at the Random Forest level. The total relevance value of each attribute for each tree is added together and divided by the overall number of trees.

$$RFfi_i = \frac{\sum_{j\ belongsalltrees}^{i} normfi_{ij}}{T} \tag{5}$$

RFfi(i) = the importance of feature i calculated from all trees in the RF model, T = total number of trees.

Whether it's data from a classification or regression model, the random forest approach can be quite helpful in a variety of situations. It finds an accurate representation of the decision trees it employs and is simple to use and quick to train.

2.2.1 Classification of Extracted Features

Most related work have focused on the extraction of low-level acoustic features for speech recognition. Feature extraction is a key mechanism in audio processing to capture the information in a voice signal. Speaking from different speakers can be done in part by using feature representation. The term "MFCC" refers to a group of properties based on fundamental frequency, energy, ZCR, and MFCC. The pause and climb of the speech signal are reflected in the signal energy, which also simulates the voice strength, volume, or loudness. In our experiment, we divided the characteristics derived from the speaker dataset into training and testing data, with 65% of the training data and 35% of the testing data respectively. Using the MFCC extraction technique, our feature extraction model of 400 speech wav files creates a number of rows. The performance of the learning algorithms in relation to a specific collection of features has been evaluated using the industry-standard criteria of accuracy and precision.

2.3 Visual Model

The first and most important step in the processing of video signals is to import all the necessary libraries required. Once the libraries are loaded into MATLAB, the next step is to organize the data set in the required manner. Two different directories were created for the Training and Testing, respectively. These lists are used dynamically to access the video samples saved in the respective directory. Figure 3 shows the directory containing both training and testing files.

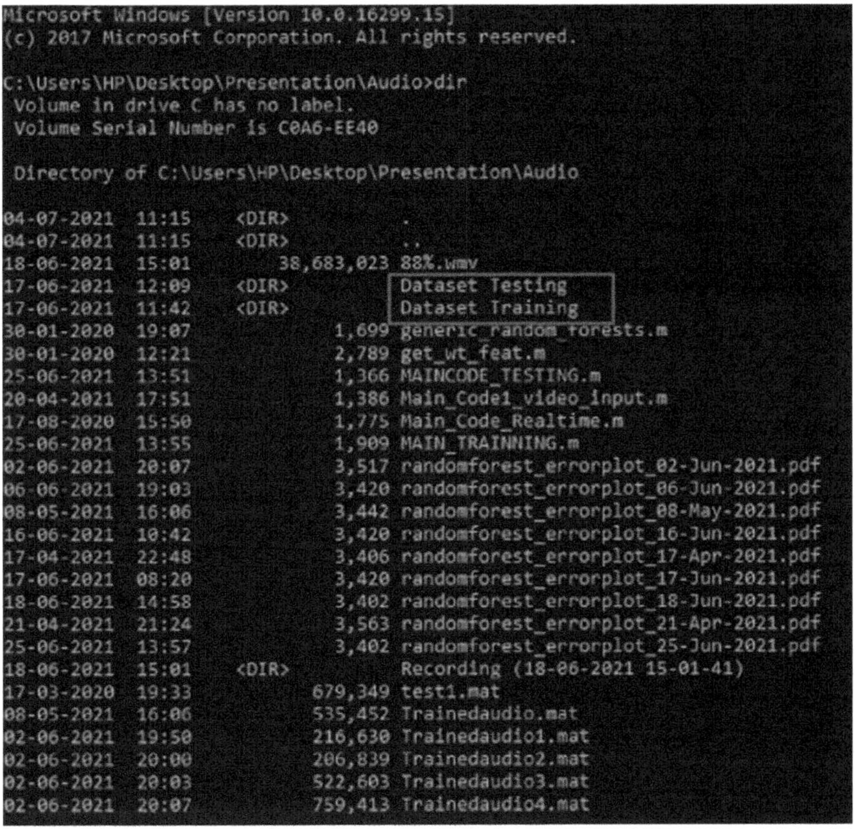

Fig. 3 Directory containing both training and testing files

Fig. 4 ROI extraction using the Viola-Jones algorithm

After the lip region has been extracted from the video using the Viola-Jones method, the lip region identification (also known as lip localization) is made. The FFNN technique is then used to train the extracted set. The next step is to arrange the data set in the necessary way after the libraries have been loaded into MATLAB. For the testing and the training, two separate directories were made. The video samples kept in each directory are accessed dynamically using these lists.

Visual features can be recognised using the object-recognition framework of the Viola-Jones algorithm. As a result, the Viola-Jones algorithm has been applied in this proposed methodology to extract the lip region and detect faces. The Region of Interest (RoI) extraction using the Viola-Jones technique is shown in Fig. 4.

Five steps are involved in the Viola-Jones Algorithm:

Step 1: Generate the Haar features

Step 2: Produce the integral image

Step 3: Boosting with the Adaptive Boosting (AdaBoost)

Step 4: Execute the cascading classifier

Step 5: Execute the Feed Forward Neural Network (FFNN)

The Haar Features were created in step 1. These characteristics aid in the machine's comprehension of the image. Finding the total of all the image pixels in the darker area of the haar characteristic and the total of all the image pixels in the lighter area of the haar characteristic is the goal here. then determine how they differ from one another. Now, the haar feature will be towards 1 if the image has a section separating light pixels on one side from dark pixels on the other. It signifies the discovery of an edge. Think about how a table's edge would seem in black and white. It will have that edge-like black-and-white appearance because one side will be lighter than the other. The horizontal and vertical features of the two crucial Face Detection characteristics specify how the system perceives the eyebrows and the nose.

The value for the region (x, y) on the image in step 2 is equal to the sum of all the pixels above and to the left of the area on the original image. This is known as a fundamental picture. Because it enables the quick calculation of rectangular spots, this intermediary illustration is necessary. Instead of looping through the pixels in that specific region to validate each one, the pixels for that region can be determined using this. This integral image would still speed up the process because haar feature recognition involves calculating the sum of the dark and light regions. We can rapidly do these laborious computations using the integral picture to determine whether a feature or set of features satisfies the requirements.

Step 3 involves applying Adaboost to boost performance. A set of guidelines known as AdaBoost is used in machine learning to learn how to choose the best subset of features and functions out of those that are available. A classifier known as a "Strong Classifier" is the result of the collection of rules. Linear combinations of "Weak Classifier" make up a strong classifier. The approach runs for T iterations from a high degree to detect such weak classifiers, where T is the number of weak classifiers to find and is selected by the user. The method displays the error rate for each feature in each loop, and the lowest error rate feature is then selected for iteration. The algorithm can identify false positives and true negatives in the data, helping it to be more accurate. It learns from the photographs that are sent to it as input. We consider all potential positions and combinations of those features and generate a very accurate model. For instance, training can be extremely time-consuming because to the numerous potential outcomes and combinations that are examined for each frame or image.

The cascade classifier is utilised in step 4. A cascade classifier, which includes a robust/strong classifier generated by an AdaBoost algorithm, is a multi-stage classifier that performs detection quickly and accurately at every level. The robust classifier created by the AdaBoost algorithm would be present at every level. An input is assessed sequentially (i.e., stage by stage). A classifier's input is instantly rejected if its output for a certain level is negative. The input is passed on to the next stage if the output is positive. In line with Viola & Jones, this multi-stage approach enables the creation of less complex classifiers that can subsequently be utilised to swiftly reject the greatest amount of negative information while devoting more time to positive data. Another "trick" to improve the speed and accuracy of our model is cascading. In order to find out if our most significant or greatest feature is present in the image within the sub-window, we first take a sub-window. We don't even look at the sub-window if it's not there; we just throw it out. The second feature in the sub-window is then examined if it is present. If it's absent, the sub-window is rejected. We continue until we run out of features, at which point we reject the sub-windows that lack that feature. Although evaluations may just take a few milliseconds, doing so for each feature could take a long time. This process is greatly accelerated by cascading, and the machine can provide results much more quickly.

Step 5 involves applying FFNN. A feed-forward neural network is essentially an algorithm with biological inspiration. It has several layer-organized processing units that resemble basic neurons. All of the units on the layer before this one are connected to it by this single unit. Each of these links has a unique strength or weight; they are not all equal. These connections' weights encode the network's information. The components of a neural network are frequently referred to as nodes as well. Data enters the inputs and moves layer by layer through the network until it reaches the output. When it performs the function of a classifier, there may occasionally be no feedback among the layers. They are called "feedforward neural networks" because of this. Machine learning algorithms use feed-forward neural networks. They are used specifically in techniques for image processing. Since all input only travels in one direction—forward—they are often referred to as feed-forward neural networks. Data enters the input nodes first, travels through the hidden layers, and then leaves the output nodes with the outcome. The output node's data cannot be fed back into the network via any of the network's links. Figure 5 shows a feed-forward neural network.

Where W = Weight of a model; I = input node; H = Hidden node; O = output node; OA = output node activated; B = bias node (Fig. 6).

The calculation of the sigmoid activation function is shown in Fig. 7, where e is the total discrepancy between the network's output and the desired value (s).

The modeling of the feed-forward neural network is described in the following. To calculate hidden node values, Eqs. (6) and (7) are used.

$$H_1 = I_1 W_1 + I_2 W_3 + B_1 W_5 \tag{6}$$

$$H_2 = I_1 W_2 + I_2 W_4 + B_1 W_6 \tag{7}$$

To select an activation function layer for the hidden layer, the following equation is executed.

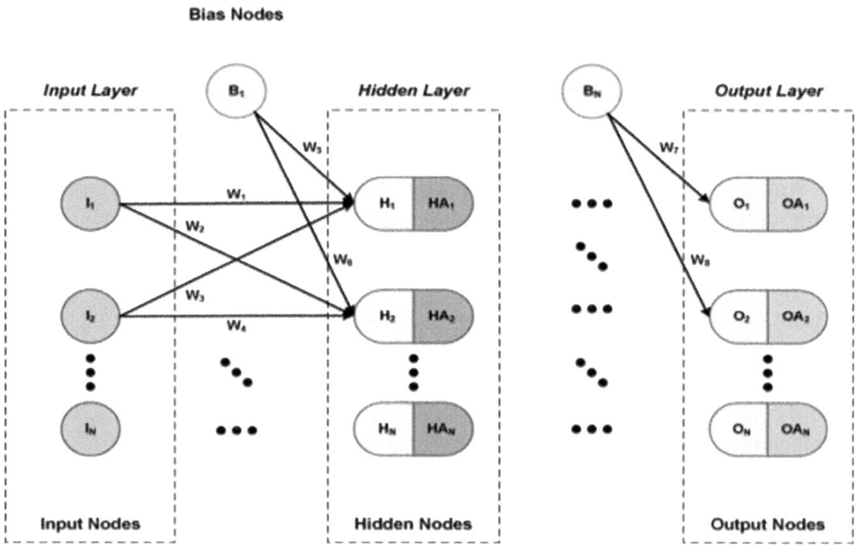

Fig. 5 The structure of the feed-forward neural network

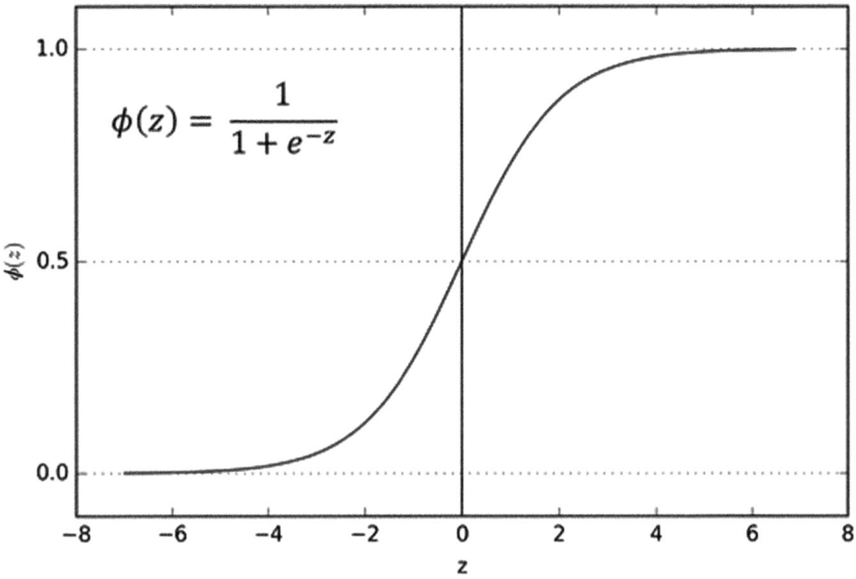

Fig. 6 Sigmoid activation function

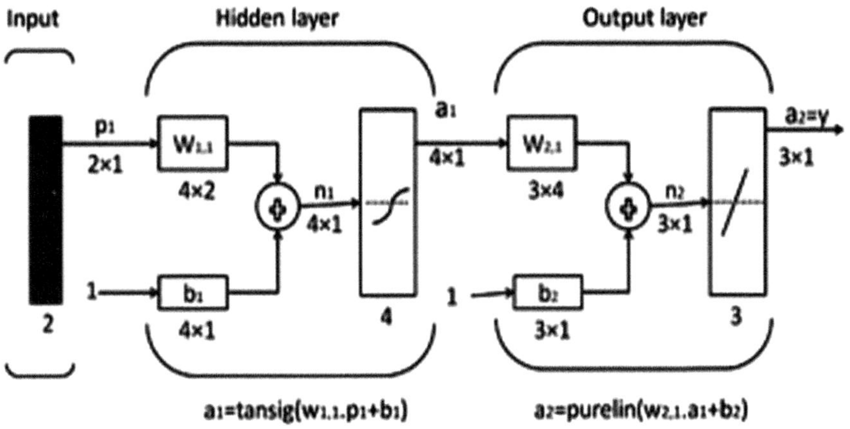

Fig. 7 A 2-layer feed-forward network

$$S(x) = \left(\frac{1}{1 + e^x}\right) \tag{8}$$

To calculate hidden node activation values, the following equations are performed (Fig. 8).

$$HA_1 = \frac{1}{1} + e^{-H_1} \tag{9}$$

$$HA_2 = \frac{1}{1} + e^{-H_2} \tag{10}$$

To select an activation function for the output layer, the following equation is executed.

$$Y = f(x) = x \tag{11}$$

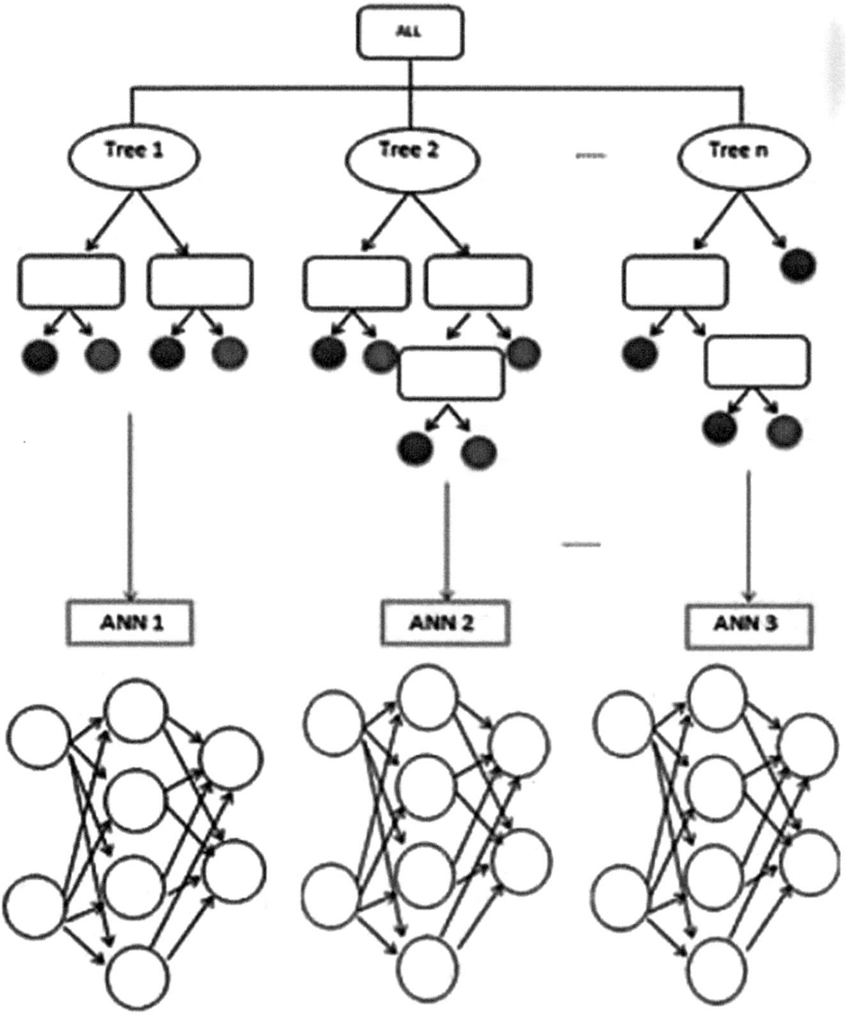

Fig. 8 Fusion model using random forest and artificial neural network

To calculate output node activation values, the following equations are performed.

$$OA_1 = O_1 \tag{12}$$

$$OA_2 = O_2 \tag{13}$$

The weight update formula is listed as follows:

$$W_j = w_j - A(del(E(w))/del(wt)) \tag{14}$$

Based on weights, delta is the learning rate and the remaining is the error estimation.

$$Loss = Yi' - yi \tag{15}$$

Here, yi is the actual value and 〖Yi〗 ^′ is the predicted value.

2.4 Fusion Model

Artificial neural network modelling is a relatively new technique that is based on how learning algorithms work in the human brain. It places a high value on adaptability and the capacity to understand intricate nonlinear/linear interactions. The most widely used networks are multi-layer feed-forward (MLF) networks that were trained using the back-propagation algorithm. To create ANN models, MLF networks are used. Input, hidden, and output layers are the three different types of neuron layers that make up the fundamental architecture. Signals strictly follow a feed-forward direction as they pass through feed-forward ANN networks from input to output units. Different numbers of neurons make up the buried layers. The neuron's output ("a") and input ("p") are parameters in Eq. (16). All the parameters are matrices and are designated by the letters "w" and "p," which stand for weight and bias, respectively:

$$\alpha = f(net) = f(n) = f(w^T.p + b) = f\left(\sum_{i=1}^{R} PW_R^T.P_R + b\right) \tag{16}$$

Continue and repeat the procedure of calibrating and optimising wi, bi to train the function as intended (s). The Mean Square Error (MSE) value will be minimised when wi, bi(s) is optimal. This process will go on until the necessary precision is achieved. Every time the process is performed in the next step, weights and biases will vary. The feed-forward convolutional neural network is depicted in Fig. 7.

$$W_{ij}^{(l+1)} = W_{ij}^{(l)} - \alpha \frac{\delta_e(w, b)}{\delta w^t_{(i,j)}} \tag{17}$$

2.4.1 Random Forest Modelling

A predictor called a random forest is made up of many randomized base regression trees.

$$rn(X, \theta_m, D_n), m >= 1 \tag{18}$$

The aggregated regression estimate is created by combining these random trees.

$$(X, D_n) = E_\theta[X, \theta, D_n] \tag{19}$$

This stands for expectation in relation to the random parameter in this instance. Theta controls the order of cuts when building individual trees, including the decision of which coordinate to divide and where to split it. Theta, a model variable, is believed to be unconnected to X and the training sample Dn. This specifically excludes any step involving bootstrapping or resampling in the training set. This eliminates any data-dependent tree-building techniques, such as finding the best splits by maximizing a criterion based on the actual observations. We do, though, permit theta to be predicated on a second sample that is separate from yet distributed as Dn.

2.4.2 Hybrid Model

In this study, a random forest and an artificial neural network are combined to create the proposed model. The algorithm is used to select the random values for the random forest decision trees that are the most effective. The features and data that must be chosen are then determined for each decision tree. After entering these numbers, the artificial neural network method replaces each tree's decision with the result of the neural network with the highest number of relevant neurons. A majority of votes was then used to decide on the label.

Then the equations that achieved the fusion of both classifiers are listed in Eqs. 18 and 19.

$$\mathrm{Z(i)} = \frac{1}{2\mathrm{W_L}} \sum_{\mathrm{n=1}}^{\mathrm{W_L}} |\mathrm{sgn}[\mathrm{x_i(n)}] - \mathrm{sgn}[\mathrm{x_i(n-1)}] \tag{20}$$

where sgn(.) is the sign function

$$\text{i.e,}\quad \text{sign}[x_i(n)] = \begin{cases} 1 & x_i(n \geq 0), \\ -1 & x_i(n < 0). \end{cases} \tag{21}$$

3 Results and Discussions

The result obtained from the proposed system using MFCC with the custom dataset for audio model processing is at an accuracy of 88% when compared with the other algorithms such as MFCC, ANN & GMM for the local dataset [1], which performed at an accuracy of 77% and for Spectral Subtraction with Voice Activity Detection (SSVAD) with TIMIT Speech dataset [2] gave an accuracy of 82.36%.

Similarly, in the visual model processing, the proposed system with the use of FFNN and HMM for the custom dataset successfully gave an output of 57% when compared to other models such as LSTM & CNN combined for an OLUV'S dataset [6] with an accuracy of 62.70% and CNN with Alexnet containing Miracle-VCI dataset [4] with an accuracy of 64.6% is underrated this difference is balanced by in the part of the integration.

In the fusion model, a hybrid model combining ANN and RF for the custom dataset was performed at an accuracy of 86%, which, when compared with a TCD-TIMIT & LRS2 dataset with an Attention-based Seq 2 Seq networks [7] with an accuracy of 46.52% and for Digit Dataset with the HMM and weighting in decision fusion model [8] with an accuracy of 48.4% is better.

3.1 *Audio Mode Result*

The audio recognition was conducted separately for 5 different Kannada words like "Guruthu", "Howdu", "Kelida", "Illa" and "Neenu" and the results are compared original dataset. A total of 400 datasets have been used, of which 250 are used for training and 150 are used for testing.

The results of audio models utilising random forests are shown in this section. The performance validation plot for the audio model is displayed in Fig. 9. Figure 10 displays the audio model's gradient descent. The confusion matrix for the audio model is displayed in Fig. 11, and the audio model's regression is given in Fig. 12.

3.2 *Visual Model Result*

A custom dataset is used to perform visual speech recognition using a feed-forward neural network. Figures 13 and 14 illustrate the gradient descent plot of the visual model and the performance plot of the visual speech recognition, respectively.

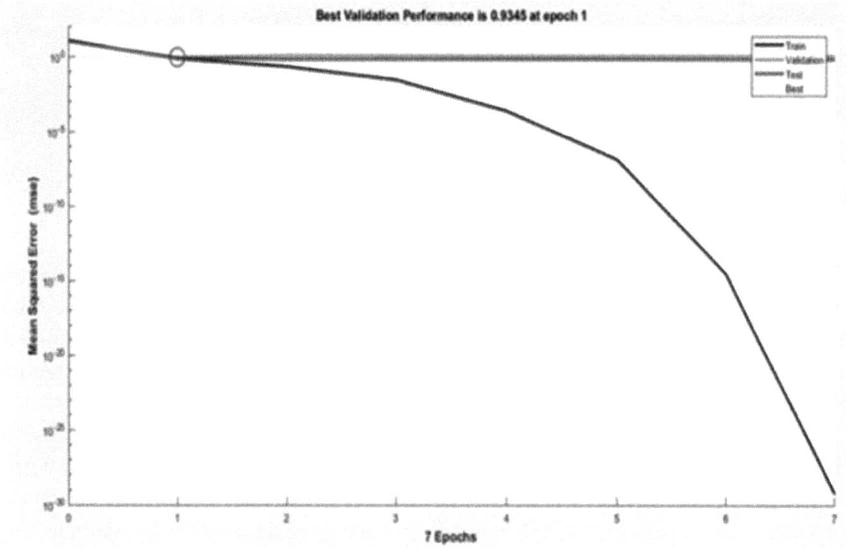

Fig. 9 Performance validation plot of the audio model

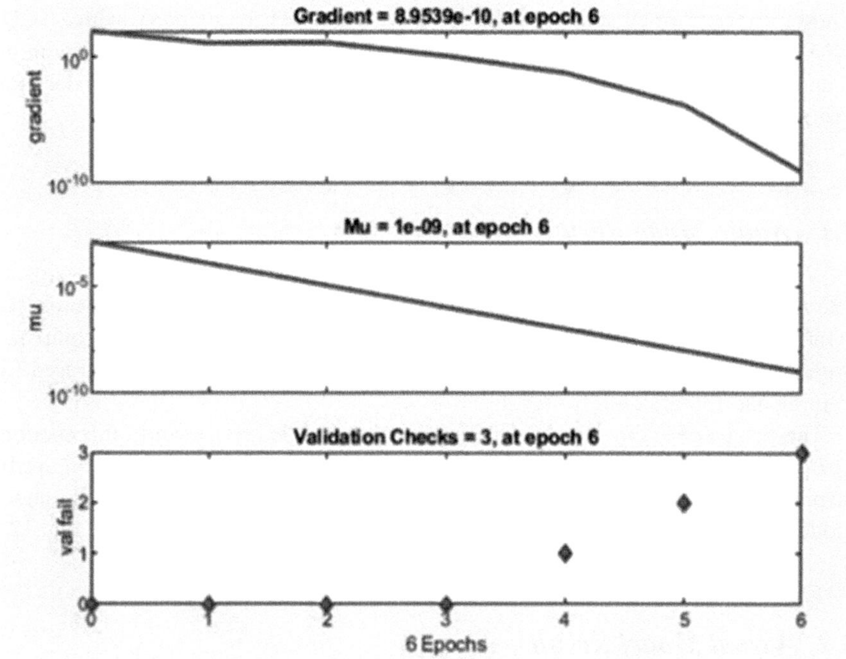

Fig. 10 Gradient descent plot of the audio model

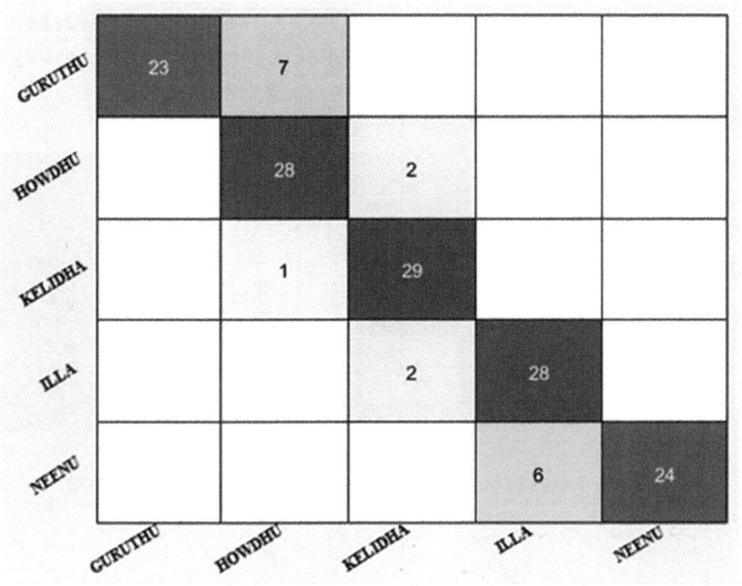

Fig. 11 Confusion plot of the audio model

Figures 15 and 16 depict the visual model's regression plot and the visual speech processing's confusion matrix plot, respectively.

3.3 *Fusion Model Result*

The following section explains how a hybrid model for the Kannada specific dataset integrated audio and visual inputs. Figure 16 displays the regression plot for the fusion model, and Fig. 17 displays the gradient descent plot for the fusion model. Figures 18 and 19 depict the fusion model's confusion matrix and regression plot, respectively. The Performance validation plot of the fusion model is displayed in Fig. 20.

Table 2 compares the results of the proposed model in terms of accuracy with the existing art of techniques. As can be seen that [4] achieved an accuracy of 46.52% and used attention-based seq 2 seq networks. In [5], authors proposed a hybrid HMM and adaptive weight techniques and achieved 48.4% accuracy. Similarly, in [17] and [18], the authors used VGG16 and HMM techniques and obtained 76 and 46% accuracy, respectively. The proposed model has achieved the highest accuracy of 89%.

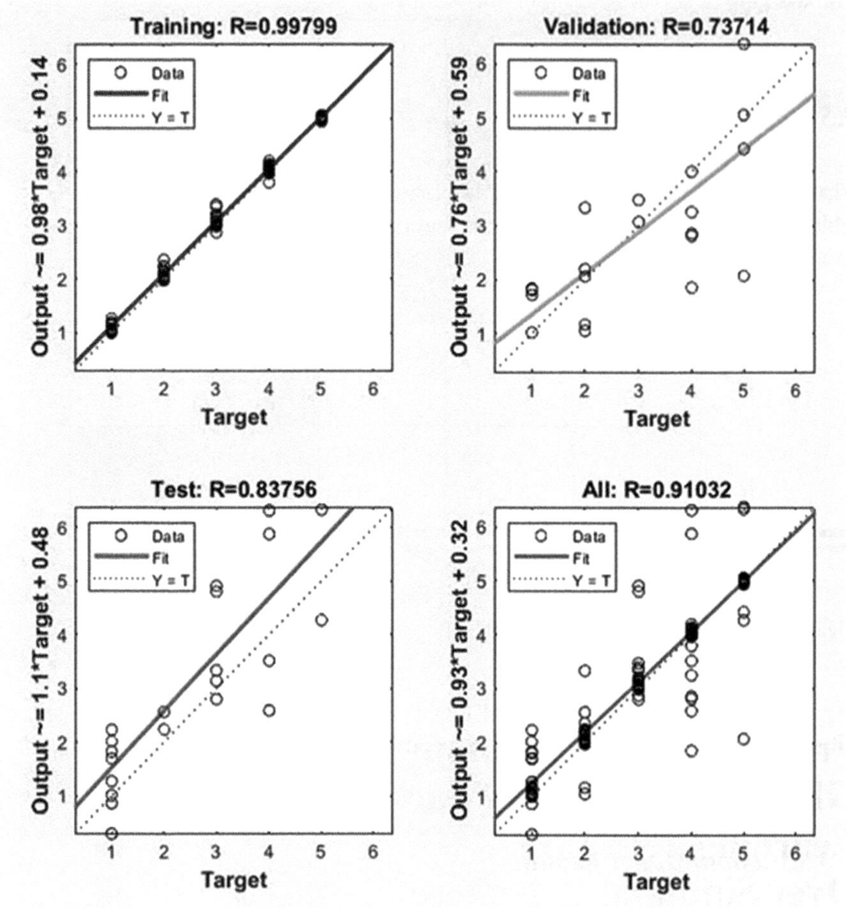

Fig. 12 Regression plot of the audio model

4 Conclusions and Future Works

A hybrid approach for Audio Visual Speech Recognition using Kannada Language Custom Datasets is proposed in this manuscript. The three main components of audiovisual speech recognition are as follows. Visual speech recognition comes in second, audio-visual speech recognition comes in third. Audio-speech recognition is the first. An accuracy of 88% was achieved in the audio feature extraction using MFCC and classification using the random forest machine learning technique. The accuracy of visual feature extraction using the feed-forward neural network and classification using the Haar-cascade detection approach was 57%. After integrating the auditory and visual information, a hybrid algorithm named Random Forest and ANN was used to classify the features, and the accuracy obtained was 89%.

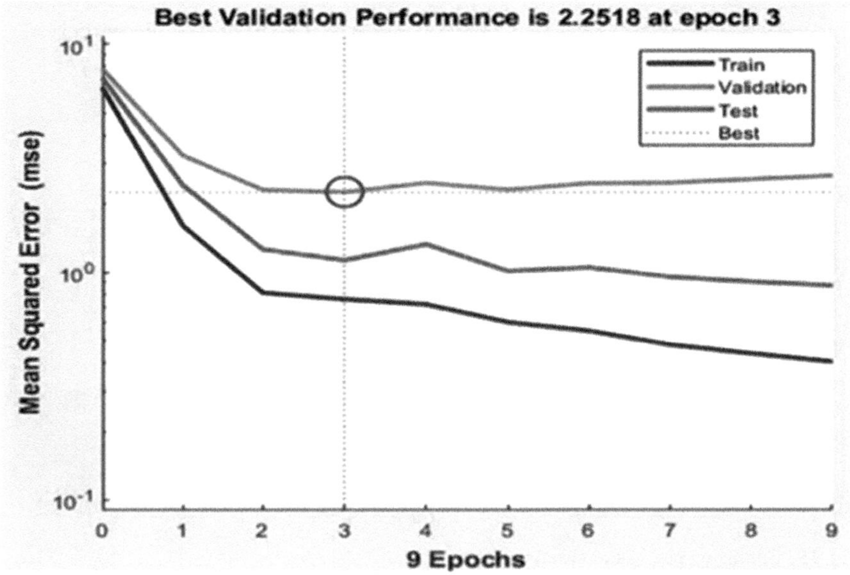

Fig. 13 Performance validation plot of the visual processing

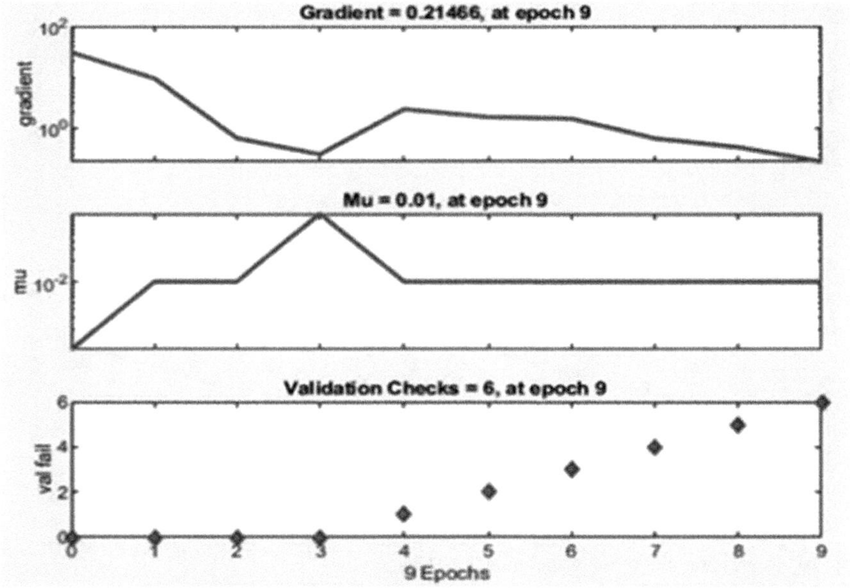

Fig. 14 Gradient descent plot of the visual model

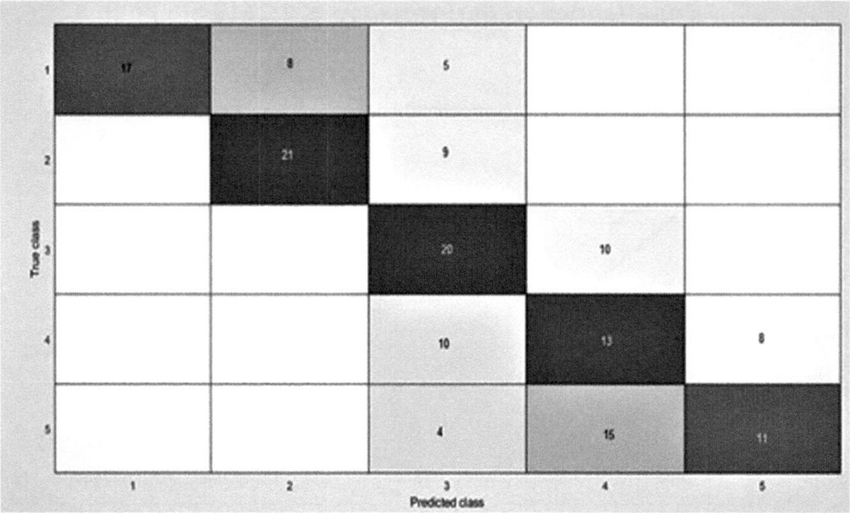

Fig. 15 Confusion plot of the visual processing

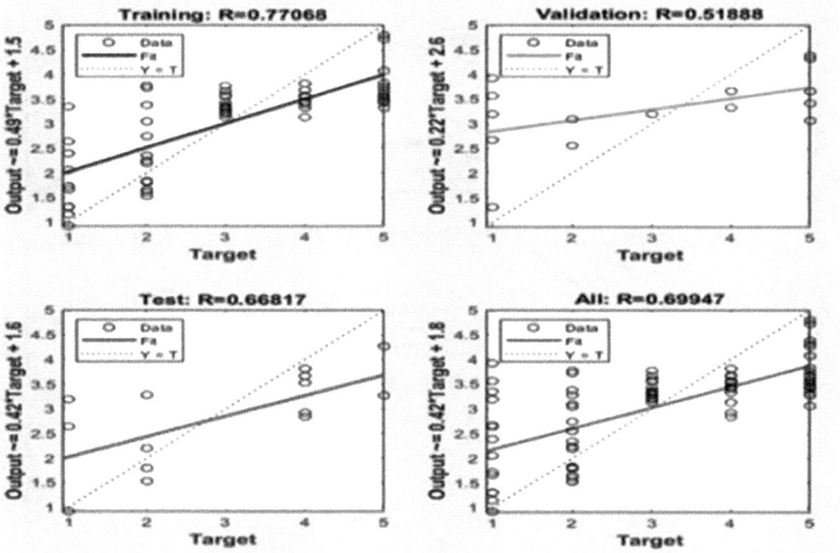

Fig. 16 Regression plot of the visual model

Kannada audio-visual speech recognition technology is still in its infancy. Future studies may make use of numerous datasets and various machine learning and deep learning models. Future development of the suggested model involves Processing with just the lip moment and no audio sample acquisition at all. AVSR can be used

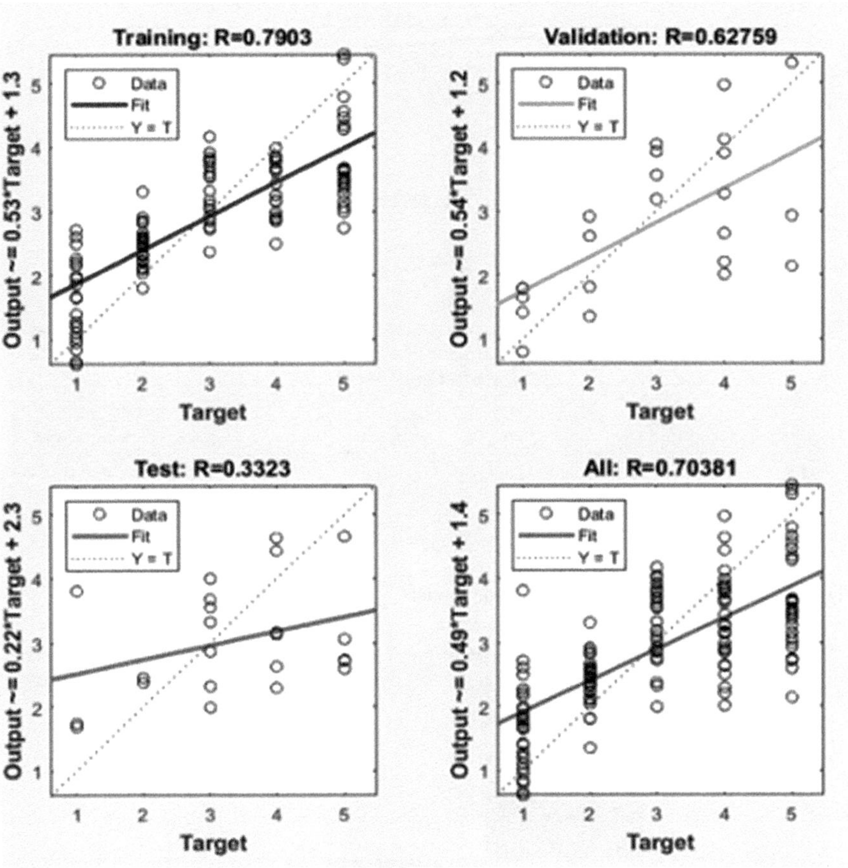

Fig. 17 Regression plot of the fusion model

to predict sentences in natural settings. Better performance can be attained and real-time data inputs can be used for processing with the use of efficient algorithms and a useful compact dataset.

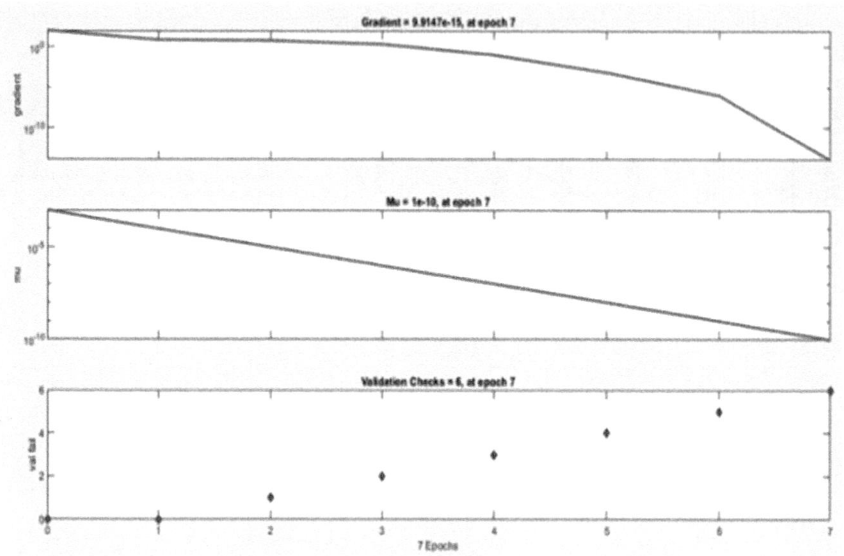

Fig. 18 Gradient descent plot of the fusion model

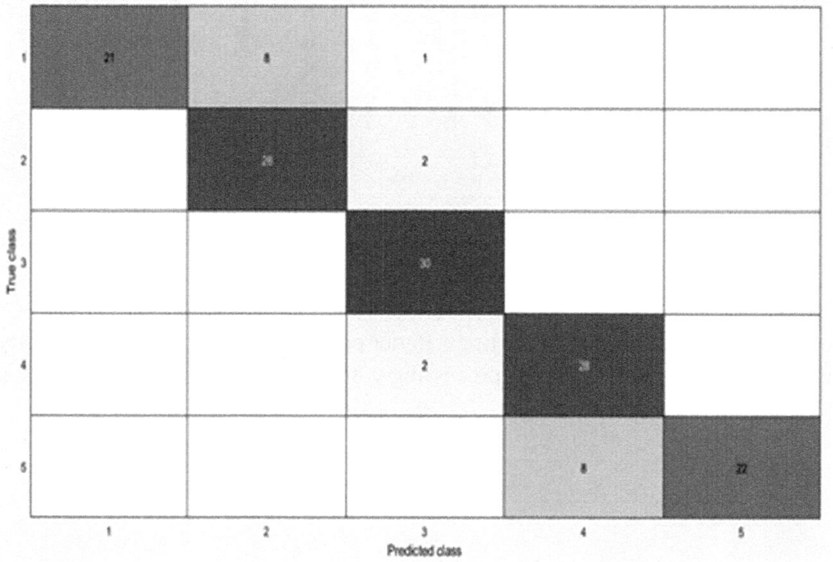

Fig. 19 Confusion plot of the fusion model

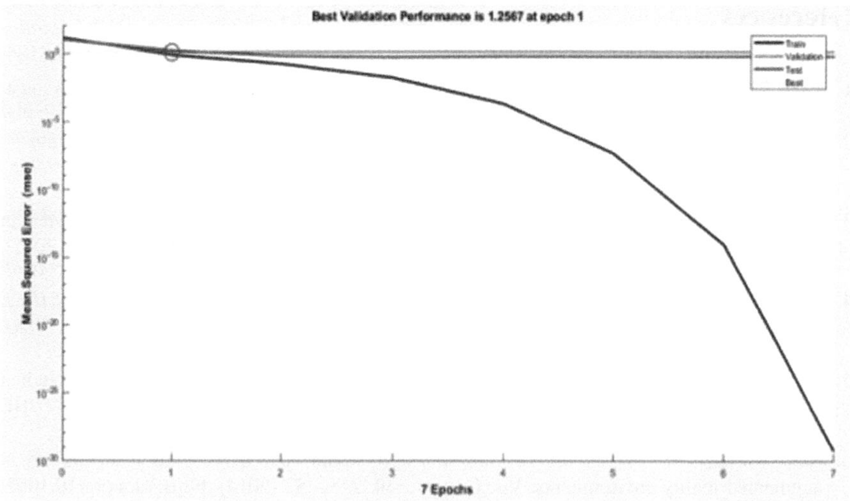

Fig. 20 Performance validation plot of the fusion model

Table 2 Comparison of the proposed model with the existing methods

Method	Accuracy (%)	Dataset
Attention-based seq 2 seq networks [4]	46.52	TCD-TIMIT
HMM & adaptive weight [5]	48.4	Digit database
VGG16 [17]	76	Custom
HMM [30]	46	Custom
Proposed model	89	Custom

Funding: The authors declare that no funds and grants were received during the preparation of this manuscript.

Data Availability: Based on the request to corresponding author.

Declarations

Conflict of Interest: We wish to confirm that there are no known conflicts of interest associated with this publication and there has been no significant financial support for this work that could have influenced its outcome.

References

1. W.H. Sumby, I. Pollack, Erratum: visual contribution to speech intelligibility in noise. J. Acoust. Soc. Am. **26**(2), 212–215 (1954). https://doi.org/10.1121/1.1907309
2. A.J. Goldschen, O.N. Garcia, E.D. Petajan, *Continuous Automatic Speech Recognition by Lipreading* (George Washington University, 1993), pp. 321–343. https://doi.org/10.1007/978-94-015-8935-2_14
3. J. Son Chung, A. Senior, O. Vinyals, A. Zisserman, Lip reading sentences in the wild, in *Proceedings of the IEEE Conference on Computer Vision and Pattern Recognition* (2017), pp. 6447–6456. https://doi.org/10.1145/3579654.3579677
4. E.D. Petajan, Automatic lipreading to enhance speech recognition, in *Proceedings of IEEE Communication Society Global Telecommunications Conference* (Atlanta, Georgia, 1984), pp. 26–29. https://doi.org/10.1109/IJCNN.1992.226994
5. A.P. Kandagal, V. Udayashankara, Visual speech recognition based on lip movement for Indian languages. Int. J. Comput. Intell. Res. **7**(2), 2029–2041 (2017). https://doi.org/10.17577/IJERTV9IS040509
6. M.R. Mirzaei, S. Ghorshi, M. Mortazavi, Audio-visual speech recognition techniques in augmented reality environments. Vis. Comput. **30**, 245–257 (2014). https://doi.org/10.1007/s00371-013-0841-1
7. L.R. Aran, F. Wong, L.P. Yi, A review on methods and classifiers in lip-reading, in *IEEE 2nd International Conference on Automatic Control and Intelligent Systems (I2CACIS)* (2017), pp. 196–201. https://doi.org/10.1109/I2CACIS.2017.8239057
8. T. Pao, W. Liao, T. Wu, C. Lin, Automatic visual feature extraction for Mandarin audio-visual speech recognition, in *IEEE International Conference on Systems, Man and Cybernetics* (2009), pp. 2936–2940. https://doi.org/10.1109/ICSMC.2009.5346011
9. T. Thein, K.M. San, Lip localization technique towards an automatic lip reading approach for Myanmar consonants recognition, in *International Conference on Information and Computer Technologies (ICICT)* (2018), pp. 123–127. https://doi.org/10.1109/INFOCT.2018.8356854
10. A. Garg, J. Nnoyola, S. Bagadia, Lip reading using CNN and LSTM, in *Proceedings Amit2016LipRU* (2016). https://api.semanticscholar.org/CorpusID:22889293
11. M.H. Rahmani, F. Almasganj, Lip-reading via a DNN-HMM hybrid system using combination of the image-based and model-based features, in *3rd International Conference on Pattern Recognition and Image Analysis (IPRIA)* (2017), pp. 195–199. https://doi.org/10.1109/PRIA.2017.7983045.
12. G. Lv, D. Jiang, R. Zhao, X. Jiang, H. Sahli, Multi-stream asynchrony dynamic bayesian network model for audio-visual continuous speech recognition, in *14th International Workshop on Systems, Signals and Image Processing and 6th EURASIP Conference focused on Speech and Image Processing, Multimedia Communications and Services* (2007), pp. 418–421. https://doi.org/10.1109/IWSSIP.2007.4381130.
13. E. Marcheret, V. Libal, G. Potamianos, Dynamic stream weight modeling for audio-visual speech recognition, in *IEEE International Conference on Acoustics, Speech and Signal Processing—ICASSP'07* (2007), pp. IV-945–IV-948. https://doi.org/10.1109/ICASSP.2007.367227
14. T. Saitoh, R. Konishi, Profile lip reading for vowel and word recognition, in *20th International Conference on Pattern Recognition* (2010), pp. 1356–1359. https://doi.org/10.1109/ICPR.2010.335
15. K.K. Nawas, M.K. Barik, A.N. Khan, Speaker recognition using random forest, in *International Conference on Innovative Technology for Sustainable Development (ICITSD-2021)*, vol. 37 (2021), pp. 1–5. https://doi.org/10.1051/itmconf/20213701022
16. M.S. Rao, G.B. Lakshmi, P. Gowri, K.B. Chowdary, Random forest based automatic speaker recognition system. Int. J. Anal Exp Modal Anal **XII**(IV), 526–535. ISSN NO:0886-9367 (2020)

17. S. Nivetha, A. Rathinavelu, S. Gayathri, Speech recognition system for isolated tamil words using random forest algorithm. Int. J. Recent Technol. Eng. (IJRTE) **9**(1), 2431–2435 (2020). ISSN: 2277-3878. https://doi.org/10.35940/ijrte.A1467.059120
18. P. Dhakal, P. Damacharla, A.Y. Javaid, V. Devabhaktuni, A near real-time automatic speaker recognition architecture for voice-based user interface. Mach. Learn. Knowl. Extr. **1**(1), 504–520 (2019). https://doi.org/10.3390/make1010031
19. L. Zheng, Q. Li, H. Ban, S. Liu, Speech emotion recognition based on convolution neural network combined with random forest. Chin. Control Decis. Conf. (CCDC) (2018). https://doi.org/10.1109/CCDC.2018.8407844
20. N. Damodar, H.Y. Vani, M.A. Anusuya, Voice emotion recognition using CNN and decision tree. Int. J. Innov. Technol. Exploring Eng. (IJITEE) **8**(12), 4245–4249, ISSN: 2278-3075. https://doi.org/10.35940/ijitee.L2698.1081219
21. R. Shashidhar, S. Patilkulkarni, S.B. Puneeth, Audio visual speech recognition using feed forward neural network architecture, in *IEEE International Conference for Innovation in Technology (INOCON)* (2020), pp. 1–5. https://doi.org/10.1109/INOCON50539.2020.9298429
22. R. Shashidhar, S. Patilkulkarni, Visual speech recognition for small scale dataset using VGG16 convolution neural network. Multimed. Tools Appl. **80**, 28941–28952 (2021). https://doi.org/10.1007/s11042-021-11119-0
23. V. Sooraj, M. Hardhik, S. Nishanth, C. Sandesh, R. Shashidhar, Lip-reading techniques: a review. Int. J. Sci. Technol. Res. **9**(2), 4378–4383 (2020). https://doi.org/10.3390/acoustics5010020
24. B.P. Yuhas, M.H. Goldstein, T.J. Sejnowski, Integration of acoustic and visual speech signals using neural networks. IEEE Commun. Mag. **27**(11), 65–71 (1989). https://doi.org/10.1007/11
25. K. Thangthai, R. Harvey, S. Cox, B.-J. Theobald, Improving Lipreading performance for robust audiovisual speech recognition using DNNs, in *The 1st Joint Conference on Facial Analysis, Animation, and Auditory-Visual Speech Processing* (Vienna, Austria)
26. G.I. Chiou, J.N. Hwang, Lip-reading by using snakes, principal component analysis, and hidden Markov models to recognize color motion video. IEEE Trans. Image Process. **6**(8), 1192–1195 (1997). https://doi.org/10.1142/S0218001418560074
27. V. Sooraj, M. Hardhik, N.S. Murthy, C. Sandesh, R. Shashidhar, Lip-reading techniques: a review. Int. J. Sci. Technol. Res. **9**, 4378–4383 (2020)
28. S. Rudregowda, S. Patil Kulkarni, G. HL, V. Ravi, M. Krichen, Visual speech recognition for kannada language using VGG16 convolutional neural network. Acoustics **5**(1), 343–353 (2023). https://doi.org/10.3390/acoustics5010020
29. S.S. Morade, S. Patnaik, Comparison of classifiers for lip reading with CUAVE and TULIPS database. Optik—Int. J. Light Electron Opt. **126**(24), 5753–5761 (2015). https://doi.org/10.1016/j.ijleo.2015.08.192
30. Lu., Yuanyao, J. Yan, Gu., Ke, Review on automatic Lip reading techniques. Int. J. Pattern. Recognit. Artif. Intell. **32**(7), 1856007 (2018). https://doi.org/10.1142/S0218001418500222
31. S. Dupont, J. Luettin, Audio-visual speech modeling for continuous speech recognition. IEEE Trans. Multimed. **2**(3), 141–151 (2000). https://doi.org/10.1186/1687-6180-2012-51
32. R. Shashidhar, S. Patilkulkarni, S.B. Puneeth, Combining audio and visual speech recognition using LSTM and deep convolutional neural network. Int. J. Inf. Tecnol. 1–12 (2022). https://doi.org/10.1007/s41870-022-00907-y
33. G. Pu, H. Wang, Review on research progress of machine lip reading. Vis. Comput. (2022). https://doi.org/10.1007/s00371-022-02511-4
34. M. Rashid, S.A.R. Abu-Bakar, M. Mokji, Human emotion recognition from videos using spatio-temporal and audio features. Vis. Comput. **29**, 1269–1275 (2013). https://doi.org/10.1007/s00371-012-0768-y
35. A.A. Shaikh, D.K. Kumar, J. Gubbi, Automatic visual speech segmentation and recognition using directional motion history images and Zernike moments. Vis. Comput. **29**, 969–982 (2013). https://doi.org/10.1007/s00371-012-0751-7

36. R. Shashidhar, S. Patilkulkarni, Audiovisual speech recognition for Kannada language using feed forward neural network. Neural Comput. Applic. **34**, 15603–15615 (2022). https://doi.org/10.1007/s00521-022-07249-7

Analysis of Machine Learning Approach for Spamming Electronic Mail Detection

Sajan Shrestha, A. Mohammed Owais Khan, Parth Jain, Vinayakumar Ravi, and H. L. Gururaj

Abstract Individuals and organizations use electronic mail (E-mail) to send and receive digital messages over the internet. In 2023, 347.3 billion emails are sent and received per day among which are spam emails. These emails can lead to communication overload, waste of time, irritation, loss of important emails, and potential exposure to malware. Irrelevant or unsolicited messages, malware, advertisements etc. are common features of spam emails. In this study we are employing various machine learning methods such as logistic regression, decision trees, random forest, SVM, naive bayes and BERT to identify spam emails. We transform email data into a specialized format known as a sparse matrix using techniques like TFIDF. Then, we apply these methods, encompassing logistic regression, decision trees, random forests, and naive Bayes, to aid us in the identification process. Additionally, the text is tokenized using BERT Transformer, serving as input for BERT. Considering multiple parameters, Logistic Regression and SVM Algorithms 99% accuracy was achieved in spam detection.

Keywords Cyber security · Spam · Machine learning · Deep learning · BERT

1 Introduction

The exponential growth of email communication globally has revolutionized how individuals and organizations connect and disseminate information. However, this technological advancement has brought about significant challenges, including the

S. Shrestha · A. M. O. Khan · P. Jain · H. L. Gururaj
Manipal Institute of Technology Bengaluru, Manipal Academy of Higher Education, Manipal, India
e-mail: gururaj.hl@manipal.edu

V. Ravi (✉)
Center for Artificial Intelligence, Prince Mohammad Bin Fahd University, Khobar, Saudi Arabia
e-mail: vravi@pmu.edu.sa

B. Pradhan and S. Mukhopadhyay (eds.), *IoT Sensors, ML, AI and XAI: Empowering A Smarter World*, Smart Sensors, Measurement and Instrumentation 50,
https://doi.org/10.1007/978-3-031-68602-3_21

proliferation of fraudulent content, malware dissemination, and inundation of unsolicited messages, collectively known as spam. Spam emails encompassing text bugs and programmed viruses, posing threats to both users and network security.

As evidenced by the staggering increase in daily spam volume from 293.6 billion in 2019 to 333.2 billion in 2022, representing 48.63% of total emails [1] depicted in Fig. 2, the urgency to combat this escalating menace is undeniable. Projections suggest a continued upward trajectory, with estimates indicating a potential surge to 400 billion spam emails per day by 2025 if current trends persist.

In response to this growing threat, considerable resources have been allocated by organizations and researchers worldwide to develop robust spam filtering solutions. using the latest improvements in artificial intelligence AI and machine learning ML some standout algorithms include logistic regression support vector machine (SVM) naive Bayes decision tree and random forest although these methods have proven useful in spotting spam achieving perfect accuracy is still a challenge because requires extensive training data incorporating a range of aspects such as website credibility URL analysis and user feedback.

This study aims to push forward the current state of the art in spotting email spam by integrating refined data preprocessing techniques. Extending beyond traditional approaches, additional features including sender reputation, phishing website identification, and header analysis are incorporated, enhancing the discriminatory power of the models. Using TF-IDF vectorization for feature extraction, a thorough assessment of different machine learning algorithms is carried out, resulting in a comparative analysis utilizing confusion matrices and classification reports covering accuracy, recall, precision, and F1-score metrics. Structured methodically, this paper unfolds with an exploration of related works in Sect. 2, delineating the existing landscape of spam filtering methodologies. Section 3 elucidates the intricacies of the proposed methodology, detailing the innovative approach adopted in feature engineering and model construction. Subsequently, Sect. 4 unveils the empirical findings derived from extensive experimentation, offering insights into the performance of diverse machine learning algorithms. Finally, Sect. 5 synthesizes the research findings, presenting a comprehensive conclusion and outlining avenues for future research endeavors. Through rigorous investigation and meticulous analysis, this study endeavors to make significant contributions to the ongoing efforts in combating the scourge of email spam.

2 Literature Review

In this section, related work has been depicted with existing works. Billions of people are using emails and the major problem they are facing is inevitable spam emails which contains malware and unwanted information. This results in loss of time and money, and it is estimated that by 2025 there will be loss of $10.5 trillion because of spam emails [2]. Such spam emails can be detected using spam filters which differentiate the spam emails with ham emails based on the content of emails, email header details, the route that it takes from its original sender.

Some major types of spam filter [3] for email spam detection are given below:

Bayesian Filters. This filter is based on Bayes theorem which determines the probability that a message is spam or ham based on the words and phrases it contains.

Content-Based Filters. Content-based filters examine the actual content of messages for certain keywords, phrases, or patterns commonly associated with spam.

Header Analysis Filters. Header Analysis filters analysis the email header information like sender's email address, IP address, subject line etc. to distinguish between spam and ham emails.

Machine learning Filters. In this filters, different Machine learning algorithms are used for detection of spam mails. Those algorithms include decision trees, random forests, logistic regression, neural networks, support vectors etc. Machine learning filters are trained using data sets and along with time the accuracy of these filters can be improved.

Sender Policy Framework (SPF). SPF is an email authentication protocol that helps to verify the legitimacy of the sender's domain. Spam filters can check SPF records to determine if an email is coming from an authorized source.

Although there are many other spam filters, the spam traps will still be there. Numerous organizations and researchers are struggling to create sturdy and effective filters to address this issue. Many models utilizing machine learning algorithms demonstrate impressive performance in discerning whether an email is spam or legitimate. Many researchers have got above 90% accuracy in their related algorithms along with that they have consider features like email text, header information, DNS record. They have used popular machine learning algorithms and Deep learning algorithms for training their model.

Spam email can also be detected based on the link present on it. For example, if mail contains phishing websites, then it is obvious that the email is spam. Detecting a phishing website can be determined based on the features like address bar, abnormality, HTML and java script and Domain information.

Lokesh et al. [4] Saleh et al. [5] presents spam filters that are based on Negative Selection Algorithms (NSA) that can detect their own type (normal or self-pattern) but not the new types (different pattern). In this study, Negative Selection works same as T-cells in biological system as it recognize its own kind normal pattern) but attack other external entities (different pattern). Ahmed, Naeem et al. [6] uses machine learning technique for spam filters and classify different types of spam filter (standard spam filter, client side spam filter, enterprise level spam filter and case based spam filter). Thashina et al. [7] had developed a different user interface which will detect whether the email is spam or ham. In this interface, users must log in and can use it for sending and receiving email. Now, the received email goes through various pipelines where machine learning models like Naïve Bayes will detect whether the email is spam or ham. Similarly, focusing on data preprocessing procedure, Nikhil and Sonowal [8] used different steps like data cleaning, data integration, data transformation and data reduction to transform the input data into model understandable data. The flow chart of the model clearly depicts the working mechanism of the whole research. Different algorithms like adaboost classifier, KNN, SVM were used in this research paper. Yaseen [9] used BERT, BiLSTM, KNN and Naïve Bayes for such spam detection. In this paper, Deep Learning Algorithms are used where each input text is converted into the vectorized form that can be given as an input to all the models. Here, pre-trained BERT is used which can convert the whole text into a meaning full vectorized text that will result proper classification of text. Similarly, Kingshuk and Kar [10] have used enron email data set in deep learning models like LSTM and BERT. NLP was used to analyze data preprocessing of text of email.

3 Methodology

In this section, the detailed methodology has been explained, including data collection, data preprocessing, modeling, and testing techniques.

3.1 Data Collection

In this research, a dataset sourced from Kaggle was utilized, consisting of email messages. The dataset included 5570 unique emails, containing 5169 distinct words. The Dataset can be accessed through following link https://www.kaggle.com/datasets/shantanudhakadd/email-spam-detection-dataset-classification.

3.2 Data Pre-processing

In the field of machine learning data preparation is paramount. This process entails converting unfiltered data into a format conducive to training and testing models. The initial step in this process is data cleaning, aimed at rectifying missing values, eliminating erroneous data entries, and adapting the dataset to its specific requirements [11]. This typically involves replacing missing values with statistical measures such as the mean, median, or mode of the corresponding feature, and removing rows containing inconvenient data.

Following data cleaning, the next step involves converting the cleaned data into a format understandable by machine learning models. This transformation is facilitated by transformers such as TF-IDF, Count Vectorizer, Bert Transformer, or Word2Vec. In this research, TF-IDF is employed to convert the cleaned data into a vectorized form, enabling machine learning algorithms such as Linear Regression, Logistic Regression, Support Vector Machines (SVM), Naïve Bayes, Decision Trees, and

Random Forests to process the input data effectively. Similarly, the Bert Tokenizer is utilized to ensure that the BERT model comprehends the input data appropriately.

TFIDF

TFIDF stands for Term Frequency- Inverse Document Frequency, a numerical value defining the significance of a word in each document. The numerical significance of a word is calculated using vectorization and feature extraction in which TF-TDF is calculated for each word using formula as shown in Eq. (1). Using this TF-IDF, a sparse matrix is calculated which will have n row (distinct types of word in the document) and m column (total number of text) [12]. In addition, we add more columns to this sparse matrix which give the other information about email like sender reputation, phishing website etc. For example, additional column is made for phishing websites in which the range of given phishing website is calculated: the most popular phishing website, the value is high and safe one, the value is almost zero.

The mathematical expression for TF-IDF is given below [13]:

$$\mathrm{TF-IDF:}\ \mathrm{W(i,j)} = \mathrm{F(i,j)}\ \mathrm{x}\ \log(\mathrm{N}/\mathrm{F(i)}) \tag{1}$$

W(i,j): Word i in document j.

F(i,j): Frequency of i in document j.

N: Total number of documents.

F(i): Total frequency of word i in all documents.

BERT Tokenizers

BERT Tokenizers is a tool used to preprocess the text data before feeding it into BERT-based model. It involves breaking down the text into smaller unit, typically words or sub words, to create a sequence of tokens that can be understand by the model. BERT tokenizer adds special tokens to the input text. These tokens include:

[CLS]: This token is added at the beginning of the input text and represents the classification token.

[SEP]: This token is used to separate sentence or input sequences. It is added between pairs of sentences or at the end of a single sentence.

[MASK]: During the initial training phase of a masked language model particular input tokens are randomly obscured and the model learns to predict them.

Once the input text is tokenized and special tokens are added, BERT Tokenizers converts each token into a vector consisting of input ids, token type ids and attention mask.

Input_ids. It represents the input text converted into integer indices using the BERT tokenizer's vocabulary. Each token in the input text is replaced by its corresponding index in the tokenizer's vocabulary.

Token_type_ids. It indicates which segment each token belongs to. It will separate each segment by its identical number called token_type_ids. For example, in a question-answering task, tokens in the question segment might be assigned a token type ID of 0 and tokens in the passage segment might be assigned a token type ID of 1.

Attention_mask. It indicates which tokens should be attended to by the model and which ones should be ignored. Generally, value of 1 indicates that the token should be attended to, while a value of 0 indicates that the token should be ignored (masked).

Overall,'input_ids' represent the tokens in the input text as integer indices, 'token_type_ids' distinguish between different segments of the input (if applicable), and 'attention_mask' indicates which tokens should be attended to by the model. These three components together form the input representation used by BERT-based models.

3.3 Modeling

Machine learning model is made for spam detection which separates ham from spam emails based on the information about that email. Before Model training, the dataset is distributed into 80% of initial data for training model and remaining data are used for testing purposes. The training dataset is further divided into 10 folds for cross validation which will make sure that our model is training properly or not.

3.3.1 Logistic Regression

Logistic regression is a helpful math tool we use to guess if something will happen or not in various fields. It tries to figure out the chance of something falling into one of two groups. Logistic regression models the probability that Y = 1 as a function of the predictors X. This is done using the logit function (also called sigmoid function or logistic function) as shown in Eq. (2), which ensure that the predicted probabilities lie between 0 and 1.

$$P(Y = 1|X) = \frac{1}{1 + e^{-\beta_0 + \beta_1 X_1 + \cdots + \beta_n X_n}} \tag{2}$$

P(Y = 1|X) It is the probability that shows how likely it is for the outcome to be 1 based on the predictors.

$\beta_0, \beta_1, \beta_2, \ldots, \beta_n$ is the coefficients associated with each predictor $X_0, X_1, X_2, \ldots, X_n$.

e represents the base of the natural logarithm.

At the beginning of training a model with logistic regression, we start with a random value or zero for θ. We then use Gradient Descent to find the best value for θ that makes the model work well. And while we're doing Gradient Descent, we also use regularization to make sure the model doesn't become too specific to the training data.

Similarly, cross validation is done in every fold. This overall method of finding the parameter is known as modeling. Furthermore, the determined optimum value of parameter θ is used to decide whether an email is spam or legitimate using Sigmoid (Logistic) function Eq. (3) which provides the value between 0 to 1 [14]. If the value obtained from the sigmoid function is less than 0.5 then it is considered as ham otherwise it is spam.

For modelling, optimal value of θ is calculated using gradient descent:

$$h(x) = \frac{1}{1 + e^{-\theta X}} \tag{3}$$

$$\theta = \theta - \frac{1}{m} \sum_{i=1}^{m} [h(X_i) - Y_i).X_i]. \tag{4}$$

h(x) shows the Probability of the dependent variable Y being 1 given the predictors X.

X is vector representation of predictor.

θ is vector of coefficient associated with each predictor.

m is the total number of parameters taken into the account.

3.3.2 Naïve Bayes Classifier

This method depends on Bayes' Theorem (Eq. 5) which evaluates the possibility of the event a taking place following the happening of event b [15].

$$P(A|B) = \frac{P(A).P(B|A)}{P(B)} \tag{5}$$

P(A/B): Possibility of event A occurring given that event B has occurred.

P(B/A): Possibility of event B occurring given that event A has occurred.

P(A): Possibility of event A happening.

P(B): Possibility of event b happening.

Consider a classification task where probability of one training set of being in class label C is to determine with some features$x_0, x_1, x_2, \ldots, x_n$. According to Bayes theorem, the possibility of class C given the features is given as:

$$P(C|x_0, x_1, x_2, \ldots, x_n) = \frac{P(x_0, x_1, x_2, \ldots, x_n|\mathrm{C})P(C)}{P(x_0, x_1, x_2, \ldots, x_n)} \tag{6}$$

Naïve Assumption. Naïve Bayes assumes feature independence, making it easier to find the combined probability by splitting it into smaller parts—individual probabilities. This aids classification by reducing complexity into manageable parts.

$$P(x_0, x_1, x_2, \ldots, x_n|C) = P(x_0|C).P(x_1|C)\ldots.P(x_n|C) \tag{7}$$

Prior Probabilities P (C) is calculated by counting the occurrences of each class the training data and normalizing to obtain probabilities that sum up to 1.

Conditional Probabilities $P(x_i|C)$ represent the probabilities of observing a particular feature value given the class. For discrete features, such as word counts in text data, these probabilities are calculated by counting the occurrences of each features value within each class and normalizing. During the training phase, these parameters are updated based on the distribution of features and class labels in the training data. Once trained, these parameters are then used to make predictions on new instances.

3.3.3 SVM

SVM (support vector machine) is a supervised learning algorithm which is particularly effective in high dimensional spaces, and it is widely used for text classification like spam email detection. In this algorithm, a hyperplane is determined which separates different classes in the feature space. In addition, two extreme hyper planes (positive and negative) are also determined to separate the classes. In SVM, kernel function is applied in each data set to manipulate data set which makes it easier to find the margin line for that given data set.

Hyperplane is a decision boundary that divides the classes in the feature space. For example, in two-dimensional feature space, the hyperplane is a line, in three-dimensional feature space, it is a plane and for higher dimensional space the decision boundary is called hyperplane. The main aim of SVM is to find a hyperplane which has maximum distance between hyperplane and the nearest data points from each class. These points are called support vectors.

The margin is the distance between the hyperplane and the nearest data points from each class (support vector). SVM seeks to maximize the margin because a larger margin generally leads to better generalization of unseen data and improved robustness.

The kernel function is used to handle nonlinear decision boundaries by mapping the input features into a higher- dimensional space. Common kernel functions include linear, polynomial, radial basis function (RBF) and sigmoid kernels.

Regularization parameter (C) helps to balance the margin maximization and classification error. A small C value allows for a large margin but may lead to more misclassifications, while a large C value results in a smaller margin but fewer misclassifications.

Linearly separable data

In linearly separable data, the hyperplane is determined with maximum margin such that all the datapoints are correctly classified. Mathematically, it can be formulated a constrained optimization problem.

Mathematically, it can be written as Eq. (8).

$$min_{w,b}\frac{1}{2}||w^2|| \tag{8}$$

Constraints: $y_i(w.x_i + b) > 1$ for i = 1..., N

W is the weight vector that defines the hyperplane.

b is the bias term.

x_i are the feature vectors.

y_i are the class label (+1 or -1).

N is the number of data points.

Nonlinearly separable data (Soft Margin SVM).

In Nonlinearly separable data, soft margin is used which introduces a slack variables (ξ) and a regularization parameter (C) to control the trade-off between maximizing the margin and minimizing the classification error. The optimization problem becomes:

$$min_{w,b}\frac{1}{2}||w^2|| + C\sum_{i=1}^{N}\xi_i \tag{9}$$

Constraints: $y_i(w.x_i + b) > 1 - \xi_i$ for 1..., N

ξ_i>0 for i = 1... N.

C is the regularization parameter.

ξ_i are slack variables that allow for misclassifications.

The first term $\frac{1}{2}||w^2||$ represents the margin as before.

The second term $C\sum_{i=1}^{N}\xi_i$ penalize misclassifications.

3.3.4 Decision Tree

Decision Tree is a popular and intuitive machine learning algorithm used for both classification and regression tasks. It models decisions as a tree structure where each internal node represents a decision based on a feature, each branch represents the outcome of that decision, and each leaf node represents the final decision or prediction.

The decision tree algorithm grows the tree recursively by selecting the best split at each node until a stopping criterion is met. Stopping criteria may include reaching a maximum tree depth, having a minimum number of samples in a node, or no further improvement in impurity reduction. Decision trees make decisions by recursively splitting the dataset based on the values of input features. At each node, the algorithm selects the feature that best splits the data into homogeneous subsets, aiming to maximize information gain (for classification) or minimize variance (for regression). For classification tasks, common splitting criteria include Gini impurity, entropy, chi-square and variance, which measure the impurity or disorder of the data. Once the tree is built, predictions are made by traversing the tree from the root node to a leaf node based on the values of input features. For classification, the majority class in the leaf node is assigned as the predicted class.

Entropy. Entropy serves as a metric to gauge the disorder or uncertainty present within a dataset. When we talk about decision trees entropy helps us understand how unsure we are about the different labels in a group of data.

Mathematically, the entropy of a node is calculated using Eq. 10.

$$Entropy = -\sum_{i-1}^{C} p_i log_2(p_i) \tag{10}$$

C is the number of classes or categories.

p_i is the probability of the class i in the node.

Information Gain. Information gain is a metric used to measure the effectiveness of a feature in reducing entropy when splitting a dataset. It quantifies the reduction in uncertainty (or increase in purity) achieved by partitioning the dataset based on a particular feature. Mathematically, the information gained for splitting a dataset on a feature is calculated using Eq. 11.

$$InformationGain = Entropy(parent) - \sum_{i=1}^{n} (\frac{N_i}{N}) Entropy(Child_i) \tag{11}$$

Entropy(parent) is the entropy of the parent node before the split.

n is the number of child nodes resulting from the split.

N_i is the number of samples in the i-th child node.

N is the total number of samples in the parent node.

$Entropy(Child_i)$ is the entropy of the i-th child node after the split.

Variance. Variance is often used as a measure of impurity in decision tree algorithms for regression tasks. When selecting the best split for a node in a regression tree, the split that minimizes the variance of the target variable within the resulting child nodes is chosen.

$$Varience = \frac{\sum (X - \mu)^2}{N} \tag{12}$$

X: actual value of element.

μ: mean value of element.

N: Number of elements.

Chi-Square(χ2). The chi-square test is a statistical tool we use to find out if there's a significant relationship between different categories of data. In the context of decision trees, chi-square test is commonly used in feature selection for classification tasks, particularly in the construction of categorical splits.

$$Chi - Square = \sqrt{\frac{(expected - actual)^2}{expected}}. \tag{13}$$

3.3.5 Random Forest

Random Forest is a method within collaborative learning approaches, which builds a collection of decision trees and then aggregates their predictions to make a final prediction. In other words, random forests construct many individual decision trees at training and the result of random forest is calculated by averaging all the predicted values of all decision trees. During training by this method, features are selected using feature importance and some decision tree is removed to prevent overfitting and import the generalize the performance.

Feature importance is a metric utilized in machine learning to assess the relevance of individual input features in prediction-making. Random Forest assists in determining the most influential features affecting the model's output, offering insights into the underlying connections within the data. Feature importance is particularly useful for understanding the relative contribution of different features to the model's performance and for feature selection. It is a common metric used to decrease the impurity or information gained at each node of the decision trees in the forest.

Mean Decrease in Impurity (MDI)

It quantifies the contribution of each feature to the reduction of impurity during the construction of decision trees. Impurity refers to the degree of uncertainty or disorder in a dataset, and reducing impurity is a key goal in building decision trees. Formula for mean decrease in impurity is given in Eq. 14.

$$MDI = \frac{\sum_{t \in Trees} ImpurityDecrease_t}{NumberofTrees} \tag{14}$$

Trees represent the set of decision trees in Random Forest.

$ImpurityDecrease_t$ represents the decrease in impurity achieved by splitting on a particular feature in tree t.

The impurity decrease at each node of the decision tree is typically calculated using the formula given in Eq. (15).

$$ImpurityDecrease_t = Impurity_{parent} - \frac{N_{left}}{N_{parent}} \times Impurity_{left} + \frac{N_{right}}{N_{parent}} \times Impurity_{right} \tag{15}$$

$Impurity_{parent}$ is the impurity of the parent nodes before splitting.

$Impurity_{left}$ and $Impurity_{right}$ are the impurities of the left and right child nodes after splitting, respectively.

N_{parent} represents the combined count of samples within the main node.

N_{left} and N_{right} denote the quantity of samples in the child nodes to the left and right, correspondingly.

When considering feature importance, a higher MDI value signifies increased significance, whereas lower values suggest reduced importance.

Mean Decrease in Accuracy

It is a measure of feature importance which quantifies each input feature's contributions to the model's overall accuracy.

The baseline accuracy $ACC_{baseline}$ of the model is computed on validation set or through cross validation, this serves as the reference accuracy against which the importance of each feature will be measured. The model is re-evaluated on the perturbed dataset (dataset that is created by randomly permuting the feature in validation set while keeping the target variable unchanged) and the accuracy $Acc^{j}_{permuted}$ of the model is computed. This accuracy represents the model performance when the values of feature j are randomly permuted.

The decrease in accuracy MDA_j caused by permuting feature j is calculated as the difference between the baseline accuracy and the accuracy on the perturbed dataset:

$$MDA_j = ACC_{baseline} - Acc^j_{permuted} \tag{16}$$

This equation calculates the decrease in accuracy when feature j is permuted, indicating the importance of that feature in contributing to the model's predictive performance. Features with lower MDA value are considered more important, as they have a greater impact on the model's accuracy when their values are permuted.

3.3.6 BERT (Bidirectional Encoder Representations from Transformers)

BERT, also known as Bidirectional Encoder Representations from Transformers, is a model used for understanding language, See Fig. 1. It learns from a big bunch of text without being told exactly what to do. Instead, it guesses missing words and tries to predict the next sentence. BERT is unique because it looks at words before and after to understand their meanings better.

BERT uses a transformer design, enabling it to grasp connections between distant words in text information. more effectively compared to traditional sequence models like recurrent neural networks (RNNs). BERT consists of multiple layers of transformer encoders, enabling it to learn contextual representations of words or tokens in a bidirectional manner.

Let's dive into how BERT works:

Input Representation or Tokenization: BERT breaks down input sequences into tokenized representations, segment features, and positional encodings. Token embeddings are obtained by looking up the token IDs in an embedding table. Segment embeddings distinguish between different segments or sentences in the input. Position embeddings encode the position of each token in the sequence to capture positional information. All this conversion of input text to the vectorized text is done by BERT Tokenizers.

Pre-training Objectives: BERT's initial training involves two main objectives: the Masked Language Model (MLM) and Next Sentence Prediction (NSP). With MLM, BERT fills in missing words in sentences based on context. NSP helps BERT understand the order of sentences in text. These tasks equip BERT with a strong foundation in language understanding, crucial for its performance across different language tasks.

Fine-Tuning Process: Following its initial training, BERT can be fine-tuned for specific tasks using supervised learning. This involves adjusting the pre-trained

Fig. 1 BERT architecture

BERT model to suit tasks like text classification, named entity recognition, or sentiment analysis. During fine-tuning, specialized layers may be added onto the pre-trained BERT model. The entire model is then trained on labeled data specific to the task at hand to improve its performance. In this study, the customized BERT architecture depicted in Figure BERT Architecture is utilized for spam email detection.

Here, in this architecture the message of email is given as input of BERT Tokenizer which will convert the text message to vectorized one which contain all the information about that text. The vectorized text contains input_ids, token_ids and masked_ids which will uniquely identify each word in the given input text. Now the vectorized data is given to the BERT which will understand each message in proper way using its pre-defined parameters which was determined during pre-training of BERT. It will give some meaningful data which will represent each message uniquely and this will be given to Artificial neural network which will predict whether the messages are spam or ham. Here, Backpropagation is continued for proper training of the model.

3.4 *Optimization*

3.4.1 Regularization

L1 Regularization (LASSO—Least Absolute Shrinkage and Selection Operator)

LASSO serves as a regularization method aimed at curbing overfitting in models and conducting feature selection. It proves particularly beneficial for datasets featuring a multitude of features. The LASSO technique encourages sparsity in the coefficient vector which sets some coefficients to zero performing feature selection. Those features with coefficients zero are excluded during classification. It improves the performance of the machine learning model since the features with less significant impact on the classification are excluded including only the relevant features.

LASSO regression model also called regression model with L1 regularization is a technique which is used to prevent over-fitting by adding a penalty term to the standard cost function which also helps in shrinking the some of the coefficient to zero performing feature selection.

$$loss_{regularized} = loss_{original} + \lambda \sum_{i=0}^{n} |w_i| \tag{17}$$

λ is the regularization parameter

w_i is model weight.

L2 Regularization (Ridge Regularization)

Ridge regularization, also known as L2 regularization, is a regularization technique used to prevent overfitting of the regression model which is done by adding a penalty term to the standard cost function. It is similar to LASSO regularization. The difference that lies between them is the penalty term added to the cost function. Also, L2 regularization does not shrink the coefficient to exactly zero, it may decrease the coefficient closer to zero. So, it does not eliminate the entire feature but may mitigate the impact of the feature. So, it does not perform feature selection.

$$loss_{regularized} = loss_{original} + \lambda \sum_{i=0}^{n} {w_i}^2 \quad (18)$$

λ is the regularization parameter

w_i is model weight.

3.4.2 Hyperparameter Tuning

Hyperparameters are settings or configurations external to the model and not learned during training. The selection of hyperparameters can greatly influence how well the model performs, affecting factors like its accuracy, ability to adapt to new data, and how efficiently it operates.

Method to perform hyperparameter Tuning is given below:

Manual Tuning. Involves manually selecting hyperparameters based on domain knowledge and experimentation. This approach is time-consuming and may not always yield optimal results.

Grid Search: This method carefully tries out different combinations of hyperparameters one by one to see which works best. It tests each combination using cross-validation. However, it can take a lot of time, especially if there are many hyperparameters to test.

Random Search. Instead of trying every possible combination, random search picks hyperparameter combinations randomly from a list. It then checks how well each combination works using cross-validation. Random search is quicker than grid search and can often find good combinations without trying every possibility.

Bayesian Optimization: This approach uses a clever strategy to figure out which combinations to try next. It learns from previous tests and uses that knowledge to select new combinations. Bayesian optimization is even faster than random search and grid search, especially when there are lots of hyperparameters to consider.

Automated Hyperparameter Tuning Tools. Several libraries and platforms, such as scikit-optimize, Hyperopt, Optuna, and TensorFlow's kerastuner, offer automated hyperparameter tuning capabilities using various optimization algorithms.

When adjusting hyperparameters, it's important to select suitable measures to judge how well different setups perform. These measures can include accuracy, precision, recall, F1-score, ROC AUC, mean squared error (MSE), etc., depending on the problem type (like classification or regression). The dataset gets split into sections, and each hyperparameter setup gets tested on each section to find its average performance. This cross-validation process helps ensure that the hyperparameter tuning works well with new data too. Fine-tuning hyperparameters usually involves lots of trial and error, where you try different setups, see how they do, and then adjust your search based on what you find. You might need to repeat this process a few times until you're satisfied with your model's performance. Once you've determined the best hyperparameters, you can train your final model using the entire training dataset with these settings.

3.4.3 Tree Pruning

It is a technique for selectively removing parts of the tree that are deemed unnecessary or contribute to overfitting, with the aim of improving the model's generalization performance on unseen data.

Method of tree pruning is given below:

Evaluation of Nodes. Traverse the decision tree in a bottom-up fashion, starting from the left node. At each internal node, calculate the impact of removing its subtree on the tree's performance. Common pruning criteria include measures of impurity reduction or error reduction before and after pruning.

Make pruning Decisions. Compare the calculated pruning criterion against predefined threshold. If the criterion exceeds the threshold, pruning is considered beneficial for improving the tree's performance. If pruning is beneficial, remove the subtree rooted at the current node. Convert the internal nodes into a leaf node, representing a class label (for classification) or a predicted value (for regression).

Refine Pruned Tree. After pruning, merge adjacent left nodes if they represent similar classes or values and update the decision tree structure to reflect the pruned nodes and branches.

Evaluate Pruned trees. Evaluate the pruned decision tree on a validation set or through cross-validation and Compute performance metrics such as accuracy, precision, recall, or mean squared error. Now, Compare the performance of the pruned tree with the performance of the original (unpruned) tree. Ensure that pruning improves or maintains the model's generalization performance.

Iterative Pruning. If the pruned tree does not meet the desired performance criteria, iteratively adjust pruning criteria and repeat the pruning process. Continue iterating until the desired level of generalization performance is achieved.

By following these steps, tree pruning helps refine decision trees to improve their generalization performance while maintaining interpretability.

3.4.4 Tree Burning

Tree burning refers to the pruning or removal of decision trees from the Random Forest model to prevent overfitting and improve generalization performance. In Random Forest, each decision tree is trained on a bootstrap sample of the original dataset and a random subset of features. While individual decision trees in a Random Forest tend to be low-bias models (i.e., they can capture complex relationships in the training data), they are prone to overfitting, especially when the trees are deep or when the dataset is noisy.

Tree burning is a technique used to address this issue by selectively pruning or removing decision trees from the random forest that contributes less to the overall predictive performance. The process typically involves:

Evaluation of Tree Performance.

Each decision tree's performance is evaluated, often using out-of-bag (OOB) samples or a separate validation set.

For each data point in the original dataset:

If the data point was not included in the bootstrap sample for tree i:

Mark the data point as an out of bag (OOB) sample for tree i.

Otherwise, the data point is not considered an OOB sample for tree i.

For each decision tree in the Random Forest:

Use the OOB samples associated with tree i to make predictions.

Calculate the error metric between the predicted values for the OOB samples and the true target values.

Lastly, calculate the mean of the error measure obtained for all decision trees in the Random Forest. This average error measure approximates how well the model performs on new data without requiring a separate validation set. This makes it a helpful method for assessing the model's ability to generalize.

Ranking trees by Importance. Decision trees are ranked based on their importance or contribution to the model's performance. Importance can be measured using metrics like Mean Decrease in Impurity (MDI) or Mean Decrease in Accuracy (MDA).

Selective Removal. Decision trees that are deemed less important or contribute less to the model's performance are pruned or removed from the Random Forest ensemble.

By selectively removing decision trees that do not significantly contribute to the model's predictive performance, tree burning helps improve the generalization ability of the Random Forest model, making it more robust to noise and improving its performance on unseen data.

3.5 Testing

Testing a model will help to determine whether the algorithm is working efficiently or not.

Confusion Matrix. Confusion matrix is a tool used in machine learning and statistics to visualize and evaluate the performance of a classification model. A confusion matrix is usually a square matrix with rows and columns corresponding to the true classes and predicted classes, respectively.

Classification Report. Classification report is the report which evaluated the efficiency of machine learning algorithms [9]. In classification report there are mainly four parameter which are explained below:

Accuracy. In machine learning, accuracy refers to the measure of how well a classification model correctly predicts the class labels of the input data points. It is a common metric to evaluate the performance of classification algorithms determined using Eq. (19).

$$Accuracy = \frac{NumberofCorrectPredictions}{TotalNumberofPredictions} \tag{19}$$

Recall. Recall, also known as sensitivity or true positive rate, is a metric that calculated using Eq. (20) and quantifies the ability of a classification model to correctly identify all positive instances from the total actual positive instances in a data set.

$$Recall = \frac{TruePositive}{TruePositive + FalseNegatives} \tag{20}$$

Precision. Precision is a tool that measures the accuracy of positive predictions made by a classification model. Precision is the ratio of True Positive and sum of True Positive and False Positive as in Eq. (21).

$$Precision = \frac{TruePositives}{TruePositives + FalsePositives} \tag{21}$$

F1-Score. F1 score is a metric that combines both precision and recall into a single value as shown in Eq. (22), providing a balanced measure of a model's performance, particularly in situations where class imbalance exists. classification report of all the machine learning algorithms

$$F1 - Score = \frac{2 \times Precision \times Recall}{Precision + Recall} \tag{22}$$

Kappa. The Kappa Coefficient, commonly referred to as Cohen's Kappa Score, is a statistic used to assess the effectiveness of machine learning classification models. Its formula, which is based on the conventional 2 × 2 confusion matrix, is used to assess binary classifiers in statistics and machine learning. The Kappa statistic (or value) is a metric that compares an Observed Accuracy with an Expected Accuracy (random chance)

$$kappa(k) = \frac{2 * (\mathrm{TP} * \mathrm{TN} - \mathrm{FN} * \mathrm{FP})}{(\mathrm{TP} + \mathrm{FP}) * (\mathrm{FP} + \mathrm{TN}) + (\mathrm{TP} + \mathrm{FN}) * (\mathrm{FN} + \mathrm{TN})} \quad (23)$$

Area Under ROC Curve (AUC). AUC provides an aggregate measure of performance across all possible classification thresholds. One way of interpreting AUC is as the probability that the model ranks a random positive example more highly than a random negative example.

AUC-ROC values range from 0 to 1 where:

AUC = 0.5 implies the classifier has no discrimination ability and performs as well as random chance.

AUC > 0.5 indicates better-than-random performance, with higher values suggesting better discrimination ability.

AUC > 0.5 indicates better-than-random performance, with higher values suggesting better discrimination ability.

4 Experimental Results and Discussion

In this section, the results obtained after testing all the machine learning algorithms are present in tabulation form. Machine learning algorithms like Logistic Regression, Naïve Bayes, Support Vector Machine (SVM), Decision Tree, Random Forest and BERT are used as a model for training and the comparative analysis of efficiency of those algorithms are tabulated as shown in Table 1 and Fig. 2.

The provided table offers a comprehensive evaluation of various machine learning algorithms employed for email spam detection, presenting key performance metrics including precision, recall, F1-score, kappa, and accuracy. Notably, algorithms like Logistic Regression, Support Vector Machine (SVM), Decision Tree, and Random Forest exhibit robust performance, achieving high precision and recall values for both ham (non-spam) and spam categories. For instance, SVM demonstrates precision values of 0.96 and 0.99 for ham and spam, respectively, alongside recall values of 1.00 and 0.75, indicating a strong ability to correctly classify both classes. Similarly, Decision Tree and Random Forest display balanced precision and recall scores for both ham and spam, highlighting their effectiveness in accurately identifying spam emails while minimizing false positives and false negatives. However, Naive Bayes shows relatively lower precision for spam (0.47), suggesting a higher tendency for misclassifying legitimate emails as spam. Additionally, BERT exhibits notable

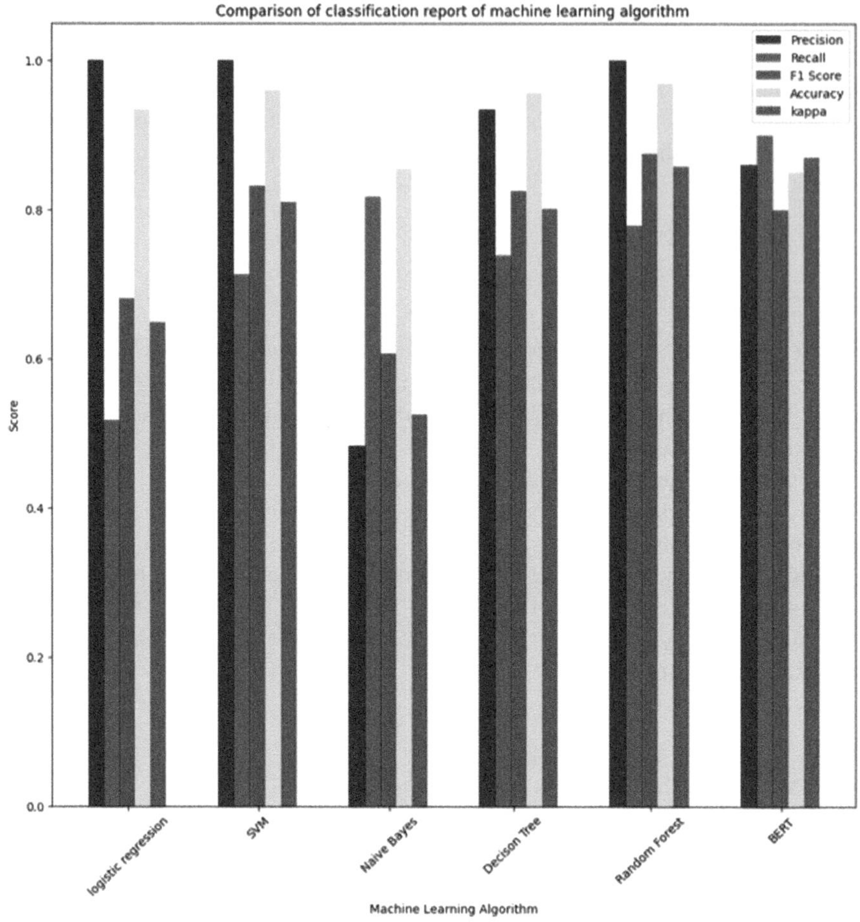

Fig. 2 Comparison of classification report of machine learning algorithms

performance with high recall for spam (1.00), albeit with slightly lower precision, underscoring its capability in capturing a significant portion of spam emails while maintaining a moderate level of precision. These insights offer valuable guidance for selecting the most suitable machine learning algorithm tailored to the specific requirements of email spam detection applications.

The AUC-ROC (Area Under the Curve of the Receiver Operating Characteristic) is a commonly used measure for assessing the effectiveness of binary classification models, including those used for email spam detection. In the context of our study, we obtained AUC values for various machine learning algorithms. Logistic Regression yielded an AUC of 0.76, indicating its moderate discriminatory power in differentiating between spam and non-spam emails, see Fig. 3. SVM performed better with an AUC of 0.86, suggesting a stronger ability to correctly classify instances. Similarly, Naive Bayes achieved an AUC of 0.84, demonstrating

Table 1 Comparison of classification report of machine learning algorithms

Metrics:		Precision	Recall	F1-Measure	Kappa	Accuracy
Logistic Regression	Ham	0.93	1	0.96	0.72	0.98
	Spam	0.99	0.54	0.7		
SVM	Ham	0.96	1	0.98	0.9	0.99
	Spam	0.99	0.75	0.85		
Decision Tree	Ham	0.97	0.99	0.98	0.75	0.96
	Spam	0.91	0.78	0.84		
Random Forest	Ham	0.97	1	0.99	0.92	0.96
	Spam	0.98	0.83	0.9		
Naive Bayes	Ham	0.98	0.84	0.91	0.58	0.85
	Spam	0.47	0.9	0.62		
BERT	Ham	1	0.8	0.9	0.87	0.85
	Spam	0.76	1	0.7		

respectable discriminatory performance. Decision Tree and BERT both attained an AUC of 0.86 and 0.80, respectively, suggesting comparable discriminatory abilities. However, Random Forest exhibited the highest AUC of 0.95, signifying superior discriminatory power in distinguishing between spam and non-spam emails. These AUC values provide valuable insights into the comparative effectiveness of the machine learning algorithms employed, aiding in the selection of the most suitable approach for email spam detection in real-world applications.

The analysis of email data reveals a notable trend in email classification based on message length. When examining the relationship between message length (on the x-axis) and the frequency of emails (on the y-axis) classified as either "ham" or "spam," a discernible pattern emerges. It is observed that shorter message lengths are predominantly associated with "ham" emails, indicative of legitimate correspondence, See Fig. 4. Conversely, as the length of messages increases, there is a notable rise in the occurrence of emails classified as "spam." This empirical observation underscores a potential correlation between message length and email categorization, suggesting that shorter messages are more likely to be genuine communications, while longer messages tend to exhibit characteristics consistent with spam content. These observations offer important direction for enhancing and perfecting email spam detection algorithms. They underscore the importance of including message length as a notable aspect in accurately distinguishing between genuine and unwanted emails.

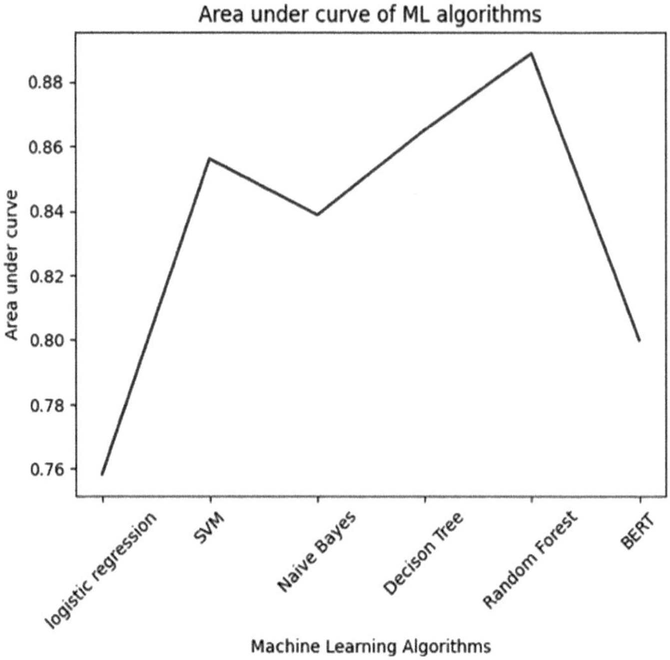

Fig. 3 Area under curve of each machine learning algorithms

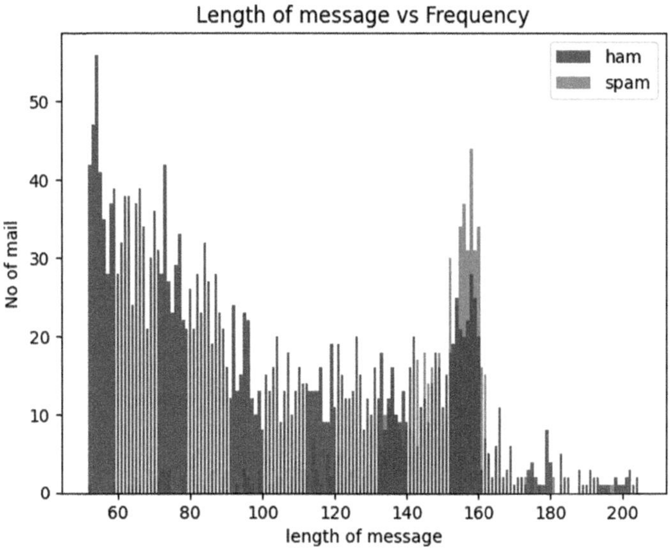

Fig. 4 Length of email message vs Frequency

5 Conclusion

In conclusion, machine learning algorithms such as Logistic Regression, SVM, Naïve Bayes, Random Forest, Decision Tree, BERT was used in this research among which this research paper highlights the effectiveness of employing advanced machine learning algorithms, such as Logistic Regression, SVM, alongside additional features, and sophisticated data preprocessing techniques, for email spam detection. Through comprehensive experimentation and comparative analysis, these algorithms demonstrated high accuracy and robust performance in distinguishing between spam and ham mails. Furthermore, the integration of features like sender reputation and phishing website identification, coupled with techniques like TF-IDF vectorization and BERT tokenization, enhanced the discriminatory power of the models. The findings underscore the importance of continuous innovation and refinement in spam filtering methodologies to address the escalating menace of spam emails. This research contributes valuable insights to the field of cybersecurity and offers practical guidance for developing more effective spam detection systems to safeguard users and organizations from the threats posed by unsolicited and malicious email content.

References

1. *Kaspersky's 2022 spam and phishing report*. Secure list English Global securelistcom
2. Y. Kho dun, *How many emails are sent per day?*. MacKeeper. Ani Petrosyan, & 8, M. (2023, March 8). Spam *e-mail traffic share 2022*.Statista (2023)
3. A. Petrosyan, & 8, M. (2023). *Spam e-mail traffic share 2022*. Statista. https://www.statista.com/statistics/420400/spam-email-traffic-share-annual
4. H.H. Lokesh, Gururaj, G.B. Gowda., Phishing website detection based on effective machine learning approach. J. Cyber Security Technol. **5**(1), 1–14 (2021)
5. J. Saleh, A. Karim, B. Shanmugam et al., An intelligent spam detection model based on artificial immune system. Information **10**(6), 209 (2019)
6. A. Naeem et al., Machine learning techniques for spam detection in email and IoT platforms: analysis and research challenges. Security Commun. Netw., 1–19 (2022)
7. S. Thashina et al., Email based spam detection. Int. J. Eng. Res. Technol. (IJERT) (2020)
8. K. Nikhil, S. Sonowal, Email spam detection using machine learning algorithms. *2020 Second International Conference on Inventive Research in Computing Applications (ICIRCA)*. IEEE (2020)
9. Q. Yaseen, Spam email detection using deep learning techniques. Procedia Comput. Sci. **184**, 853–858 (2021)
10. D. Kingshuk, N. Kar, Email spam detection using deep learning approach. *2022 International Conference on Machine Learning, Big Data, Cloud and Parallel Computing (COM-IT-CON)*. Vol. 1. IEEE (2022)
11. D. Sarah Jane, M. Buckley, D. Greene, SMS spam filtering: methods and data. Expert Syst. Appl. **39**(10), 9899–9908 (2012)
12. C. Priyanga, K. Qian, The impact of data preprocessing on the performance of a naive bayes classifier. *2016 IEEE 40th Annual Computer Software and Applications Conference (COMPSAC)*. Vol. 2. IEEE (2016)

13. S. Ksh Nareshkumar, et al., A novel approach for dimension reduction using word embedding: An enhanced text classification approach. *International Journal of Information Management Data Insights* 2.1, 100061 (2022)
14. Kumar, K. Varun, M. Ramamoorthy, Machine learning-based spam detection using naïve bayes classifier in comparison with logistic regression for improving accuracy. J. Pharmaceutical Negative Results, 548–55 (2022)
15. Z. Hai, Research of text classification based on TF-IDF and CNN-LSTM. J. Phys. Conf. Series. Vol. 2171. No. 1. IOP Publishing (2022)

An Enhanced Caries Detection and Prediction Using DentSU_Net

L. Jani Anbarasi, R. Neeraja, H. Sharen, Malathy Jawahar, and Vinayakumar Ravi

Abstract Segmenting medical images is crucial for clinical diagnosis and case analysis. Currently, most of the successful techniques are based on U-shaped encode and decoder-based convolutional neural networks (CNNs). A major drawback of these approaches is their limited capacity to establish highly relevant pattern connections and comprehensive contextual associations, leading to imprecisions during segmenting regions of interest. To address this limitation and improve the utilization of global semantic features with minimal semantic representation between encoding and decoding stages, the proposed work DentSU_Net enhanced independent channel elite features using Swin_Unet incorporated through Squeeze and Excitation (SE) attention layer to effectively segment the dental caries from panoramic X-ray images. This layer selectively emphasizes informative features and suppresses less relevant features recalibrating the importance of different channels for better learning and representation features for precisely identifying the caries region from dental images. The experimental results state that the proposed DentSU_Net attained a dice coefficient of 0.90, accuracy of 0.97, IoU of 0.83, sensitivity of 0.85, and specificity of 0.99. The results outperform the performance of prior studies and can help dentists in automatically and more effectively segmenting the caries zone.

L. J. Anbarasi · R. Neeraja · H. Sharen
School of Computer Science and Engineering, Vellore Institute of Technology, Chennai, India
e-mail: janianbarasi.l@vit.ac.in

R. Neeraja
e-mail: neeraja.r2020@vitstudent.ac.in

H. Sharen
e-mail: Sharen.h2020@vitstudent.ac.in

M. Jawahar
Leather Process Technology Division, CSIR-Central Leather Research Institute, Adyar, Chennai, India
e-mail: malathyj@clri.res.in

V. Ravi (✉)
Center for Artificial Intelligence, Prince Mohammad Bin Fahd University, Khobar, Saudi Arabia
e-mail: vravi@pmu.edu.sa

B. Pradhan and S. Mukhopadhyay (eds.), *IoT Sensors, ML, AI and XAI: Empowering A Smarter World*, Smart Sensors, Measurement and Instrumentation 50,
https://doi.org/10.1007/978-3-031-68602-3_22

Keywords Swin_Unet · Dental caries detection · Caries segmentation · CNN · Squeeze and Excitation attention model

1 Introduction

Dental caries detection is an essential component of preventive dentistry, serving the purpose of early identification and diagnosis of dental decay or cavities. Dental caries has been categorized into various grades by the American Dental Association based on the extent and progression of lesions such as normal, initial, moderate, and extensive spread [1]. Dental caries has been categorized into various grades by the American Dental Association based on the extent and progression of lesions such as normal, initial, moderate, and extensive spread [1]. According to the World Health Organization (WHO), dental caries is a widespread health issue with global implications. Dental caries is a progressive and expanding dental condition that, if not addressed, can lead to significant harm to adjacent teeth and perhaps result in tooth loss. The primary determinant is the existence of bacteria in the oral cavity, which generates acids that can damage durable structures like dentin, enamel, and cementum [2]. Diagnosing initial posterior proximal caries through routine clinical examinations poses a significant challenge in clinical practice [3]. Radiographs are commonly used in clinical practice as a diagnostic tool and standard imaging system for dental treatments due to their cost-effectiveness. The panoramic X-rays provide a comprehensive view of the maxillomandibular region, allowing for a more detailed understanding of the surrounding contextual information. Furthermore, it has a lower radiation rate compared to alternative techniques [4]. By utilizing panoramic radiographs, dentists are able to identify a range of dental issues such as bone irregularities, cavities, concealed dental structures, and fractures resulting from accidents that are difficult or practically impossible to detect through visual examination [5]. Dentists can utilize these techniques to develop a suitable treatment strategy for each patient. In certain situations, dentists may be assigned the task of analyzing radiographs, namely panoramic X-rays, because the analysis process is performed manually. This procedure is laborious and necessitates a specific level of proficiency to differentiate pertinent dental characteristics from extraneous ones such as jaw bones, nasal bones, and spine bones [6]. Inconsistencies can arise as a result of the disparity in expertise levels across different dentists. The analysis of radiographs appears to fluctuate partly due to individual expertise, biases, and competencies [7]. In certain instances, the inconsistency in professional dentists' proficiency in analyzing radiographs may lead to incorrect diagnosis and improper treatment [8, 9]. Nevertheless, this system has certain limitations. Tracing minor lesions is not feasible due to the limited ability of X-ray scans to depict the three-dimensional nature of tooth structures in a two-dimensional format.

Several automated systems have been developed to address these complexities. Some studies employed traditional machine learning approaches like active contour

[9], Bayesian methods [10], and support vector machines [11] for dental caries detection by extracting features manually. Artificial intelligence (AI) has gained significant popularity and widespread use in various medical disciplines, including hepatology [10], brain segmentation [11], mammography [12], cardiac image segmentation [13] and dentistry [14, 15]. The incorporation and fusion of AI techniques in dental care specifically Convolutional Neural Network (CNN) based models have demonstrated exceptional performance in detecting caries lesions [16, 17]. The U-Net architecture is widely recognized for its high performance in challenges involving the segmentation of medical images using CNNS due to its efficacy in pixel-level classification and its adaptability in processing both 2D and 3D medical images [18]. The primary contributions of the proposed work are as follows:

- Dental Caries from panoramic images are analyzed, specifically focusing on its automatic detection and segmentation using U-shaped CNN and transformers.
- Swin Transformer enhanced using Squeeze and Excitation attention model in the bottleneck of U-Net architecture is proposed generating a hybrid DentSU_Net for dental caries segmentation from panoramic X-Ray images
- This resultant multiscale characteristic helps in preserving essential fused features throughout the decoding process resulting in enhanced segmentation of dental caries. This DentSU_Net model efficiently extracts features and hierarchical representations for automatic panoramic X-ray segmentation of caries.

Section 2 covers related articles, Sect. 3 details the proposed architecture, and Sect. 4 elaborates on the results and discussion. Section 5 comprises the conclusion and future work.

2 Related Works

The literature articles offer a thorough overview of the latest advancements in deep learning techniques for automatic dental caries detection. Imak et al. [19] proposed Multi-Input Deep CNN Ensemble (MI-DCNNE) architecture to automatically diagnose caries in teeth using periapical images. The framework utilized a score-based ensemble scheme to enhance the performance by employing score fusion in the softmax layer. The dataset comprises of 340 periapical images which contain both non-caries X-rays and caries X-rays. The study concluded that the proposed framework effectively identified dental caries, with an 99.13% accuracy rate.

Alharbi et al. [20] utilized various U-Net models to identify caries lesions in dental X-rays using DNS X-ray dataset, comprising 1500 images acquired from Ivisionlab. The DNS dataset was enhanced by identifying the dental cavities in X-ray images and generated a binary representation of the images. Furthermore, employed U-Net3+, U-Net, U-Net++, and on the extended DNS dataset to get knowledge about the hierarchical characteristics and improve the boundary of the cavity. The findings indicate that U-Net3+surpasses the preceding U-Net versions with 95% accuracy score.

Qayyum et al. [21] proposed a highly efficient self-training based centroid cropping sampling (CCS) approach for detecting and segmenting caries with limited labelled images to train the main model and an extensive collection of unlabeled images to train the student model. The performance of the proposed method is evaluated in different models such as FCN, PSPNet, LRASPP, FPN, LinkNet and Deeplab-v3 with a backbone of pre-trained ResNet-50, ResNet-101, MobileNet-v3 networks using Dental Caries Detection (DCD2) dataset comprised of 229 dental radiographs where 141 X-rays were annotated, and 88 X-rays were unlabeled. The Deeplab-v3 model with ResNet-101 backbone achieved an average accuracy of 99.43%.

Mahran et al. [22] demonstrates the feasibility of implementing the Attention U-Net architecture for the purpose of segmenting teeth. The paper showcases the exceptional efficacy of the proposed network for segmenting teeth precisely utilizing Tufts Dental X-Ray Dataset. During tenfold cross-validation, the model showed 95.01% average dice coefficient, 90.6% intersection over union, and 98.82% pixel accuracy.

The objective of the study by Park et al. [23] was to assess the effectiveness of a deep learning algorithm in detecting caries by accurately segmenting the tooth surface using 2348 intraoral images collected from 445 participants. The dataset comprised of randomly arranged 1638 X-rays for training, 410 X-rays for validation and 300 X-rays for testing. The image segmentation, classification, and caries identification process utilized CNNs such as and Faster R-CNN, U-Net, ResNet-18. The accuracy and area under ROC (Receiver Operating Characteristic) curve for the caries image classification algorithm were enhanced to 0.813 and 0.837, respectively, from their initial values of 0.758 and 0.731. Saini et al. [24] used CNN to identify dental cavities at an early stage using digital color images in order to facilitate prompt and efficient treatment. The pre-trained networks such as Inception-V3, ResNet-50, VGG-16 and VGG-19 were trained. The binary X-ray dataset containing caries and without caries has been used for training, validation, and testing. The Inception-V3 outperformed with 99.89% and 98.95% training and validation accuracy. Additionally, Inception-V3 exhibits the lowest loss compared to the VGG-16 CNN models.

A diagnostic algorithm utilizing a neural network has been devised to diagnose dental caries in digital radiography by Geetha et al. [25]. The diagnostic system consists of window-based adaptive thresholding, morphological operations, back-propagation neural network, Laplacian filtering, and statistical feature extraction. The backpropagation method is employed for classifying the tooth surface as normal or with caries. The dataset comprised 105 X-ray images for training the network using a tenfold cross validation technique. The framework achieved 97.1% accuracy with 2.8% False Positive (FP) rates.

The proposed technique by Lakshmi et al. [26] employed the Sobel edge detection technique in conjunction with DCNNs to accurately predict the presence of cavities during their initial development. The proposed approach utilized the Sobel algorithm to identify the boundaries in dental images. This proposed system efficiently predicted caries and achieved an accuracy of 96.08%. The work by Geetha et al. [27] proposed a method for diagnosing dental caries utilizing a combination of Support Vector Machine classifiers, Laplacian filter, and adaptive thresholding. The proposed method is compared to Otsu thresholding, watershed segmentation, and active contouring method in the analysis. The Adaptive thresholding demonstrates superior performance, with 96.9% accuracy and 96.1% precision..

3 Methodology

This work examined Dental Caries pictures for caries segmentation utilizing a customized DentSU_Net model.

3.1 DentSU_Net Architecture

A U-shaped encoder-decoder network, called DentSU_Net has been analyzed to leverage the capabilities of Transformers in dental medical segmentation. The encoder employs a shifted windows based Swin-Transformer to efficiently capture distinct characteristics in dental caries images. The decoder in DentSU_Net is symmetrical and utilizes the Swin Transformer architecture. The up sampling operations are performed using patch expanding layers for maintaining the feature map's spatial resolution. The Swin Transformer is used for feature extraction all through the encoding phase. Unlike Convolutional Neural Networks (CNNs), this model generates a structured representation of features and accurately captures long-range dependencies. Swin Transformer performs self-attention within a specific window of patches, rather than across all patches. This approach is computationally advantageous, as the complexity of self-attention scales linearly with image size. Consequently, it is well-suited for addressing semantic segmentation tasks involving high-resolution images.

The input panoramic X-rays images (Ip_{pdi}) are splitted into non-overlapping patches (Pa_{pdi}) of range $\frac{H}{P}X\frac{W}{P}XC$ patches where H refers height, W refers width and P refers patch size (Eq. 1).

$$P_{\mathrm{a}\,pdi} = f_{pp}(Ip_{pdi}) \tag{1}$$

Every patch is considered an input feature and is transformed into a c-dimensional space by a linear embedding layer and applies a linear transformation to these input features resulting transformed output L_E with weight (W_i) and bias (b_i) (Eq. 2).

$$L_{Epdi} = f_{pp}(W_i(P_{\mathrm{a}\,pdi}) + b_i) \tag{2}$$

The linearly embedded features undergo a series of transformations through the attention mechanisms, feedforward networks, and normalization layers within the.

Swin Transformer resulting capture complex patterns through hierarchical feature extraction (Eq. 3).

$$ST_{pdi} = f_{pp}(L_{E\,pdi}) \tag{3}$$

The encoder includes Swin Transformer blocks followed by patch merging layers that downsamples to filter out redundant features and broaden the receptive field resulting in more comprehensive features (Eq. 4).

$$P_{\mathrm{M}\,pdi} = f_{pp}(ST_{pdi}) \tag{4}$$

The resultant comprehensive features $P_{\mathrm{M}\,pdi}$ are enhanced using the bottleneck layer which comprises convolutional layer followed by squeeze and attention layer (SE_{attn}) and convolution (Co) of filter size 3 × 3. This layer selectively emphasizes informative features and suppress less relevant features recalibrating the significance of various channels for better learning and representation features as $B_{\mathrm{L}\,pdi}$ (Eq. 5).

$$B_{\mathrm{L}\,pdi} = f_{pp}[Co^{3x3}(SE_{attn}(Co^{3x3}(P_{\mathrm{M}\,pdi})))] \tag{5}$$

The decoder part comprised of Swin Transformer blocks and patch expanding layers that increases the size of the patches to recover the spatial resolution of the generated feature maps. The patch expanding layer helps in increasing the size of the feature maps, ensuring that the final resultant output ($P_{\mathrm{E}\,pdi}$) has a higher resolution as shown in Eq. 6.

$$P_{\mathrm{E}\,pdi} = f_{pp}(B_{\mathrm{L}\,pdi}) \tag{6}$$

This high resolution features are subjected to swin transformer to model long-range dependencies suitable for dental caries segmentation tasks to relate the context and relationships between pixels to determine the region of interest efficiently $ST_{\mathrm{E}\,pdi}$ as given in Eq. 7.

$$ST_{\mathrm{E}\,pdi} = f_{pp}(P_{\mathrm{E}\,pdi}) \tag{7}$$

This resultant multiscale characteristic helps in preserving essential fused features throughout the decoding process resulting enhanced segmentation of dental caries. This DentSU_Net model efficiently extracts features, and hierarchical representations for automatic panoramic x-ray segmentation of caries. The proposed custom DentSU_Net architecture is displayed in Fig. 1.

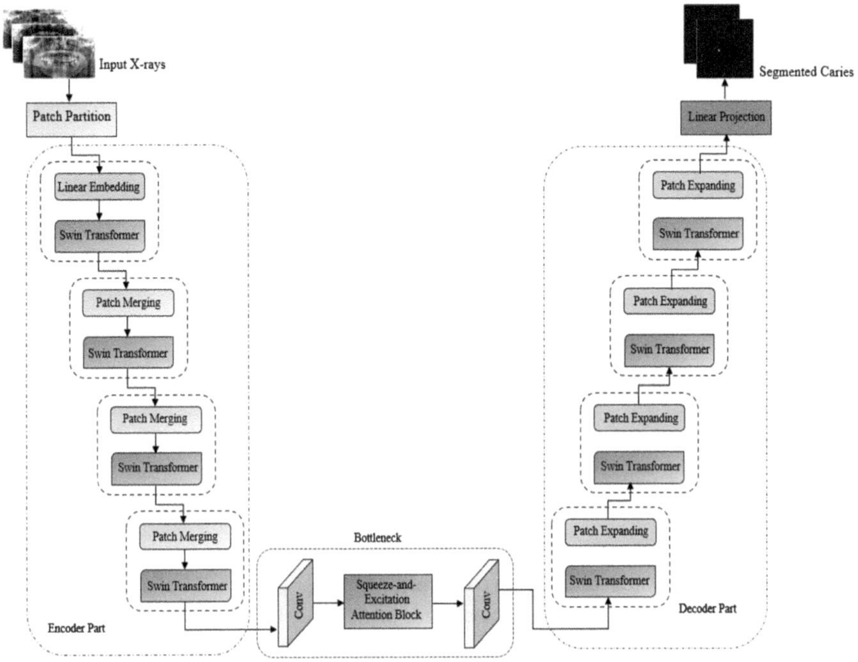

Fig. 1 The DentSU_Net architecture

3.2 *Swin Transformer*

The Swin Transformer block features a shifted windows design illustrated in Fig. 2. A Swin Transformer block consists of a multi-head self-attention module, Layer Norm (LN) layer, a residual connection, and a 2-layer Multi-Layer Perceptron (MLP) with GELU non-linearity. The first block uses a window-based multi-head self-attention module (W-MSA), and the second block uses a shifted window-based multi-head self-attention module (SW-MSA).

The continuous Swin Transformer can be represented by its input and output given in Eqs. 8 to 11.

$$\hat{o}^{l} = W_i - MS_A(L_N\left(o^{l-1}\right) + \left(o^{l-1}\right) \tag{8}$$

$$o^{l} = M_{LP}(L_N\left(\hat{o}^{l}\right)) + \hat{o}^{l} \tag{9}$$

$$\hat{o}^{l+1} = SW_i - MS_A(L_N\left(o^{l}\right)) + o^{l} \tag{10}$$

$$o^{l+1} = M_{LP}(L_N\left(\hat{o}^{l+1}\right)) + \hat{o}^{l+1} \tag{11}$$

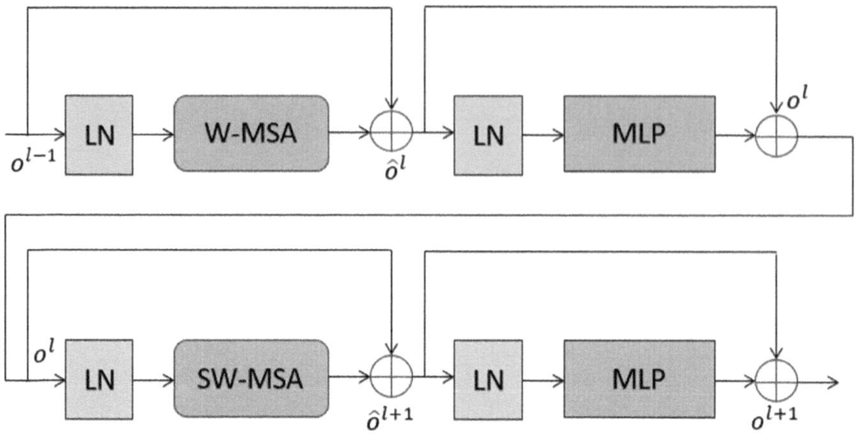

Fig. 2 Architecture of basic Swin Transformer

where $\hat{o}^l$ and o^l represent the output of Swin Transformer layer.

The encoder analyses inputs with a resolution of $\frac{H}{P}X\,\frac{W}{P}XC$ patches where H refers height, W refers width and P refers patch size and passes them into Swin Transformer blocks to acquire representations features. Simultaneously, the patch merging layer downsample the features by c-dimension and increases the feature dimension in 'n' fold. The operation is iterated four times in the encoder.

The patch merging layer combines the input patches into a unified entity. This operation reduces the feature resolution by a factor of 'n'. The input patches are concatenated together by the patch merging layer. This downsamples the feature resolution by 'nx'. A linear layer is applied to the concatenated features to unify the feature dimension to 'nx' the original dimension because the concatenation process improves the feature dimension by '2nx'. The bottleneck is formed by employing two sequential Convolutional layers that are strengthened with an SE attention model to acquire deeper feature representations. The decoder has a patch expanding layer and a Swin Transformer block to increase retrieved deep features by reconfiguring the feature maps of neighboring dimensions.

3.3 Bottleneck Layer

The bottleneck is constructed using only two consecutive Convolutional layers enhanced using Squeeze and Excitation attention model to learn deep feature representations. The feature resolution and dimension remain intact during this process. The SE attention block is composed of 2 phases, namely squeeze phase and an excitation phase. In squeeze phase, the channel dependency challenges are addressed by compressing the global spatial features into a channel descriptor for all encode swin

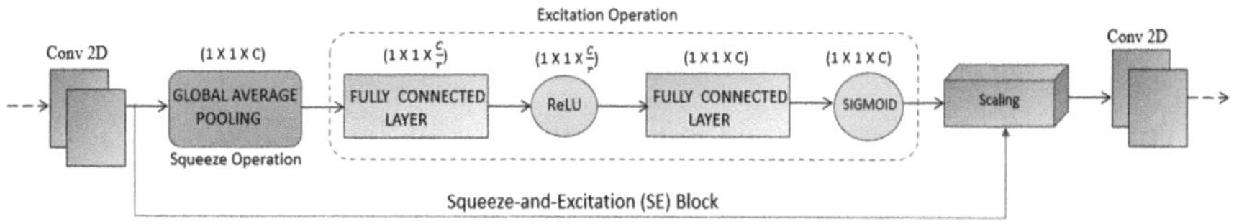

Fig. 3 Bottleneck layer with squeeze-and- excitation attention block

transformer features $P_{M\,pdi}$. The most basic kind of global average pooling operation is employed to compute channel based statistics. The input feature maps $P_{M\,pdi}$ underwent a reduction in size throughout their spatial dimensions, specifically the height (HT) and width (WD), resulting in the generation of a channel wise statistical value $z_S \in R^{CH}$. Let z_{S_e} be the eth element of z_S, as computed in Eq. (12).

$$z_{S_e} = f_{Squeeze}\left(P_{M\,pdi}\right) = \frac{1}{HT \times WD}\sum_{j=1}^{HT}\sum_{k=1}^{WD} P_{M\,pdi}(j, k) \tag{12}$$

The squeeze phase is thereafter accompanied by excitation phase in order to carry out the dynamic feature wise recalibration. The squeeze phase yields embedded global feature information, which is subsequently utilised to identify channel related dependencies in the excitation phase. The utilization of a gating operation, in conjunction with a sigmoid function activation, as demonstrated in Eq. (13).

$$sm = f_{Excitation}(z_S, WX) = \sigma(gating\,(z_S, WX)) = \sigma(WX_2 ReLU(WX_1\, z_S)) \tag{13}$$

where $WX_1 \in R^{\frac{ch}{re}XCH}$ is the dimensionality reduction layer, $WX_2 \in R^{CHX\frac{ch}{re}}$ is the dimensionality increasing layer, $ReLU$ and σ (sigmoid) are the activation functions, $'ret'$ is the reduction ratio. These features are convolved using convolutional layer resulting informative features for better learning and representation features as $B_{L\,pdi}$ as shown in Fig. 3.

4 Results and Discussion

This section offers in-depth information on the dataset used, the experiments conducted, the model training method, and the validation procedures. The paper also includes a discussion on the performance comparison of existing CNN models, and the proposed approach.

4.1 Experimental Setup

The proposed model was executed utilizing the Python TensorFlow program. The computation employed Keras library. The system setup utilized for training and testing the transfer learning and DentSU_Net model included an Intel Xeon Gold 6230 CPU running at 2.10 GHz, an NVIDIA Quadro RTX 5000 graphics card with 16 GB of VRAM, 64 GB of RAM, and a 2 TB hard drive.

4.2 Data Acquisition, Pre-Processing and Augmentation

The kaggle dataset [28] contained 100 panoramic X-ray images along with its mask for segmenting the caries region. The data is augmented to reduce overfitting by expanding the training dataset. Zoom augmentation technique enhances the model's ability to identify items irrespective of their dimensions. Furthermore, it improved the model's capacity to identify incomplete items. The rotation range was set to 2 degrees, and the images are zoomed within a range of 0.2 to 0.5. Rotation and zoom augmentation techniques are executed meticulously to make minimum changes to the image. Setting a rotation range of 2 to 5 degrees and using a zoom range of 0.2 to 0.5 for dental caries images helps to improve the model's performance and stability, while still ensuring the diagnostic accuracy remains intact. Also images were subjected to width and shifting of range 0.2 and a total of 1000 images were generated after augmentation. Figure 4 shows the different variations of a Dental panoramic x-ray images sample after the augmentation process and the mask corresponding to the sample images.

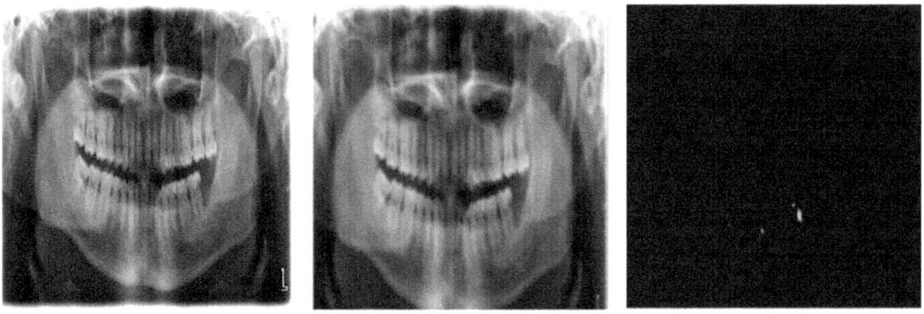

(a) (b) (c)

Fig. 4 **a** Sample original, **b** Augmented images and **c** Mask

4.3 Hyperparameter Tuning

Hyperparameter tuning adjusted key parameters like batch size, epochs, learning rate, and optimization units to enhance the model's performance in comparison to the conventional method. The analysis employed an early stopping strategy and utilized Adam optimizers, which led to improved performance. The loss function was utilized to obtain the maximum values for all the accuracy measurements. The early stopping strategy allowed training iterations to stop when the performance of the model ceased to increase on a separate validation dataset. The ReduceLROnPlateau function is used to decrease the learning rate when the model stagnates in the learning process. The learning rate initially ranged from 0.000001 to 1. The updated learning rate is increased by 0.1 factor. The patience option is configured as 10, specifying the threshold for the number of iterations without improvement that will trigger a fall in the learning rate. The model is trained for 100 epochs.

4.4 DentSU_Net Model Training and Validation

The DentSU_Net framework is trained using a dataset consisting of X-ray images with 80% used for training and 20% for testing. The proposed model provided a comprehensive analysis of the effectiveness of several backbones of U-Net and Swin Transformer [29] enhanced using squeeze and Excitation attention model in extracting regions of interest (ROI), using both qualitative and quantitative research to evaluate their performance. The image segmentation outcomes of different areas, including larger and smaller ROI_s (Region Of Interest) of the test dataset are evaluated and assessed using the validation and training methods. The dimensions of the training and test images are $256 \times 256 \times 3$. The default batch size is considered as 8. The model is trained for 100 epochs using the Adam optimizer by incorporating an early stopping, 0.000001 learning rate, ReduceLROnPlateau with 0.1 factor. The model ended at early training for 57th epoch with 1.0000e-06 learning rate. The training model reached 0.2797 loss value and 0.8646 Intersection Over Union (IoU) score..

4.5 Evaluation Metrics

An evaluation of the DentSU_Net models' performance was conducted using metrics such as IoU, Dice Coefficient, Accuracy, Sensitivity, Pixel accuracy, Specificity. The IoU or Jaccard Index is a measure that quantifies the overlap between the predicted and the ground truth segmentation. The calculation involves dividing the area of the union between the two regions by the overlap area. The Dice Coefficient is used to quantify the similarity or overlap between two sets. The recall or sensitivity for pixel classification measures the capacity to correctly recognize real positive instances, while the true negative rate or specificity evaluates the accuracy of recognizing true negative classes. The evaluation parameters used to measure skin lesion segmentation are shown in Eqs. (14) to (19) where $TP_{os}, TN_{eg}, FP_{os}, and FN_{eg}$ refers to true positive, true negative, false positive and false negative.

$$DiceScore = \frac{2 * TP_{os}}{(TP_{os} + FP_{os}) + (TP_{os} + FN_{eg})} \tag{14}$$

$$IOU = \frac{AreaofIntersection}{AreaofUnion} = \frac{TP_{os}}{TP_{os} + FP_{os} + FN_{eg}} \tag{15}$$

$$Accuracy = \frac{TP_{os} + TN_{eg}}{TP_{os} + TN_{eg} + FP_{os} + FN_{eg}} \tag{16}$$

$$Sensitivity = \frac{TP_{os}}{TP_{os} + FN_{eg}} \tag{17}$$

$$Specificity = \frac{TP_{os}}{TP_{os} + FP_{os}} \tag{18}$$

$$BalancedAccuarcy = \frac{Sensitivity + Specificity}{2} \tag{19}$$

The effectiveness of the learning progress during testing and training data across 100 epochs using DentSU_Net is shown in Table 1. The results are compared with existing model [30]. The displayed outcomes include the loss curve over 100 epochs and the IoU learning curve using the Jaccard Index for both models that employ different encoders and optimizers. The primary accuracy metric selected was IoU because of its exceptional performance in skin segmentation. By tuning loss function, optimizer, and batch size hyperparameters, the experiment yielded a notable improvement in F1 score, Accuracy and IoU and as shown in Table 2. Table 3 displays the segmented caries region along with the original mask to show the proposed DentSU_Net effectiveness.

4.6 Comparison with Existing Research Models for Dental Caries Segmentation

Qayyum et al. [21] performed centroid cropping sampling (CCS) using DL techniques attaining 99.43%. Park et al. [23] was to assess the effectiveness of a deep learning algorithm to detect tooth surface resulting in 83% accuracy. Saini et al. [24] used VGG-16, VGG-19, Inception-V3, and ResNet-50 to identify dental cavities and attained 99.89%.

Table 1 Displaying the assessment of the intersection over union (iou) score and total loss curve over epochs to show the learning curve

	IoU	Loss
Swin_Unet	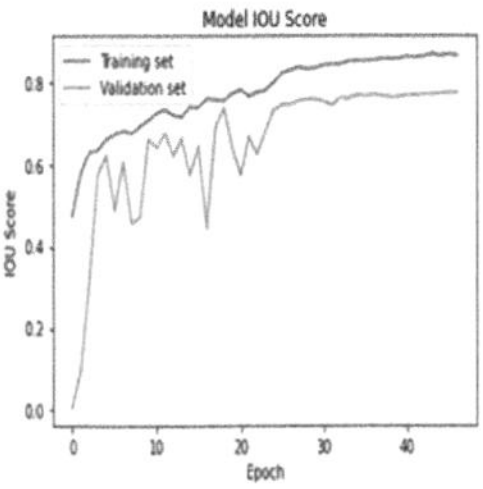	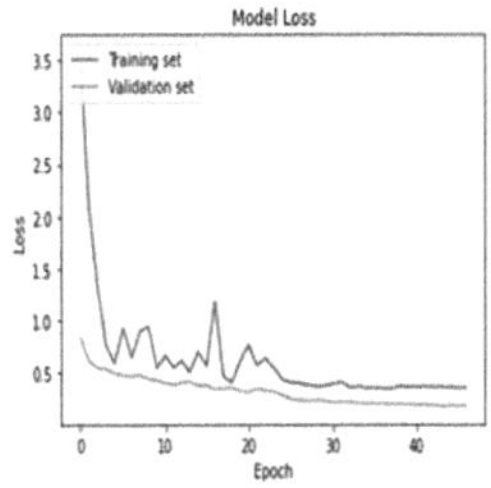
DentSU_Net (Proposed)	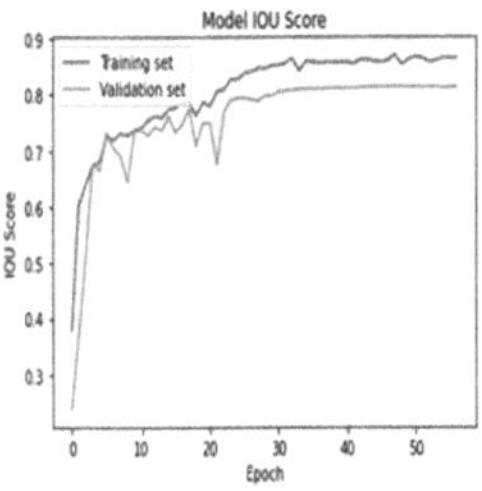	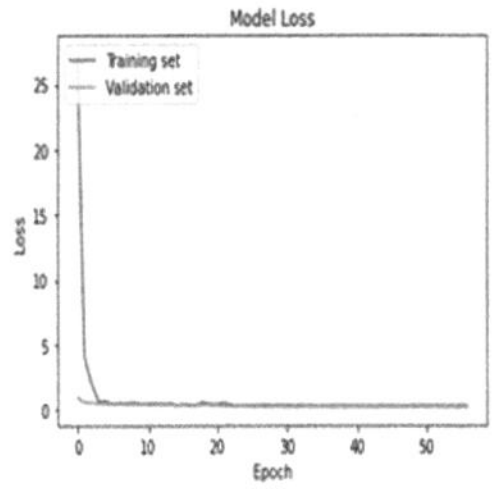

Table 2 Performance evaluation during the testing phase of the DentSU_Net model

Model	DSC	IoU	Accuracy	Sensitivity	Specificity	F1-Score
Swin_Unet	0.88	0.79	0.96	0.81	0.95	0.90
DentSU_Net (Proposed)	0.90	0.83	0.97	0.85	0.99	0.92

Geetha et al. [25] performed morphological operations, Laplacian filtering, statistical feature extraction, window-based adaptive thresholding and attained 97.1% accuracy. Lakshmi et al. [26] analyzed Sobel edge detection technique and achieved an accuracy of 96.08%. The comparison of the existing methodologies with the proposed DentSU_Net is illustrated in Table 4.

5 Conclusions and Future Works

An extensive investigation was performed on the custom DentSU_Net, a hybrid model that combines Swin_UNet with Squeeze and Excitation attention models, to evaluate their efficacy in dental caries image segmentation. The Kaggle dataset was used for an experimental study to assess the ability of a customized DentSU_Net model to segment caries regions in panoramic X-ray images. The study showed that the evaluated DentSU_Net models outperformed Swin_UNet, with similar results in

Table 3 Segmented dental caries images

Original image	Ground truth	Segmented region
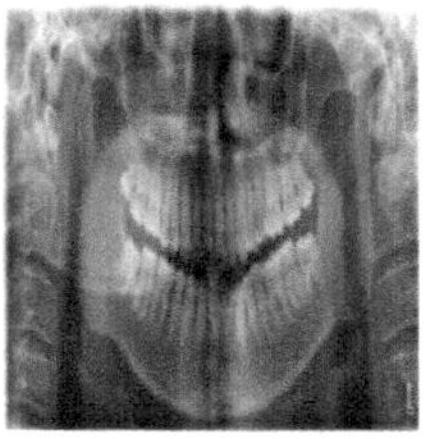		
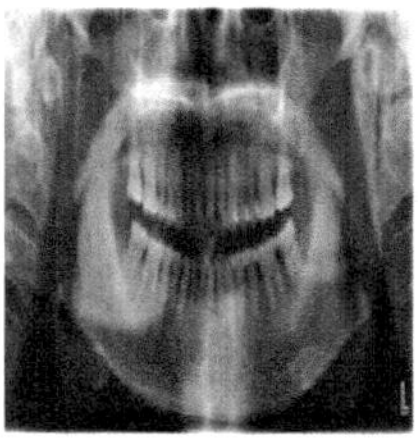	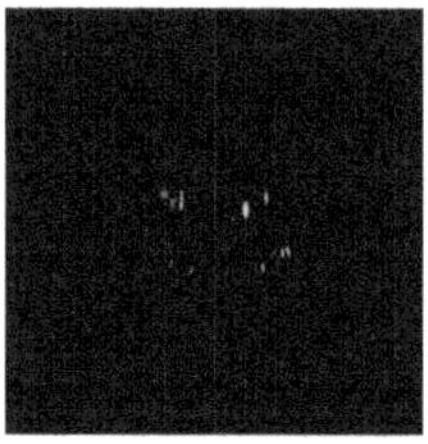	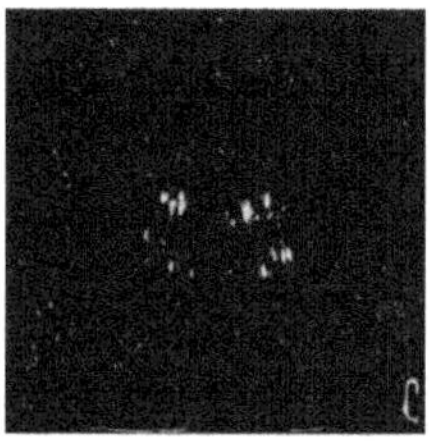
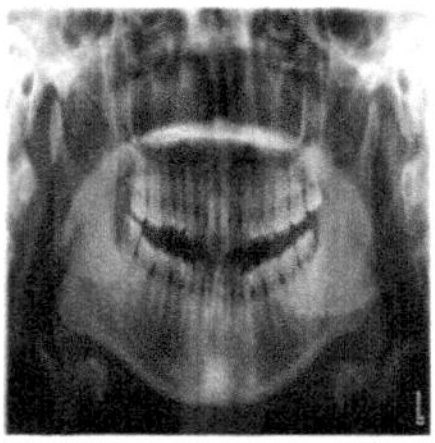		

Table 4 Comparison with the existing research work

References	Types	Network rchitecture/algorithm	Accuracy
[21]	FCN, PSPNet, LRASPP, FPN, LinkNet and Deeplab-v3	Deep Learning Algorithm	99.43%
[23]	U-Net, ResNet-18, and Faster R-CNN	Deep Learning algorithm	83%
[24]	VGG-16, VGG-19, Inception-V3, and ResNet-50	Transfer Learning Models	99.89%
[25]	Morphological operations, Laplacian filtering	Machine Learning Model	97.1
[26]	Sobel algorithm	Machine Learning Model	96.08%
[30]	Panoramic X-Rays	Swin_UNet	0.96
Proposed	Panoramic X-Rays	DentSU_Net	97%

measures including dice coefficient, IoU, accuracy, sensitivity, specificity, and F1-Score. The improved feature maps of the encoders helped to identify caries formations with sharper edges. The results of the experiment showed that the DentSU_Net model performed exceptionally well, with a dice coefficient of 0.90, accuracy of 0.97, IoU of 0.83, and sensitivity and specificity of 0.85 and 0.99, respectively. It is important to weigh the potential trade-offs among performance, speed, and complexity depending on clinical environments, research purposes, and resource limitations.

References

1. D.A. Young et al., The American dental association caries classification system for clinical practice: a report of the American dental association council on scientific affairs. J. Am. Dent. Assoc. **146**, 79–86 (2015)
2. S. Lee et al., Deep learning for early dental caries detection in bitewing radiographs. Sci. Rep. **11**, 1–8 (2021)
3. S. Williams, L. Jamieson, A. MacRae, C. Gray, Review of Indigenous oral health. Australian Indigenous Health Bulletin **11**(2) (2011)
4. Z. Kong et al., Automated maxillofacial segmentation in panoramic dental X-ray images using an efficient encoder-decoder network. IEEE Access **8**, 207822–207833 (2020). https://doi.org/10.1109/access.2020.3037677
5. C.-W. Wang et al., A benchmark for comparison of dental radiography analysis algorithms. Med. Image Anal. **31**, 63–76 (2016). https://doi.org/10.1016/j.media.2016.02.004
6. A. Wirtz, S.G. Mirashi, S. Wesarg, Automatic teeth segmentation in panoramic X-ray images using a coupled shape model in combination with a neural network. *Medical Image Computing and Computer Assisted Intervention—MICCAI 2018*, pp. 712–719 (2018). https://doi.org/10.1007/978-3-030-00937-3_81
7. S. Akmal, Y. Jin Tiau, Image quality assessment in panoramic dental radiography: a comparative study between conventional and digital systems. Quant. Imaging Med. Surgery **3**(1), pp. 43–48 (2013), www.ncbi.nlm.nih.gov/pubmed/23483085, https://doi.org/10.3978/j.issn.2223-4292.2013.02.07. Accessed 24 Nov. 2022
8. R. Fitzgerald, Error in radiology. Clin. Radiol. **56**(12), 938–946 (2001). https://doi.org/10.1053/crad.2001.0858
9. R.Ó. Brady, P. Laoide, McCarthy, R. McDermott, Discrepancy and error in radiology concepts, causes and consequences. Ulster Med. J. **81**(1), 3–9 (2012). https://www.ncbi.nlm.nih.gov/pmc/articles/PMC3609674/
10. M. Sharma, N. Kumar, Improved hepatocellular carcinoma fatality prognosis using ensemble learning approach. J. Ambient. Intell. Humaniz. Comput. **13**, 5763–5777 (2022)
11. V. Cherukuri, P. Ssenyonga, B. Warf, A. Kulkarni, V. Monga, S. Schiff, Learning based segmentation of CT brain images: application to post-operative hydrocephalic scans. IEEE Trans. Biomed. Eng. (2017)
12. E. Kumaraswamy, S. Kumar, M. Sharma, An invasive ductal carcinomas breast cancer grade classification using an ensemble of convolutional neural networks. Diagnostics **2023**, 13 (1977)
13. Y. Liu, S. Chen, L. Wee, A. Dekker, I. Bermejo, Deep learning based unpaired image-to-image translation applications for medical physics: a systematic review **68**(5), 05TR01–05TR01 (2023), https://doi.org/10.1088/1361-6560/acba74
14. C. Huang, J. Wang, S. Wang, Y. Zhang, A review of deep learning in dentistry. Neurocomputing **2023**, 126629 (2023)
15. H. Ding, J. Wu, W. Zhao, J.P. Matinlinna, M.F. Burrow, J.K. Tsoi, Artificial intelligence in dentistry—a review. Front. Dent. Med. **4**, 1085251 (2023)

16. D.L. Duong, M.H. Kabir, R.F. Kuo, Automated caries detection with smartphone color photography using machine learning. Health Inform. J. **27**, 14604582211007530 (2021)
17. Z. Zheng, H. Yan, F.C. Setzer, K.J. Shi, M. Mupparapu, J. Li, Anatomically constrained deep learning for automating dental cbct segmentation and lesion detection. IEEE Trans. Autom. Sci. Eng. **18**, 603–614 (2020)
18. R. Wang, T. Lei, R. Cui, B. Zhang, H. Meng, A.K. Nandi, Medical image segmentation using deep learning: A survey. IET Image Process. **16**, 1243–1267 (2022)
19. Imak, A. Celebi, K. Siddique, M. Turkoglu, A. Sengur, I. Salam, Dental caries detection using score-based multi-input deep convolutional neural network. IEEE Access **10**, 18320–18329 (2022), https://doi.org/10.1109/ACCESS.2022.3150358.
20. A. Albahbah, H. El-Bakry, S. Abdelghany, Detection of caries in panoramic dental x-ray images using back-propagation neural network. Int. J. Electron. Commun. Comput. Eng. **7**, 250–256 (2016)
21. A. Qayyum, A. Tahir, M.A. Butt et al., Dental caries detection using a semi-supervised learning approach. Sci. Rep. **13**, 749 (2023). https://doi.org/10.1038/s41598-023-27808-9
22. H. Mahran, A.M. Walid Hussein, S. El Din Mohammed Saber, *Automatic Teeth Segmentation Using Attention U-Net*. Preprints (2023). https://doi.org/10.20944/preprints202306.1468.v2
23. E.Y. Park, H. Cho, S. Kang, S. Jeong, E.K. Kim, Caries detection with tooth surface segmentation on intraoral photographic images using deep learning. BMC Oral Health. **22**(1), 573 (2022). Published 2022 Dec 7. https://doi.org/10.1186/s12903-022-02589-1
24. D. Saini, R. Jain, A. Thakur, Dental caries early detection using convolutional neural network for tele dentistry. *2021 7th International Conference on Advanced Computing and Communication Systems (ICACCS)*, Coimbatore, India, pp. 958–963 (2021), https://doi.org/10.1109/ICACCS51430.2021.9442001
25. V. Geetha, K.S. Aprameya, D.M. Hinduja, Dental caries diagnosis in digital radiographs using back-propagation neural network. Health Inf. Sci. Syst. **8**(1), 8 (2020). Published 2020 Jan 3. https://doi.org/10.1007/s13755-019-0096-y
26. M.M. Lakshmi, P. Chitra, Classification of dental cavities from X-ray images using Deep CNN algorithm. *2020 4th International Conference on Trends in Electronics and Informatics (ICOEI)(48184)*, Tirunelveli, India, pp. 774–779 (2020), https://doi.org/10.1109/ICOEI48184.2020.9143013
27. V. Geetha, K. Aprameya, Analysis of image segmentation techniques for diagnosis of dental caries in X-ray images. *World Academy of Science, Engineering and Technology, Open Science Index 146*. Int. J. Comput. Inform. Eng. **13**(2), 30–33 (2019)
28. https://www.kaggle.com/datasets/thunderpede/panoramic-dental-dataset?select=labels
29. L. Ze, et al., Swin transformer: Hierarchical vision transformer using shifted windows. *Proceedings of the IEEE/CVF International Conference on Computer Vision* (2021)
30. C. Hu et al., Swin-unet: Unet-like pure transformer for medical image segmentation. *European Conference on Computer Vision*. Cham: Springer Nature Switzerland (2022)

Skin Lesions Classification of Dermoscopy Images Using Deep Learning Technique

Abhinav Mishra, Akshaj Tammewar, Akshay Jaiswal, Aman Ali Shaikh, Shilpa Gite, and Biswajeet Pradhan

Abstract Skin lesions are a severe disease and the most predominant type of cancer worldwide. It is becoming more prevalent in modern society, with rising cases every year. The World Health Organization (WHO) claims melanoma is from the most severe forms of skin cancer, affecting well over 100,000 people worldwide each year. While early-stage lesions are frequently treatable, the prognosis worsens dramatically as the illness progresses, enforcing the need to identify cancer correctly and as early as possible. The computer-based analysis of dermoscopy images helped in the early detection of melanoma which had a significant impact on increasing the survival rate of the people who are infected from this. However the order of accurately recognizing the types of melanoma is quite challenging because of the small deviations between the lesions and skin, and there are lots of visual similarities between melanoma and non-melanoma lesions. However, images with a lower resolution than the original skin image are produced as a result of segmenting the skin lesions. In this paper, by using a dermoscopic image of a skin tumor, we present a novel deep learning-based method for resolving the issues with skin lesion analysis aimed at enhancing accuracy and efficiency. The models which being proposed are *inception ResNet-V2, EfficientNet, InceptionNet, and MobileNet*, which are trained and then evaluated on HAM 10000 datasets to compare the performance and accuracy. The suggested strategies produced a robust accuracy in the validation sets. The experimental studies conducted on a clinical dataset reveal that deep learning-based features fare better for classification than conventional machine learning methods.

A. Mishra · A. Tammewar · A. Jaiswal · A. A. Shaikh · S. Gite
Computer Science Engineering Department, Symbiosis Institute of Technology, Symbiosis International (Deemed University), Pune 412115, India

S. Gite (✉)
Symbiosis Centre of Applied AI (SCAAI), Symbiosis International (Deemed) University, Pune 412115, India
e-mail: shilpa.gite@sitpune.edu.in

B. Pradhan (✉)
Centre for Advanced Modelling and Geospatial Information Systems (CAMGIS), School of Civil and Environmental Engineering, Faculty of Engineering and IT, University of Technology Sydney, Ultimo, NSW 2007, Australia
e-mail: Biswajeet.Pradhan@uts.edu.au

B. Pradhan and S. Mukhopadhyay (eds.), *IoT Sensors, ML, AI and XAI: Empowering A Smarter World*, Smart Sensors, Measurement and Instrumentation 50,
https://doi.org/10.1007/978-3-031-68602-3_23

Keywords Skin lesions segmentation · HAM10000 dataset · Deep learning · Multi-class classification · Transfer learning

1 Introduction

In today's world, diseases like cancer are becoming more predominant, and people of every age group suffer from it. There are various types of cancer, with skin cancer being one of the most common [1]. Skin is one of the largest organs in the body, which covers all body parts such as muscles, bones, etc. The use of the skin in the human body is critical as, even minor changes in their functioning, may impact other organs or body parts. In the outside environment, the skin is mostly exposed thus, the chances of disease and infection are very high [2]. A lesion refers to a region on the skin that has suffered damage through injury or disease. Skin lesions are initial clinical indications of chickenpox, melanoma, and others [3]. One of the most fatal among them is Skin Cancer, with about 1.04 million melanoma cases in 2018 alone and in 2022, an estimated 330,000 new cases of melanoma were diagnosed worldwide and almost 60,000 people died from the disease. These facts show that skin cancer is climbing up the cancer list [4]. Skin cancer is classified mainly into melanoma and non-melanoma. If left undiagnosed, the most dangerous form of skin cancer in the world is melanoma, as it causes significant damage to the part of the skin. Melanomas are skin cancers that develop from melanocytes, which basically make melanin. It primarily appears as an abnormal mole on the skin, and the cell division at a much-uncontrolled rate invades the nearby tissues and infects that tissue. So basically, the primary indicator is the change in the appearance of the infected area. Visual detection features like the ABCDE (asymmetry, border, color, dermoscopic structure, and evolution) rule and the CASH (color, architecture, symmetry, and homogeneity) rule are utilized to diagnose melanoma [5]. If these features are present in the ABCDE rule and CASH rule, then it is more likely to be a malignancy. Because it has a great potential of being cured if caught early, early diagnosis and a comprehensive body checkup are essential in its therapy. Physicians frequently use dermoscopy or Epiluminescence Microscopy (ELM) to assess whether skin lesions are cancerous or benign. Computerized non-invasive dermatology significantly impacts skin cancer detection thanks to artificial intelligence (AI), computer vision (CV), and image analysis tools that have advanced technologically. This greatly lowers the cost of therapy while also raising the rate of survival. The rest of the paper is organized as follows: a comprehensive literature review and classification of skin-related diseases is present in Sect. 2. In Sect. 3, the comparative methodology is presented in a stepwise manner, and exploratory data analysis of the dataset is also shown. In Sect. 4, the results, comparative analysis, and discussions of implemented architecture are presented whereas Sect. 5 discusses the architectures used. Lastly, Sect. 6 concludes this paper with future scope.

2 Previous Work

This literature review of the multi-class classification of skin cancer is divided into three sections, namely: (i) image processing, (ii) application of conventional machine learning classifiers, and (iii) deep learning models for segmentation and classification.

2.1 Image Processing Techniques in Skin Cancer Dataset

In a paper, Victor et al. [6] proposed the model that image processing removes the noise by using the mean of the median filter. To find the region of interest (ROI), the histogram equalized phase is fed as the input for the segmentation. Meanwhile, for extracting the images, support vector machine (SVM), K-nearest neighbors (KNN), Decision tree (DT), and backtracking (BT) were used. In a separate work, Raza et al. [7] focused on developing and improving an active learning-based data screening framework based on a probabilistic strategy. A watershed merged mask was used to classify the segmented legion SVM and a neural classifier to generate segments for the mole part.

Dubal et al. [8] proposed a way that effectively detects and categorizes various skin lesions as benign or cancerous using cameras to generate images. Hasan et al. [9] proposed a process divided into two stages. In the first stage, image enhancement and augmentation were used. In the second stage, they proposed a 46-layered structure for the U-Net to segment the lesion and used U-Net 32 and U-Net 46 to make their architecture. Burdick et al. [10] investigated the impact of enlarging the segmentation boundary, including extra pixels that neighbor the lesion. Their results indicated that segmentation border enlargement produces better results across all metrics.

Xu et al. [11] presented a method that first reduced color images into intensity images and performed approximate segmentation on the image by intensity thresholding. Image edges are used to refine segmentation and find the legion's potential position; double thresholding is used to focus on images. Lynn et al. [12] proposed a method of first removing unwanted noise in images, followed by the segmentation process. The mean-shift algorithm is used to detect skin cancer, and the underlying ABCD dermatology rules perform feature extraction.

Sujitha et al. [13] employed pre-processing techniques combining median filter and Karhunen-Loeve Transform (KLT). In order to provide more precise results, they introduce a combination of watershed transformation and active contours algorithms. Using the ABCD rule, Dubal et al. [8] suggested a technique to extract characteristics. The skin lesion segmentation problem was approached from two different angles by Rajab et al. [14]. The first is a method of segmenting data based on regions, and the second uses edge detection. The best-performing technique was found to be iterative thresholding over a range of signal-to-noise ratios.

2.2 Application of Conventional Machine Learning Classifiers

Using computer vision and image processing techniques, Taufiq et al. [15] suggested a real-time mobile-friendly healthcare system for melanoma detection. For segmentation, the GrabCut algorithm was used with the SVM for the classification. To classify skin lesions, an automatic framework, which is a union of segmentation and classification by Sumitra et al. [16], was introduced. The fusion of SVM and KNN classifiers was used with the fusion of the extracted images.

Bhimavarapu et al. [17] implemented numerous deep learning in order to diagnose melanoma from images. Convolutional neural networks (GC_SCNN) were used. Mane et al. [18] first processed the image, and then lesion was then segmented using image segmentation techniques. The classification was done using SVM with a linear kernel.

Dey et al. [19] presented techniques that are a combination of segmentation technique and SGO-based Kapur's thresholding and were found to be very effective for accuracy, sensitivity, and specificity in recognizing digital images of melanoma from dermoscopy. The watershed segmentation technique was used by Murugan et al. [20], and the produced segments were then subject to feature extraction. Shape, ABCD Rule, and GLCM are features that were extracted and were then used to classify data. In their work, SVM classifiers produced the best results.

2.3 Deep Learning Models for Segmentation and Classification

In their model for detecting accurate skin lesion segmentation, Sarker et al. [21] merged the SLSNet adversarial network (GAN) model with 1-D kernel factorized networks, as well as the multiscale aggregation method. Mohakud et al. [22] proposed a hyper-parameter optimized Fully Convolution Encoder-Decoder Network (FCEDN) for segmenting skin images of various skin cancer images. Harangi et al. [23] looked at the creation of an ensemble for deep convolutional neural networks to classify photos into three categories: seborrheic keratosis, nevus, and melanoma. The output of the suggested model is based on the member robust convolutional neural networks (CNNs) in one frame. Saeed et al. [24] employed deep learning models to evaluate medical images, including image classification, object recognition, segmentation, registration, and a variety of other topics. Alom et al. [25] suggested a distinctive deep learning framework, which includes a NABLA-N network. The model used NABLA-N and IRRCNN for the segmentation and classification of the benchmarking on the ISIC2018 dataset. Jianli et al. [26] created a model to segment cancer images of skin through the genetic neural network (GNN) model. To increase the convergence speed of the BPNN, weight and threshold optimization using a genetic algorithm is carried out.

Kaul et al. [27] proposed a representation of the feature maps produced by a different convolutional autoencoder within the convolutional neural networks. They were quickly able to integrate deep CNN with their architecture. Bibi et al. [28] suggested an innovative technique that uses the conventional deep learning framework consisting of the lesion segmentation and classification task. Filtering techniques and color transformations were used to improve contrast. Canonical correlation analysis was performed to extract characteristics from each of the pre-trained CNN models and combine them. Maximum entropy score-based selection (MESbS), a solution to the problem of the introduction of redundant features, was offered. The HAM10000 dataset's claimed accuracy was 96.7%.

Zafar et al. [29] proposed a distinctive method of separating the borders of lesions by combining U-Net and ResNet to form a new architecture called Res-UNet. Image inpainting was also used to remove hair for improved segmentation results. The model was tested on the ISIC2017 dataset. Ashraf et al. [30] used sophisticated pre- and post-processing approaches and various models, including U-Net, Res-UNet, and ResUNet++. The segmentation accuracy was increased in the post-processing stage using conditional random field (CRF) and test time augmentation (TTA). The authors achieved better results in training on the combined dataset of ISIC2016. Thapar et al. [31] presented the swarm intelligence (SI) algorithm used for the ROI segmentation with the Grasshopper Optimization Algorithm (GOA). The results of the proposed segmentation and classification techniques are evaluated using classification.

Kassem et al. [32] reviewed, summarized, and evaluated the level of evidence supporting computer-aided systems' ability to detect skin cancer with diagnostic precision. This paper reviewed around 49 deep learning papers and 53 articles employing conventional machine learning techniques in skin cancer detection. Verstockt et al. [33] proposed a literature review, which gave a brief overview of infrared cameras, thermal excitation techniques, and measurement setups that were used in skin lesion detection. Stofa et al. [34] presented a range of analyses covering classification and segmentation techniques. The most recent techniques are examined and analyzed on various performance measures. The authors suggest further research should be tested on various open datasets, and more variations in skin tone should be validated to produce a more robust system. Furthermore, a comprehensive analysis of different segmentation techniques should be carried out to achieve better accuracy. A detailed overview of the literature review and the accuracy achieved by the models in the diagnosis of skin cancer are presented in Table 1.

After an extensive literature review as mentioned above, it was found that to diagnose or detect skin cancer and other skin-related diseases, extensive CAD architectures used conventional machine learning and deep learning to achieve an accuracy of 90% and above. The accuracy of the CAD architectures can be improved using robust and flexible algorithms. During the literature review, it was found that no paper had done an extensive exploratory data analysis of skin cancer datasets nor carried out a comparative analysis of different types of architectures. Therefore, in this study, a detailed exploratory data analysis was carried out as it assists in not only identifying obvious errors but also gaining a better understanding of the patterns of the data. It also assists in the detection of outliers and anomalous events. A lot of

Table 1 Summary of the literature review in the skin cancer disease detection domain

Author	Model	Diseases	Datasets	Accuracy
Victor et al. [6]	SLSNet, GAN model	Skin cancer	(ISIC 2018), (ISBI 2017)	97.61%
Hammad et al. [7]	(ALDS) with SVM and Neural classifier	Skin cancer	PH2	80%
Dubal et al. [8]	CNN	Skin cancer	Custom dataset	76.9%
Hasan et al. [9]	Modified U-Net	Skin cancer	ISIC-2018	93%
Burdick et al. [10]	InceptionV3, VGG-16	Skin cancer	ISBI-2016	69.3%
Xu et al. [11]	Elastic curve model	Skin cancer	Custom dataset	–
Lynn et al. [12]	KNN, decision tree	Skin cancer	ISIC	78.2%
Sujitha et al. [13]	CNN and watershed transform	Skin cancer	Custom dataset	–
Rajab et al. [14]	Isodata algo, NN edge detection	Skin cancer	Custom dataset	–
Taufiq et al. [15]	CNN, OpenCV	Skin cancer	Klinik und Poliklinik	80%
Sumithra et al. [16]	CNN, GNN	Skin cancer	ISIC-2018	87%
Usharani et al. [17]	GC-SCNN	Skin cancer	PH2, ISIC	99.75%
Mane et al. [18]	SVM, BayesNet	Skin cancer	PH2	92.3%
Dey et al. [19]	Thresholding, segmentation	Skin cancer	ISBI2016	95.72%
Murugan et al. [20]	kNN, RF, SVM	Skin cancer	ISIC	89.43%
Sarker et al. [21]	(GAN), SLSNeT	Skin cancer, segmentation	ISBI 2017, ISIC 2018	97.61%
Mohakud et al. [22]	FCEDN	Skin lesion segmentation	ISIC 2016, ISIC 2017	98.32%
Harangi et al. [23]	CNNs, ResNet, VGGNet	Skin cancer	ISBI 2017	–
Saeed et al. [24]	DCNNs, GAN	Skin cancer	PH2, ISBI-2017, ISIC-2016	97.78%
Alom et al. [25]	NABLA-N, IRRCNN	Skin cancer	ISIC-2018	87%

papers used limited-sized data, but in this paper, augmentation is used to increase the size of the data. Most of the papers are binary class classifiers of skin cancer, whereas multi-class classification of skin cancer is done in this paper.

3 Data and Methodology

In this section, the dataset, pre-processing techniques, and classification architectures are presented and discussed in detail. Later, this section presents the mathematical formulas for evaluation metrics like loss, accuracy, recall, specificity, sensitivity, and precision, along with exploratory data analysis of the dataset.

3.1 Dataset

Multiple sources were used to collect this sizable dataset of 10015 dermatoscopic images of enumerable usual skin lesions. The aim of the HAM10000 dataset is to tackle the problem caused by small-size datasets. The legion cases include actinic keratoses, intraepithelial carcinoma, basal cell, and vascular lesions. Histopathology verifies the maximum number of skin lesion images of the dataset and the remaining are done by follow-up examinations. The size of the images is 600×400 and it is provided by the International Skin Imaging Collaboration (ISIC) publicly.

3.2 Pre-processing

The accuracy of the classification model is mainly dependent on the pre-processing steps. First, the dataset's images are resized to 224×224 pixels, and the image's contrast is adjusted according to the transfer learning architecture. Then, to balance the imbalance in the dataset, data augmentation was used. One-hot encoding was implemented to classify the data into multiple classes separately. The dataset was randomly divided into 80% training and 20% test data. After pre-processing, 80% of the data was used to train the model, and the remaining 20% was used to evaluate the classification model.

3.3 Classification Models

3.3.1 Inception

Inception is an architecture, as shown in Fig. 1, which is aimed at improving resource utilization while outperforming state-of-the-art models [35]. The inception network applies dimension reduction and projections whenever computational requirements become excessive. 1×1 convolutions are used to compute reductions before expensive convolutions are carried out. Moreover, apart from reductions, they also include ReLU activation. The architecture includes modules stacked upon each other and max-pooling layers with a stride of 2 to cut the resolution by half. This architecture brings about the benefits of increasing units at each stage without an uncontrolled growth in computational complexity. With careful manual input, inception networks can perform two to three times faster than similarly performing models without the inception architecture. Inception-v2 dives deeper into inception and introduces batch normalization [36]. Batch normalization addresses the internal covariate shift phenomenon. Most convolution neural networks use stochastic gradient descent (SGD). While SGD is simple and has proven to be effective, careful hyperparameter tuning of learning rates and careful selection of initial values must be carried out. Moreover, training can be complicated and slowed down as layer inputs are affected by the parameters of previous layers resulting in an amplifying effect as the network becomes deeper [37]. The change in the distribution of layers' input causes hidden layers to adapt to the new distribution resulting in the training process slowing down. This change in input distribution is referred to as the internal covariate shift. As a result, it allows for the use of higher learning rates without the risk of divergence, significantly accelerating the training of the network. It even acts as a regularizer such that the need for dropout is reduced. Batch Normalization also prevents the network from getting stuck in saturated modes. Inception V2 achieves state-of-the-art performance using only a fraction of the training steps. It is a top performer using an ensemble of batch-normalized networks, exceeding human raters on the ImageNet classification.

3.3.2 EfficientNet

EffientNet, as shown in Fig. 2, is aimed at rethinking the scaling operation of convolutional neural networks to achieve superior accuracy and efficiency [38]. EfficientNet employs a straightforward compound scaling method. Standard scaling methods only increase a single dimension of neural networks (width, depth, or resolution). The compound scaling method uniformly scales the network dimensions and resolution using a set of fixed scaling coefficients. EfficientNet-B7, in particular, exceeds GPipe's best accuracy while requiring less than 8x fewer parameters and performing over 6x faster [38]. Generally, it can be assumed that higher-resolution images will require deeper networks. Analogously, the network width should be increased when the resolution is greater to seize the fine details as pixels increase in high-resolution

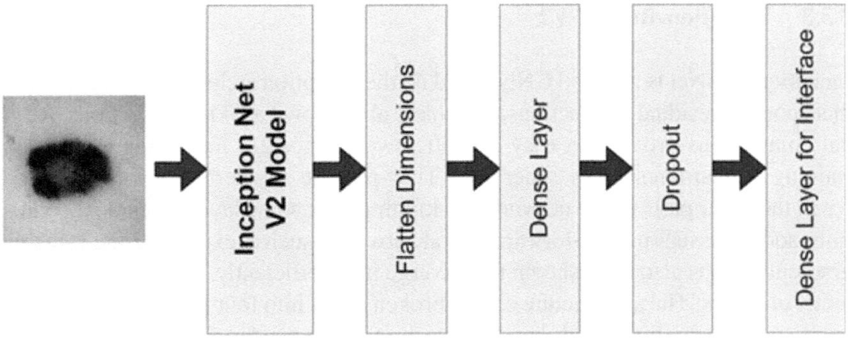

Fig. 1 Inception module architecture

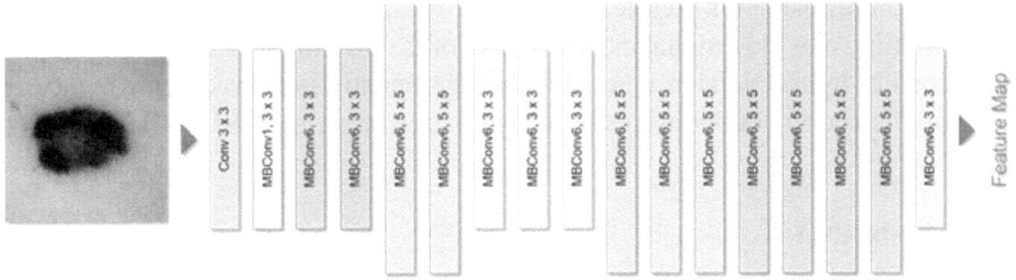

Fig. 2 EfficientNet architecture

images. This shows the need to balance different scaling dimensions against the traditional method involving single dimensions. Tan et al. [38] introduced a new compound scaling method using a compound coefficient ϕ to scale all dimensions in a monitored manner uniformly.

Depth: $d = \alpha^{\phi}$, width: $w = \beta^{\phi}$, resolution: $r = \gamma^{\phi}$.

α, β, γ are determined by a grid search and constrained such that:

$$\alpha \cdot \beta^2 \cdot \gamma^2 \approx 2 \tag{1}$$

where $\alpha \geq 1, \beta \geq 1, \gamma \geq 1$.

The EfficientNet-B0 variant's primary base is the mobile inverted bottleneck MBConv with squeeze-and-excitation optimization. EfficientNet-B0 is used as a baseline, and the compound scaling technique is applied. First $\phi = 1$ is set, and a grid search is carried out to determine α, β, γ under the given constraint. α, β, γ are set as constants, and B0 baseline is scaled up using different ϕ values producing EfficientNet-B1 to B7.

EfficientNet's compound scaling method can be applied to any baseline ConvNet under specified constraints to scaling it up while maintaining efficiency. EfficientNet models achieve state-of-the-art accuracy while requiring fewer parameters.

3.3.3 Inception-ResNet-V2

Inception-ResNet is a hybrid CNN based on the inception architecture family [39]. It incorporates residual connections, which are also known as skip connections. Residual connections provide an easy and effective method for improving and making training in neural networks easier [40]. They provide an alternative path for data to reach the latter parts of the network by skipping over some layers. Residual connections address issues in feed-forward neural networks, such as exploding and vanishing gradients. It has also been shown to converge more efficiently, even in cases of hundreds of layers. The architecture can be broken down into four parts, with the feature extractor extracting the high-level features and the encoder the mid-level features. These two are joined at the fusion layer, and then the decoder forecasts the output using the merged features. The architecture is shown below in Fig. 3.

Furthermore, Inception-ResNet employs activation scaling of residuals, stabilizing the training of wide residual inception networks. We employed Inception-ResNet V2, a costlier hybrid version that achieves significantly improved performance of 3.7% top-5 error on 50000 images of the validation set of ILSVRC [39].

3.3.4 MobileNet

The MobileNet architecture is aimed at resource-limited and mobile environments [41]. Their architecture is built on depth-wise separable convolutions. Separable convolutions can be of either two types- spatial or depthwise. Spatial separable convolutions are straightforward; however, it suffers from significant limitations and is therefore not used to a great extent. As the name suggests, it pertains to kernels' spatial dimensions, i.e., height and width. The operation involves dividing a kernel into two smaller kernels. For example, a 4×4 kernel can be divided into a 4×1 and 1×4 kernel.

The number of convolutions is dropped from 16 to 2 convolutions of 4 for a total of 8. As a result, the computational complexity is reduced for a faster-performing network. The drawback arises, as not all kernels can be divided into two smaller

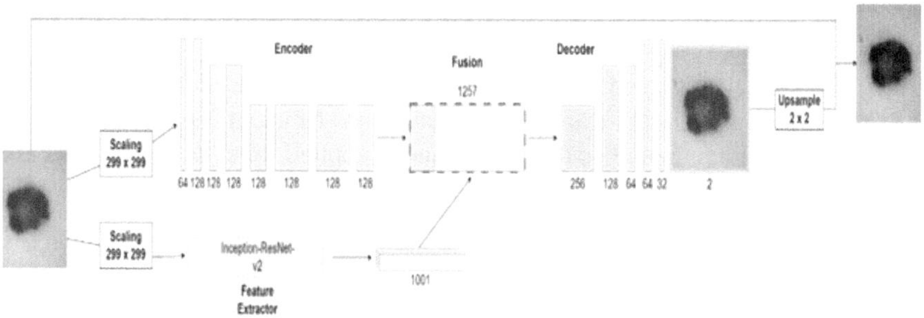

Fig. 3 Inception-ResNet V2 architecture

kernels. In practice, despite many possible kernels, only a small fraction of them can be separated into two smaller kernels.

Depthwise separable convolutions, which MobileNet uses, factorize the kernels into two kernels: depthwise convolution and pointwise convolution (1×1 convolution) [42] as shown in Fig. 4. As the name suggests, it also addresses the depth dimension by applying a filter to each input channel. While standard convolutions perform filtering and combining inputs into outputs, depthwise works by dividing it into two layers: the filtering layer and the combining layer. This factorization results in significantly reduced model size and computational requirements at the cost of a negligible loss in accuracy.

Standard convolutions have computational costs given by Eq. 2:

$$D_K \cdot D_K \cdot M \cdot N \cdot D_F \cdot D_Z \tag{2}$$

where $D_K \cdot D_K$ is the kernel size, M is the number of input channels, N is the output channels, and $D_F \cdot D_F$ is the size of the feature map.

As depth-wise convolutions can be expressed as two steps of applying the filter and then performing the combining step, the computational costs are reduced to:

$$\frac{1}{N} + \frac{1}{D_K^2} \tag{3}$$

MobileNet has a total of 28 layers. Batchnorm and ReLU nonlinearity are used. Stride convolutions handle downsampling. The last average pooling layer reduces spatial dimensions to 1, followed by the fully connected layer. MobileNet's architecture allows extraordinarily efficient operations requiring significantly reduced computational requirements while beating out state-of-the-art models despite a reasonable tradeoff of accuracy for reduced size.

After extensive and detailed research, analysis, and implementation of classification algorithms on the skin cancer dataset are presented in Fig. 5.

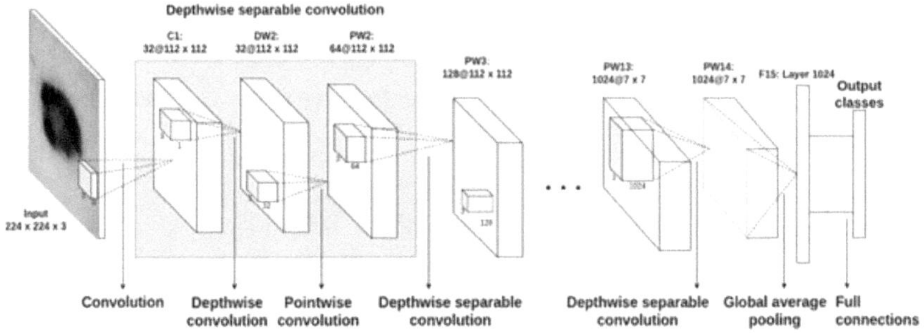

Fig. 4 MobileNet architecture

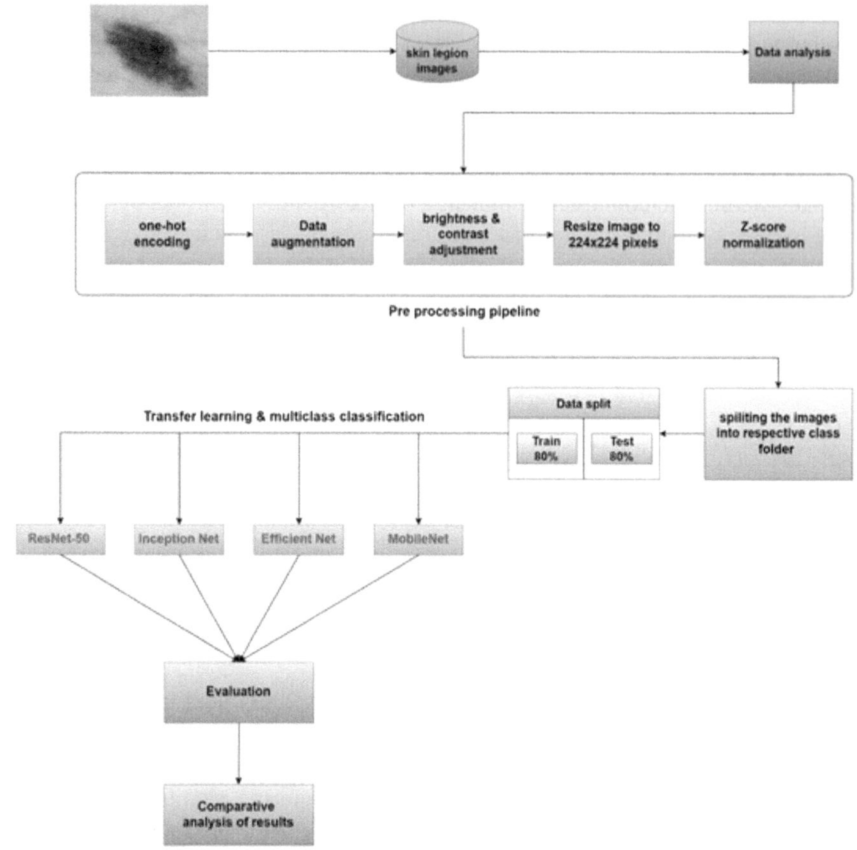

Fig. 5 Proposed architecture for the framework

3.4 Performance Matrix

$$\begin{array}{c|c} \text{True Positive (TP)} & \text{False Positive (FP)} \\ \hline \text{True Negative (TN)} & \text{False Negative (FN)} \end{array} \tag{4}$$

$$Accuracy = \frac{TP + TN}{TP + FN + FP + TN} \tag{5}$$

$$Precision = \frac{TP}{TP + FP} \tag{6}$$

$$Specificity = \frac{TN}{FP + TN} \tag{7}$$

$$Sensitivity = \frac{TP}{TP + FN} \tag{8}$$

$$Sensitivity = \frac{TP}{TP + FN} \tag{9}$$

3.5 Exploratory Data Analysis

Exploration of the data helps detect an error, understand relationships, unearthing important factors, find patterns, and provide new insights. So, looking at Fig. 6, which describes the count of cases by age—it's clear that the most significant number of cases are around 45 years of age, i.e., peaking at age 45 and going down both sides. By gender (Fig. 6), males are more susceptible to skin lesion problems. By localization on the body (Fig. 6), most skin lesions are generally detected on the back, followed by the lower extremities, trunk area, and so on (upper extremity, abdomen, face, chest, foot, etc.). By cell types of the disease, the skin lesion is primarily detected in the form of melanocytic nevi, a little bit in melanoma. Other varieties are present but are rare (including benign keratosis-like lesions, basal cell carcinoma, actinic keratoses, etc.).

Figure 7 gives us a clear idea of the types of skin lesions detected most. Melanocytic nevi is at 69.9%, exceeding everything else by a significant margin. Melanoma came in second place with 11.1% presence, followed by Benign keratosis-like lesions with 11% presence and traces of Basal cell carcinoma at 5.1%, Actinic keratoses at 3.3%, Vascular lesions at 1.4%, and Dermatofibroma 1.1%. Regarding how the lesions were detected, histopathology tops the chart by taking the first position in the detection rate with 53.3% of all cases. Next is the follow-up examination at 37% detections, the expert consensus at 9%, and finally, confirmation by in-vivo confocal microscopy at 0.7%.

Next, looking at the localization versus cell type (Fig. 8), it's clear that in the dataset, the majority is the melanocytic nevi cells causing the lesion, the body parts mainly being the back, trunk, upper extremity, abdomen, lower extremity, foot and a little bit in hand, neck, face. Then comes the benign keratosis-like lesion mainly

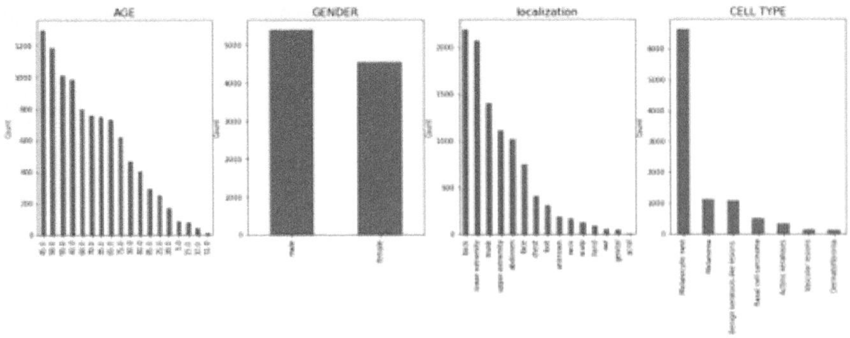

Fig. 6 Skin lesion distribution

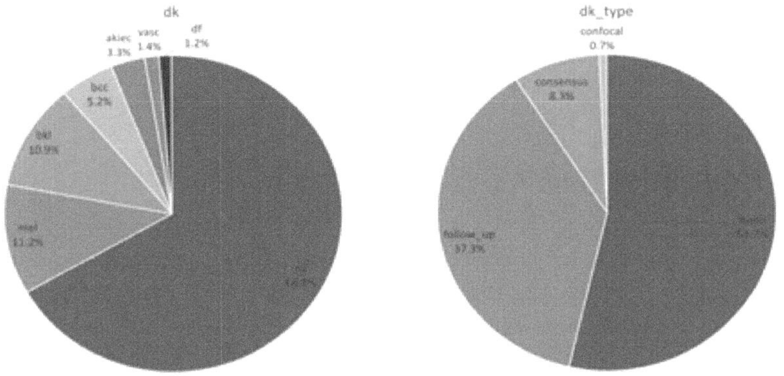

Fig. 7 Share of skin lesion

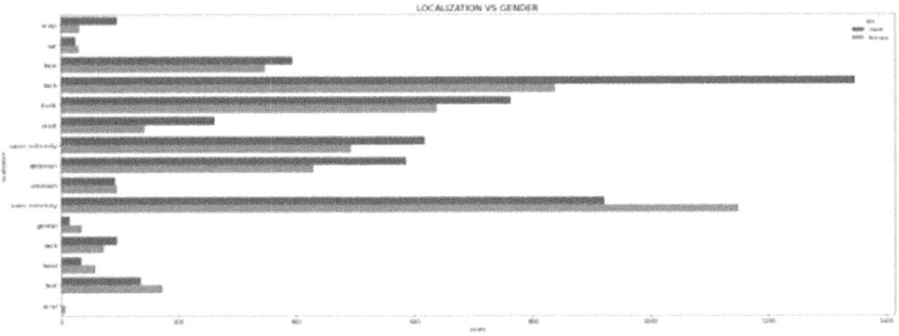

Fig. 8 Localization versus cell type

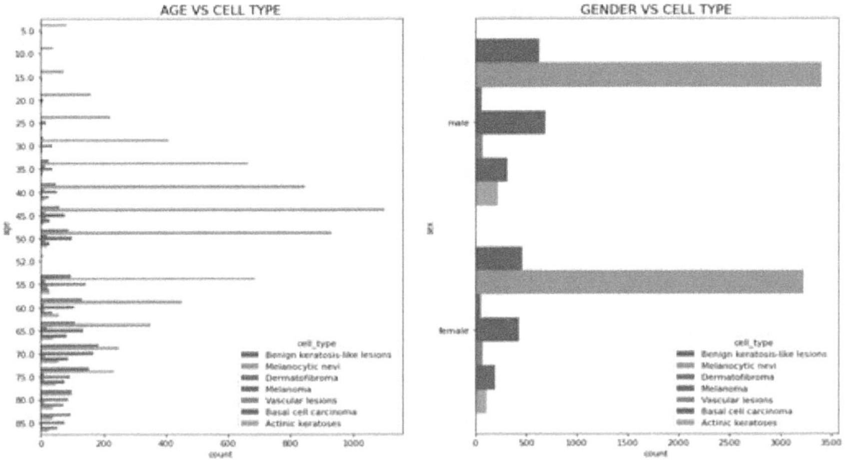

Fig. 9 Age versus cell type and gender versus cell type

occurring on the face and back and a little bit on the trunk, chest, etc. then comes the melanoma cells mainly targeting the back, face, and upper extremity. Others also show a little trace here and there, but nothing much.

Now moving on to the age versus cell types and gender versus cell types (Fig. 9). Looking at the age versus cell types first, as stated above, the age of about 45 years is most affected by skin lesion diseases, and the most common cell type found is the melanocytic nevi; it's clear that this combination is going to be on top. Followed by age around 50 years, then 40 years, 50 years, 55 years, 35 years, and so on. One peculiar factor also visible in the age vs cell type plot is that the benign keratosis-like lesions are found in older people more around the age of 70 years, and cases are decreasing on either side and nearly vanish around the ages of 30-35 years despite being the second most common type of lesion type tying with the melanoma cells. Coming to melanoma cells, this also, like the previous type, is more visible in older people of around 70 years (maximum), then decreases as age increases or decreases. Melanoma also vanishes in the ages below 20-25 years. Next in the ranks comes Basal cell carcinoma, which also maxes at the age of 70 years and vanishes in the ages below 45-40, and then there are traces of other cell types, mainly in the elderly. In gender versus cell types, melanocytic nevi still take the top place for both genders, followed by melanoma, then benign keratosis-like lesions.

4 Results

This section is divided into three parts: results of classification techniques, comparative analysis, and result discussion.

Table 2 Training and validation results of the inception ResNet-V2

Model	Loss	Accuracy	Sensitivity	Precision	Recall	Specificity
Inception ResNet-V2 training	0.4731	0.9699	0.9684	0.9723	0.9684	0.9954
Inception ResNet-V2 validation	0.8718	0.8614	0.8547	0.8715	0.8547	0.9790

4.1 *Classification Techniques Results*

This paper's multi-class classification of the skin legion uses state-of-the-art pre-processing techniques and four advanced deep-learning algorithms. A detailed comparative analysis of the four algorithms is also done in this study to check which algorithm performs best on pre-processed data. It can be deployed on a computer-aided diagnosis.

4.1.1 Inception ResNet-V2

Pre-trained Inception ResNet-V2 is imported from TensorFlow-Hub. Hyperparameter tuning of the model is done on the HAM10000 dataset. Table 2 presents the best results of validation and training.

Figures 10, 11, and 12 show the performance of the InceptionRest-V2. Accuracy shows how good a model is in the classification of the classes accurately. The maximum accuracy achieved by inceptionResNet-V2 is 86%. Other parameters used in this study are sensitivity, specificity, precision, and recall. Their values for this model can be seen in Table 2. The confusion matrix for this model is shown in Fig. 13.

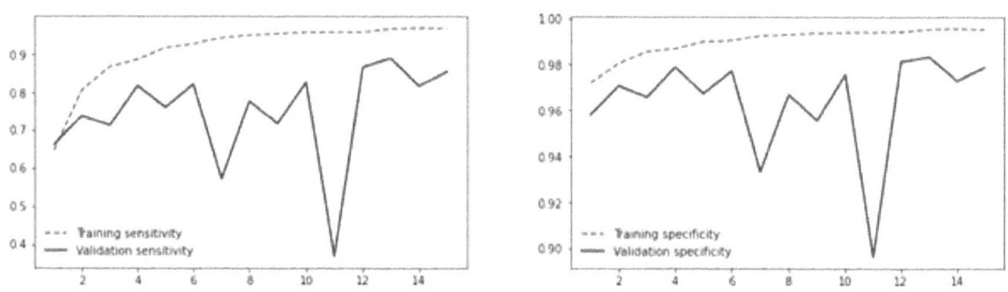

(a) Sensitivity (b) Specificity

Fig. 10 Depiction of training and validation sensitivity and specificity

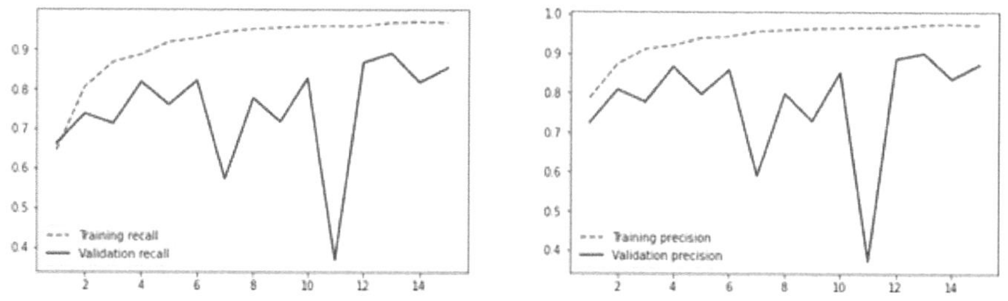

(a) Recall (b) Precision

Fig. 11 Depiction of training and validation recall and precision

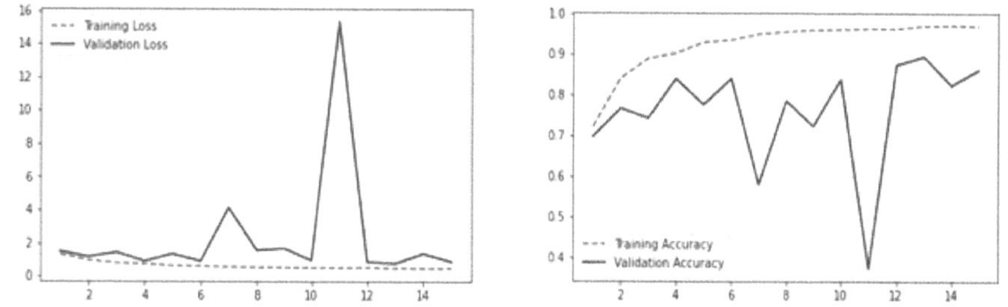

(a) Loss (b) Accuracy

Fig. 12 Depiction of training and validation of loss and accuracy

Fig. 13 Confusion matrix of inception ResNet-V2

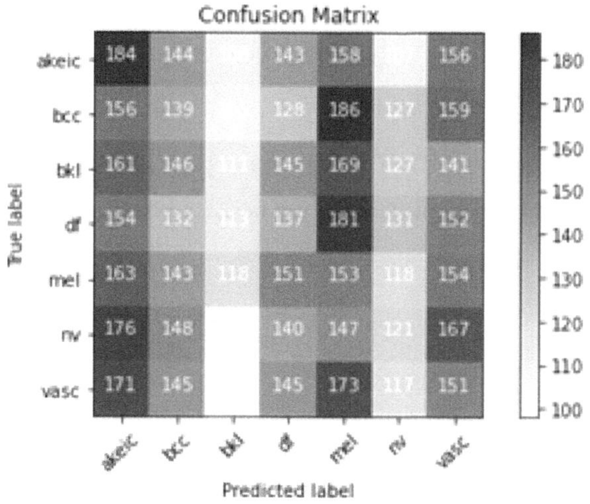

Table 3 Training and validation parameters of the EfficientNet

Model	Loss	Accuracy	Sensitivity	Precision	Recall	Specificity
EfficientNet training	0.2053	0.9813	0.9803	0.9825	0.9803	0.9971
EfficientNet validation	0.4410	0.8931	0.8832	0.9054	0.9253	0.9885

4.1.2 EfficientNet

EfficientNet is trained on the Imagenet dataset by Google. Trained parameters of the EfficientNet are used for the multi-class classification of skin lesions. The best-performing results of the model are presented in Table 3.

The accuracy achieved after hyperparameter tuning on the HAM10000 dataset by EfficientNet is 89%. This model's maximum sensitivity and specificity are 88% and 89%, which are presented in Table 3. The performance of the EfficientNet on other parameters during the training and validation process can be seen in Figs. 14, 15, 16. The confusion matrix of the EfficientNet is shown in Fig. 17.

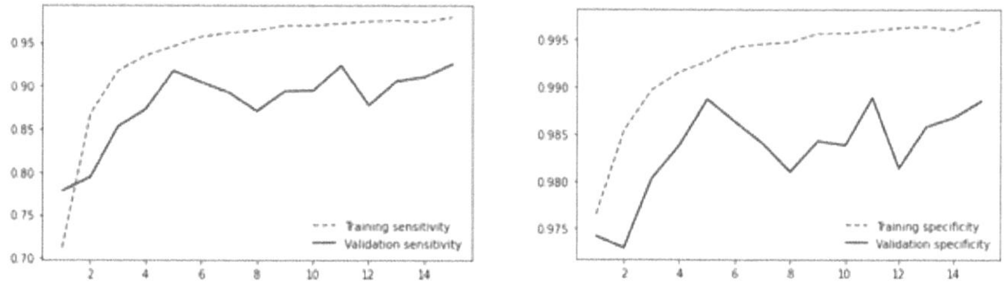

(a) Sensitivity (b) Specificity

Fig. 14 Depiction of training and validation sensitivity and specificity

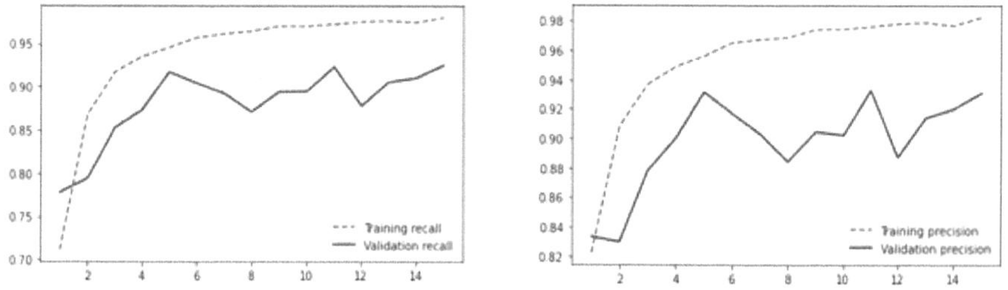

(a) Recall (b) Precision

Fig. 15 Depiction of training and validation recall and precision

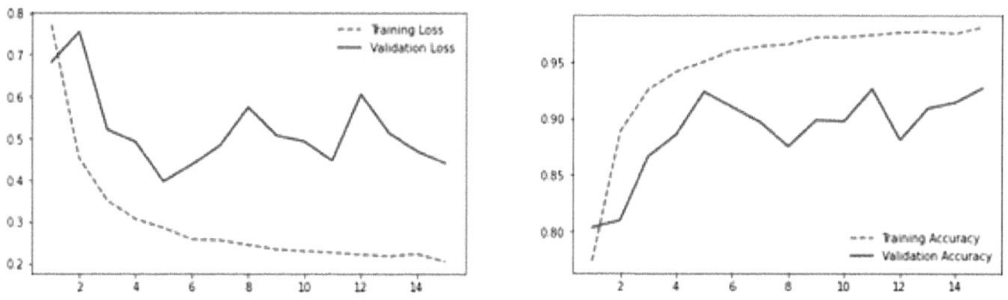

(a) Loss (b) Accuracy

Fig. 16 Depiction of training and validation of loss and accuracy

Fig. 17 Confusion matrix of efficientNet

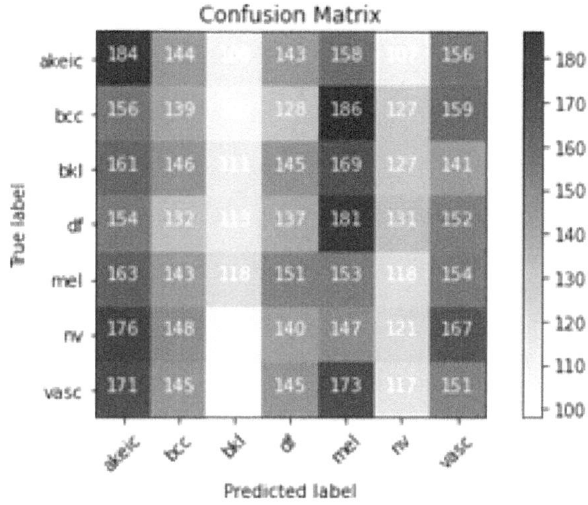

Table 4 Training and validation parameters of the InceptionNet

Model	Loss	Accuracy	Sensitivity	Precision	Recall	Specificity
InceptionNet training	0.3534	0.9600	0.9560	0.9651	0.9560	0.9942
InceptioNet validation	0.5938	0.8820	0.8708	0.8962	0.8708	0.9832

4.1.3 InceptionNet

Pre-trained InceptionNet is trained on the ImageNet dataset and tuned on the HAM10000 dataset for multiclass classification of skin legion. Table 4 presents the best performance of the InceptionNet.

InceptionNet performance is presented in the Figs. 18, 19, 20. The accuracy achieved by the model is shown in Fig. 20. Specificity means how well the model is in predicting the true negative, and sensitivity shows how accurately the model

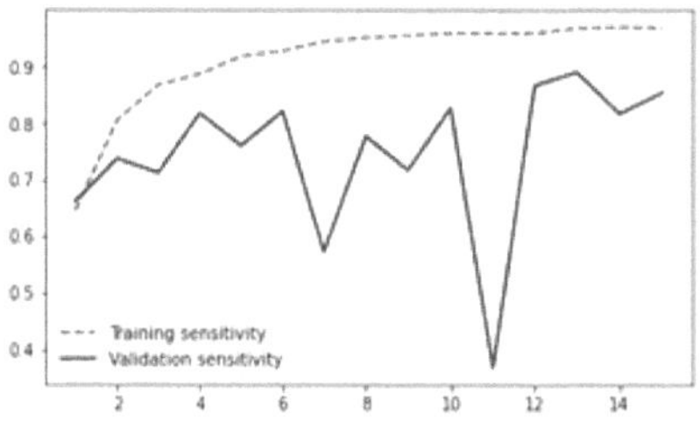

(a) Sensitivity

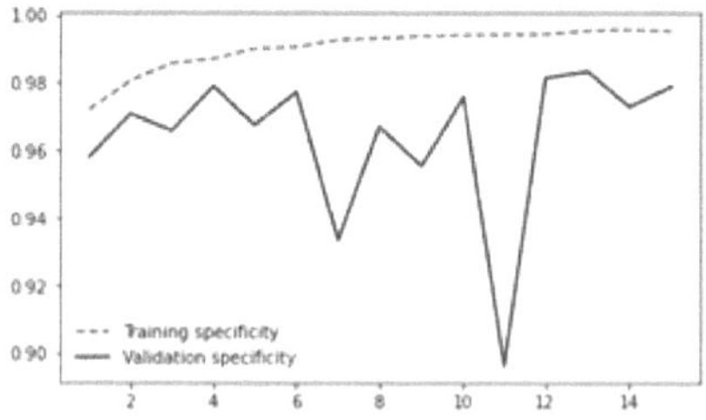

(b) Specificity

Fig. 18 Depiction of training and validation sensitivity and specificity

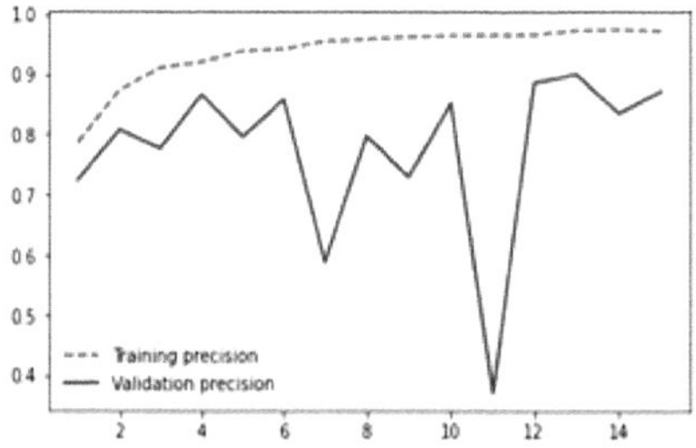

(a) Recall

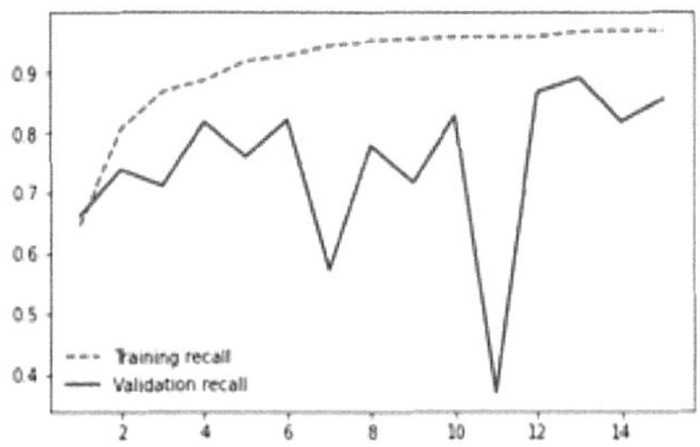

(b) Precision

Fig. 19 Depiction of training and validation recall and precision

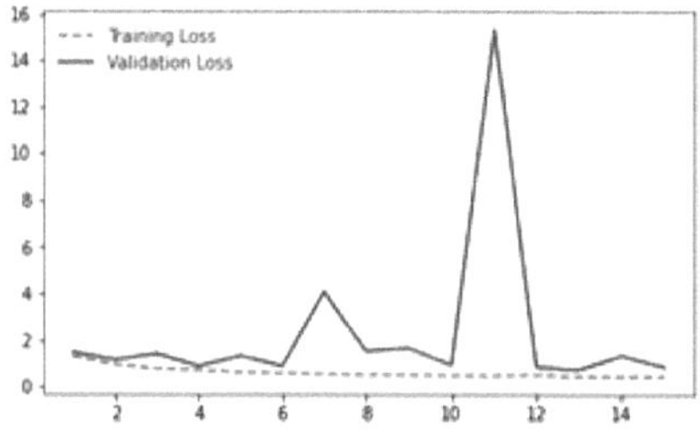

(a) Loss

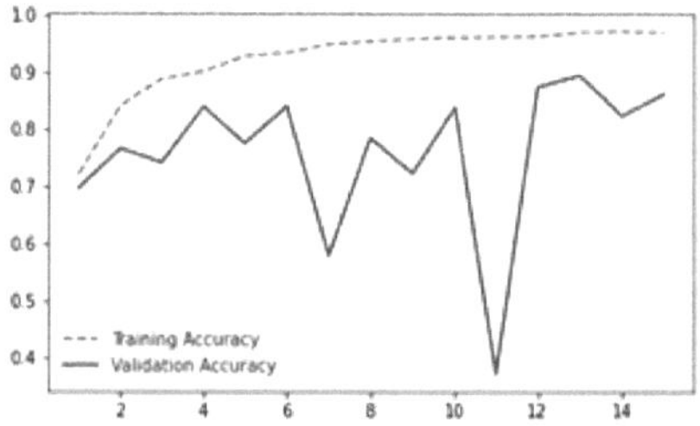

(b) Accuracy

Fig. 20 Depiction of training and validation of loss and accuracy

predicts the true positive; values achieved by the model are 87% and 98%, and values of other parameters that the model achieves are presented in Table 4. The confusion matrix of the model is shown in Fig. 21.

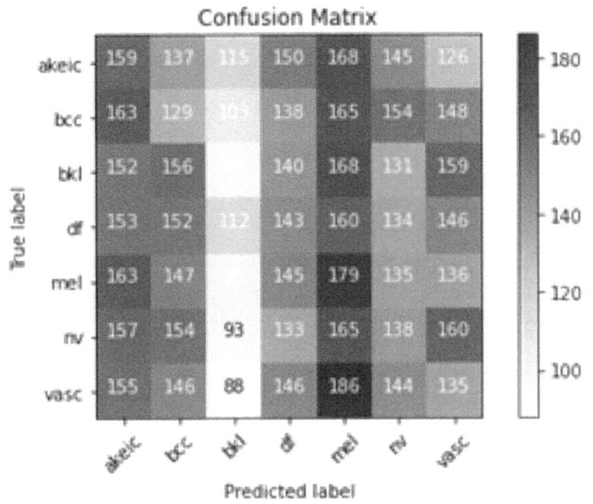

Fig. 21 Confusion matrix of InceptionNet

Table 5 Training and validation Results of MobileNet

Model	Loss	Accuracy	Sensitivity	Precision	Recall	Specificity
MobileNet training	0.3534	0.9600	0.9560	0.9651	0.9560	0.9942
MobileNet validation	0.5938	0.8820	0.8708	0.8962	0.8708	0.9832

4.1.4 MobileNet

Pre-trained MobileNet is tuned on train and test data extracted from the HAM10000 dataset for multiclass classification of skin legion. The performance of the model is presented in Table 5.

Figures 22, 19 and 24 show the performance of the pre-trained MobileNet on the training and validation dataset. Accuracy shows how good a model is in the classification of the classes accurately. The maximum accuracy achieved by inceptionResNet-V2 is 88%. Other parameters used in this study are sensitivity, specificity, precision, and recall. Their values for this model can be seen in Table 5. The confusion matrix for this model is shown in Figs. 25 and 23.

5 Discussion

The problem faced by traditional convolutional neural networks was the selection of the kernel size because the location of the information can vary widely in the image. The problem is solved by Inception net by using multiple size filters on the

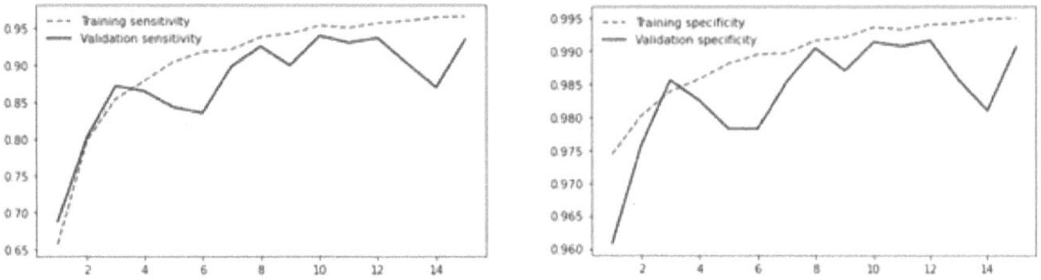

(a) Sensitivity (b) Specificity

Fig. 22 Depiction of training and validation sensitivity and specificity

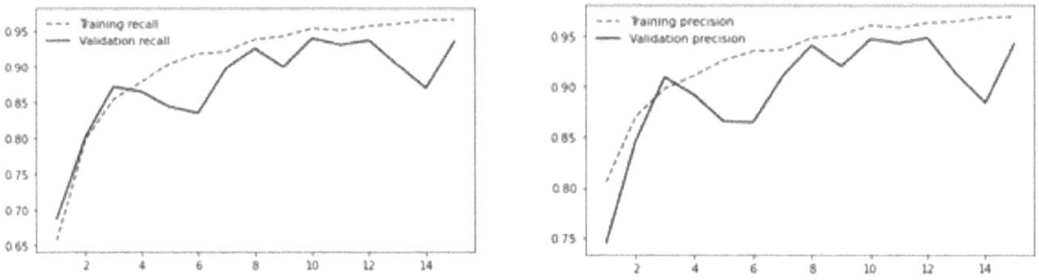

(a) Recall (b) Precision

Fig. 23 Depiction of training and validation recall and precision

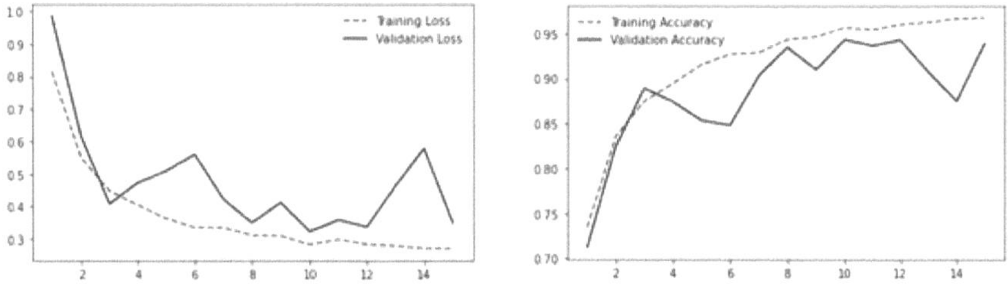

(a) Loss (b) Accuracy

Fig. 24 Depiction of training and validation of loss and accuracy

same level, and the outputs of these are sent to the next inception module. Inception Resnet V2 is a combination of inception architecture and residual connection. The concatenation part is replaced with the residual connection in this model filter. These residual connections not only avoid degradation problems caused by deep structures but also reduce the training time; because of this, awesome architecture achieved an accuracy of 86.14%.

Mobilenet was developed to be used in mobile applications, and it was TensorFlow's first mobile computer vision model. The technique used by this fantastic model to reduce the number of parameters is depthwise separable convolutions. Furthermore, this consists of two integral processes: Depthwise convolutional and

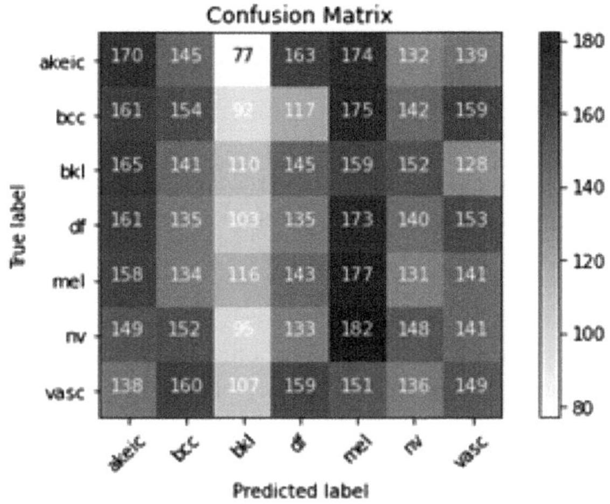

Fig. 25 Confusion matrix of MobileNet

Table 6 Training parameters of the models

Model	Loss	Accuracy	Sensitivity	Precision	Recall	Specificity
InceptionNet	0.3534	0.9600	0.9560	0.9651	0.9560	0.9942
MobileNet	0.2571	0.9617	0.9586	0.9652	0.9586	0.9942
Inception ResNet-V2	0.4731	0.9699	0.9684	0.9723	0.9684	0.9954
EfficientNet	0.2053	0.9817	0.9803	0.9825	0.9803	0.9971

pointwise convolution. In this study, the mobile net achieved an accuracy of 88.20%. EfficientNet is developed using neural architecture search using the AutoML MNAS framework. This model builds so that it gets penalized if the network becomes computationally heavy for slow inference time. In contrast, the model inverted bottleneck architecture, similar to mobile net V2, is used in the architecture but larger due to increasing FLOPs to make it light and mobile. So these are things responsible for the outstanding performance of EfficientNet; also, it achieved an accuracy of 92.73%.

5.1 *Comparative Analysis*

The performance of the four models is compared in this report section. Four pretrained advanced deep-learning algorithms are tuned on the HAM10000 dataset. The results generated by these algorithms are evaluated using accuracy, sensitivity, precision, recall, and sensitivity, as shown in Tables 6 and 7.

Accurate classification of skin cancer is crucial as medical misdiagnosis can have irrevocable consequences. In this study, it was found that Inception ResNet-V2 produced the poorest results with respect to all performance metrics and achieved only

Table 7 Validation parameters of the models

Model	Loss	Accuracy	Sensitivity	Precision	Recall	Specificity
Inception ResNet-V2	0.8718	0.8614	0.8547	0.8715	0.8547	0.9790
InceptionNet	0.5938	0.8820	0.8708	0.8962	0.8708	0.9832
MobileNet	0.4637	0.8931	0.8832	0.9054	0.8832	0.9846
EfficientNet	0.4910	0.9273	0.9253	0.9308	0.9253	0.9885

an accuracy of 86.14%. On the other hand, it was found that EfficientNet is the best performer achieving an accuracy of 92.73%, a sensitivity of 92.53%, a precision of 93.08%, and a specificity of 98.85%. These results can be verified in Tables 6 and 7. EfficientNet's compound scaling operation's ability to uniformly scale the neural network dimensions using fixed scaling coefficients results in it performing faster and requiring significantly fewer parameters. Despite a relatively high resolution of input images, EfficientNet's capability of balancing various scaling dimensions results in its strong performance across all performance metrics compared to all other architectures discussed.

6 Conclusions

A framework is presented in this study which is based on pre-trained advanced deep-learning models like InceptionNet, Inception Resnet, EfficientNet, and MobileNet. Ensemble learning is also explored in this study by implementing the Inception Resnet to classify skin lesions. Correct classification of skin cancer is critical as misdiagnosis of cancer results in worsening the existing condition or even death. Four of the most advanced deep learning models for the CAD system are implemented on the skin-legion dataset to explore and compare the results of the pre-trained models in this study. In future work, segmentation will be implemented before classification, custom deep-learning models will be implemented, and a Keras tuner will be used for hyperparameter tuning. Explainable Artificial Intelligence (XAI) frameworks, like GRAD-CAM can also be implemented for better understanding of the parameters.

Funding Statement
The study is supported by the Centre for Advanced Modelling and Geospatial Information Systems (CAMGIS), University of Technology Sydney.

Conflicts of Interest
The authors declare that they have no conflicts of interest to report regarding the present study.

Data availability
The data used in this work are open-access data.

References

1. P.T. Bradford, Skin cancer in skin of color. Dermatol. Nursing/Dermatology Nurses' Assoc. **21**(4), 170 (2009)
2. L. Schielein, L. Tizek, T. Biedermann, A. Zink, Tick bites in different professions and regions: pooled cross-sectional study in the focus area Bavaria, Germany. BMC Public Health. **22**(1), 2022 (2022)
3. O. Jones, C. Ranmuthu, P. Hall, G. Funston, F. Walter, Recognising skin cancer in primary care. Adv Ther. **37**(1), 603–16 (2020)
4. R. Kaur, N. Kaur, Analysis of Melanoma cancer detection techniques. SN Comput Sci. **3**, 93 (2022)
5. D. Jensen, B. Elewski, The ABCDEF rule: combining the "ABCDE Rule" and the "Ugly Duckling Sign" in an effort to improve patient self-screening examinations. J Clin Aesthet Dermatol. **8**(2), 15 (2015)
6. A. Victor, M. Ghalib, Detection and classification of human malignant melanoma using CAD approach. Int. J. Scientific Technol. Res. **9**(4), 2501–5 (2020)
7. M. Farooq, M. Azhar, R. Raza, Automatic lesion detection system (ALDS) for skin cancer classification using SVM and neural classifiers, in *2016 IEEE 16th International Conference on Bioinformatics and Bioengineering (BIBE)*, pp. 301–8 (2016)
8. P. Dubal, S. Bhatt, C. Joglekar, S. Patil, Skin cancer detection and classification, in *2017 6th International Conference on Electrical Engineering and Informatics (ICEEI)*, pp. 1–6 (2017)
9. S. Hasan, M. Gezer, R. Azeez, S. Gülseçen, Skin lesion segmentation by using deep learning techniques, in *2019 Medical Technologies Congress (TIPTEKNO)*, pp. 1–4 (2019)
10. J. Burdick, O. Marques, J. Weinthal et al., Rethinking skin lesion segmentation in a convolutional classifier. J Digit Imaging. **31**(3), 435–40 (2018)
11. L. Xu, M. Jackowski, A. Goshtasby, D. Roseman, S. Bines, C. Yu et al., Segmentation of skin cancer images. Image Vis. Comput. **17**(1), 65–74 (1999)
12. N. Lynn, Z. Kyu, Segmentation and classification of skin cancer Melanoma from skin lesion images, in *2017 18th International Conference on Parallel and Distributed Computing, Applications and Technologies (PDCAT)*, pp. 117–22 (2017)
13. S. Sujitha, M. Priya, J. Premaladha, K. Ravichandran, A combined segmentation approach for melanoma skin cancer diagnosis, in *2015 IEEE Seventh National Conference on Computing, Communication and Information Systems (NCCCIS)*, pp. 11–6 (2015)
14. M. Rajab, M. Woolfson, S. Morgan, Application of region-based segmentation and neural network edge detection to skin lesions. Comput. Med. Imaging Graph. **28**(1–2), 61–8 (2004)
15. M. Taufiq, N. Hameed, A. Anjum, F. Hameed, m-skin doctor: a mobile enabled system for early melanoma skin cancer detection using support vector machine. eHealth 360°, pp. 2501–5 (2017)
16. R. Sumithra, M. Suhil, D.S. Guru, Segmentation and classification of skin lesions for disease diagnosis. Procedia Comput. Sci. **45**, 76–85 (2015)
17. U. Bhimavarapu, G. Battineni, Skin lesion analysis for Melanoma detection using the novel deep learning model fuzzy GC-SCNN. Healthcare. **10**(5), 962 (2022)
18. S. Mane, S. Shinde, A method for Melanoma skin cancer detection using dermoscopy images, in *2018 Fourth International Conference on Computing Communication Control and Automation (ICCUBEA)*, pp. 1–6 (2018)
19. N. Dey, V. Rajinikanth, A. Ashour, J. Tavares, Social group optimization supported segmentation and evaluation of skin Melanoma images. Symmetry. **10**(2), 51 (2018)
20. A. Murugan, S. Nair, K. Kumar, Detection of skin cancer using SVM, random forest and kNN classifiers. J. Med. Syst. **43**(8), 269 (2019)
21. M.M.K. Sarker, H.A. Rashwan, F. Akram, V.K. Singh, S.F. Banu, F.U.H. Chowdhury et al., SLSNet: Skin lesion segmentation using a lightweight generative adversarial network. Expert Syst. Appl. **183**, 115433 (2021)
22. R. Mohakud, R. Dash, Skin cancer image segmentation utilizing a novel EN-GWO based hyper-parameter optimized FCEDN. J. King Saud Univ. Comput. Inf. Sci. (2022)

23. B. Harangi, Skin lesion classification with ensembles of deep convolutional neural networks. J. Biomed. Inf. **86**, 25–32 (2018)
24. J. Saeed, S. Zeebaree, Skin lesion classification based on deep convolutional neural networks architectures. JASTT. **2**(01), 41–51 (2021)
25. M.Z. Alom, T. Aspiras, T.M. Taha, V.K. Asari, Skin Cancer Segmentation and Classification with NABLA-N and Inception Recurrent Residual Convolutional Networks. arXiv preprint arXiv:1908.02877 (2019)
26. L. Liu, Y.Y. Tsui, M. Mandal, Skin lesion segmentation using deep learning with auxiliary task. J Imaging. **7**(4), 67 (2021)
27. C. Kaul, S. Manandhar, N. Pears, Focusnet: an attention-based fully convolutional network for medical image segmentation, in *2019 IEEE 16th International Symposium on Biomedical Imaging (ISBI 2019)*, p. 455–8 (2019)
28. A. Bibi, M.A. Khan, M.Y. Javed, U. Tariq, B.G. Kang, Y. Nam et al., Skin lesion segmentation and classification using conventional and deep learning based framework. Comput. Mater Contin. **71**(2), 2477–95 (2022)
29. K. Zafar, S. Gilani, A. Waris, A. Ahmed, M. Jamil, M. Khan et al., Skin lesion segmentation from dermoscopic images using convolutional neural network. Sensors (Basel). **20**(6), 1601 (2020)
30. H. Ashraf, A. Waris, M. Ghafoor et al., Melanoma segmentation using deep learning with test-time augmentations and conditional random fields. Sci Rep. **12**(1), 3948 (2022)
31. P. Hapar, M. Rakhra, G. Cazzato, M.S. Hossain, A novel hybrid deep learning approach for skin lesion segmentation and classification. J. Healthcare Eng., 1–21 (2022)
32. M. Kassem, K. Hosny, R. Damaševičius, M. Eltoukhy, Machine learning and deep learning methods for skin lesion classification and diagnosis: a systematic review. Diagnostics **11**(8), 1390 (2021)
33. J. Verstockt, S. Verspeek, F. Thiessen, W. Tjalma, L. Brochez, G. Steenackers, Skin cancer detection using infrared thermography: measurement setup, procedure and equipment. Sensors (Basel) **22**(9), 3327 (2022)
34. M.M. Stofa, M.A. Zulkifley, M.A.A.M. Zainuri, Skin lesions classification and segmentation: a review. Int. J. Adv. Comput. Sci. Appl. **12**(10) (2021)
35. C. Szegedy, W. Liu, Y. Jia, P. Sermanet, S. Reed, D. Anguelov et al., Going deeper with convolutions 2015 IEEE conference on computer vision and pattern recognition (CVPR), in *2015 IEEE Conference on Computer Vision and Pattern Recognition (CVPR)*, pp. 1–9 (2015)
36. C. Szegedy, V. Vanhoucke, S. Ioffe, J. Shlens, Z. Wojna, Rethinking the inception architecture for computer vision, in *2016 IEEE Conference on Computer Vision and Pattern Recognition (CVPR)*, p. 2818–26 (2016)
37. S. Ioffe, C. Szegedy, Batch normalization: accelerating deep network training by reducing internal covariate shift, in *ICML'15: Proceedings of the 32nd International Conference on International Conference on Machine Learning*, p. 448–56 (2015)
38. M. Tan, Q. Le, EfficientNet: rethinking model scaling for convolutional neural networks, in *Proceedings of the 36th International Conference on Machine Learning*. vol. 97; p. 6105–14 (2019)
39. C. Szegedy, S. Ioffe, V. Vanhoucke, A. Alemi, Inception-v4, Inception-ResNet and the impact of residual connections on learning. *AAAI Conference on Artificial Intelligence*, p. 31 (2016)
40. K. He, X. Zhang, S. Ren, J. Sun, Deep residual learning for image recognition, in *2016 IEEE Conference on Computer Vision and Pattern Recognition (CVPR)*, p. 770–8 (2016)
41. A. Howard, M. Zhu, B. Chen, D. Kalenichenko, W. Wang, T. Weyand et al., MobileNets: Efficient Convolutional Neural Networks for Mobile Vision Applications. In: arXiv preprint arXiv:1704.04861 (2017)
42. F. Chollet, Xception: deep learning with depth-wise separable convolutions, in *2017 IEEE Conference on Computer Vision and Pattern Recognition (CVPR)*, p. 1800–7 (2017)

GPSR Compliance

The European Union's (EU) General Product Safety Regulation (GPSR) is a set of rules that requires consumer products to be safe and our obligations to ensure this.

If you have any concerns about our products, you can contact us on ProductSafety@springernature.com

In case Publisher is established outside the EU, the EU authorized representative is:

Springer Nature Customer Service Center GmbH
Europaplatz 3
69115 Heidelberg, Germany

Batch number: 08998949

Printed by Printforce, the Netherlands